Iptycenes Chemistry

Chuan-Feng Chen • Ying-Xian Ma

Iptycenes Chemistry

From Synthesis to Applications

Springer

Chuan-Feng Chen
Institute of Chemistry
Chinese Academy of Sciences
Beijing
China

Ying-Xian Ma
Institute of Chemistry
Chinese Academy of Sciences
Beijing
China

ISBN 978-3-642-32887-9 ISBN 978-3-642-32888-6 (eBook)
DOI 10.1007/978-3-642-32888-6
Springer Heidelberg Dordrecht London New York

Library of Congress Control Number: 2012954546

Printed on acid-free paper

Springer is part of Springer Science+Business Media (www.springer.com)

Preface

In 1942, triptycene was synthesized by Bartlett and his coworkers, which was served as the first and simplest member of iptycene family. In the next 40 years, iptycene family had not attracted much attention until Hart first formally proposed the concept "iptycene" in 1981. Since then, iptycene chemistry has truly established. Iptycenes are a class of aromatic compounds with arene units fused to bicyclo[2.2.2]octatriene bridgehead system. These unique three-dimensional rigid structures make them promising candidates for more and more applications in molecular machines, supramolecular chemistry, material science, coordination chemistry, sensor applications in the last decade. Iptycene chemistry is walking into its golden age with great opportunities and challenges. However, during the 70 years of iptycene chemistry, there were only some relevant reviews. Thus, a comprehensive book, which reviews the retrospections and prospects of iptycenes chemistry, is not only an urgent need but also of great significance. In addition, this year also marks the 70th anniversary for the development of iptycene chemistry (1942–2012). This current situation inspired us to prepare this comprehensive book *Iptycene Chemistry: From Synthesis to Applications*.

Overall, this book contains three parts. Part I includes a brief introduction of the basic naming rules and general properties of iptycenes and their derivatives. Part II details the various methods for synthesis and functionalized reactions of triptycenes, pentiptycenes, other higher iptycenes, heterotriptycenes, and iptycene-based polymers. Chapter 2 aims at the synthetic methods and the reactions of triptycene and its derivatives, as well as the synthesis of the extended triptycenes containing fused rings. In Chap. 3, the synthesis and reactions of pentiptycenes and their derivatives are discussed in details. In addition, the method for the preparation of extended pentiptycenes is also provided. The preparation of the other iptycenes, including the heptipycene, noniptycene, and other iptycene members containing the more complicated framework will be talked about in Chap. 4. According to the different positions of the hetero atom(s), Chap. 5 is divided into three subsections: (1) the bridgehead-substituted heterotriptycenes, (2) heterotriptycenes with heterocycles, and (3) miscellaneous heterotriptycenes and their derivatives. The last chapter (Chap. 6) in this part mainly describes the methods for the preparation of various iptycene-derived polymers. After that, the applications of iptycenes in different areas are discussed in Part III. First, the varied molecular machines capable of mimicking the behaviors of macroscale objects, including gearings, brakes, ratchets, compasses, gyroscopes, wheelbarrows, and the rotaxane-based molecular machines are described in Chap. 7. In Chap. 8, we talk about the iptycene-based materials, involving the long alkyl or alkoxy-substituted triptycenes and iptycenes (and iptycene-based polymers) with the unusual internal free volumes (IFVs) on the applications of liquid crystals. Then, the materials containing good optical and electrical properties, and the porous materials based on iptycene moieties with adsorption and separation capacities are also shown in this chapter. In Chap. 9, we mainly depict the design and synthesis of various novel triptycene- and pentiptycene-derived hosts, and their complexation behaviors with different kinds of guests. Three aspects of iptycenes in self-assembly, including self-assembly in crystal with multiple supramolecular interactions, the construction of self-assembled monolayers with iptycenes and surface modification, and the self-assembly in solution based on the novel iptycene-derived synthetic host, are discussed in Chap. 10. Chapter 11 describes the iptycene molecules served as the building blocks for the metal complexes and the varied and novel complexes with special properties. Then, the varied chemosensors and biosensors based on the iptycene derivatives, especially, the iptycene-based conjugated polymers are discussed in Chap. 12. The varied triptycene-based molecules served as molecular balances to offer the attractive platforms for the study of noncovalent interactions are depicted in details in Chap. 13. Finally, another four applications are shown in Chap. 14, including the different drug activities, especially, antitumor activities of the iptycenes and their derivatives; the iptycenes act as models for Jahn–Teller effect systems and artificial photosynthesis, as well as the iptycenes applied in preparation of carbene. We sincerely hope this book will not only be useful and helpful for the researchers in iptycene chemistry, but can stimulate and facilitate future researches also.

Finally, we would like to acknowledge June Tang, associate editor from Springer, for her kind invitation, and her help and suggestions during the preparation of this manuscript. We also thank Dr. Xiao-Zhang Zhu, Dr. Yi Jiang, Yun Shen, and Ying Han from our research group for their review of the manuscript.

Chuan-Feng Chen
Ying-Xian Ma

Contents

Abbreviations

1D	One dimension
2D	Two dimension
3D	Three dimension
6FDA	4,4-(hexafluoroisopropylidene) diphthalic anhydride
A	Electron acceptor
AFPs	Amplifying fluorescent polymers
Am	Pentyl
AM1 Calculations	Austin Model 1 calculations
BET	Brunauer–Emmett–Teller
Boc	*t*-butyloxy carbonyl
BPAPC	Poly(bisphenol A carbonate)
BQ	Benzoquinone
CAN	Ceric ammonium nitrate
CBPQT	Tetracationic cyclobis(paraquat-*p*-phenylene)
COD (cod)	1,5-cyclooctadiene
Cp	Cyclopentadienyl
CPK	Corey–Pauling–Koltun
D	Electron donor
DATRI	2,6-di-aminotriptycene
DAU	Daunomycin
DB24C8	Dibenzo-[24]crown-8
DB30C10	Dibenzo-[30]crown-10
DBU	1,8-diazabicyclo[5.4.0]undec-7-ene
DCC	Dicyclohexylcarbodiimide
DCE	1,2-dichlorocthane
DCM	Methylene dichloride
DDQ	2,3-dichloro-5,6-dicyano-1,4-benzoquinone
DFT Calculations	Density functional theory calculations
DIBAL-H	Diisobutylaluminium hydride
DIBAH	Diisobutylaluminium hydride
DIPEA	Diisopropylethylamine
DLS	Dynamic light-scattering

DMA	*N, N′*-dimethylacetamide
DMAD	Dimethyl acetylenedicarboxylate
DMAP	*N, N*-4-dimethylaminopyridine
DME	1,2-dimethoxyethane
DMF	*N, N*-dimethylformamide
DMS	Dimethyl sulfide
DMSO	Dimethyl sulfoxide
DNA	Deoxyribonucleic acid
DNT	2,4-dinitrotoluene
DR	Dichroic ratio
DTT	Dithiothreitol
EDTA	Ethylene diamine tetraacetic acid
EPR	Electron paramagnetic resonance
ESI-MS	Electrospray ionization mass spectra
FIrpic	Ir^{III} bis(4,6-difluorophenyl-pyridinato)-picolinate
GPC	Gel permeation chromatography
HFIP	Hexafluoro-2-propanol
HMPA	Hexamethyl-phosphoramide
HOPG	Highly oriented pyrolytic graphite
HPLC	High-performance liquid chromatography
IEC	Ion-exchange capacity
IFV	Internal free volume
IMFV	Internal molecular free volume
IR	Infrared spectroscopy
IUPAC	International Union of Pure and Applied Chemistry
K_a	Association constant
LbL	Layer-by-layer
LC	Liquid crystal
LCs	Liquid crystal solutions
LDA	Lithium diisopropylamide
LiDBB	Lithium 4,4′-di-*t*-butylbiphenilide
LIF	Laser-induced fluorescence laser
MALDI-TOF MS	Matrix assisted laser desorption ionization-time of flight mass spectrometry
MCPBA (*m*-CPBA)	*Meta*-chloroperbenzoic acid
MDR	Multidrug-resistant
MOFs	Metal-organic frameworks
MP	Methyl propiolate
MS	Mass spectrometry
NBS	*N*-bromosuccinimide
NMP	*N*-methyl-2-pyrrolidone
NMR	Nuclear magnetic resonance
o-DCB	1,2-dichlorobenzene
PC	Polycarbonate
PDAC	Poly(diallyldimethylammonium chloride)

PIMs	Polymers of intrinsic microporosity
Pn	Pentacene
PPDs	Poly(*p*-phenylenebuta diynylene)s
PPEs	Poly(phenyleneethynylene)s
PPVs	Poly(phenylene vinylene)s
PS	Polystyrene
PTP	Permeability transition pore
Py	Pyridyl
RH	Relative humidity
ROMP	Ring-opening olefin metathesis polymerization
ROX	Carboxy-X-rhodamine
S_NAr	Nucleophilic aromatic subsititution
SPR	Surface plasmon resonance
STM	Scanning tunneling microscopy
SWCNTs	Single-walled carbon nanotubes
TBA	Tributylamine
TBAF	Tetra-*n*-butylammonium fluoride
TBAF	Tetrabutylammonium fluoride
TEM	Transmission electron microscope
Tf	Trifluoromethanesulfonyl
TFA	Trifluoroacetic acid
T_g	Glass-transition temperature
THF	Tetrahydrofuran
TIPS	Triisopropylsilyl
TLC	Thin-layer chromatography
TMEDA	*N, N,N′,N′*-tetramethylethylenediamine
TMS	Trimethylsilyl
TNT	2,4,6-trinitrotoluene
Tp	9-triptycyl
TPC	Triptycene
TPQ	Triptycene quinone
TP-TCNQ	9,10-dihydro-9,10-*o*-benzeno-1,4-bis(dicyanomethyl-ene)anthracene
Ts (Tos)	*p*-toluenesulfonyl
Upy	Ureidopyrimidinone
UV	Ultraviolet

Part I
Introduction and Background

Chapter 1
Introduction and Background

1.1 Introduction

Iptycenes are a class of aromatic compounds with arene units fused to bicyclo[2.2.2]octatriene bridgehead system. The first and simplest member of this family, triptycene was obtained by Bartlett et al. [1] in 1942. In the next 40 years, iptycene chemistry had not attracted much attention, and the studies almost focused on the synthesis and reactions of triptycene and its derivatives. In 1981, Hart et al. [2] first proposed the concept "iptycene" based on triptycene, denoting the number of arene planes separated by a bridgehead system. Since then, the door of iptycene chemistry has truly been opened, and the potential applications of iptycenes and their derivatives with unique three-dimensional rigid structures have been gradually developed. In recent years, iptycenes and their derivatives, especially triptycenes [3–10] and pentiptycenes [4, 5, 11, 12] have drawn much attention, and more and more applications in molecular machines, supramolecular chemistry, material science, coordination chemistry, sensor applications, and many other research areas have been discovered. It can be believed that iptycene chemistry is winning more and more chemists' attentions, and walking into its golden age with great opportunities and challenges. In addition, this year also marks the 70th anniversary for the development of iptycene chemistry (1942–2012).

During the 70 years of iptycene chemistry, there were some relevant reviews [3–18]. However, most of the reviews were focused on the limited area(s), and reported in recent several years. Swager [4] reviewed his group's work on synthesis of iptycene-based conjugated polymers and their applications in materials science in 2008. In the same year, Yang and Yan [11] reviewed the synthesis and applications of central-ring functionalized pentiptycenes. In 2009, Chong and MacLachlan [5] briefly described the applications of iptycenes in supramolecular and materials chemistry. Some mini-reviews on the applications of iptycene derivatives in medicine chemistry [14], microporous polymers as potential hydrogen storage materials [16], and physical organic chemistry [6, 17] have also been reported in the past several years. Moreover, a recent feature article on the synthesis of novel triptycene-derived hosts and their applications in molecular recognition and molecular assemblies [8] has been published. More recently, Jiang and Chen [18] also highlighted the synthesis

C.-F. Chen, Y.-X. Ma, *Iptycenes Chemistry*,
DOI 10.1007/978-3-642-32888-6_1, © Springer-Verlag Berlin Heidelberg 2013

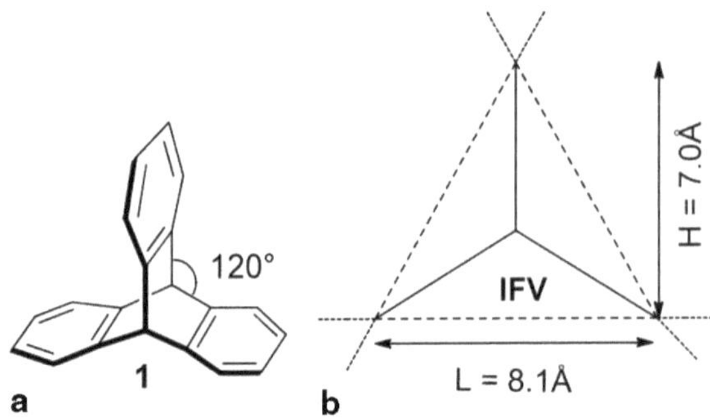

Fig. 1.1 **a** Structure of triptycene **1**. **b** Dimension of **1** with definition of "IFV"

and applications of triptycene and pentiptycene derivatives during last 10 years. In this book, we will discuss in detail not only the synthesis but also the applications of iptycenes and their derivatives. Firstly, we will introduce various methods for the synthesis and functionalization reactions of triptycenes, pentiptycenes, other higher iptycenes, heterotriptycenes, and homotriptycenes, as well as the methods for the preparation of iptycene-based polymers. Then, the applications of iptycenes in different areas will be discussed, according to the different disciplines and molecular behaviors.

1.2 Structure Properties

Triptycene has D_{3h} symmetry with a unique Y-shaped rigid structure (Fig. 1.1a), like "the triptych of antiquity, which was a book with three leaves hinged on a common axis" [1]. In this molecule, the three phenyl ring "panels" are connected with the bridgehead carbons. With dimensions calculated from the centroids of the hydrogen atoms at the extremities, the length of triptycene molecule is 8.1 Å, whereas the height is 7.0 Å (Fig. 1.1b) [19]. The [2.2.2] bridgehead system keeps the angle between aromatic rings at 120°, due to the twisting or deformation caused by high energy barrier. In this rigid three-dimensional molecular structure, this three-bladed geometry can hinder efficient packing; the resulting free volumes in the clefts between the aromatic faces were defined as the "IFV (Fig. 1.1b)" by Long and Swager [20]. In addition, triptycene with three benzene rings contains three open electron-rich cavities with rich reactive positions to provide supramolecular interactions with other molecules. These unique structural features with richer reaction positions make them useful for a wide field of applications with good prospects.

As another most common member of iptycene family, pentiptycene with five phenyl rings has two structural isomers: *para*-pentiptycene and *ortho*-pentiptycene (**2** and **3**, Fig. 1.2). For *para*-pentiptycene, it is in D_{2h} symmetry with an H-shaped structure, whereas *ortho*-pentiptycene has a quite different C_{2v} symmetry. There is a "sterically shielded" central benzene ring and four side ones in *para*-pentiptycene scaffold. These unique structural features with richer reaction positions make them a wide field of applications with good prospects.

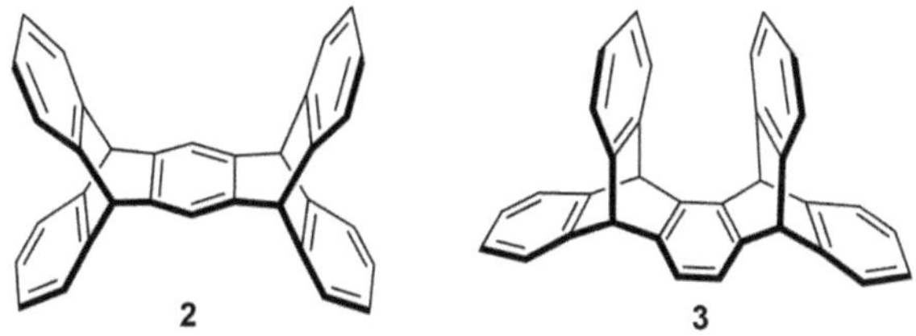

Fig. 1.2 Structures of *para*-pentiptycene **2** and *ortho*-pentiptycene **3**

Compared with triptycene and pentiptycene, the higher iptycenes, like heptiptycene, noniptycene, and nonadecaiptycene, contain more phenyl rings with complicated and diversified rigid structure, along with more than one structural isomers (Table 1.1) [21]. Likewise, the iptycenes with more phenyl rings are in a rigid, nonplanar structure with π-electron-richness character.

1.3 Physical and Chemical Properties

Triptycene is a fine white rhomboid, which is crystallized from methanol–water system. It is very soluble in benzene, soluble in ethyl alcohol, ether, acetone, chloroform, and slightly soluble in methanol. More detailed physical and chemical properties are listed in Table 1.2.

In comparison with triptycene, pentiptycene and the higher iptycene have the poorer solubility in common organic solvent, and face the comparatively difficult situation in synthesis. Thus, comprehensive basic physical–chemical property data are absent.

1.4 Spectral Properties

Triptycene is in a unique D_{3h} symmetry rigid conformation, thus its spectral properties exhibit some special characteristics. In the ^{13}C NMR spectrum, there are only four different peaks: 145.27 ppm for C_1, 125.08 ppm for C_2, 123.53 ppm for C_3, and 54.11 ppm for C_4 (Fig. 1.3). The mass spectrum of triptycene exhibits a parent ion at *m/z* 254. There is a group of doubly charged ions spacing at half-mass units between *m/z* 123 and 128 (Fig. 1.4).

There is no recession from aromatic properties in the absorption spectrum of triptycene. It shows a group of strong bands at 211, 264, 271, and 279 nm (Fig. 1.5). It is noteworthy that the near ultraviolet spectrum of triptycene exhibits a differentiable smaller (by a factor of 5) bathochromic shift than that of barrelene [23].

Triptycene can be served as a trimer of three *o*-xylene molecules with a D_{3h} symmetry, covalently bound by methine bridgehead carbons of the type found in

Table 1.1 The possible isomers of iptycenes

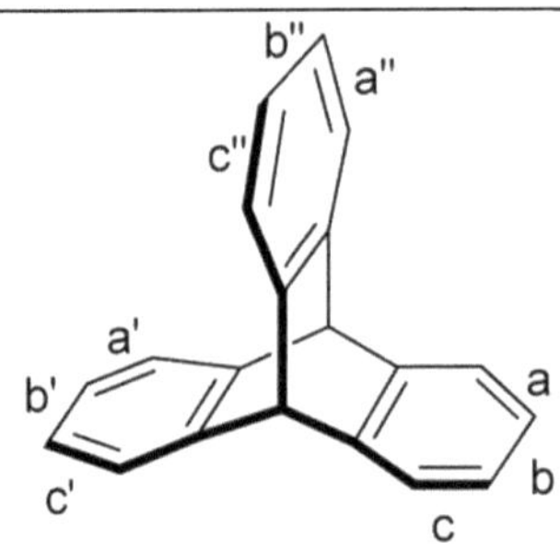

Entry	Compounds	Number of isomers	Fusion bonds	Point group
1	Triptycene	1	–	D_{3h}
2	Pentiptycene	2	a	C_{2v}
			b	D_{2h}
3	Heptiptycene	5	ac	D_{3h},
			aa′	C_s
			ab′	C_1
			ac′	C_2
			bb′	C_{2v}
4	Noniptycene	8	aca′	C_1
			acb′	C_s
			aa′a″	C_{3v}
			aa′b″	C_s
			aa′c″	C_s
			ab′b″	C_s
			ab′c″	C_2
			bb′b″	D_{3h}
5	Undecaiptycene	5	aca′c′	C_{2v}
			aca′a″	C_s
			aca′b″	C_1
			aca′c″	C_2
			acb′b″	C_{2v}
6	Tridecaiptycene	2	aca′c′a″	C_s
			aca′c′b″	C_{2v}

Iptycenes derived from triptycene by 9,10-anthradiyl fusions

Table 1.2 Basic properties of triptycene

Empirical formula of triptycene	$C_{20}H_{14}$
CAS Number	477-75-8
Molecular weight (g/mol)	254.3252
Density (g/cm^3)	1.197
Flash point (°C)	171.7
Boiling point (°C)[a]	371.8
Melting point (°C)	252–254

[a]At 760 mmHg

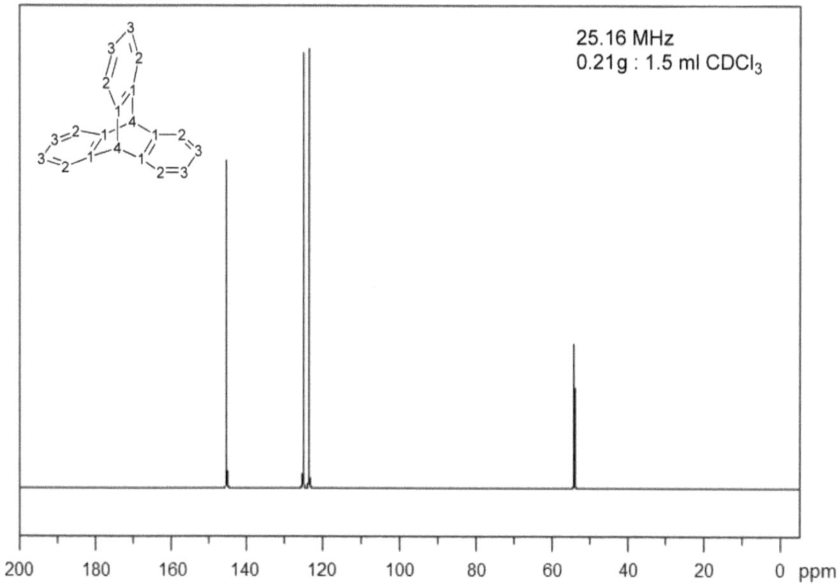

Fig. 1.3 ^{13}C NMR spectrum of triptycene. ([22])

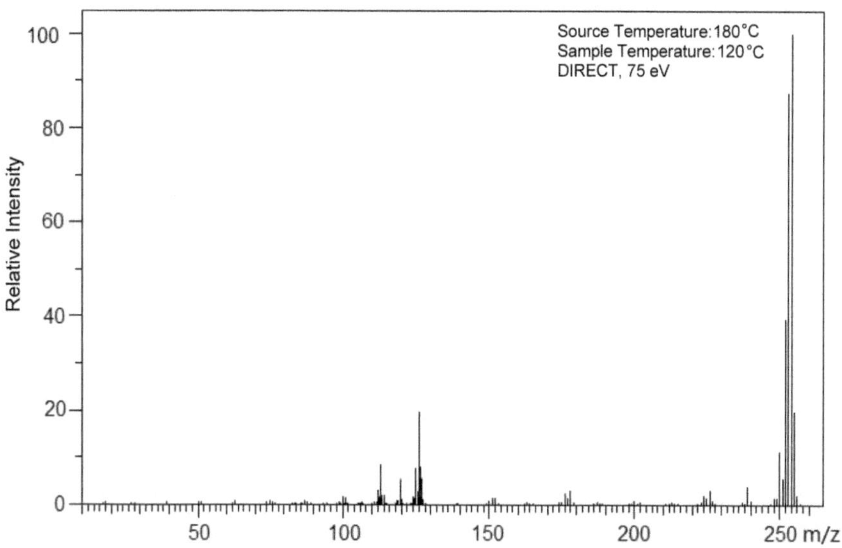

Fig. 1.4 Mass spectrum of triptycene. ([22])

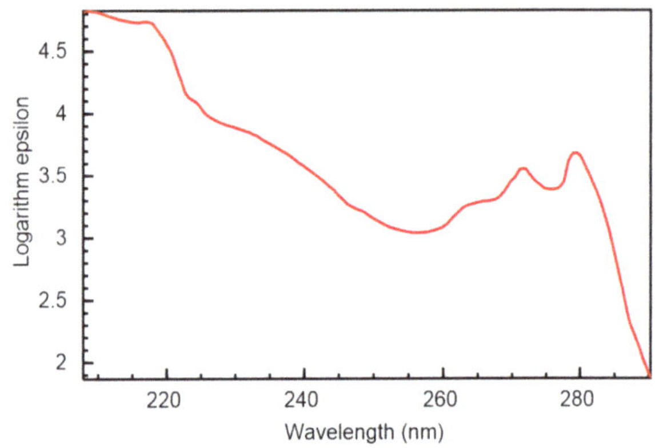

Fig. 1.5 UV/visible spectrum of triptycene. ([24])

[2.2.2]bicyclooctane [25]. It is found that the ground-state vibrational spectrum of triptycene is fit to the low frequency vibrational modes, which is revealed by fluorescence emission, IR, and Raman spectroscopies (Table 1.3).

1.5 Nomenclature

In order to simplify the IUPAC nomenclature, a nomenclature is introduced, which regards triptycene compound as a derivative of dihydroanthracene, and the carbon atoms numbered as shown in Fig. 1.6a or b. However, in some early publications, they numbered the carbon atoms as in Fig. 1.6c, in which triptycene molecules were regarded as substituted bicyclo[2,2,2]octanes. We use the former one in this book.

Table 1.3 Experimental frequencies and intensities of triptycene in the 0–1,000 cm^{-1} range

Entry	Mode	Sym	Frequencies (cm^{-1})			Intensities	
			Raman	IR	LIF	Raman	IR
1	ν_1	e′	62	–	64	0.96	–
2	ν_2	e″	144	–	–	0.35	–
3	ν_3	a_2′	–	–	211		
4	ν_4	e′	354	350	348	0.70	0.11
5	ν_5	a_1′	364	–	–	1.52	–
6	ν_6	e′	478	485	486	0.26	0.55
7	ν_7	e″	495	–	–	0.39	–
8	ν_8	e′	628	626	–	1.00	1.00
9	ν_9	a_1′	648	–	655	0.83	–
10	ν_{10}	a_2″	–	687	–	–	0.02
11	ν_{11}	e′	–	740	748	–	1.51
12	ν_{12}	a_1′	803	–	–	2.22	–
13	ν_{13}	e′	796	797	–	0.91	0.38

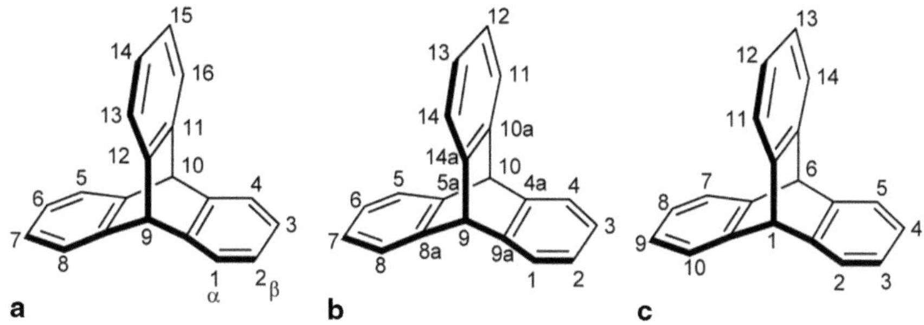

Fig. 1.6 Nomenclature of triptycene

2 3

Fig. 1.7 Nomenclature of pentiptycenes **2** and **3**

In 1980s, Hart et al. [2] proposed the concept of "iptycene" originated from the basic unit triptycene. The iptycene molecules are named after the number of arene planes separated by a bridgehead system, for example, triptycene, pentiptycene, and heptiptycene are the system with three, five, and seven phenyl rings, respectively, in which the prefix indicates the number of independent arenes. However, for these larger iptycenes, there is more than one structural isomer. In order to precisely define the structure of iptycenes, Hart et al. had suggested a set of descriptors. The example of pentiptycene illustrates this regulation, both compounds **2** and **3** are pentiptycene, it is named **2** as $[1.1.1^{b}.1.1]$pentiptycene and **3** as $[1.1.1^{a}.1.1]$pentiptycene (Fig. 1.7), respectively, where "the 1's indicate that each ring is benzenoid and the superscripts (a and b) refer to the bond to which the sp^3 carbons are attached" [11].

References

1. Bartlett PD, Ryan MJ, Cohen SG (1942) Triptycene (9,10-*o*-benzenoanthracene). J Am Chem Soc 64:2649–2653
2. Hart H, Shamouilian S, Takehira Y (1981) Generalization of the triptycene concept. Use of diaryne equivalents in the synthesis of iptycenes. J Org Chem 46(22):4427–4432
3. Skvarchenko VR, Shalaev VK, Klabunovsk EI (1974) Advances in the chemistry of triptycene. Russ Chem Rev 43(11):951–966
4. Swager TM (2008) Iptycenes in the design of high performance polymers. Acc Chem Res 41(9):1181–1189
5. Chong JH, MacLachlan MJ (2009) Iptycenes in supramolecular and materials chemistry. Chem Soc Rev 38(12):3301–3315

6. Mati IK, Cockroft SL (2010) Molecular balances for quantifying non-covalent interactions. Chem Soc Rev 39(11):4195–4205
7. Zhao L, Li Z, Wirth T (2010) Triptycene derivatives: synthesis and applications. Chem Lett 39(7):658–667
8. Chen CF (2011) Novel triptycene-derived hosts: synthesis and their applications in supramolecular chemistry. Chem Commun 47(6):1674–1688
9. Han T, Jiang Y, Chen CF (2007). Developments in synthesis and applications of triptycene derivatives. Chin Sci Bull 52(12):1349–1361
10. Jiang XQ, Guan X, Zhang H (2011) New progress of researches in triptycene derivatives. Chin J Org Chem 31(7):949–963
11. Yang JS, Yan JL (2008) Central-ring functionalization and application of the rigid, aromatic, and H-shaped pentiptycene scaffold. Chem Commun 7(13):1501–1512
12. Cao J, Jiang Y, Chen CF (2011) Advances on synthesis and applications of iptycenes and their derivatives. Prog Chem 23(11):2200–2214
13. Beyeler A, Belser P (2002) Synthesis of a novel rigid molecule family for the investigation of electron and energy transfer. Coord Chem Rev 230(1–2):29–39
14. Asche C (2005) Antitumour quinones. Mini Rev Med Chem 5(5):449–467
15. Bouffard J, Eaton RF, Mueller P, Swager TM (2007) Iptycene-derived pyridazines and phthalazines. J Org Chem 72(26):10166–10180
16. McKeown NB, Budd PM, Book D (2007) Microporous polymers as potential hydrogen storage materials. Macromol Rapid Commun 28(9):995–1002
17. Andrew TL, Swager TM (2011) Structure-property relationships for exciton transfer in conjugated polymers. J Polym Sci Polym Phys 49(7):476–498
18. Jiang Y, Chen CF (2011) Recent developments in synthesis and applications of triptycene and pentiptycene derivatives. Eur J Org Chem 43(13):6377–6403
19. Tsui NT, Paraskos AJ, Torun L, Swager TM, Thomas EL (2006) Minimization of internal molecular free volume: a mechanism for the simultaneous enhancement of polymer stiffness, strength, and ductility. Macromolecules 39(9):3350–3358
20. Long TM, Swager TM (2001) Minimization of free volume: alignment of triptycenes in liquid crystals and stretched polymers. Adv Mater 13(8):601–604
21. Hart H, Bashirhashemi A, Luo J, Meador MA (1986) Iptycenes—extended triptycenes. Tetrahedron 42(6):1641–1654
22. SDBSWeb National Institute of Ad-vanced Industrial Science and Technology, SDBS No. 2494. http://riodb01.ibase.aist.go.jp/sdbs/. Accessed 12 June 2012
23. Wilcox CF (1960) The ultraviolet spectrum of triptycene. J Chem Phys 33(6):1874–1875
24. Talrose V, Yermakov AN, Usov AA, Goncharova AA, Leskin AN, Messineva NA, Trusova NV, Efimkina MV, "UV/Visible Spectra" in NIST Chemistry WebBook, NIST Standard Reference Database Number 69, Eds. Linstrom PJ and Mallard WG, National Institute of Standards and Technology, Gaithersburg MD, 20899. http://webbook.nist.gov. Accessed 12 June 2012
25. Furlan A, Fischer T, Fluekiger P, Gudel HU, Leutwyler S, Luthi HP, Riley MJ, Weber J (1992) Low-frequency vibrations of triptycene. J Phys Chem 96(26):10713–10719

Part II
Synthesis and Reactions of Iptycenes and Their Derivatives

Chapter 2
Synthesis and Reactions of Triptycenes and Their Derivatives

2.1 Synthesis of Triptycenes and Their Derivatives

In order to verify if the triptycyl (the analog of triphenylmethyl in which the three phenyl groups were united to a CH) free radical **1a** (Fig. 2.1) should be more instable than the triphenylmethyl radical itself, Bartlett et al. [1] first synthesized triptycene **1** in a low total yield by a multistep route starting from the Diels–Alder addition reaction of anthracene and benzoquinone (Scheme 2.1) in 1942. Several years later, Bartlett et al. [2] also reported the synthesis of 9-bromotriptycene (**2**) from the 9-bromoanthracene and benzoquinone according to the similar synthetic strategy. However, it was noted that the target compound **2** could be immediately afforded by the deamination of the diamine with $NaNO_2$ in the presence of 50 % H_3PO_2 at −3 °C (Scheme 2.2).

Craig and Wilcox [3] provided a new route to the synthesis of triptycene **1** based on the reduction of triptycene monoquinone, the adduct between anthracene and *p*-benzoquinone, with $LiAlH_4$ or $NaBH_4$. The resulting reduction product was treated with ethanolic hydrochloric acid, and then followed by the chromatography on acid alumina to give the target product **1** in 15 % overall yield.

In 1956, Wittig and Ludwig [4] developed a convenient and one-pot method for the synthesis of triptycene **1** by the addition reaction between benzyne and anthracene. During the next few years, various routes to triptycene via benzyne had been reported. For example, Wittig et al. [4–6] synthesized the triptycene in 30 % yield by the addition reaction between anthracene and organomagnesium reagent generated by the reaction of *o*-bromochloro benzene with magnesium in a THF or ether solution. Soon after, they also reported the synthesis of triptycene **1** by the reaction of chlorobenzene with anthracene in the presence of butyl lithium [7] or $LiSb(C_6H_5)_6$ [8], respectively. Similarly, Sharp and co-workers [9, 10] synthesized triptycene by treating halogenated benzene with potassium *t*-butoxide, followed by the treatment of anthracene [11].

By the addition reaction between tetrafluorobenzyne and anthracene, in 1968, Heaney and co-workers [12, 13] reported the synthesis of tetrahalogenated triptycene **3** in a yield of 42 % (X = F) or 54 % (X = Cl), along with the subsidiary product **4** (Scheme 2.3).

C.-F. Chen, Y.-X. Ma, *Iptycenes Chemistry*,
DOI 10.1007/978-3-642-32888-6_2, © Springer-Verlag Berlin Heidelberg 2013

Fig. 2.1 Structure of triptycyl free radical **1a**

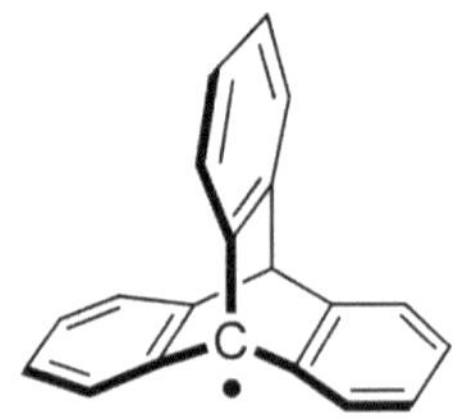

In 1964, Cadogan and Hibbert [14] reported the synthesis of *t*-butyl-substituted triptycene **5** by the reaction of anthracene with the corresponding aryne, which was formed by the heterolytic cleavage of *o*-*t*-butyl-*N*-nitrosoacetanilide in benzene (Scheme 2.4a). Followed by the similar synthetic route, Cadogan and co-workers [15] further synthesized the di-*t*-butyl-substituted triptycene **6** in a 16 % yield by the addition reaction of the corresponding di-*t*-butyl-substituted aryne with anthracene in benzene (Scheme 2.4b).

Depending on the different processes of generating benzyne or arynes, triptycene (Table 2.1) or substituted triptycenes could be obtained in different yields, which might also be regarded as a standard for testing the reactivity of benzyne or arynes.

In 1960, Stiles and Miller [21] reported the synthesis of triptycene **1** in 30 % yield by using the *o*-diazonium benzoate as the precursor of benzyne, which provided a more convenient and efficient method for the synthesis of triptycene than the previous ones. Soon after, Friedman and Logullo [25] described another type of precursor for

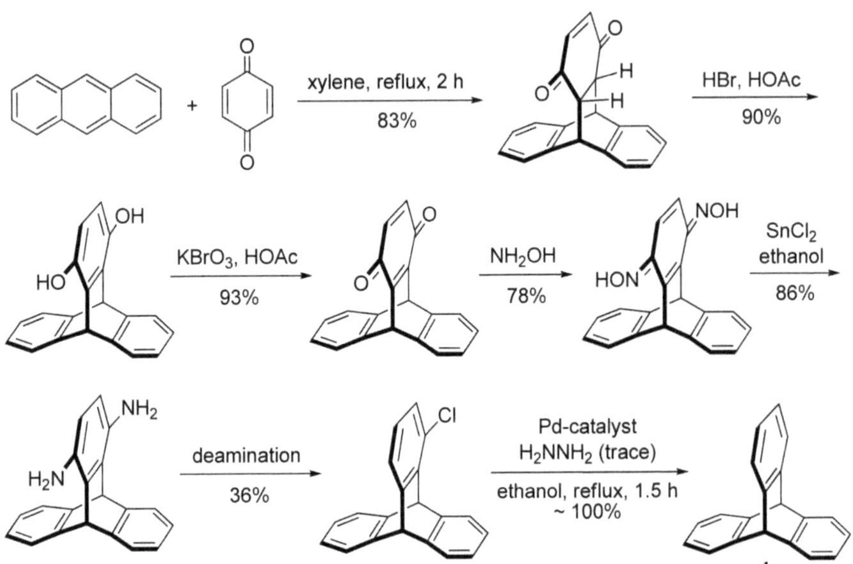

Scheme 2.1 Synthesis of triptycene **1** from anthracene with benzoquinone

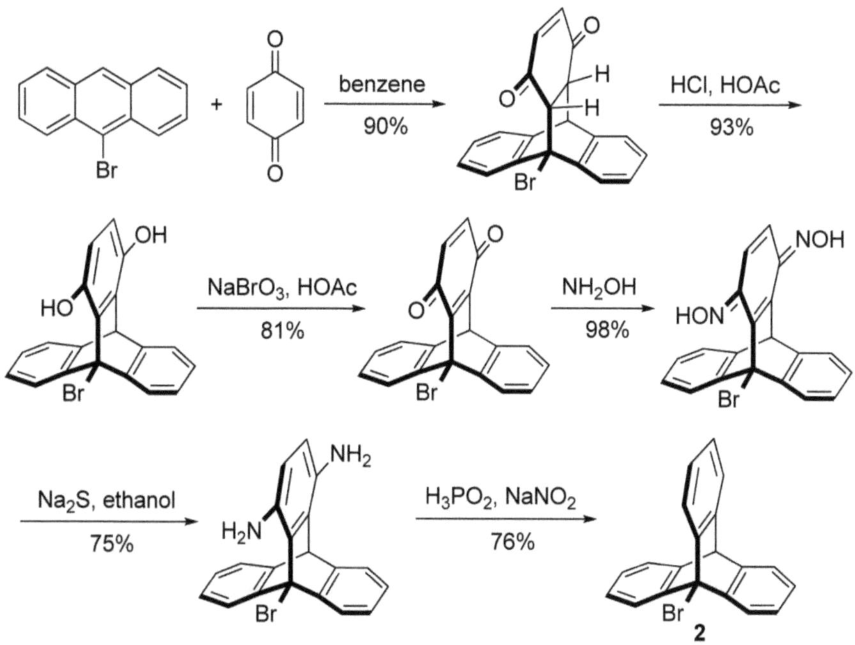

Scheme 2.2 Synthesis of 9-bromotriptycene **2**

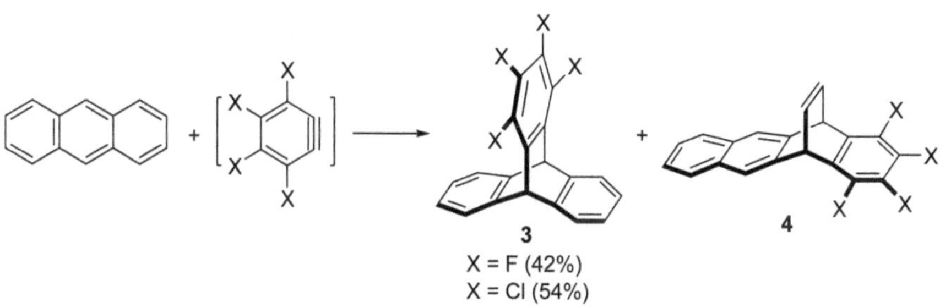

Scheme 2.3 Synthesis of tetrahalogenated triptycene **3**

releasing benzyne in situ. As shown in Scheme 2.5, the diazotization of anthranilic acid in nonprotonic solvent in the presence of amyl nitrite could form benzyne in situ; then, the freshly prepared benzyne reacted with anthrancene and gave triptycene in 50–60 % yield. Moreover, the yield could be further improved up to 70–80 % by using a molar excess of anthracene or anthranilic acid.

To generate benzyne with high efficiency under mild conditions, Kitamura et al. [26] described a novel type of hypervalent iodine benzyne precursor. This hypervalent iodine precursor ((phenyl)[[2]-(trimethylsilyl)phenyl]iodo-niumtriflate) reacted with anthracene to afford the Diels–Alder adduct **1** in 86 % yield at room temperature (Scheme 2.6).

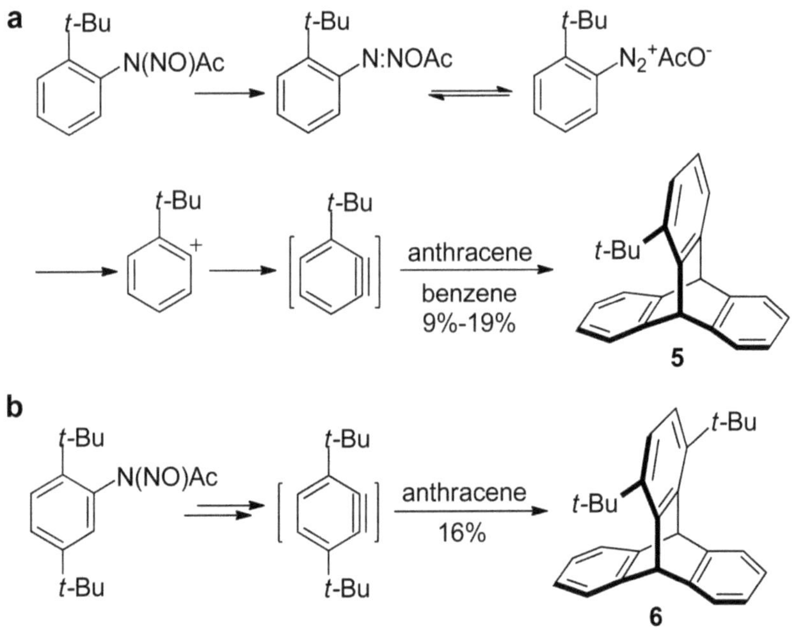

Scheme 2.4 Synthesis of *t*-butyl-substituted triptycene **5** and di-*t*-butyl-substituted triptycene **6**

On the basis of Logullo and Friedman's methodology [25], Kornfeld et al. [27] investigated the reactivity of benzyne toward 9-substituted anthracenes. As shown in Table 2.2, it was found that the different substituents in the anthracene markedly influenced the yields of the triptycene derivatives. According to these results, it was obvious that the yields were relatively high for the anthracenes with electron-donating substituents. By contrast, for the anthracene derivatives with electron-withdrawing groups the yields were unsatisfactory. Particularly, for some strong electron-withdrawing groups (such as R = CHO), there seemed to be no reactions between the 9-substituted anthracenes and benzyne. Although the acetals (Table 2.2, entries 4 and 5) had closed electronic factors, the results were different from their varied steric effect.

Similar to the method developed by Le Goff [24] (Table 2.1, entry 11), Wilcox and Roberts [28] reported the synthesis of diphenyl-substituted triptycene (**7**) in 65 % yield by the reaction between 1,4-diphenylanthracene and benzyne, which was generated in situ from diphenyliodonium-2-carboxylate (Scheme 2.7). According to the procedure of Friedman and Logullo [25], the target compound **7** could be also obtained in 57 % yield with anthranilic acid as a precursor. It was noted that the crude product **7** could be crystallized from the CCl_4 or cyclohexane solution to form complexes, in which the former was more stable.

In 1965, Klanderman [29] investigated the reaction between benzyne and disubstituted anthracenes in detail. As shown in Scheme 2.8, benzyne tended to react in Diels–Alder fashion with the B ring of anthracene **8** to form triptycene **9** or A ring to

Table 2.1 Synthesis of triptycene **1** by the reaction of anthracene with benzyne

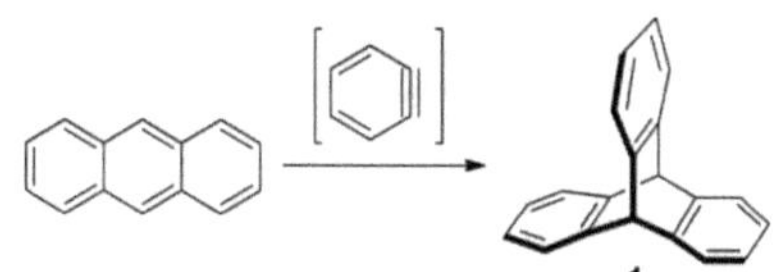

1

Entry	Precursors	Conditions	Yield (%)
1	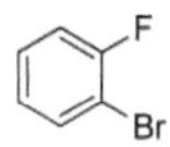	Mg, heat	30 [7]
2	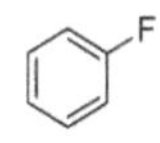	BuLi, r.t.	10 [7]
3	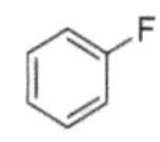	$LiSb(C_6H_5)_6$, r.t.	23 [8]
4	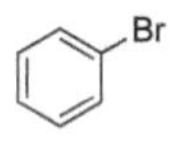	*t*-BuOK, reflux	21 [9]
5	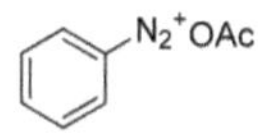		10 [16, 17]
6	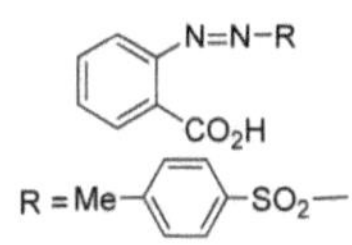	heat	13 [18]
7	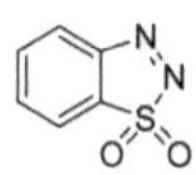		21 [19, 20]
8	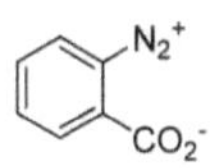	heat	30 [21]
9	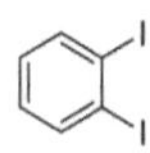	Zn, 550° C	8 [22]
10	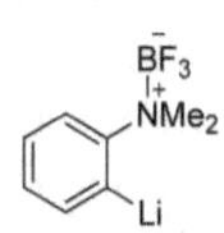	heat	11 [23]
11	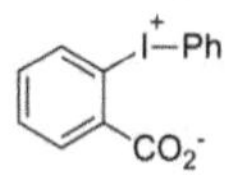	heat	23 [24]

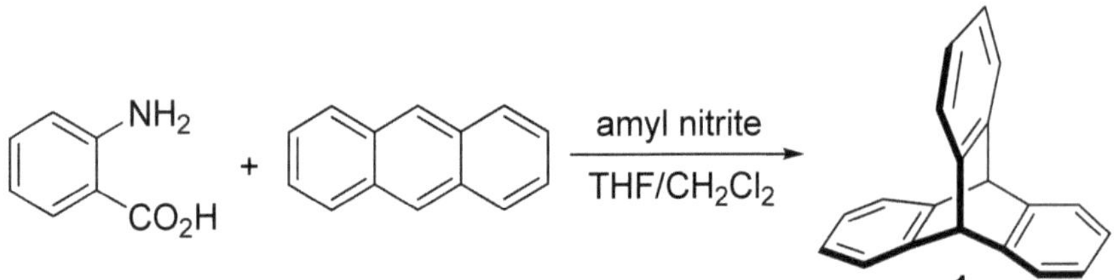

Scheme 2.5 Synthesis of triptycene **1** by using the *o*-diazonium benzoate as the precursor of benzyne

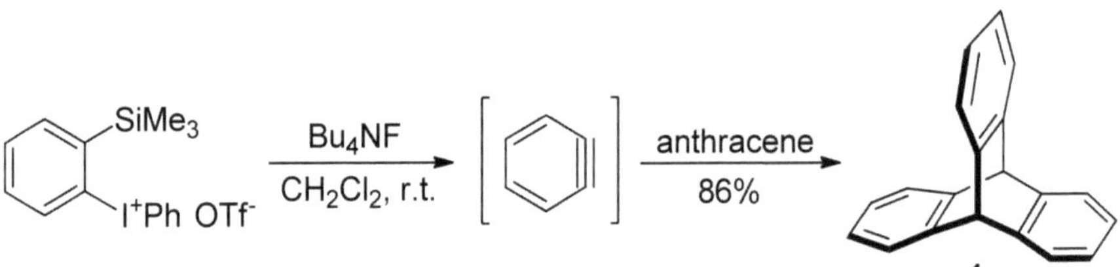

Scheme 2.6 Synthesis of triptycene **1** with hypervalent iodine benzyne precursor

form adduct **10**. The regioselectivity could be controlled by changing the substituents on the anthracene. For the reactions of anthracenes with electron-drawing groups, like cyano group, the ratio of A-ring to B-ring adducts would obviously increase, although, the total yield of the adducts decreased considerably.

Soon after, Klanderman and Criswell [30, 31] also studied the relative reactivity of a number of anthracenes with end-ring substituents toward benzyne. The results showed that benzyne would react in Diels–Alder fashion with all the A, B, and C rings of anthracenes (Scheme 2.9). The characteristics and locations of substituents in the anthracene system greatly influenced the rate of reactions. In general, the electron-donating substituents enhanced the reactivity of the substituted rings to benzyne, whereas the electron-withdrawing substituents lowered it. Furthermore, the steric requirements of benzyne were also considered. However, it was found that except 9,10-diphenyl anthracene, substituents in one ring had no or little effects on the reaction between the other ring and benzyne.

Afterward, Kadosaka and co-workers [32] reported the synthesis of a series of chloro substituted triptycenes by the treatment of benzyne or chlorobenzyne with various chloroanthracene. The mono chlorotriptycene (**15**, Scheme 2.10), dichlorotriptycenes (**16**, **17**, Scheme 2.11) and trichlorotriptycenes (**18**, **19**, Scheme 2.12) could be obtained via the corresponding Diels–Alder reaction of benzyne with chloroanthracene, or chlorobenzyne with anthracene, followed by the Logullo and Friedman's procedure [25].

Table 2.2 Triptycenes prepared from 9-substituted anthracenes

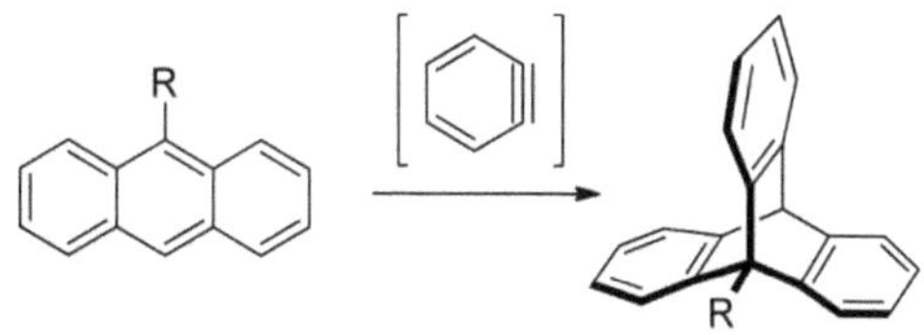

Entry	R	Yield (%)
1	NO_2	15
2	CH_2Cl	70
3	CH_2OH	6
4	$CH(OCH_3)_2$	4
5	$HC(OCH_2)_2$	53
6	CHO	-
7	CH_2CH_2Cl	49
8	$CH_2CH_2CO_2CH_3$	50

Particularly, the addition of 3-chlorobenzyne to the 9-substituted anthracene would give a mixture of two isomers: quasi-*cis*- and quasi-*trans*-isomer (**20**, Scheme 2.13). In this case, it was believed that the inductive effect of the substituents would take the leading role in determining the ratios of the isomers, whereas the steric effects had almost no or little contributions.

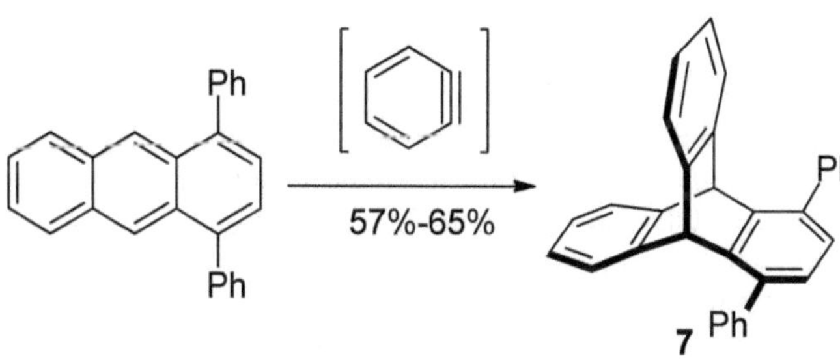

Scheme 2.7 Synthesis of compound **7**

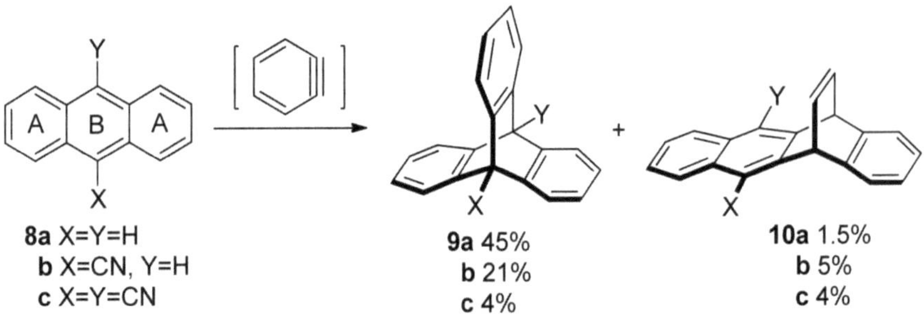

Scheme 2.8 Reaction between benzyne and disubstituted anthracenes **8**

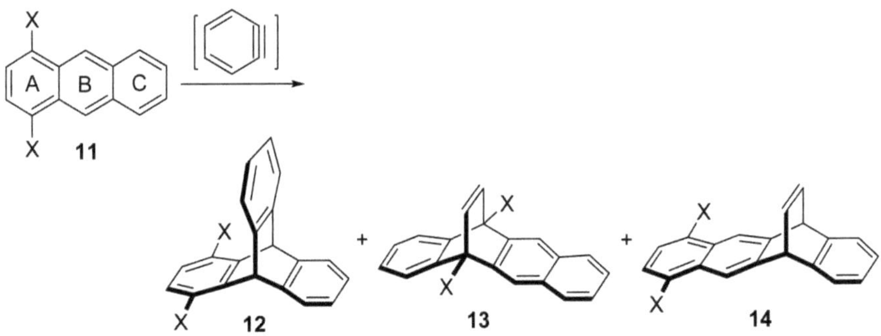

Scheme 2.9 Reaction between benzyne and anthracenes **9** with end-ring substituents

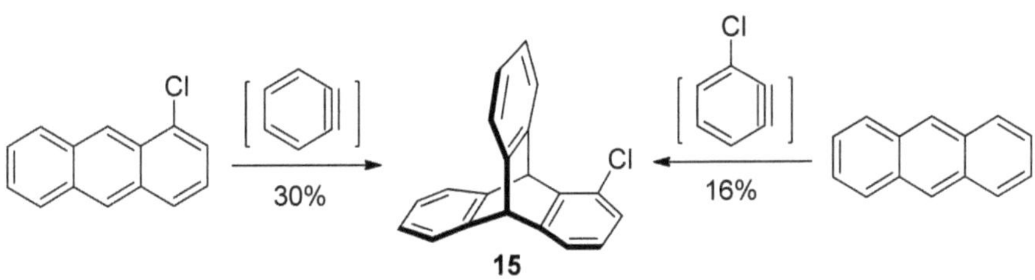

Scheme 2.10 Synthesis of monochlorotriptycene **15**

Subsequently, Rogers and Averill [33] synthesized a series of 1,8,13- and 1,8,16-trisubstituted triptycenes by the treatment of 1,8-disubstituted anthracenes with monosubstituted benzynes, based on Kadosaka's synthetic strategy [32]. As shown in Table 2.3, the nature of the substituents on the anthracene and benzyne units had obviously influenced the ratio of *syn/anti* isomers. However, it was further found that this stereo selectivity mainly depended on the substituents of benzynes.

Following the route of Friedman and Logullo [25], Klanderman and Faber [34] reported the synthesis of 9,10-bis(acetoxymethyl)triptycene (**21**) from 9,10-bis(acetoxymethyl)anthracene, which was generated by the reaction of 9,10-bis(chloromethyl)anthracene with potassium acetate and acetic acid (Scheme 2.14a).

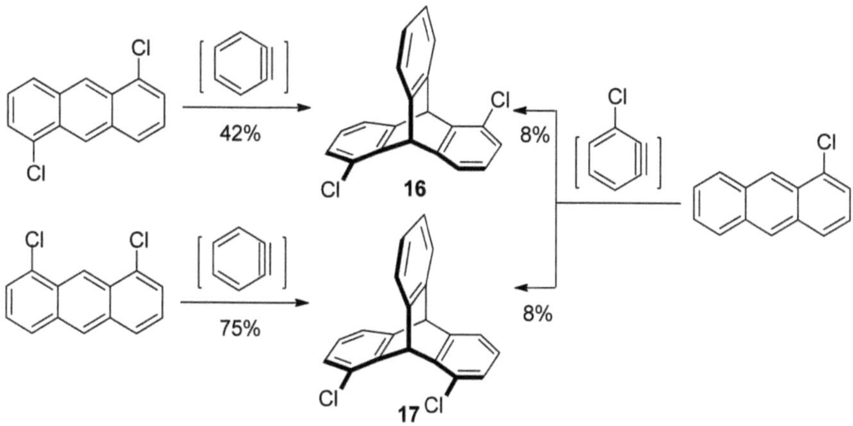

Scheme 2.11 Synthesis of dichlorotriptycenes **16** and **17**

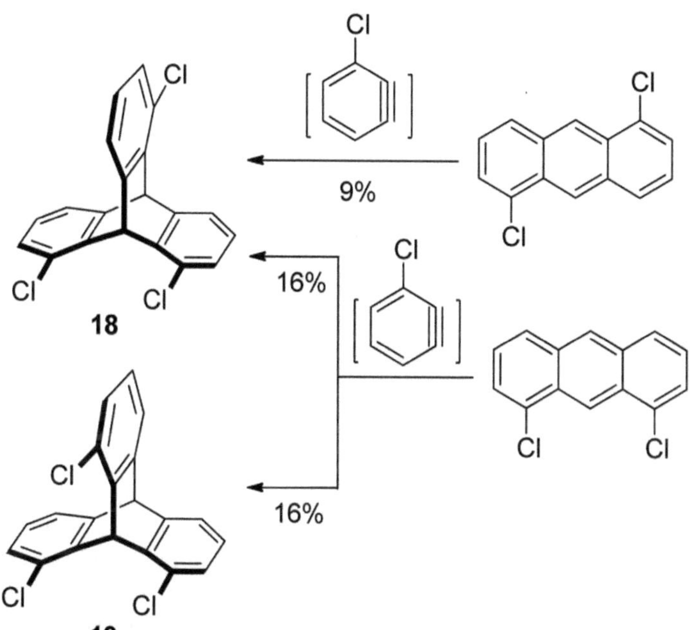

Scheme 2.12 Synthesis of trichlorotriptycenes **18** and **19**

According to the similar procedure, Hoffmeister et al. [35] also obtained 9,10-bis(acetoxymethyl)triptycene (**21**) and its hydrolysis product, 9,10-bis(hydroxymethyl) triptycene (**22**) in high yield (Scheme 2.14b). Moreover, they found that compound **22** was a key precursor for the preparation of other bisubstituted triptycene. Consequently, the hydroxymethyl group in compound **22** could effectively convert into other groups by further reactions (Scheme 2.15).

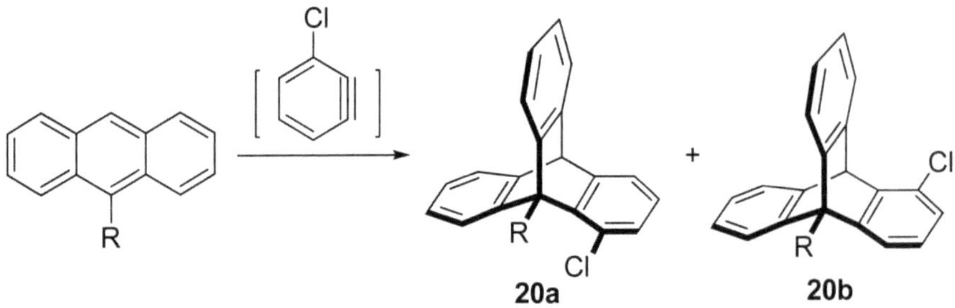

Scheme 2.13 Synthesis of quasi-*cis*- and quasi-*trans*-isomers **20a** and **20b**

Table 2.3 Ratios of *anti/syn* isomers of trisubstituted triptycenes

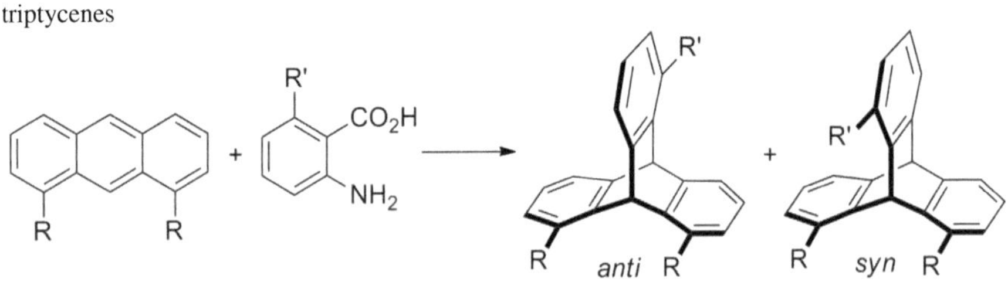

Entry	R	R′	*Anti* (%)	*Syn* (%)	Yield (%)
1	Cl	Me	25	75	74
2	CN	Me	28	72	57
3	CO_2Me	Me	31	69	58
4	Cl	Cl	77	23	27
5	CO_2Me	Cl	73	27	20
6	Cl	CO_2Me	44	56	47
7	CN	CO_2Me	99	1	38
8	CO_2Me	CO_2Me	76	24	62

Recently, Mitzel and co-workers [36] studied the influence of substituent of C-10 position in a series of 1,8-dichloroanthracene precursor molecules on the selectivity of *syn*- and *anti*-trichlorotriptycenes in depth. As shown in Table 2.4, it was found that in the case of no substituent (R = H, entry 1), the reaction of 1,8-dichloroanthracene with 6-chloroanthranilic acid gave the trichlorotriptycenes in a mixture of 21 % *syn* and 79 % *anti* form, which was similar to the result reported by Rogers and Averill [33]. For a small C-10 substituent like methyl group at the 1,8-dichloroanthracene, the reaction led the ratio of *syn*- to *anti*-forms to be 37:63 (entry 5). However, for a large group (R = *t*-Bu, entry 3), the target product was unexpectedly obtained in 100 % *anti* form, which might relate to the marked distortion of the planarity of backbone. In contrast, the results of calculations showed that C-10 substituents always changed the orbital coefficients in a way to favor the *anti* over the *syn* form, and the experiments demonstrated that such electronic effects clearly override steric repulsion effects during the reaction.

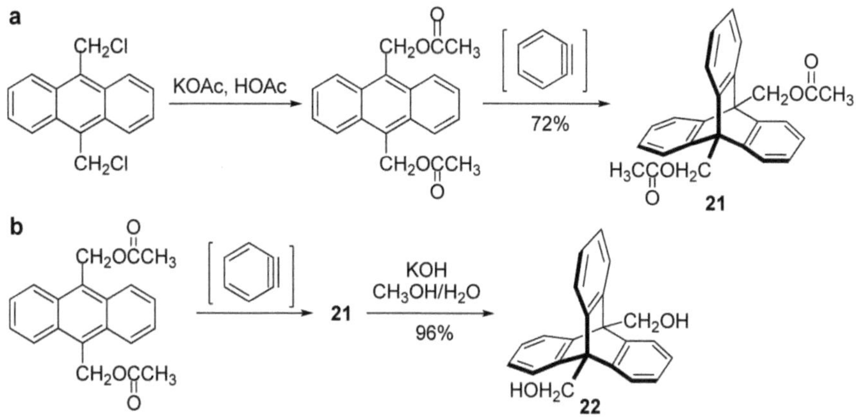

Scheme 2.14 Synthesis of 9,10-bis-(acetoxymethyl)triptycene **21** and 9,10-bis(hydroxymethyl) triptycene **22**

Anthracene derivatives with substituents in the end-rings reacted with benzyne to give a racemic mixture of optical enantiomers. Thus, stereo isomeric disubstituted triptycenes (**31a, b**) were obtained from 1,5-disubstituted anthracenes [37] (Scheme 2.16a). Similarly, Kricka and Vernon [38] utilized the addition between benzyne generated in situ and the 1,4,9,10-tetramethylanthrace to give the tetrasubstituted triptycene **32** (Scheme 2.16b).

With more and more species of benzynes being discovered, various functionalized triptycenes could be easily and efficiently formed via the widely used Diels–Alder approach. In 2002, Marks et al. [39] reported the synthesis of compound **33** by the addition of benzyne or 3,6-dimethylbenzyne to 1,2,3,4,5,6,7,8-octaethylanthracene (Scheme 2.17). According to the data of low-temperature NMR spectra, crystal structure, and MM3 calculation, the ethyl groups in both of the target compounds might arrange in a perfectly alternated up-down mode to achieve a low energy conformation. Almost at the same time, Pascal and co-workers [40] obtained a highly crowded triptycene derivative: 1,2,3,4,5,6,7,8,13,14,15,16-dodecaphenyltriptycene (**34**) by the cycloaddition of tetraphenyl benzyne to octaphenyl anthracene (Scheme 2.18). The yield was only 11 %, due to the relatively large size of phenyl groups.

In 2005, Zhu and Chen [41] reported the synthesis of hexamethoxyltriptycene **35** in 65 % yield by the reaction of 2,3,6,7-tetramethoxy-9,10-dimethylanthracene and 4,5-dimethoxybenzenediazonium-2-carboxylate in 1,2-dichloroethane in the presence of 2-methyloxirane. Furthermore, the demethylation of **35** with boron bromide in dichloromethane gave triptycene tri(catechol) **36** in 98 % yield, which could be regarded as a new building block for the construction of supramolecular systems (Scheme 2.19).

By modifying the reaction conditions described by Cadogan et al. [10], Anzenbacher and co-workers [42] synthesized di(thien-2-yl)triptycene **37** in 87 % yield by the reaction between the corresponding 3-halo-1,4-bis(thien-2-yl)benzene and anthracene in the presence of potassium *t*-butoxide (Scheme 2.20).

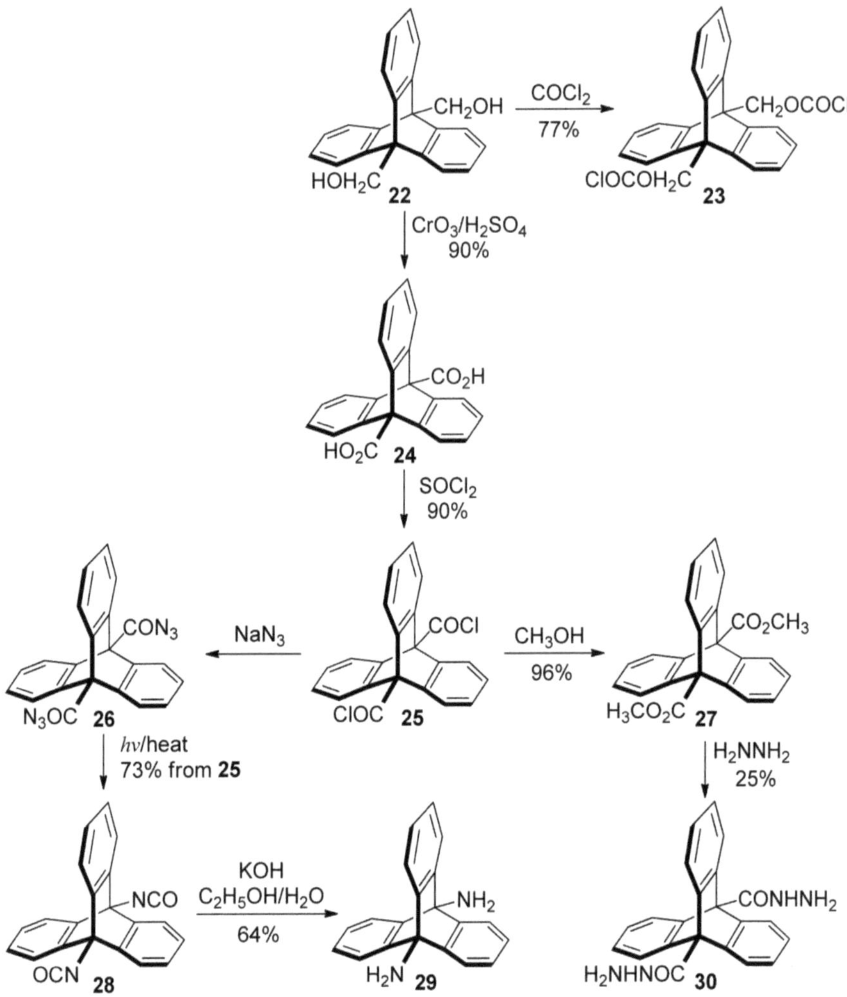

Scheme 2.15 Synthesis of disubstituted triptycenes from compound **22**

To avoid the isomeric forms in the Diels–Alder reaction, Závada and co-workers [43] developed a strategy for a stepwise formation of the third aromatic ring in the

Table 2.4 Ratio of *syn*- and *anti*-triptycene isomers depending on the C-10 substituent

Entry	R on C-10	*Anti* (%)	*Syn* (%)	Yield (%)
1	H	79	21	16
2	*c*-C_6H_{11}	79	21	60
3	*t*-Bu	100	0	43
4	*i*-Pr	70	30	40
5	Me	63	37	42
6	*n*-Bu	60	40	22
7	Ph	75	25	28

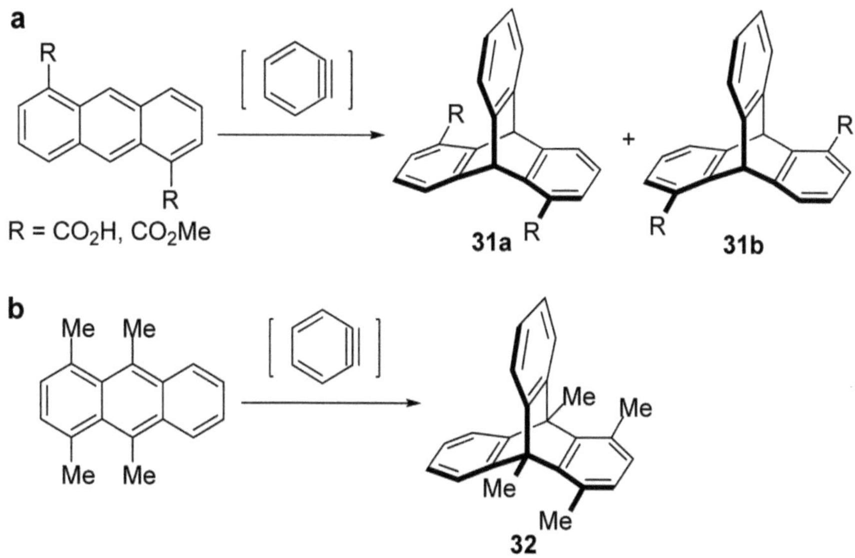

Scheme 2.16 Synthesis of stereo isomeric disubstituted triptycenes **31a**, **b** (**a**) and tetrasubstituted triptycene **32** (**b**)

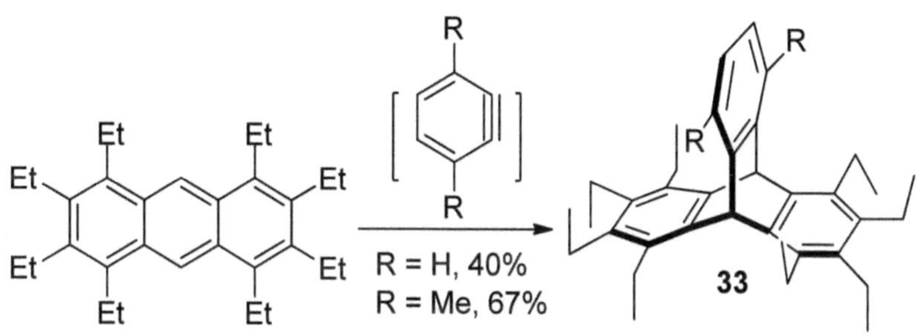

Scheme 2.17 Synthesis of compound **33**

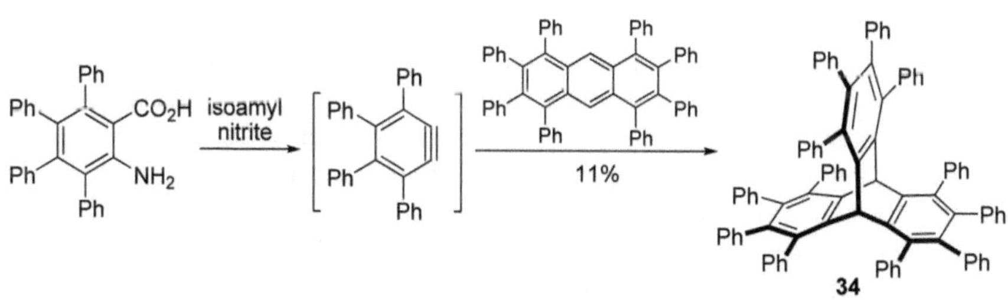

Scheme 2.18 Synthesis of 1,2,3,4,5,6,7,8,13,14,15,16-dodecaphenyltriptycene **34**

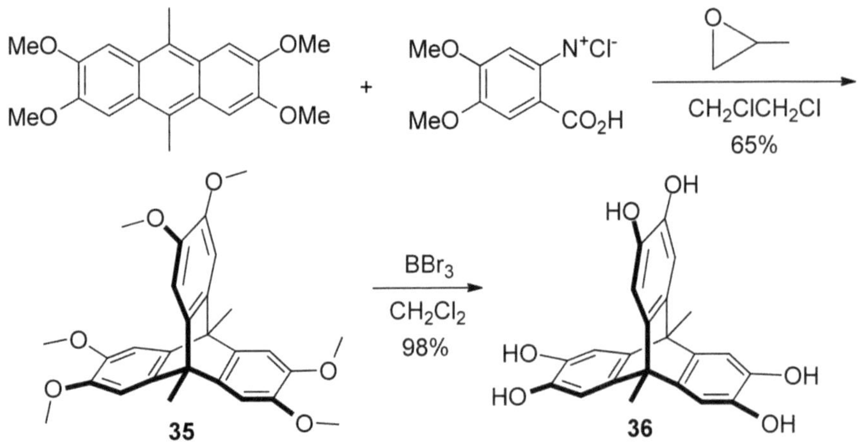

Scheme 2.19 Synthesis of triptycene tri(catechol) **36**

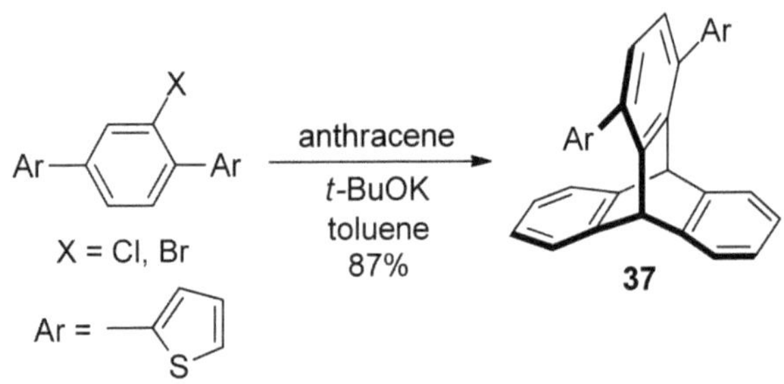

Scheme 2.20 Synthesis of di(thien-2-yl)triptycene **37**

triptycene skeleton. As shown in Scheme 2.21, although the process was longer than that of the direct aryne addition method, the target products **38a**, **b** were in the good overall yields. Moreover, the Diels–Alder addition of the resulting *exo*-diene to dimethyl acetylenedicarboxylates and the subsequent aromatization were of the utmost importance.

The Diels–Alder reactions between anthracene and benzyne or quinone are the most important methods for the synthesis of triptycene and its derivatives, whereas the strategy without the process of the Diels–Alder addition reaction provides another opportunity to synthesize triptycene and its derivatives. In 1968, Walborsky and Bohnert [44] discovered that treatment of the 9-phenyl-9-ethyl-10-methylene-9,10-dihydroanthracene with polyphosphoric acid could give 9-ethyl-10-methyl triptycene **39** via the compound **41** in an excellent yield (Scheme 2.21). In addition, the precursor **41** of **40** was treated with polyphosphoric acid to give **39** in good yield as well. Although the scope of this strategy was limited, it provided a new design idea to the synthesis of triptycene which totally differed from that of the Diels–Alder type.

Almost at the same time, Taylor and Swager [45] investigated the synthesis of 9,10-triptycenediols by metal-catalyzed [2 + 2 + 2] cyclotrimerization reaction's

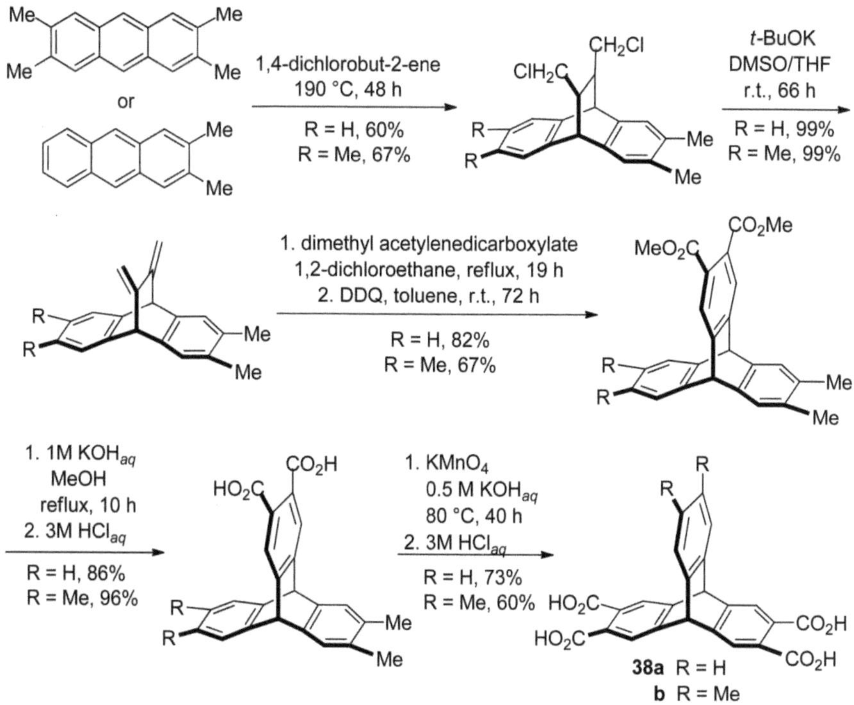

Scheme 2.21 Synthesis of compounds **38a, b**

cycloaddition. To identify the catalytic efficiency, a series of ruthenium- and rhodium-based catalysts were evaluated by the reaction of 1-hexyne. The results showed that the yields of products depended on the catalyst. This method could afford a wide scope of multisubstituted triptycene derivatives in high yields via only two or three steps from anthraquinones and alkynes (Table 2.5). To some extent, it could also take the place of Diels–Alder fashion in the synthesis of multisubstituted triptycene derivatives.

In 2008, Asao and co-workers [46] reported that the reaction between *ortho*-alkynylbenzaldehydes and benzenediazonium 2-carboxylate could undergo in the presence of AuCl catalyst via a zwitterionic intermediate at 80 °C for 0.5 h, which gave triptycyl ketones **49a** and **49b** in 51 and 35 % yield, respectively (Scheme 2.23).

2.2 Synthesis of Triptycenequinones and Their Derivatives

As early as 1931, Clar [47] first reported triptycene monoquinone which was obtained by the reaction between anthracene and *p*-benzoquinone. Later, Bartlett et al. [1, 48] improved the Clar's route, by the reaction between anthracene and *p*-benzoquinone in xylene under reflux for 2 h which gave product **50** in 83 % yield after recrystallization

Table 2.5 Synthesis of substituted 9,10-triptycenediols by rhodium-catalyzed [2 + 2 + 2] cycloaddition

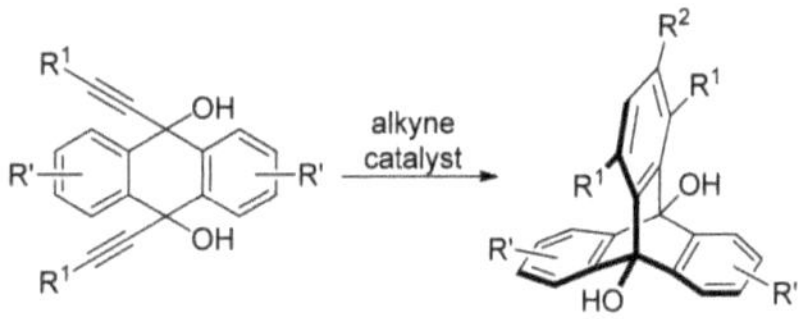

TP

Entry	Diyne	Alkyne	TP	Yield (%)
1	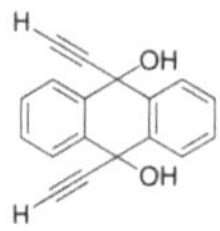	H—≡—n-Bu	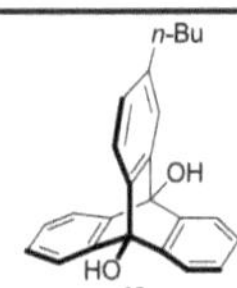 **42**	95
		H—≡—$C_{10}H_{21}$	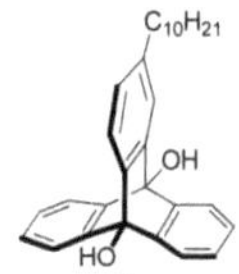 **43**	86
		H—≡—Ph	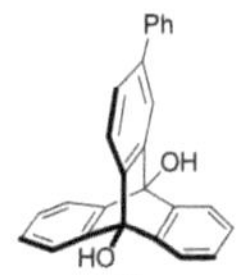 **44**	73
		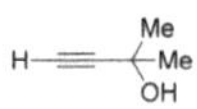	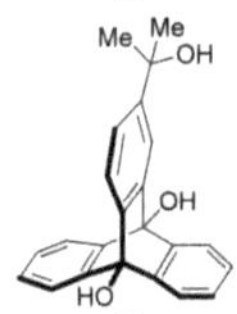 **45**	95
2	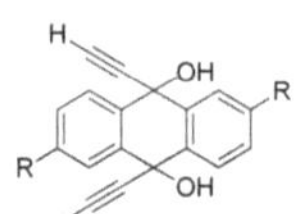 R = OC_8H_{17}	H—≡—n-Bu	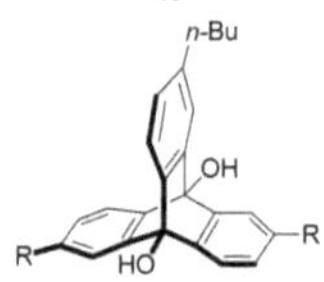 **46** R = OC_8H_{17}	81
3	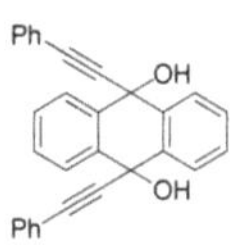	H—≡—n-Bu	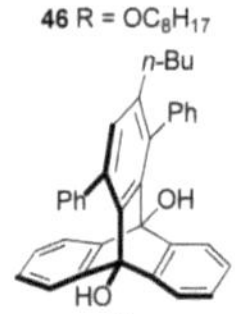 **47**	99
		norbornadiene	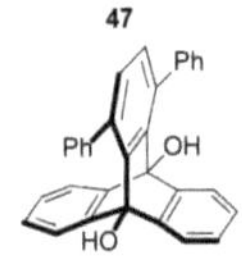 **48**	91

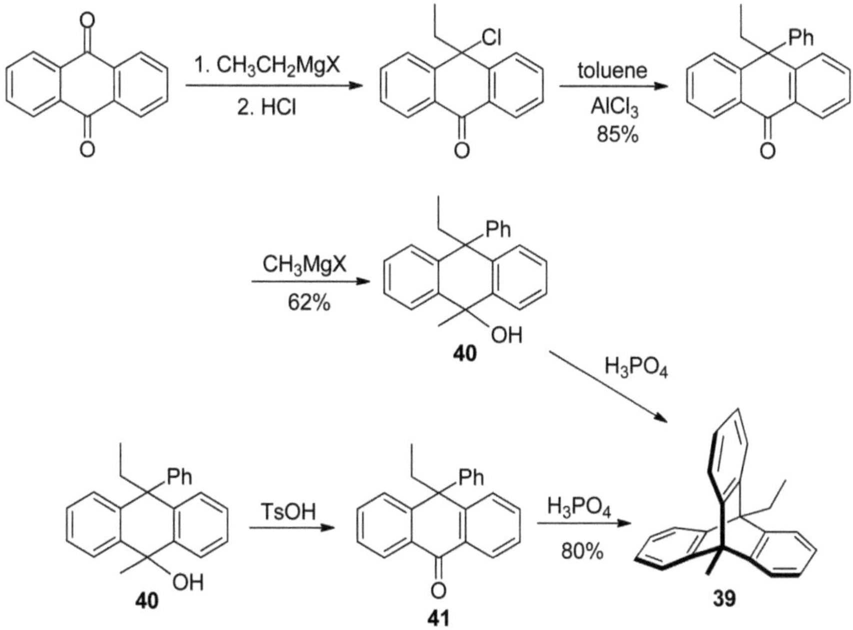

Scheme 2.22 Synthesis of 9-ethyl-10-methyl triptycene **39**

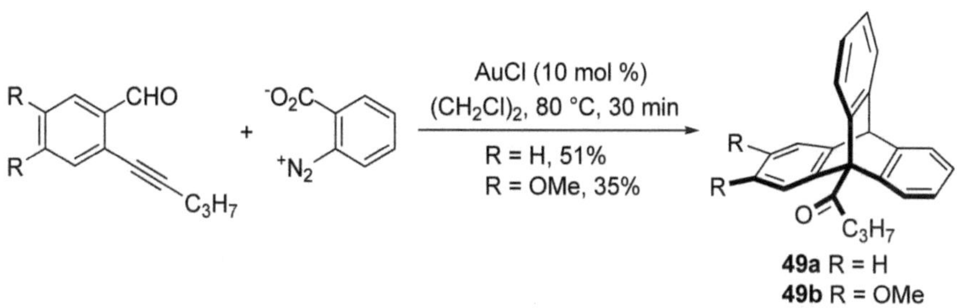

Scheme 2.23 Synthesis of triptycyl ketones **49a, b**

from xylene (Scheme 2.24). This method is simple and practical, even today the preparation of triptycene monoquinone almost follows this route.

Sonoda et al. [49] tried the reaction between 2-carboxyanthracene **51** with *p*-benzoquinone in several solvents, and found that the results were beyond their expectations. The products were produced by the addition of **52** (the homologous methyl ester of **51**) with *p*-benzoquinone instead of compound **51**. Moreover, there were isomeric issues in forming the substituted triptycene quinone. By the reaction between **52** and *p*-benzoquinone, the isomeric adducts (**53a, b**) in both *exo*- and *endo*-configuration were obtained (Scheme 2.25). Furthermore, the isomers **53a, b** were all sensitive under the base conditions; thus, the matching carboxylic acids could be obtained by the hydrolysis of **52** in aqueous NaOH solution. When these

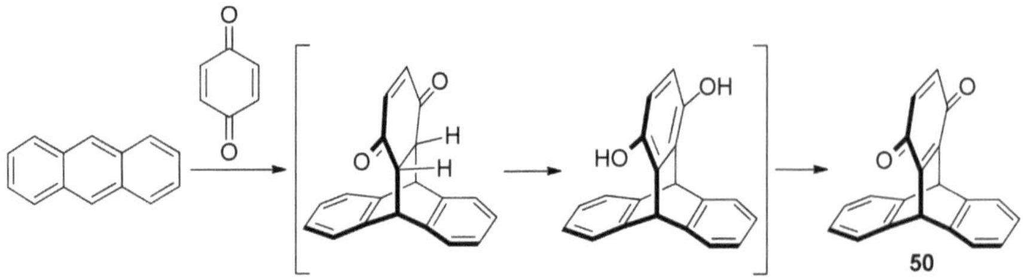

Scheme 2.24 Synthesis of triptycene quinone **50**

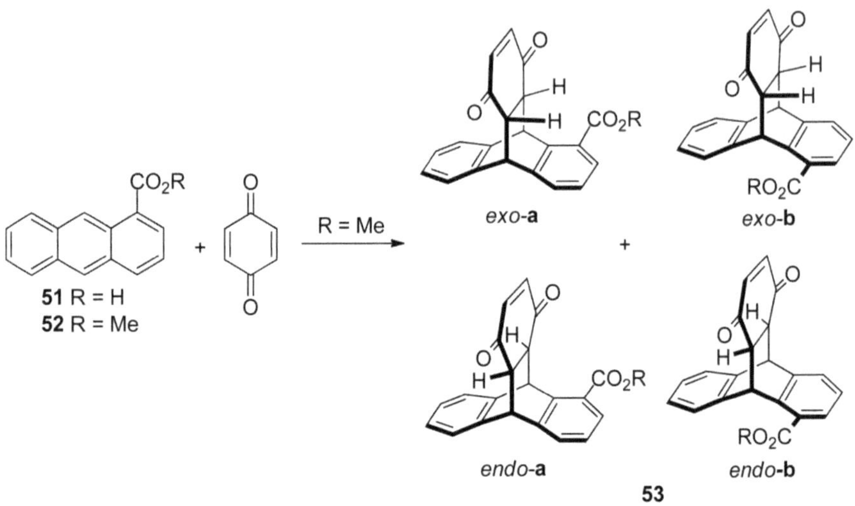

Scheme 2.25 Addition of 2-carboxyanthracene **51** and its homologous methyl ester **52** with *p*-benzoquinone

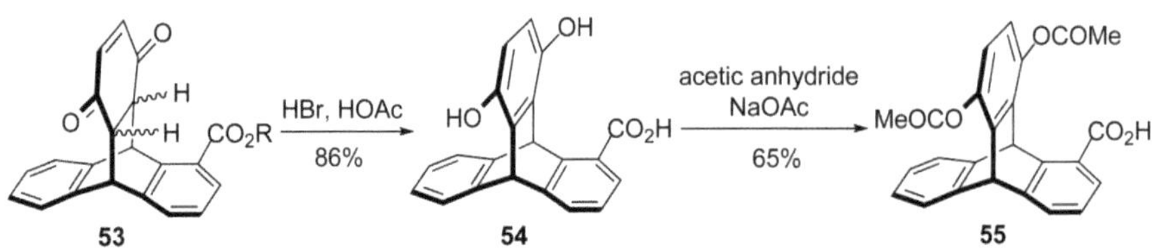

Scheme 2.26 Synthesis of compound **55**

isomers **53a**, **b** were treated with HBr in an acetic acid solution, the corresponding product **54** was afforded in a crude yield of 86 %. As the compound **54** was unstable to oxygen, the following esterification of **54** gave **55** in 65 % yield (Scheme 2.26).

In 1978, Iwamura and Makino [50] reported that the Diels–Alder reaction of 1,4-dimethoxy anthracene with *p*-benzoquinone in acetonitrile under reflux temperature for 10 h gave a mixture of *endo*- and *exo*-9,10-adducts in 86 % yield. Treatment of the stereoisomers with potassium hydroxide and silver oxide would

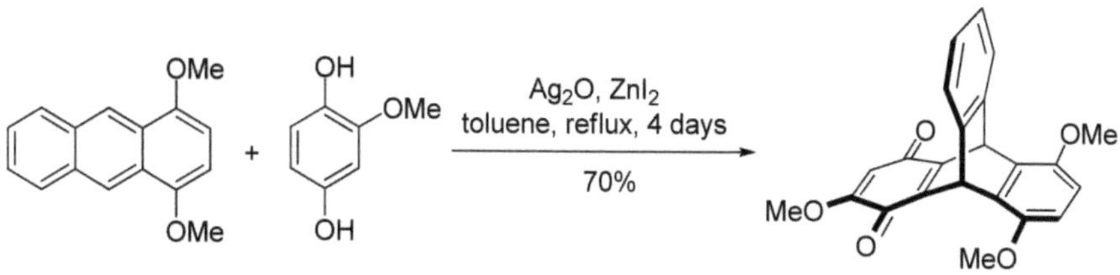

Scheme 2.27 Synthesis of triptycene monoquinone **56**

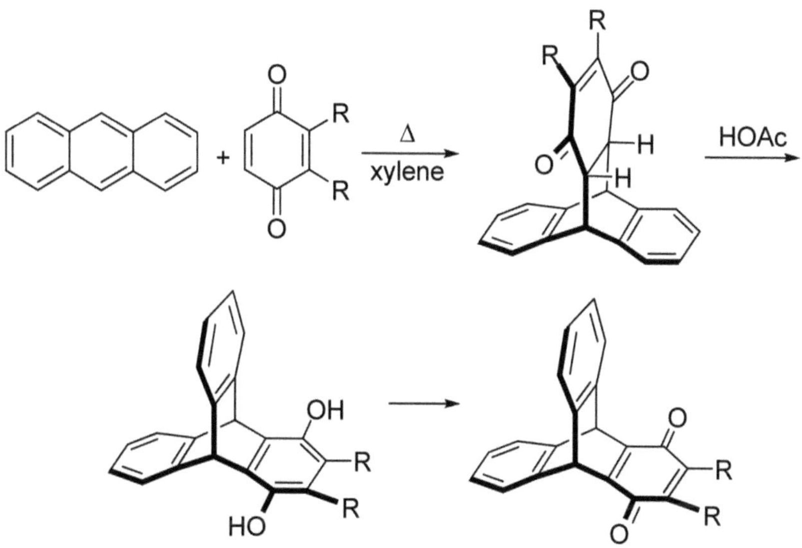

Scheme 2.28 Synthesis of disubstituted triptycene quinone **57**

produce 5,8-disubstituted triptycene monoquinones. Not long ago, Hua et al. [51] described a one-pot synthesis of the triptycene monoquinone **56** (2,5,8-trimethoxy-9,10-dihydro-9,10-[[1],[2]]benzeneanthracene-1,4-dione) by heating a mixture of 1,4-dimethoxyanthracene and methoxy-hydroquinone in toluene in the presence of silver oxide and zinc iodide (Scheme 2.27).

One decade ago, Wiehe et al. [52, 53] reported a new synthetic strategy involving a classical Diels–Alder reaction between an anthracene derivative and excess of quinone in acetic acid to give substituted triptycene quinone. In a depth analysis, they believed that the initial [4 + 2] cycloadduct was interconversed in situ to a triptycene hydroquinone, then quickly oxidized to the simple disubstituted triptycene quinone **57** in 81 % (R = Me) and 47 % (R = OMe) overall yield, respectively (Scheme 2.28). According to the same approach, a series of alkyl- and aryl-substituted triptycene quinones (**58–64**, Table 2.6) had been obtained in good yields by the

Table 2.6 Synthesis of triptycene quinones

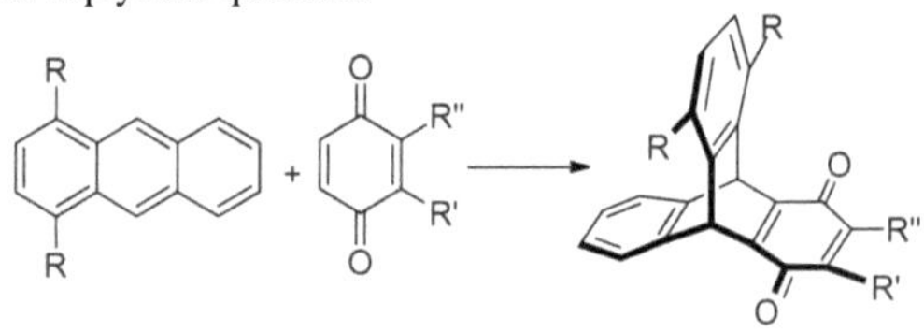

Entry	anthracene	quinone	Product
1	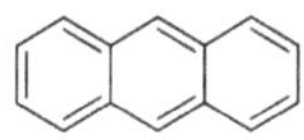	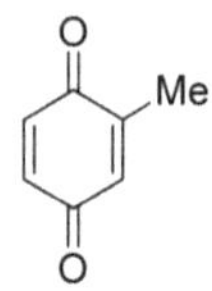	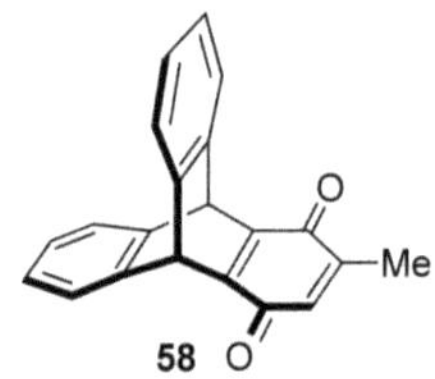 **58**
2	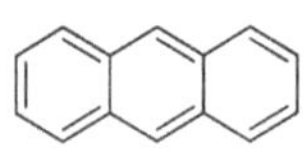	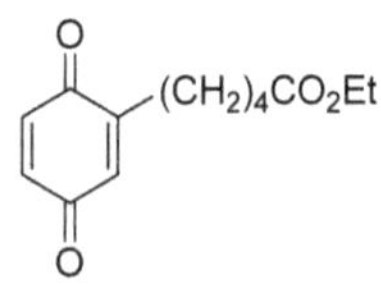	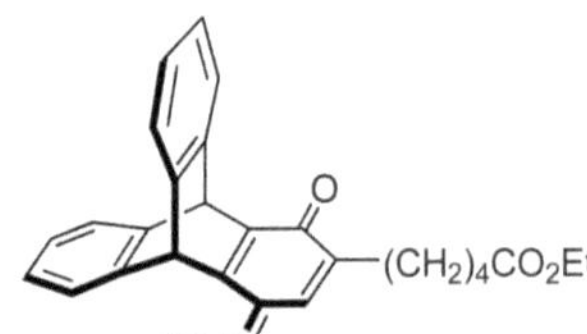 **59**
3	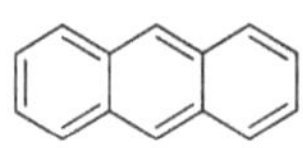	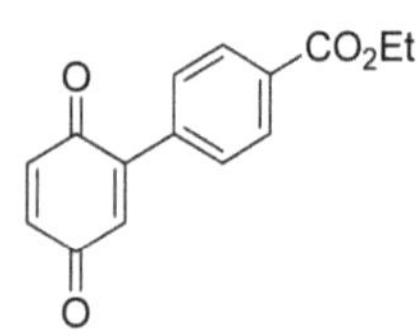	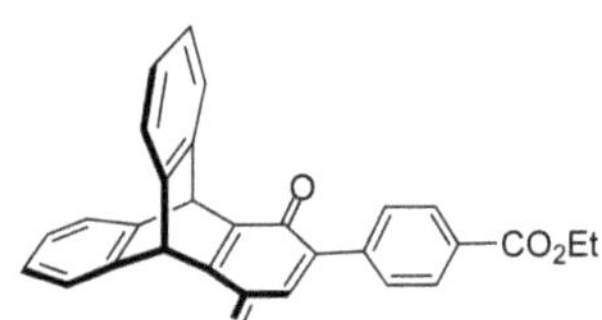 **60**
4	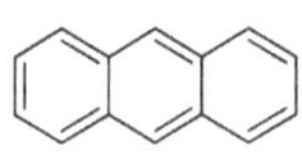	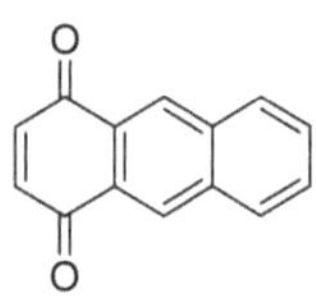	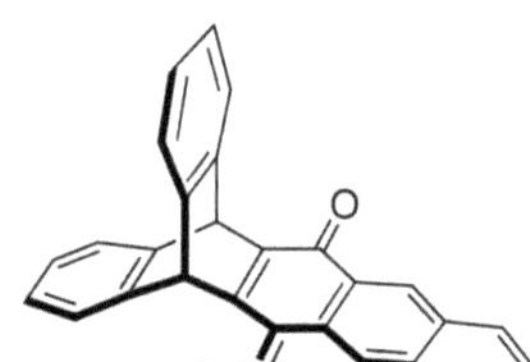 **61**
5	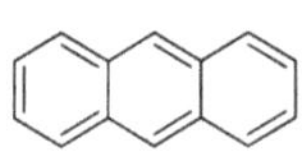	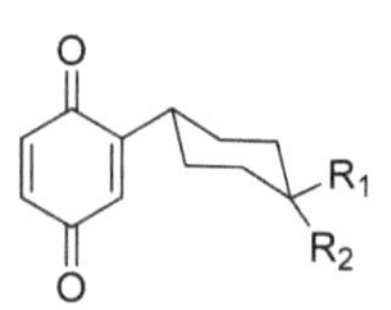	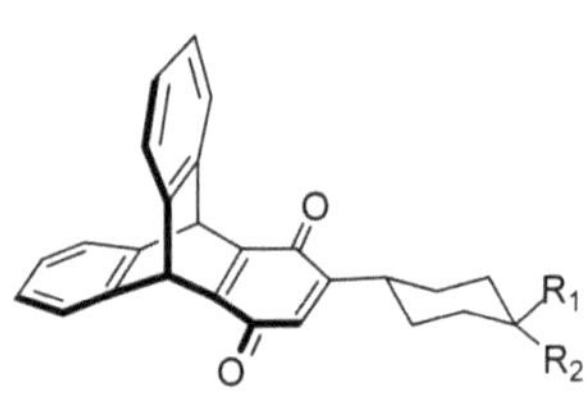 **62** $R_1 = R_2 = H$ **63** $R_1 = H, R_2 = CO_2Me$ **64** $R_1 = CO_2Me, R_2 = H$

Table 2.6 (continued)

6	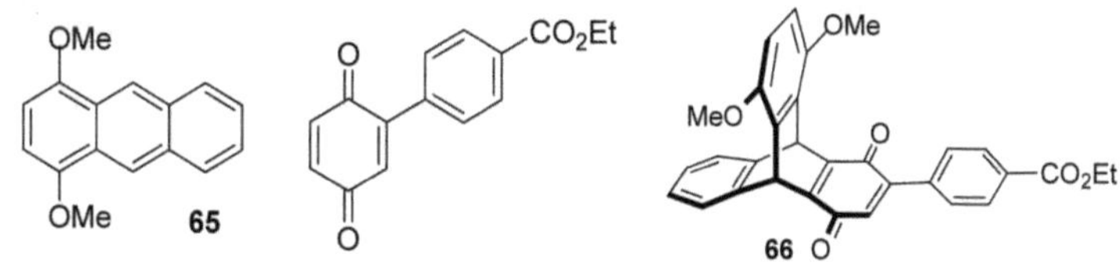	65, 66
7	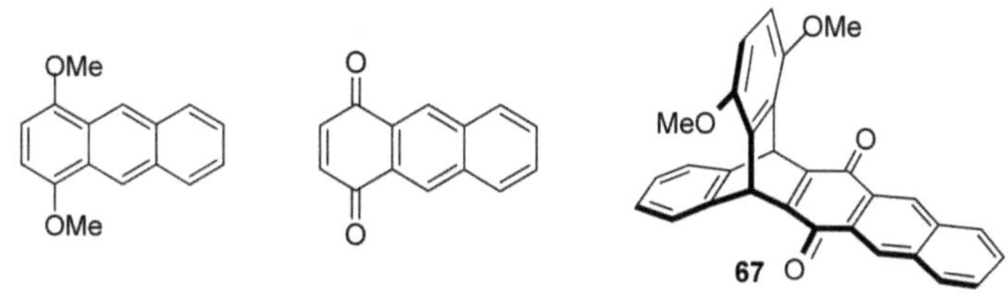	67
8	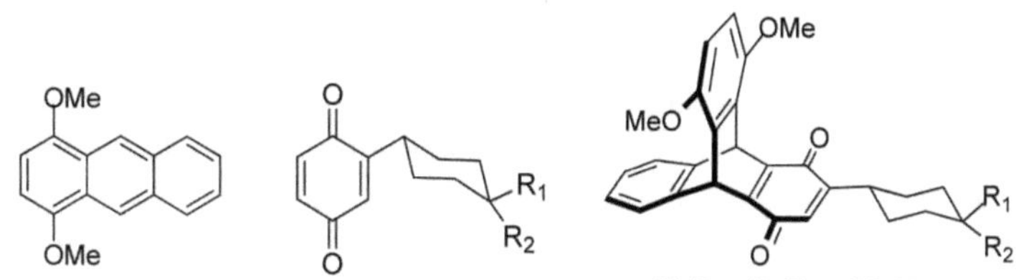	**68** R_1 = H, R_2 = CO_2Me **69** R_1 = CO_2Me, R_2 = H
9	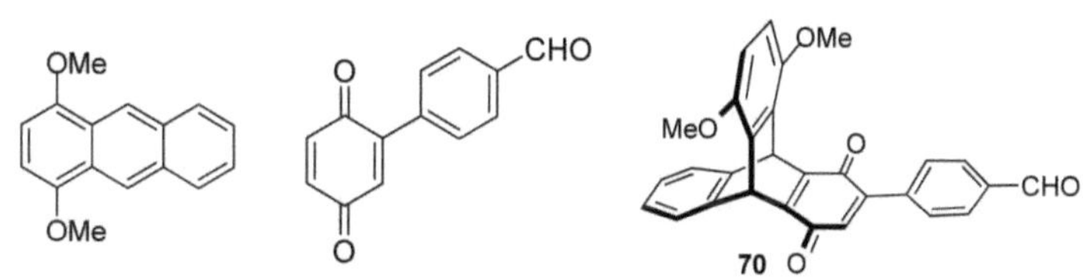	70

reaction of anthracenes and excess quinones in acetic acid. Similarly, the more complex triptycene quinones **66–70** were obtained in 71–90 % yield by the reaction between1,4-dimethoxyl substituted anthracene **65** and a variety of *p*-benzoquinone derivatives.

Generally, the Diels–Alder reactions could easily occur with strong electron acceptors as dienophiles, whereas hardly take place with electron-rich ones [54–59]. However, Fukuzumi and Okamoto [60] found that the Diels–Alder reactions between anthracene derivatives and various inert or weak dienophiles (*p*-benzoquinone derivatives) could proceed successfully in the presence of $Mg(ClO_4)_2$ in acetonitrile (Scheme 2.29). By the comparison of a series of metal ion catalysts, it was further found that scandium triflate [$Sc(OTf)_3$] was most effective to accelerate the Diels–Alder reactions [54–59].

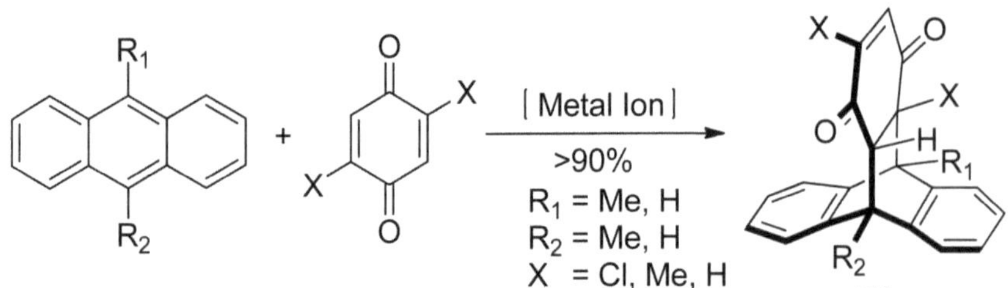

71

Scheme 2.29 Synthesis of compound **71**

2.3 Reactions of Triptycenes and Their Derivatives

2.3.1 Nitration and Amination

In 1969, Klanderman and Perkins [61] first reported the nitration of triptycene. Treatment of triptycene with concentrated nitric acid in glacial acetic acid under 27–29 °C for 3 h could afford 2-nitro triptycene in a 44 % yield, whereas 2,6- and 2,7-dinitro, and trace of the 1-nitro triptycene (**72**, 2.3 %) were also produced. It was noteworthy that the mononitration of triptycene occurred almost at the β-position (97.7 %), the further reaction also preferred to occur at the β-position to form β, β'-dinitro-triptycene. Klanderman and Perkins presumed the nitration to be first order in the stoichiometric concentration of nitric acid. Unfortunately, the result of kinetic analysis was not ideal on account of the purity of acetic anhydride, and it might be third-order [62]. Nevertheless, Rees [63] pointed out that the bond fusing the aromatic and aliphatic rings mainly determined the reactivity of the β-position. According to the theory of the Mills–Nixon effect, Homes and co-workers [64] reported that the transannular stabilization of the electron-deficient transition states for substitution in triptycene was negligible on the basis of deuteration in triptycene.

Subsequently, Shigeru and Ryusei [65] reported the reaction between triptycene and three equivalents of acetyl nitrate solution in glacial acetic acid with the pressure of acetic anhydride at 0 °C for 2 h, which gave a mixture of nitrated triptycenes (**73**–**77**) including mononitro-, dinitro-, and a small amount of trinitro-(2,6,14- and 2,7,14-trinitro-) product (Scheme 2.30). As mentioned before, the β-positions were the most reactive sites to electrophilic aromatic substitution in triptycene. In the case of nitration, a nitro group was substituted in one ring of triptycene, the further nitro substitution would occur on the β-positions of the other two ring, which resulted in a mixture of products rather than a sole one.

In 2006, Chong and MacLachlan [66, 67] attempted to improve the conditions of nitrification to achieve a certain kind of nitrotriptycene as main product, along with decreasing the side products as much as possible. They found that the reaction time and the amount of nitrating agent were important to reach the optimization conditions. In the case of the mononitration of triptycene, this reaction could quickly

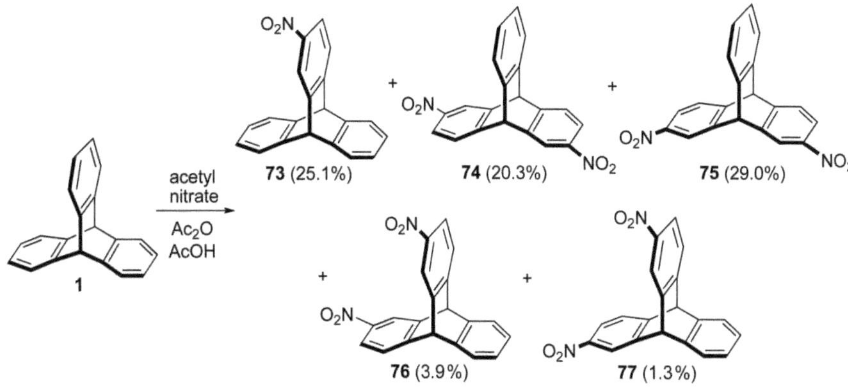

Scheme 2.30 Nitration of triptycene

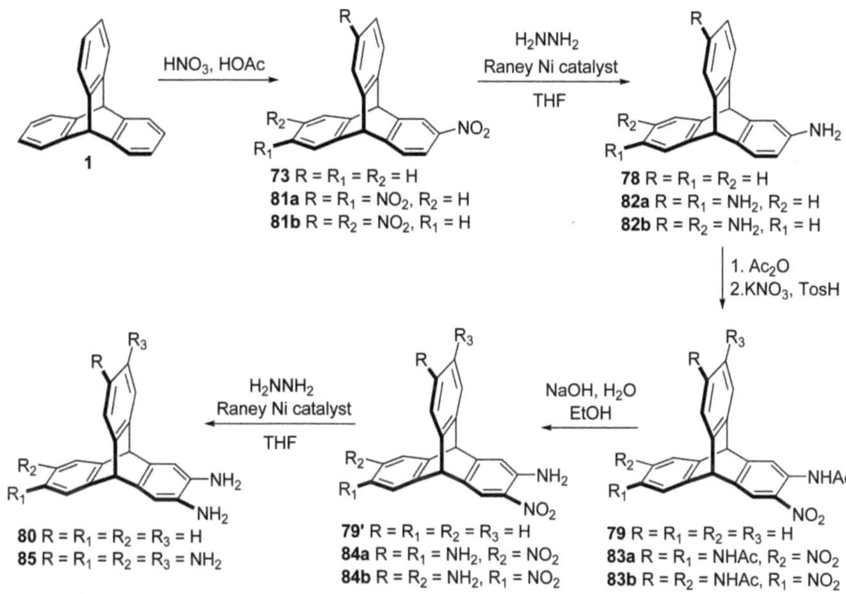

Scheme 2.31 Synthesis of polyaminotriptycenes **80** and **85**

complete after 6 h with concentrated HNO_3 in acetic acid under heating to give 2-nitrotriptycene **73** as the principal product. Then, the mononitration product **73** could be reduced into 2-aminotriptycene **78**, which could convert into an acetamide in situ. Following the selective nitration with stoichiometric nitric acid generated in situ, gave the compound **79** with an acetate protecting group. After deprotection, the target compound **80** could be afforded by the followed reduction in a yield of 53 % (Scheme 2.31). Although, the triple nitration of triptycene with concentrated HNO_3 gave a pair of trinitrotriptycene isomers (**81a, b**) in a ratio of 3:1, this result revealed that the electronic effects had negligible impact on the location of nitration. By purification,

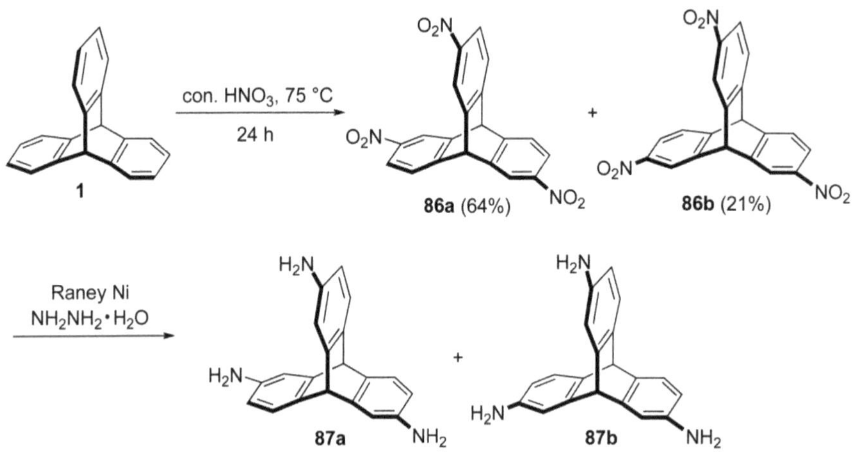

Scheme 2.32 Synthesis of triaminotriptycenes **87a, b**

2,6,14-trinitrotriptycene **81a** could be obtained as the sole isolated product with a yield of 33 %. It was noted that both of these isomers could afford the 2,3,6,7,14,15-hexa-aminotriptycene **85** by a sequence of double nitration/reduction route. They tried to directly synthesize the target hexa-amino compound without intermediate isomeric separation, followed by the route similar to 2,3-diaminotriptycene **80**, but failed due to the difficulty in the separation and purification of the final product **82**. Thus, the separation of the two amine isomers (**82a, b**) was important to ensure a better yield of the target 2,3,6,7,14,15-hexa-aminotriptycene **85**.

Soon after, Zhang and Chen [68] found that by increasing the temperature (75 °C) and prolonging the time (24 h) of the reaction, not only 2,6,14-trinitrotriptycene (**86a**) was obtained in 64 % yield, but also 2,7,14-trinitrotriptycene (**86b**) could be isolated in 21 % yield (Scheme 2.32). More importantly, the trinitrotriptycenes could be easily reduced by Raney Ni in the presence of hydroazine to afford the corresponding triaminotriptycenes (**87a, b**) in almost quantitative yields. The triaminotriptycenes could be considered as useful precursors for the preparation of other trisubstituted triptycene derivatives by Sandmeyer reaction.

Zhang and Chen [68] further found that by the treatment of the aminotriptycenes with aqueous solution of sodium nitrite in concentrated sulfuric acid, the trihydroxytriptycene **88a** could be obtained in 75 % yield. Treatment of compound **87a** with a solution of sodium nitrite in concentrated hydrobromic acid and CuBr gave tribromotriptycene **89a** in 68 % yield. Similarly, the reaction of the diazonium salt derived from **87a** with KI would give triiodotriptycene **90a** in 71 % yield (Scheme 2.23a). Under the same conditions as above, compounds **88–90b** were synthesized in 81, 62, and 80 % yield, respectively (Scheme 2.23b).

According to the procedure described as above, Chen and Swager [69] synthesized 2,6-diaminotriptycene **91** by the reduction of the corresponding dinitrotriptycene with hydrazine and Raney Ni with a yield of 95 %, which was then transformed into

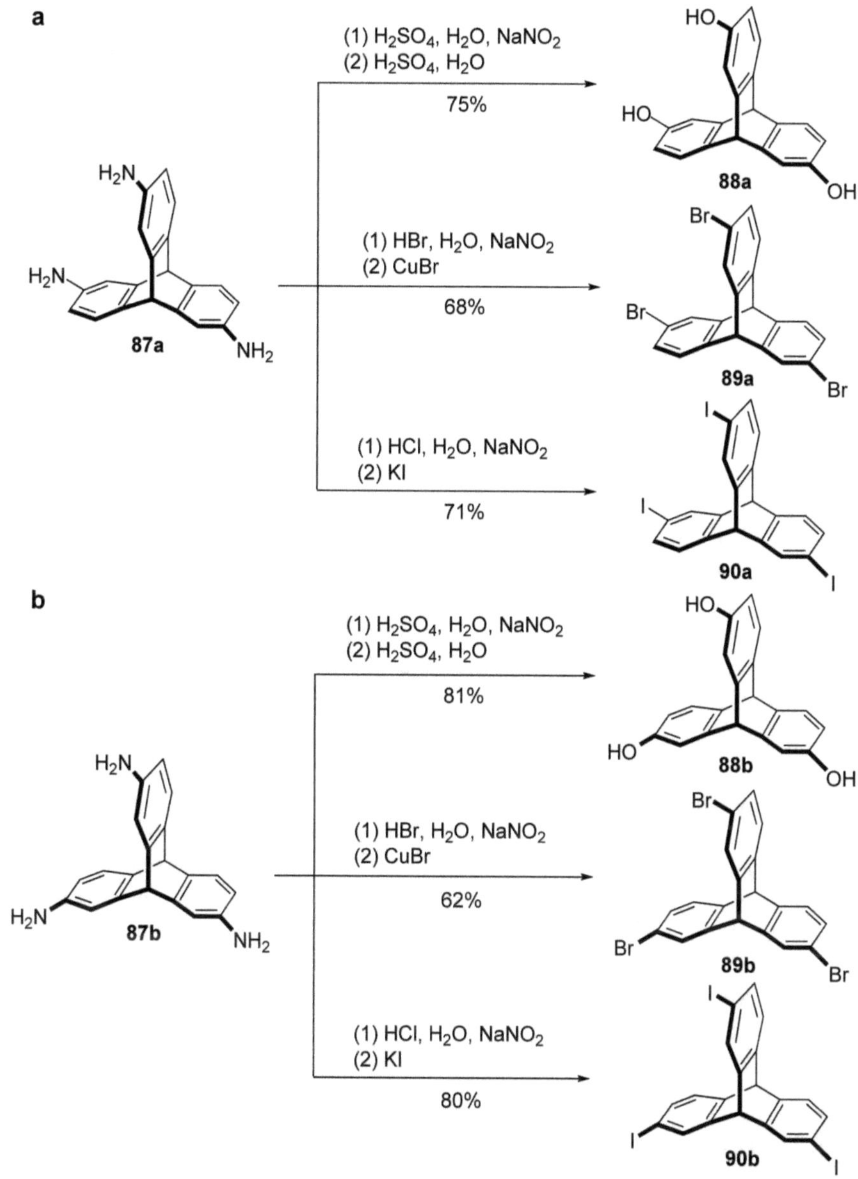

Scheme 2.33 Synthesis of 2,6,14- and 2,7,14-trisubstituted triptycene derivatives

dibromotriptycene **92** by the Sandmeyer reaction in an aqueous HBr solution in the presence of $NaNO_2$ and CuBr. Under the similar conditions, diiodotritpycene **93** was obtained in a satisfactory yield as well by the treatment of **91** with HCl, $NaNO_2$, and KI (Scheme 2.34).

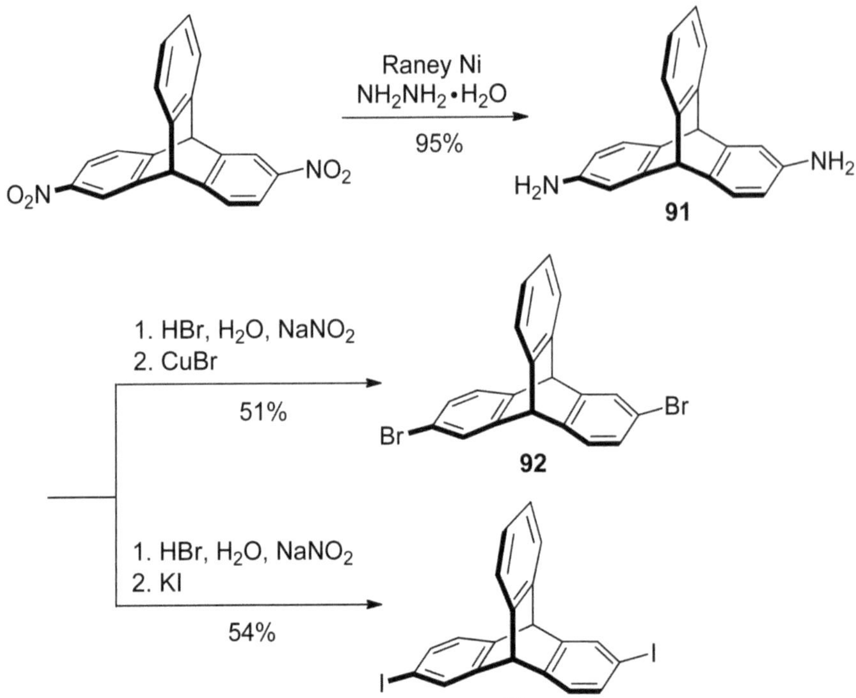

Scheme 2.34 Synthesis of dibromotriptycene **92** and diiodotritpycene **93**

Recently, Oppel and co-workers [70] introduced a two-step synthesis of an air-stable hexa-ammonium-triptycene **94**. As shown in Scheme 2.35, the hexanitrotriptycene **95** was initially obtained in 18 % yield by a one-pot reaction of triptycene with fuming nitric acid, which was then transformed into the compound **94** in almost quantitative yield by the reduction of **95** with tin(II) chloride in a mixture solution with aqueous hydrochloride and ethanol in place of Pd/C [71] and H_2 or Raney Ni/hydrazine. It was noteworthy that the reaction could perform under air without precaution.

2.3.2 *Acylation*

Compared with the nitration of triptycene, only a few reports on the acylation of triptycene were known, although the acyl-substituted triptycenes are very important precursors for the further derivatization of triptycene as well.

In 1965, Parget and Burger [72] reported the acetylation of triptycene, which was also the first example of the direct electrophilic substitution of triptycene. The reaction of triptycene with acetyl chloride in tetrachloroethane in the presence of aluminum chloride yielded 2-acetyltriptycene **96** in 97 % yield (Scheme 2.36). In

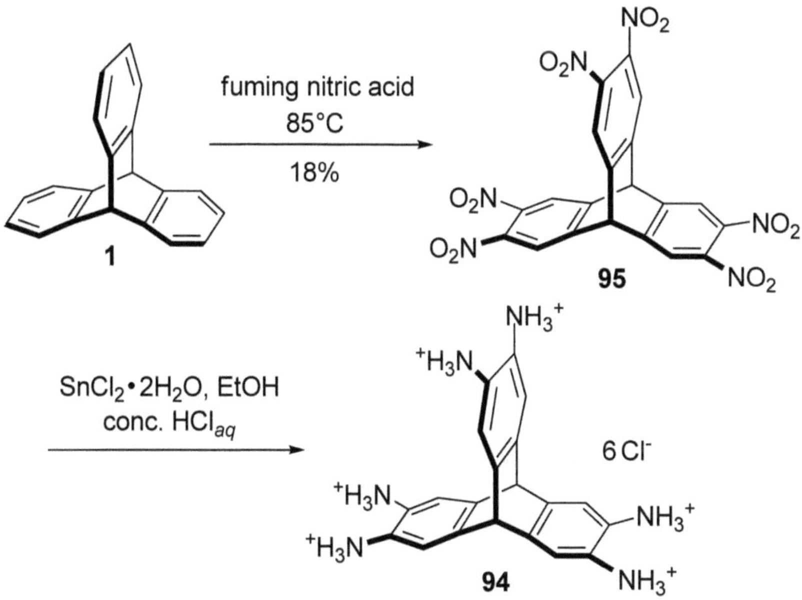

Scheme 2.35 Synthesis of hexa-ammonium-triptycene **94**

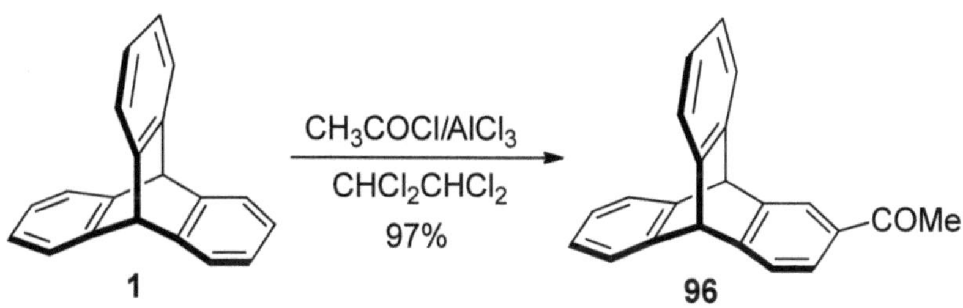

Scheme 2.36 Synthesis of 2-acetyltriptycene **96**

common with the nitration of triptycene, the acetylation also preferred to occur at β-position of the aromatic ring.

Similarly, the treatment of triptycene with benzoyl or fumaroyl chlorides in the presence of nitromethane also gave the sole β-substituted product (Scheme 2.37), which could further form the corresponding ketone **98** [73, 74].

The direct formylation of triptycene [74] with the excess of dichloromethyl methyl ether at −20 °C in the presence of aluminum chloride could provide compound **99** in 51 % overall yield by a few steps [74, 75]. Moreover, the triptycene derivatives **100** [75] and **101** [74, 76] could be further obtained from **99** (Scheme 2.38).

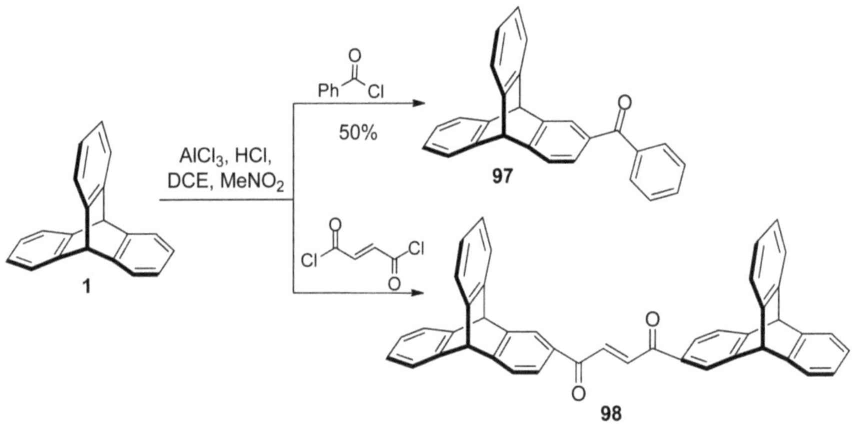

Scheme 2.37 Synthesis of compounds **97** and **98**

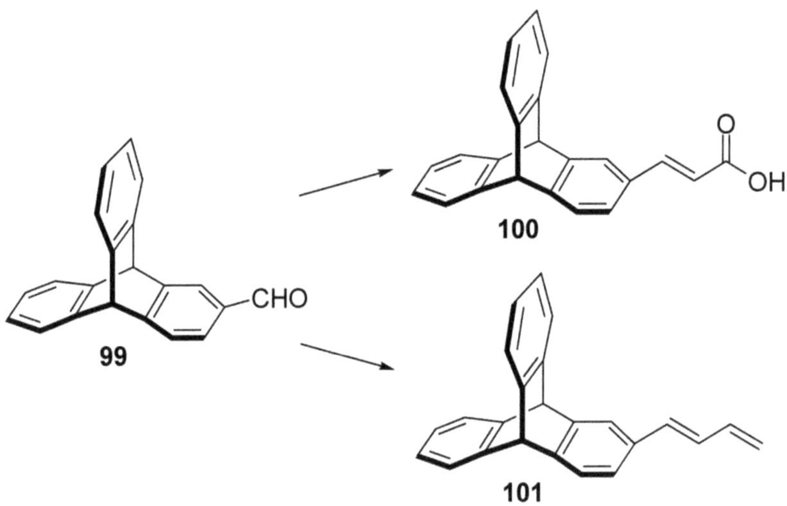

Scheme 2.38 Synthesis of compounds **100** and **101**

2.3.3 Halogenation

As mentioned before, halogenated triptycenes could easily be obtained by the Diels–Alder reaction approach starting from halogenated anthracene or halogenated benzyne. However, this addition reaction was sometimes accompanied with side reaction to form isomers, which resulted in the difficulty of separation. Thus, the direct halogenation of triptycene would provide another opportunity for the synthesis of halogenated triptycenes.

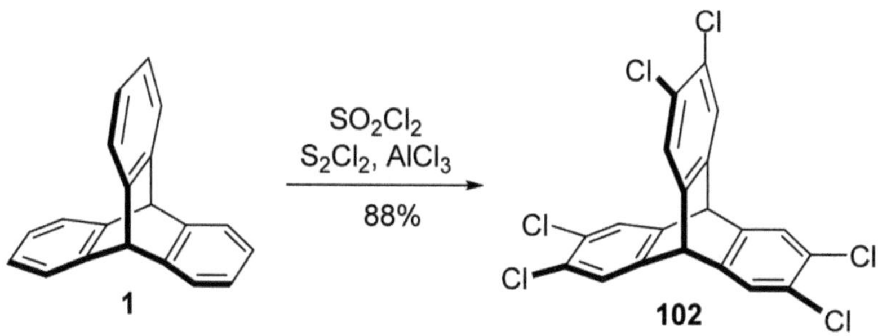

Scheme 2.39 Synthesis of dodecachlorotriptycene **102**

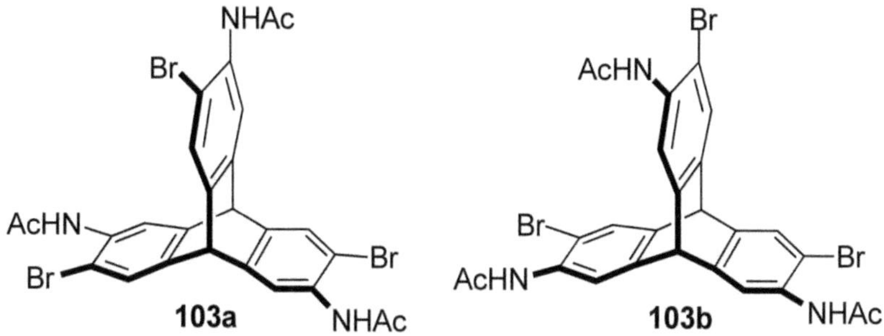

Fig. 2.2 Structures of triacetamido-tribromo-triptycenes **103a, b**

In 1970, Ballester et al. [77] reported the reaction of triptycene with sulfuryl chloride in the presence of sulfur monochloride and aluminum chloride would give dodecachlorotriptycene **102** in a good yield (Scheme 2.39)

A mixture of isomeric triacetamido-tribromo-triptycenes (**103a**, **b**, Fig. 2.2) [74, 78] could be obtained by the bromination of isomeric triacetamidotriptycenes, which were obtained by the reduction of trinitrotriptycenes in acetic acid. This result revealed that the β-positions on triptycene were also the most reactive sites to bromination reaction.

2.3.4 Oxidation

As early as 1942, Bartlett et al. [1] found that triptycene could be easily oxidized by strong oxidizing agents, such as chromium trioxide or potassium permanganate in acetic acid to form anthraquinone (**105**) and carbon dioxide. Moreover, compound **105** and 1,4-diphenylanthraquinone **104** could be obtained by oxidation of diphenyltriptycene **7** [28] with CrO_3 at the same time, whereas the ratio was uncertain (Scheme 2.40). Actually, the oxidation of substituted triptycene quinones with ammonium persulfate even led to cleavage of the quinonoid ring to give substituted 9,10-dihydro-9,10-etheno-anthracenes [79].

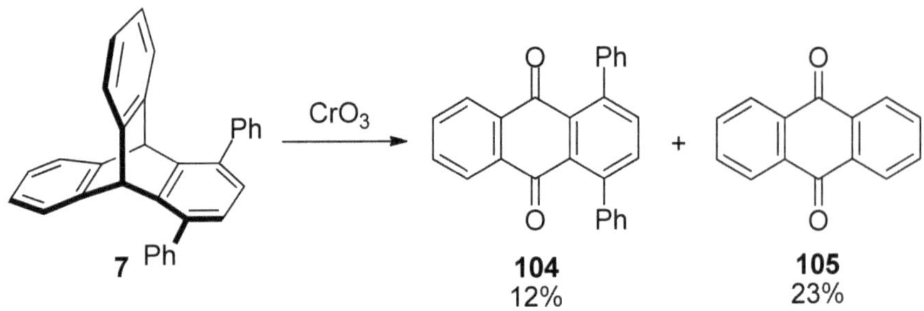

Scheme 2.40 Oxidation of diphenyltriptycene **7** with CrO_3

To keep the skeleton of triptycene intact, mild oxidizing conditions should be utilized. In fact, in the route of synthesis of triptycene, Bartlett et al. [1] carried out the oxidation of triptycenequinol to triptycene quinone in a reasonable yield by adding hot potassium bromate in a minimum amount of glacial acetic acid. As mentioned earlier, Hoffmeister et al. [35] obtained triptycene-9,10-dicarboxylic acid (**24**) by the oxidation of 9,10-bis(hydroxyl-methyl) triptycene (**22**) with chromium trioxide in the pressure of concentrated sulfuric acid in a yield of 90 %. Moreover, triptycene-2-carboxylic acid [73, 74] and triptycen-2-al [74, 75] could be obtained by the oxidation of 2-acetyl-triptycene and triptycen-2-ylmethanol, respectively.

In 2006, Závada and co-workers [80] reported the synthetic route to triptycene carboxylic acids. The reaction between appropriate aryne and dimethylanthracene in a mixture of 1,2-dichloroethane and THF (or diglyme) could conveniently generate a series of di- or tetra-methyltriptycenes (**106**) in 41–69 % yields. Then, the oxidation of **106** by potassium permanganate gave triptycene di- and tetra-carboxylic acids (**107**) in 86–96 % yields.

According to the similar approach, the highly symmetrical triptycene tetra- and hexacarboxylates (**108**, **109**) could be synthesized as well [43]. Consequently, the Diels–Alder reaction between tetramethylanthracenes and aryne formed in situ, then followed oxidation by potassium permanganate in pyridine and water, the targets **108** (Scheme 2.41), **38a**, and **109** (Scheme 2.42) could be obtained in 94 and 53 % yield, respectively. However, this procedure for synthesis of **38a** and **109** would be accompanied by the side reaction to form naphthabenzobarrelene isomers **109′** (Scheme 2.42).

Under the similar conditions, Zonta et al. [81] attempted to synthesize the desired hexacarboxylic acid **109** from hexamethyl-triptycene **110**. When potassium permanganate was only used as oxidant, the compound **110** would be partially oxidized. However, **109** and a certain amount of over-oxidation side-products were produced with strongly basic potassium permanganate as oxidant. Finally, they found that by the oxidation of **110** by potassium permanganate in pyridine and a water solution of NaOH, the target product **109** could be obtained in 78 % yield (Scheme 2.43).

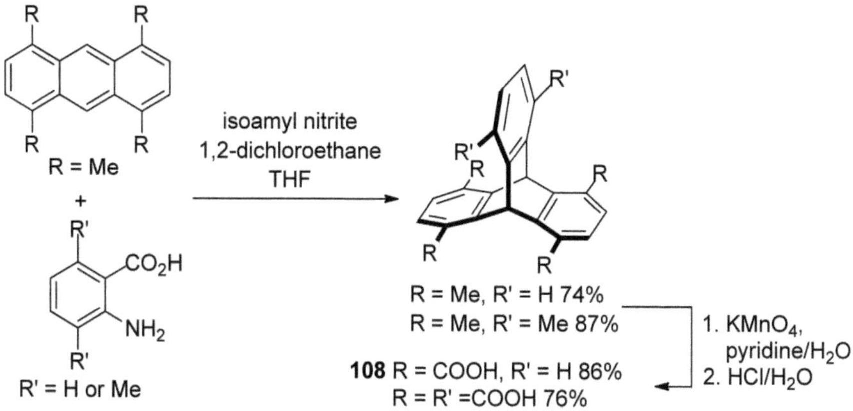

Scheme 2.41 Synthesis of compound **108**

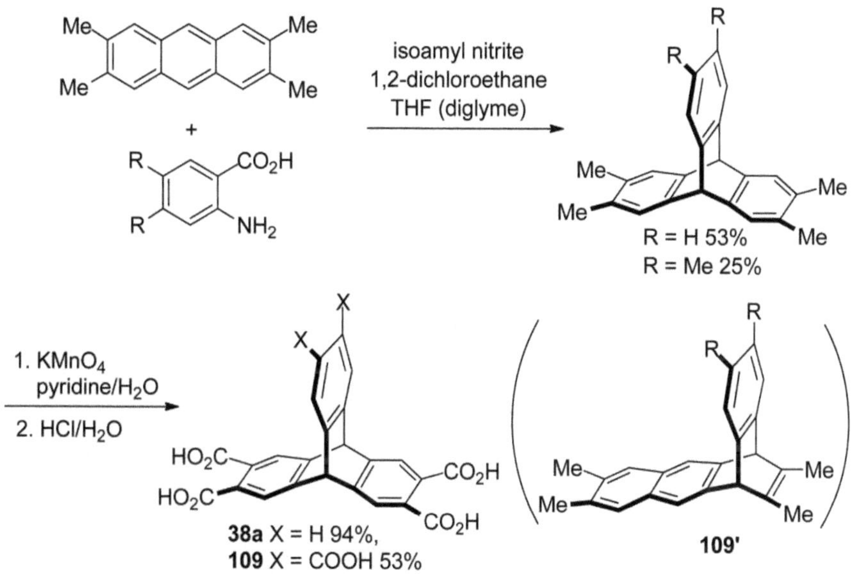

Scheme 2.42 Synthesis of compounds **38a** and **109**

2.3.5 Reduction

In common with the oxidation reaction, the skeleton of triptycene could be damaged under the extreme reduction conditions. Thus, Clar [47] found that when triptycene-1,4-quinol and triptycene quinone underwent distillation in the presence of zinc dust, the skeleton of triptycene destroyed to form 9,10-dihydroanthracene.

In 1962, Theilacker and Möllhoff [82] reported that triptycene could undergo an interesting cleavage reaction with K (in K/Na alloy) in ether to produce compound

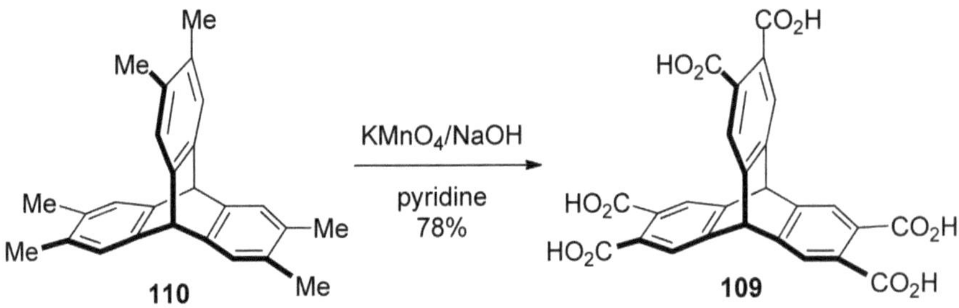

Scheme 2.43 Synthesis of hexacarboxylic acid **109**

Table 2.7 Synthesis of methyltriptycene derivatives (**106**) and triptycene carboxylic acids (**107**)

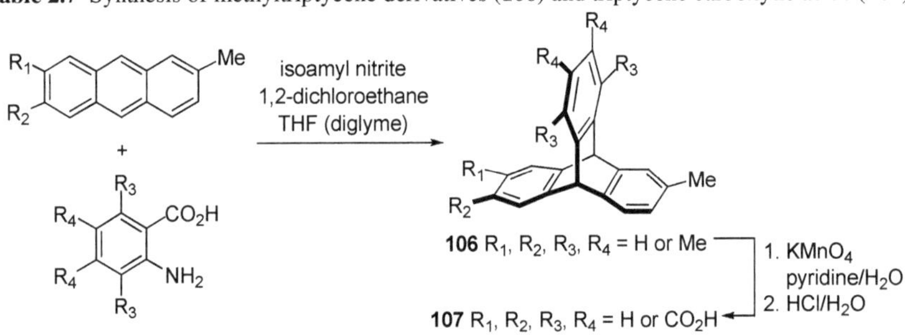

Entry	Dimethyl anthracene	Aryne	106 R	106 Yield (%)*	107 R	107 Yield (%)
1	$R_1 = H, R_2 = Me$	$R_3 = R_4 = H$	$R_1 = R_3 = R_4 = H, R_2 = Me$	51	$R_1 = R_3 = R_4 = H, R_2 = CO_2H$	87
2	$R_1 = Me, R_2 = H$	$R_3 = R_4 = H$	$R_2 = R_3 = R_4 = H, R_1 = Me$	53	$R_2 = R_3 = R_4 = H, R_1 = CO_2H$	95
3	$R_1 = H, R_2 = Me$	$R_3 = Me, R_4 = H$	$R_1 = R_4 = H, R_2 = R_3 = Me$	69	$R_1 = R_4 = H, R_2 = R_3 = CO_2H$	97
4	$R_1 = Me, R_2 = H$	$R_3 = Me, R_4 = H$	$R_2 = R_4 = H, R_1 = R_3 = Me$	61	$R_2 = R_4 = H, R_1 = R_3 = CO_2H$	94
5	$R_1 = H, R_2 = Me$	$R_3 = H, R_4 = Me$	$R_1 = R_3 = H, R_2 = R_4 = Me$	23; 41**	$R_1 = R_3 = H, R_2 = R_4 = CO_2H$	94
6	$R_1 = Me, R_2 = H$	$R_3 = H, R_4 = Me$	$R_2 = R_3 = H, R_1 = R_4 = Me$	52**	$R_2 = R_3 = H, R_1 = R_4 = CO_2H$	86

*Solvent: THF; ** diglyme.

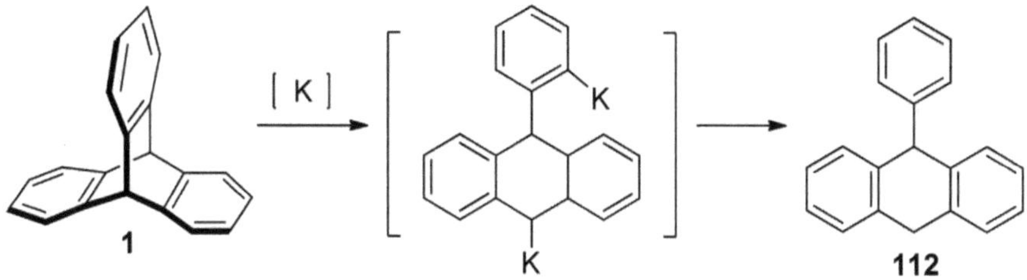

Scheme 2.44 Cleavage reaction of triptycene **1** with K (in K/Na alloy)

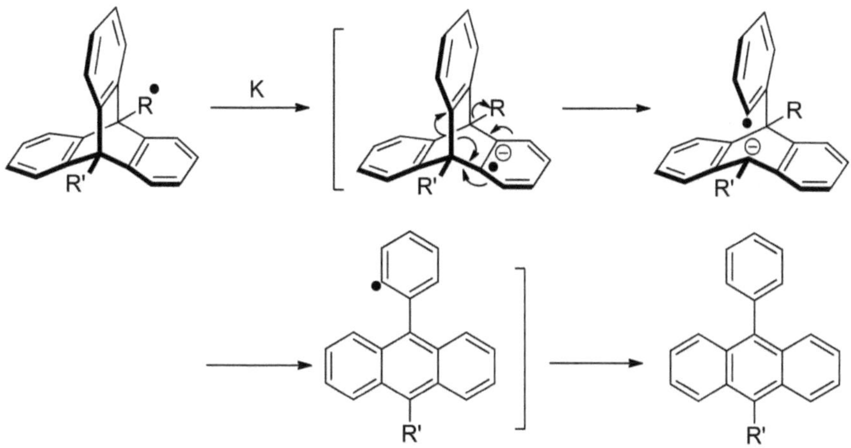

Scheme 2.45 A possible mechanism for the cleavage of triptycene with potassium in oxolan

112 (Scheme 2.44), whereas Na in the same alloy had no effect on the cleavage of triptycene **1**.

Similarly, Walsh and Ross [74, 83] described the cleavage of triptycene with potassium in oxolan. They considered that the reaction could occur via a radical-anion intermediate, and thus, suggested a possible mechanism as shown in Scheme 2.45. In this mechanism, the degree of stability for the intermediate radical-anion with the different group was the key factor considered. Apart from this, the leaving rate of group had to be considered. Among the common substituent, the phenyl group showed the lower leaving rate than methyl, H, and benzyl group. In addition, the metal agent itself could influence the cleavage. For example, there was little or almost no cleavage of 9-phenyltriptycene with sodium.

To avoid the destruction of the triptycene skeleton, the reduction could occur under mild conditions. Bartlett and Lewis [84] reported the reaction of 9-bromotriptycene with magnesium in ether, and found that the triptycen-9-yl magnesium bromide could be unexpectedly hydrolyzed to form triptycene. No reaction between 9-bromotriptycene and sodium occurred in boiling toluene; whereas, the reaction of 9-bromotriptycene with metal in mineral oil gave triptycene in approximately 100 % yield when it was heated to 140 °C or in boiling isooctane. Ginsburg and co-workers [85] also investigated the reaction of 9,10-dibromotriptycene with metals, and they

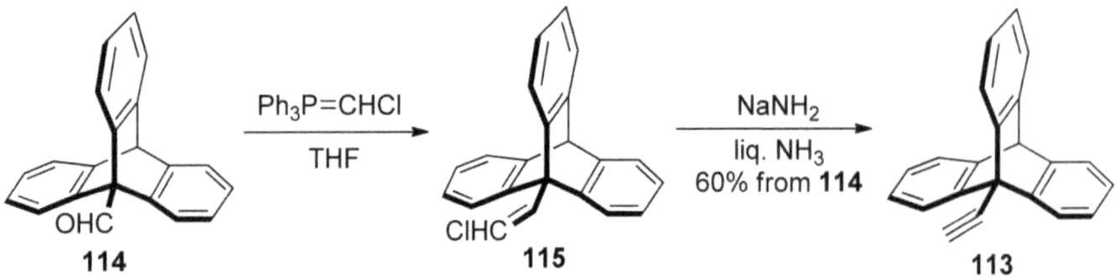

Scheme 2.46 Synthesis of 9-ethynyltriptycene **113**

found that the treatment of dibromotriptycene with Na–K alloy in reflux *n*-octane produced 9-bromotriptycene. Moreover, treatment of 9-bromotriptycene with NaH in HMPA could further give triptycene **1** in 92 % yield.

By the reduction of the diethyl ester of 9,10-triptycenediacetic acid with lithium aluminum hydride in THF equipped with a Soxhlet extractor, Hoffmeister et al. [35] obtained 9,10-bis(β-hydroxyethyl)triptycene. In 1971, Akiyama et al. [86] reported the synthesis of 9-ethynyltriptycene **113** starting from 9-triptycene carboxaldehyde **114**. As shown in Scheme 2.46, the reaction of **114** with chloromethylene triphenylphosphorane in THF could yield the chlorovinyl triptycene **115**, which then proceeded the dehydrochlorination by the sodium amide in liquid ammonia to give the target compound **113**.

Afterward, Skvarchenko and co-workers [74, 87, 88] reported that 2-acetyl triptycene was treated with PCl_5 to give the intermediate compounds (**117** and **118**), which was then followed by the reduction with $NaBH_4$ to afford another unsaturated hydrocarbons substituted triptycene **116a** (Scheme 2.47a). In contrast, the reduction of 2-acetyltriptycene by $LiAlH_4$ via two steps to obtain unsaturated hydrocarbons substituted triptycene **116b** (Scheme 2.47b).

In 1979, Rabideau et al. [89] reported the reduction of triptycene by metal–ammonia. The reduction of triptycene could be conveniently accomplished with metal in the presence of NH_3 in a THF solution at −33 °C, and form the products **119a, b** without ring opening. They further studied this reduction with several metals in various alcohols, and the results are shown in Table 2.8. On the basis of the results, it was found that the yields might be controlled by the acidity of the protonating agent (alcohol) to a great extent. The more acidic alcohols the metal consumed, the lower the yields were. It was noteworthy that the reaction of triptycene showed considerably unusual regioselectivity. In most of the cases the major product was the isomer **119b**. In addition, it was also found that metal would influence the yield of reduction product, and the reduction with Li afforded product **119b** in a higher yield than that with metal Na.

2.3.6 *Photochemical Reactions*

In 1969, Walsh [90] investigated the direct irradiation of triptycene in dilute ether or acetone solution, and obtained a single certain isomer. The further analyses confirmed

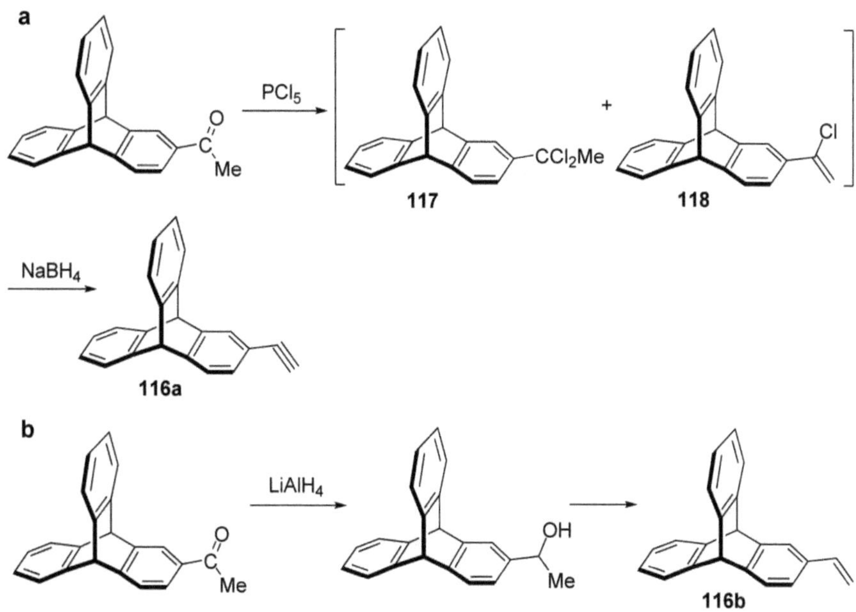

Scheme 2.47 Synthesis of unsaturated hydrocarbon-substituted triptycenes **116a, b**

Table 2.8 Metal–ammonia reduction of triptycene

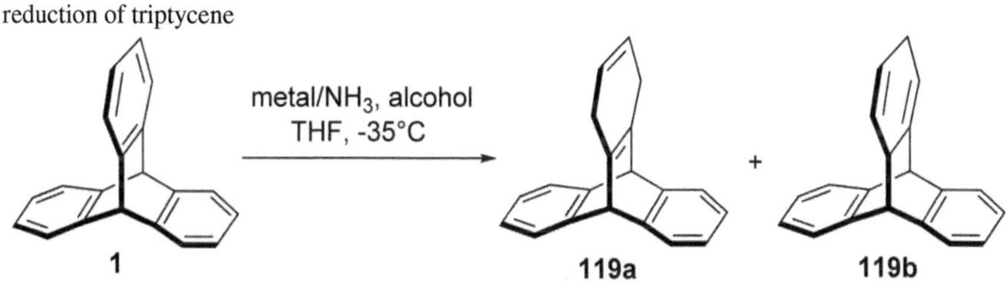

Entry	Metal	Alcohol[a]	Yield of	
			119a (%)	**119b** (%)
1	Li	$(CH_3)_3COH$	27	55
2	Li	$(CH_3)_2CHOH$	23	53
3	Li	CH_3CH_2OH[b]	18	33
4	Li	CF_3CH_2OH	16	22
5	Li	$(CF_3)_2CHOH$	10	9
6	Na	$(CH_3)_3COH$	6	11

Conditions: THF
[a]Temperature: −35 °C
[b]Temperature: −75 °C

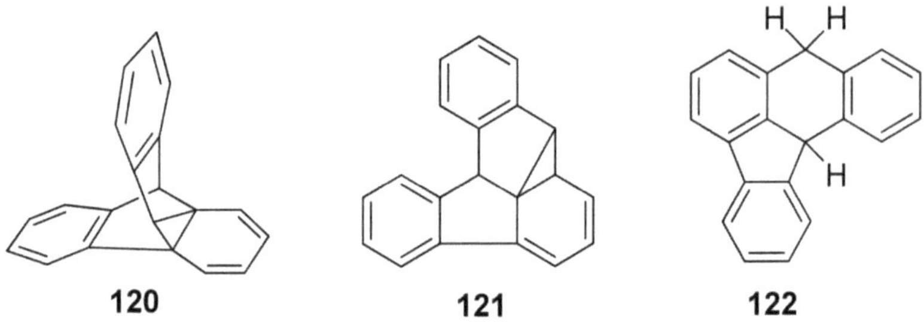

Fig. 2.3 Structures of the compounds **120–122**

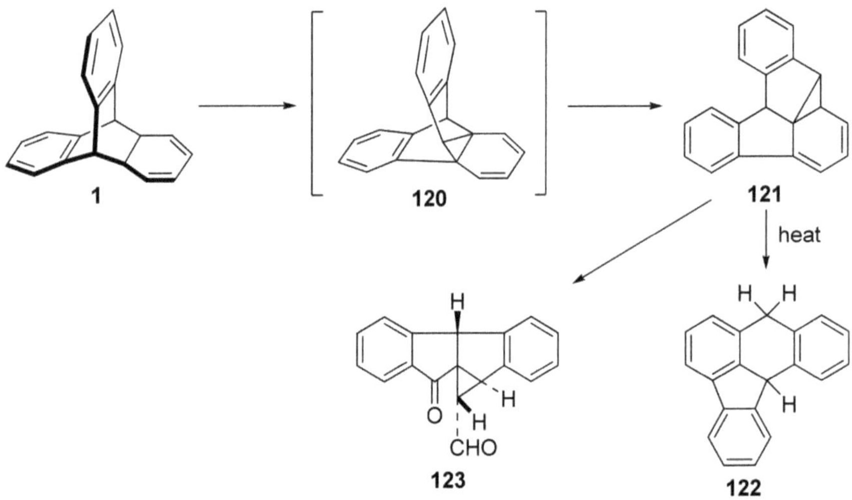

Scheme 2.48 Irradiation of triptycene **1**

that compound **121** (Fig. 2.3) was the correct structure of the isomer. This result was verified by three aspects: (1) the photoproduct further could convert to form the thermal isomerization product, 8,12b-dihydrobenz[a]-fluoranthene **122** in a quantitative yield under the melting temperature; (2) the cleavage of the cyclopropane and hydrogen migration were necessary for the formation of the ring skeleton, whereas only **121** might fit this requirement; and (3) in contrast to the photochemical behavior of the related barrelene and benzobarrelene, the anticipated semibullvalene **120** had already been confirmed instead of the product, but **121** could be obtained by the Cope rearrangement from **120**. In other words, **120** might be the initial photoproduct which finally rearranged to the more stable **121**. Moreover, molecule **121** was the most strain-free of all the possibilities based on the molecular models.

At the same time, Turro et al. [91] also reported the irradiation of triptycene **1** to give the product **121** (Scheme 2.48). They considered that **121** was yielded by the σ-tropic rearrangement of **120** as well. Furthermore, they obtained the compound **122** by heating **121**, and **123** (ketoaldehyde) which could be converted to a diol with $LiAlH_4$ by the ozonolysis of **121**. In addition, they deduced that the conversion of

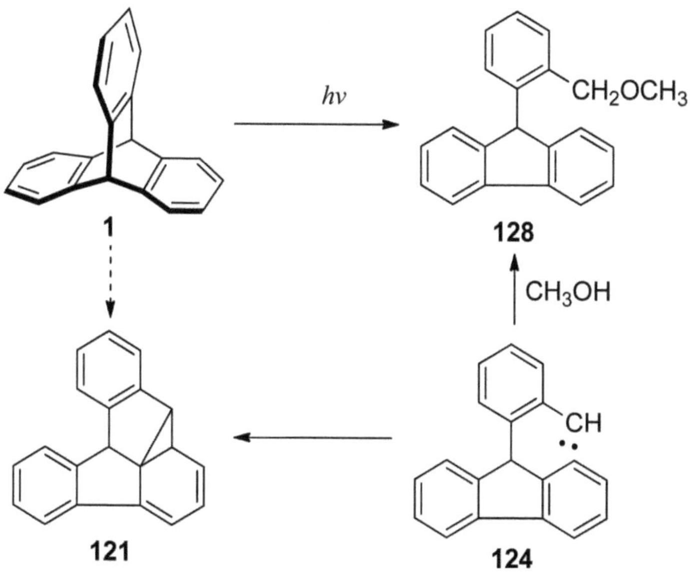

Scheme 2.49 Irradiation of triptycene **1** via carbine **124**

1–121 might be a very fast triplet rearrangement. The formation of **121** could not be quenched by 1,3-pentadiene, and the conversion occurred smoothly in benzene and acetone which are photosensitive solvents.

Soon after, Iwamura and co-workers [92–94] studied a series of photochemical reactions of triptycene and its derivatives. At first, they obtained a kind of yellow crystalline solid from the irradiation of triptycene with Pyrex-filtered UV light in methanol, and the yield was about 75 %, which was based on the consumed starting material [92]. However, it was found that the product was 9-(2-methoxymethylphenyl)-fluorene instead of the product **121** [90, 91] reported before. Moreover, they deemed that the product was generated probably by the attack of solvent molecules to carbine **124** (Scheme 2.49). The compound **124** underwent the intramolecular addition to form photoproduct **122**. At the same time, they further expounded the carbene mechanism for the photolytic rearrangement of **1** in another literature [94]. The photoproduct, which was identified as 9-(2-methoxymethylphenyl)fluorene not the compound **121** reported before [90, 91], was obtained in 75 % isolated yield after TLC on silica gel by the irradiation of a 0.003 M solution of **1** in methanol with Vycor-filtered UV light for 2 h. Under the similar conditions, the irradiation of 9-acetyl-triptycene **125** gave the photoproduct **126** (Scheme 2.50), and there was no trace of the normal isomerization products (**127**) in this process. It was noteworthy that two conformers of **126** were in a ratio of 2.2:1 discriminated by the NMR spectrum, and the results might to some extent support the carbene mechanism. Thus, as mentioned before, they inferred that the capture of **124** with reactive solvent gave **129** in a reasonable yield. At the other side, the normal photoisomer **128** could be obtained by the intramolecular addition of **124**.

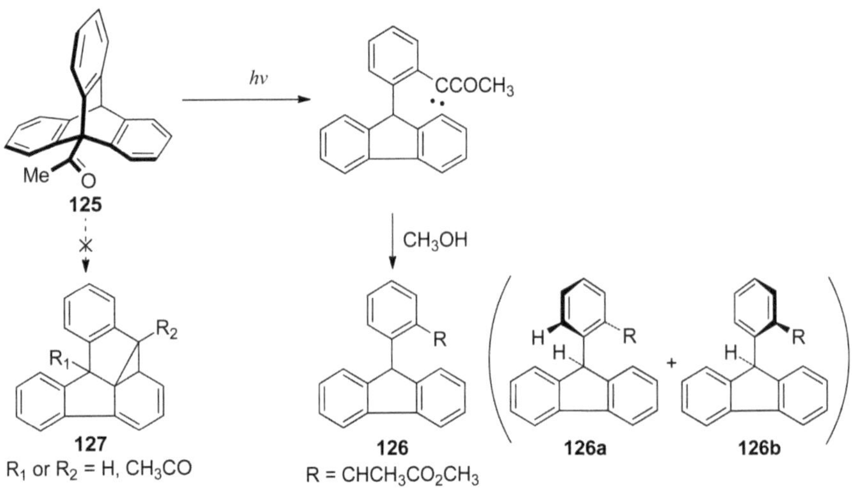

Scheme 2.50 Irradiation of 9-acetyl-triptycene **125**

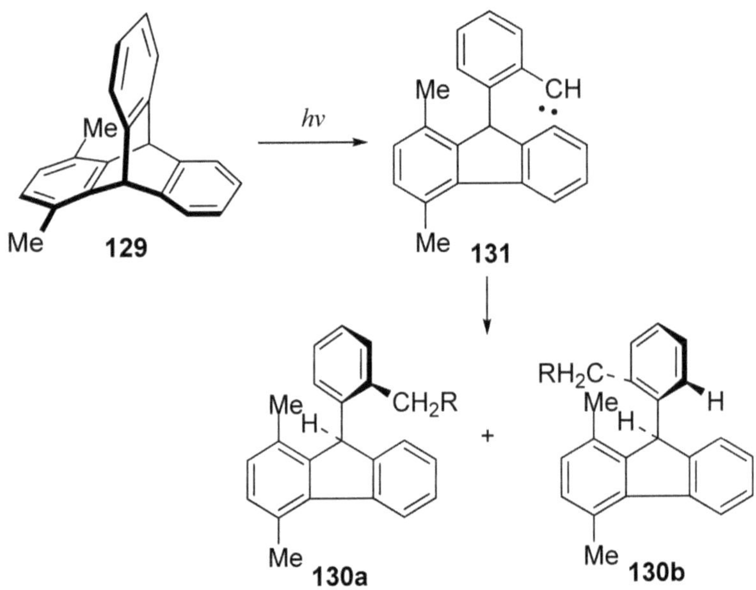

Scheme 2.51 Photolysis of 1,4-dimethyltriptycene **129**

The photolysis of 1,4-dimethyltriptycene **129** was investigated as well [93]. Consequently, it was found that two products **130a** and **130b** were obtained by the irradiation of **129** with Vycor-filtered UV light in methanol. Iwamura and co-workers presumed that the O–H bond of the solvent inserted by carbene **131** gave the main product **130a** in 89 % yield; simultaneously the minor product **130b** was obtained in about 5 % yield by the hydrogen abstraction of **131** (Scheme 2.51).

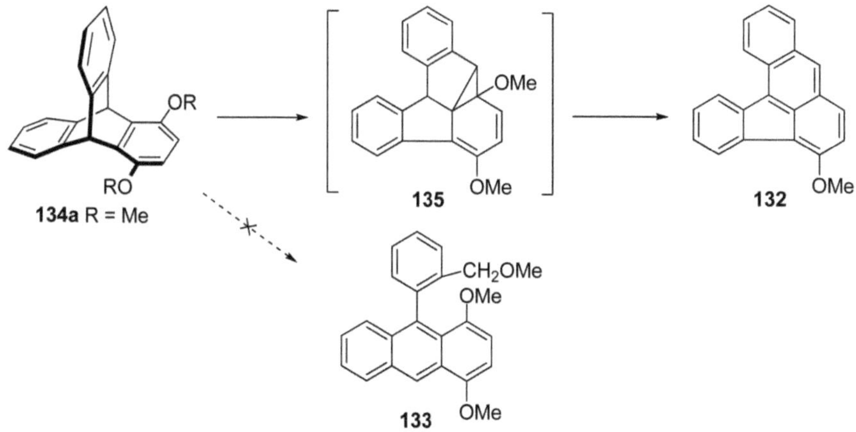

Scheme 2.52 Irradiation of dimethoxytriptycene **134a**

Wheeler and co-workers [95] soon questioned the carbene mechanism, because they found that 5-methoxybenz[a]aceanthrylene **132** rather than **133** was obtained by the irradiation with UV light of the dimethoxytriptycene **134a** in benzene or methanol (Scheme 2.52). They deemed that the reaction preferred the pathway via the intermediate **135** which was suggested by Turro et al. [90, 91], but they failed to reconcile the results of Iwamura as well.

Iwamura and Tukada [96] also reported the Pyrex-filtered irradiation of 1,4-dihydroxytriptycene (**134b**) in methanol, which gave the photoproduct **135**. By the reduction with $LiAlH_4$ and dehydration over P_2O_5, it was found that the epoxy ketone in **135** could be converted to the triptycene-like ketone **137** (Scheme 2.53). This process for the formation of compound **137** could be expounded by the carbene mechanism. Although, the photoproduct **132** reported by Wheeler et al. [95] was actually inconsistent with this carbene mechanism, the intermediate **136** could be trapped either by the aromatic double bond to give **138** (Fig. 2.4), as the result of the epoxide ring formation, or by the carbonyl group of the photoketonized hydroquinone chromophore. Thus, they guessed that the double bond of the triptycene with methoxyl substituent might be activated to undergo the intermolecular addition to give **138b** rather than being trapped by the solvent molecule such as methanol.

Afterward, Wheeler and co-workers [97] investigated the same irradiation of 1,4-dihydroxytriptycene in methanol with UV light. However, they found that the reaction did not give **136** (reported by Iwamura) but **139** in about 70 % yield based on the recovered starting material. Besides **139**, they also isolated the second product based on the spectroscopic data, especially, the strong M-84 peak in the mass spectrum: *m/z* 286 (58 %, M^+) and 202 (100 %), which suggested that the structure of this second product was **140** (Fig. 2.5). Therefore, they considered that the existence of photoproduct **140** confirmed the rearrangement of **134a** to **139** via the intermediate **140**, which was obtained by the di-π-methane rearrangement of **134a**.

In 1978, Iwamura and Tukada [96] studied the Vycor-filtered irradiation of a series of 9-alkoxytriptycenes (**141**) in alcohol, and obtained the corresponding acetals

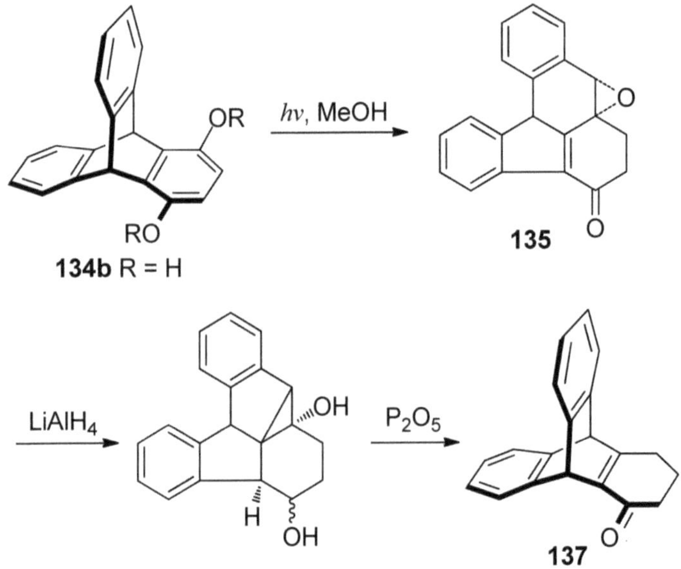

Scheme 2.53 Pyrex-filtered irradiation of 1,4-dihydroxytriptycene **134b** in methanol

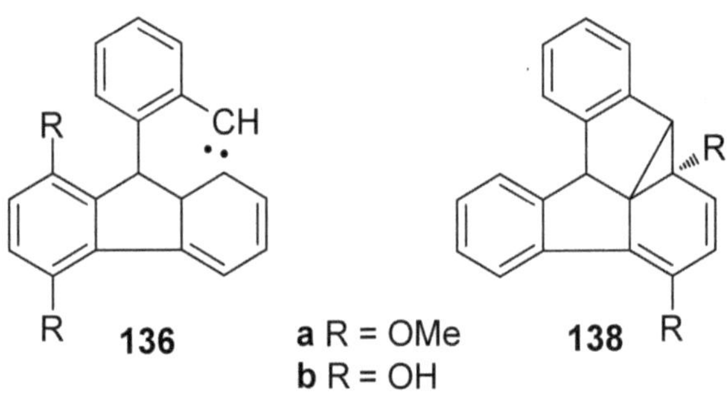

Fig. 2.4 Structures of the compounds **136** and **138**

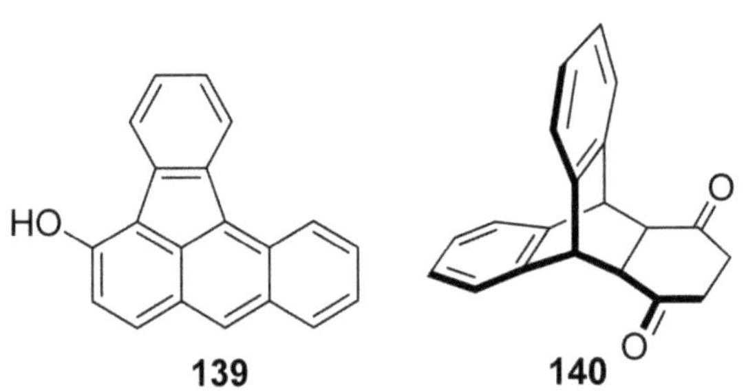

Fig. 2.5 Structures of the compounds **139** and **140**

Table 2.9 The photoreaction of 9-alkoxytriptycenes

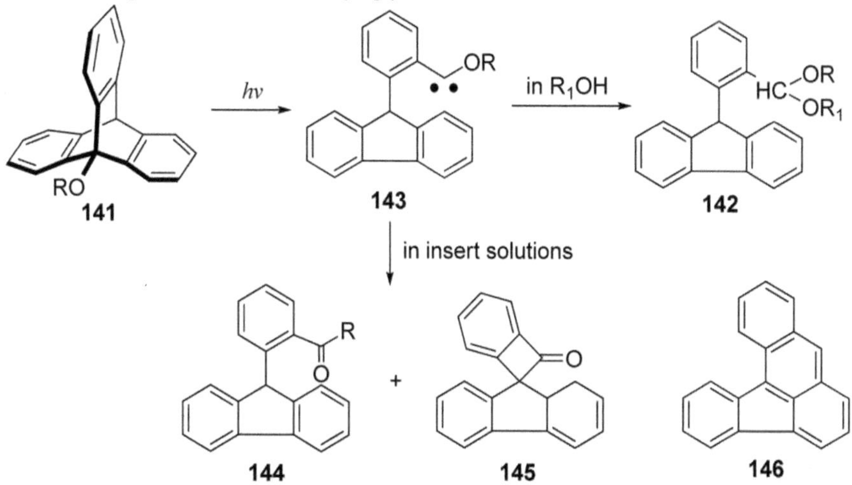

Entry	R in **141**	Solvent and/or additive	Product	Yield (%)
1	Me	CH_3OH	**142** (R = R_1 = Me)	62
2	Me	*t*-C_4H_9OH	**142** (R = Me, R_1 = *t*-C_4H_9)	43
3	Me	Cyclohexane	**144** (R = H)	73
			145	≤5
4	Me	Cyclohexane[a]	**144** (R = H)	98
5	$CH_2CH{=}CH_2$	*n*-Pentane	**144** (R = $CH_2CH{=}CH_2$)	62
			144 (R = H)	4
6	$CH_2CH{=}C(CH_3)_2$	*n*-Pentane	**144** (R = $CH_2CH{=}C(CH_3)_2$)	78
			144 (R = H)	15
7	CH_2Ph	CH_3OH	**142** (R = CH_2Ph, R_1 = Me)	29
8	CH_2Ph	*n*-Pentane	**144** (R = CH_2Ph)	66
9	Ph	Cyclohexane	**146**	31
			142 (R = R_1 = Ph)	29
			144 (R = H)	~1

[a] with 35 mM dimethyl fumarate

(**142**) via the intermediary formation of alkoxycarbenes (**143**). Moreover, the similar photoreactions occurred well in inert solvents. The detailed results are shown in Table 2.9. According to the results, they deemed that the reactions in inert solvents could be rationalized as the homolysis and rearrangement of α-alkoxycarbenes **143** which was similar to the Wittig rearrangement. Furthermore, they [98] also found that the ability to donate or accept electrons was not much related to the tendency of bridging on the site near to the substituent.

Soon after, Hemetsberger and Neustern [99] investigated the irradiation of triptycene monodeuterated at the bridgehead position. The result showed that the deuterium effect could cause the difference in the reaction rate; however, the radiationless processes in the rearrangement were unaffected.

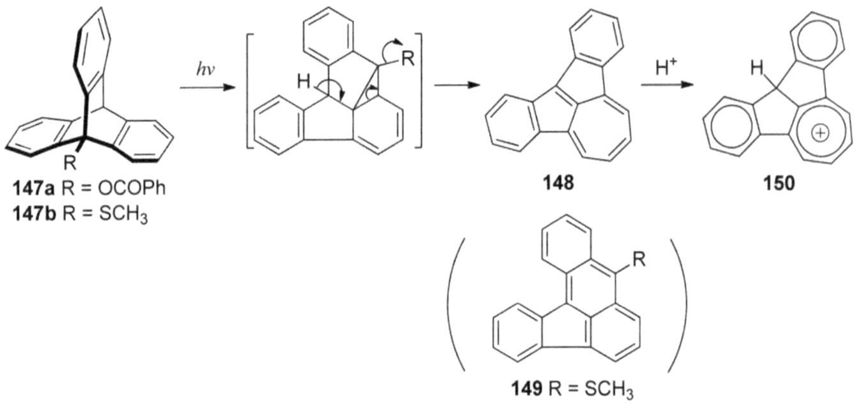

Scheme 2.54 Irradiation of triptycene derivatives **147a, b**

Iwamura et al. [100] further discovered that triptycenes (**147a–c**) with different leaving groups on the bridgehead position underwent photochemical reactions under the irradiation of the low-pressure mercury lamp to give the compound **148** in good-to-moderate yields (Scheme 2.54). For triptycene derivative **147a**, most (> 90 %) of the substrate could be consumed in 2 h to give the compound **148** with a yield of 70 %. Although, the photolysis of triptycene **147b** only gave the compound **148** in an approximate yield of 28 %, along with the side-products **146** and **149** in 15 and 8 % yield, respectively. In addition, the compound **148** could convert to a cation **150** under the acidic condition, like trifluoromethanesulfonic acid.

In 1989, Kitaguchi [101] reported the photoinduced reduction of triptycene quinone and its analogs with xanthene as a hydrogen atom donor to give the corresponding hydroquinones and the adducts with two xanthenyl radicals. Several years later, Scheffer and co-workers [102] further investigated the photochemical reaction of triptycene-1,4-quinone **50** in non-hydrogen atom-donating solvents. They found that in carefully deoxygenated acetone, ethanol or acetonitrile, the Pyrex-filtered irradiation of **50** gave the di-π-methane structure **151** as the sole product (Scheme 2.55a). However, a novel photoproduct **152** could be obtained in about 20 % yield by the irradiation of **50** in an air-saturated acetone (Scheme 2.55b). If the solvent was turned to deoxygenated carbon tetrachloride, another photoproduct, 6-chlorotriptycene-1,4-quinone **153**, would be produced by the photolysis (350 nm, Rayonet photo reactor) of **50** (Scheme 2.55c). They considered that all three photoproducts were obtained via the same intermediate compound **154**, and the cyclopropyldicarbinyl diradical **154** was "the more stable diradical in which aromaticity had been lost in only one of the two benzenoid rings and the second odd electron was stabilized by being next to a carbonyl group" [102]. However, irradiation of **50** in the solid state showed no photoreaction, which was probably attributed to the effect of intermolecular interactions [103].

Soon after, Borecka et al. [104] extended the work to the photolysis of the 9,10-dialkyl-substituted triptycene quinones. Compared with the photoproducts obtained from **50**, the photoproducts of **155a, b** showed the unusual structure and color,

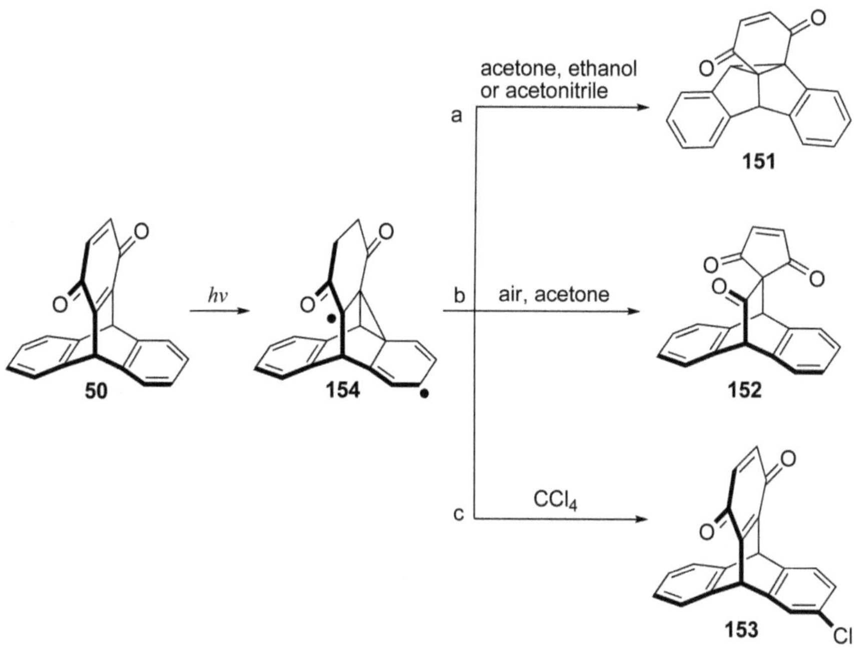

Scheme 2.55 Photochemical reaction of triptycene-1,4-quinone **50** in non-hydrogen atom-donating solvents

and these features made them favorable to be observed. The irradiation of **155a** in deoxygenated acetonitrile for 3 h gave the yellow di-π-methane photoproduct **157** (needles) and the blue–black norcaradiene-containing derivative **156a** (prisms) in 28 and 26 % isolated yield, respectively. However, the photolysis of **155b** afforded the significantly different products. Besides a small amount of the dark-blue norcaradiene compound **156b** and the dihydrobenzofuran derivative **158**, another dark orange unidentified photoproduct **159** was also isolated in 18 % yield which might be a phenolic aldehyde with the benz[a]aceanthrylene carbon skeleton (Scheme 2.56a). The carbene mechanism established by Iwamura and co-workers [94, 98] could well expound the formation of **156a, b**. Nevertheless, for quinone **155a**, both norcaradiene **156a** and di-π-methane photoproduct **157** might be formed from the plausible common intermediate biradical **160** via the di-π-methane rearrangement (Scheme 2.56b).

2.3.7 *Other Reactions*

As mentioned in the early works, the halogen at the bridgehead in 9-halogenated triptycenes showed to be unreactive with nucleophile. For example, Bartlett and co-workers [2, 48, 84] reported that the 9-bromotriptycene or 9-iodotriptycene failed

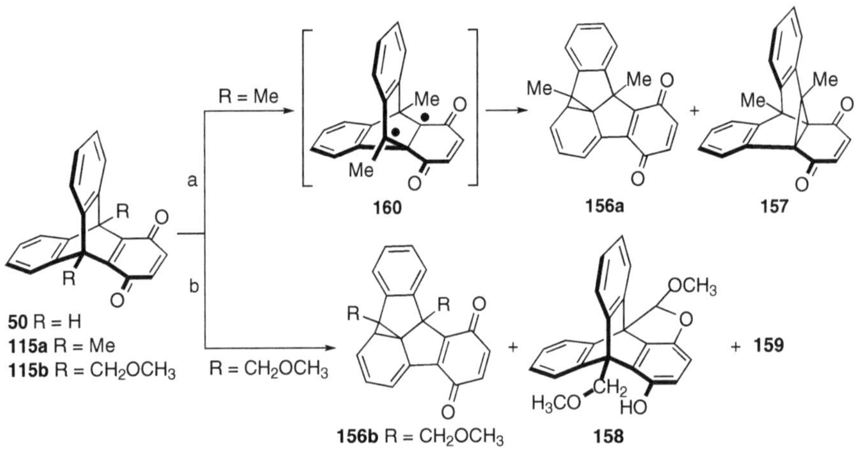

Scheme 2.56 Photolysis of the 9,10-dialkyl-substituted triptycene quinones

to react with sodium ethoxide, tin(II) chloride, sodium sulfide, or silver nitrite even under extreme conditions. Skvarchenko et al. [74] revealed that the structural reasons led the bridging carbon atom to be inert. Firstly, the intermediate formation of a carbonium cation required the transition of the bridging carbon to a sp^2-hybridized state and a coplanar arrangement of the aliphatic carbon–carbon bonds in the already strained bicyclic fragment of triptycene. Secondly, the carbonium cation could not be stabilized by conjugation with the aromatic rings but, on the contrary, be destabilized by the negative inductive effect of the three *o*-phenylene substituents [84, 105]. Thus, the S_N1 reactions would not occur. Moreover, the S_N2 reactions occurred with difficulty as well, as the carbon atoms at a bridgehead were also shielded from the attack by nucleophilic reagents [84, 105].

Moreover, Skvarchenko and co-workers [74, 75] described that in the extreme conditions, triptycene could be chloromethylated to give a mixture of 2-chloromethyl-substituted triptycene **161** in 15 % yield, along with 52 % yield of 2,6-bichloromethyltriptycene **162a**, and 16 % yield of 2,7-bichloromethyl-triptycene **162b** (Scheme 2.57).

Nevertheless, Wittig and coworkers [106, 107] reported that 9-bromotriptycene could react with *n*-butyl-lithium to give triptycen-9-yl-lithium **163**, which actually could serve as the precursor for other triptycene derivatives (Scheme 2.58). As shown in Scheme 2.59, it was also found that the reaction between **163** and mercury(II) chloride could give di-triptycen-9-yl-mercury in 76 % yield. Similarly, di-triptycen-9-yl-selenium and di-triptycen-9-yl-diselenium were obtained by the reaction of **163** and selenium chloride in 15 and 37 % yield, respectively.

In 1981, Oki and co-workers [108] reported the facile ring formation of triptycene derivatives in the presence of a cationoid center at the 2-position of a 9-substituent. Consequently, a cyclic ether **164** was obtained by the reaction of 9-(2-hydroxypropyl)-1,4-dimethoxytriptycene **165** with sulfuric acid. If 9-allyl-1,4-dimethoxytriptycene **166** was used, the same product **164** was obtained in the

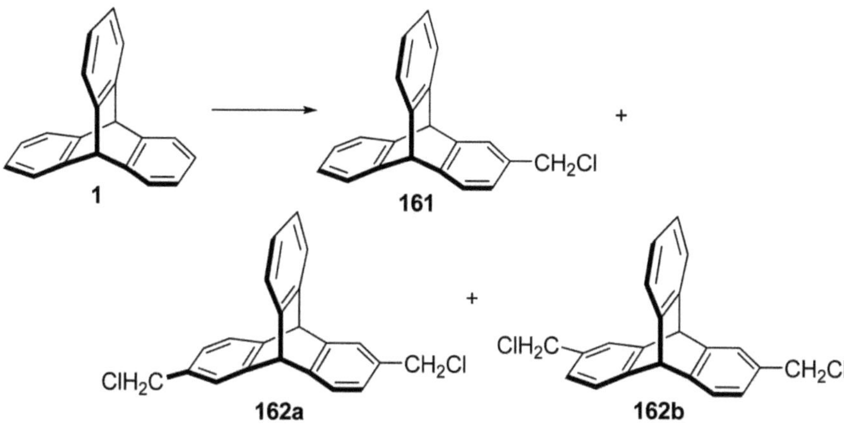

Scheme 2.57 Chloromethylation of triptycene

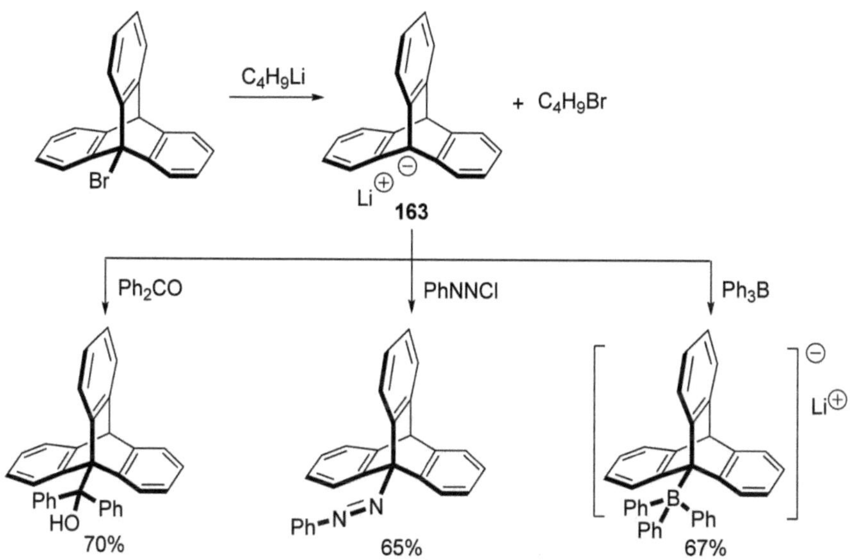

Scheme 2.58 Reaction of 9-bromotriptycene with *n*-butyl-lithium

presence of trifluoroacetic acid (Scheme 2.60). Moreover, the treatment of **166** with bromine, and 9-(2-hydroxyethyl)-1,4-dimethoxytriptycene **167** with thionyl chloride also gave the corresponding cyclic ether **168** and **169**, respectively (Scheme 2.61). Thus, they deduced that the reaction occurred via both the S_N1 and S_N2 processes with two steps: the formation of the cationoid **170** at the 2-position by an electrophilic addition to the olefin, then it cyclized to an oxonium ion **171** which would be attacked by an anion at the least hindered methyl to give the cyclic ether **172** (Scheme 2.62).

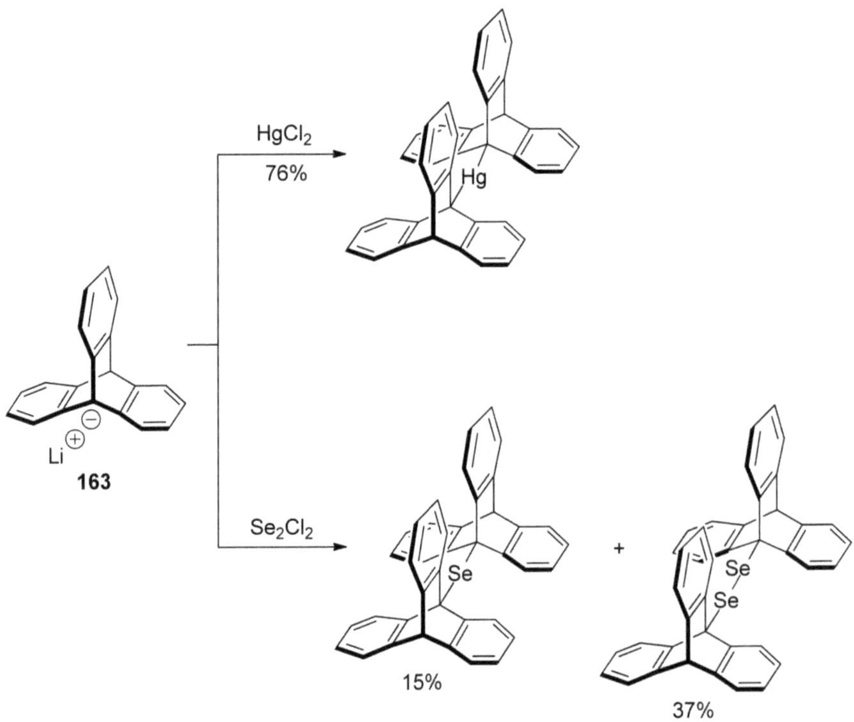

Scheme 2.59 Reaction of compound **163** with mercury(II) chloride and selenium chloride

2.4 Synthesis of Extended Triptycene Derivatives

In the early works, a lot of benzotriptycenes were obtained by the Diels–Alder addition between anthracene and arynes. For example, benzotriptycene **173** could be obtained by the reaction of anthracene with aryne from 1-nitro-naphthalene [109]. Similarly, by the use of the arynes obtained from 9-nitrophenanthrene or 9-bromophenanthrene in situ, extended triptycene **174** containing phenanthrene could be obtained. Moreover, the pyrolysis of benzil with naphthacene gave the benzotriptycene **175** (Scheme 2.63) [74, 110].

Analogously, Sugihash et al. [111] also synthesized the monobenzo-derivatives of extended triptycene **175** with a yield of 11 % by the Diels–Alder reaction between anthracene and aryne, which was generated from 3-amino-2-naphthoic acid and isoamyl nitrite in situ through the method of Friedman and Logullo [25] (Scheme 2.64a). Under the same conditions, the addition gave dibenzotriptycene **176** in a yield of 38 % when tetracene took place of the anthracene (Scheme 2.64b). Similarly, naphthotriptycenes **177** and **178** could also be synthesized by the addition of the aryne from 3-aminoanthraquinone-2-carboxylic acid with anthracene and tetracene, respectively (Scheme 2.65).

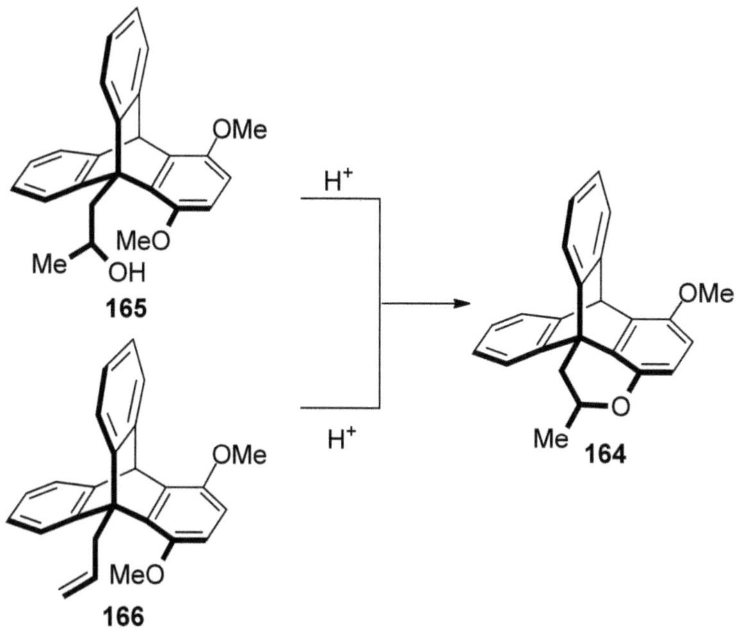

Scheme 2.60 Synthesis of triptycene derivative **164**

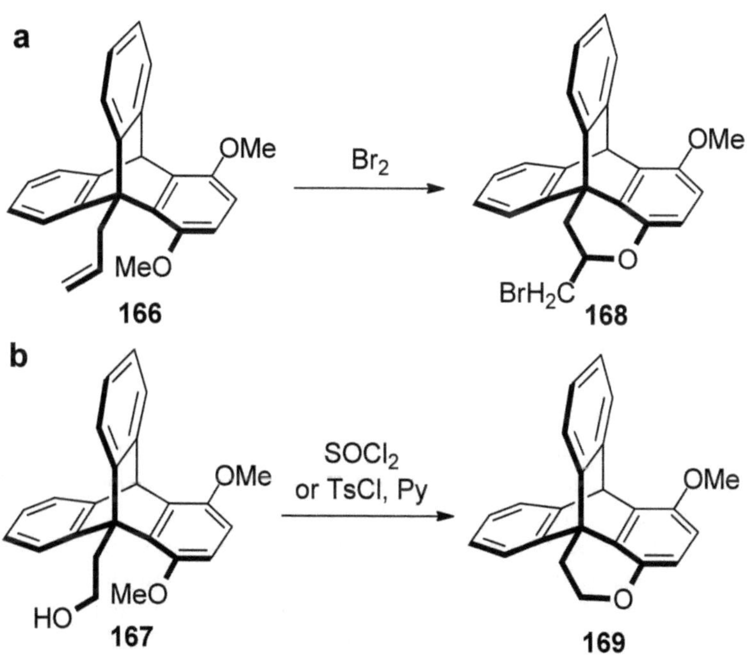

Scheme 2.61 Facile ring formation of triptycene derivatives **168** and **169**

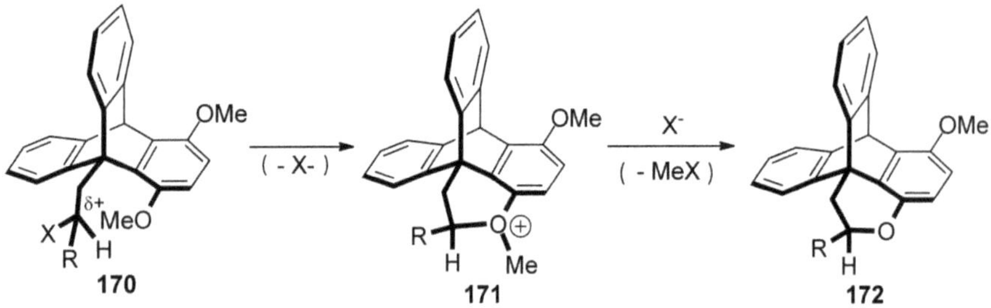

Scheme 2.62 Formation of cyclic ether **172**

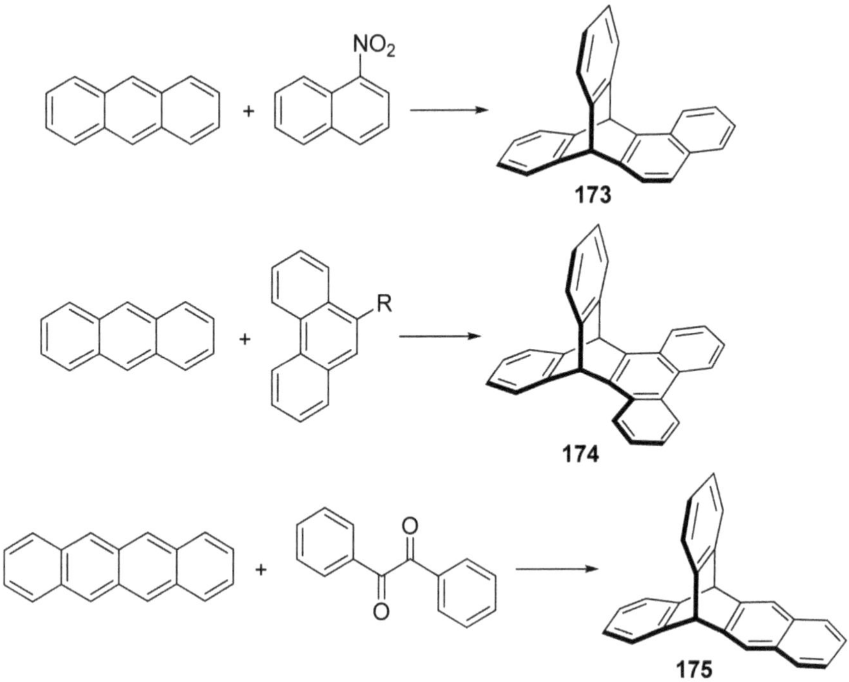

Scheme 2.63 Synthesis of extended triptycenes **173–175**

In 1974, Skvarchenko et al. [74] reported the synthesis of the extended triptycene derivatives (**179**). As shown in Scheme 2.66, the reaction between triptycene quinone and the substituted butadiene, followed by the reduction of the carbonyl groups, gave the target products (**179**).

In 1967, Regan and Miller [112] synthesized 2,3-benzotriptycene derivative **180** with substituents by the addition of the corresponding anthracene to 1,4-epoxy-1,4-dihydronaphthalene in the refluxing xylene, as shown in Scheme 2.67a. It was noteworthy that this route was originally suggested by Wittig et al. [113]. According to this strategy, tribenzotriptycene **181** [113] could also be synthesized by the reaction of pentaphene with 1,4-epoxy-1,4-dihydronaphthalene in xylene (Scheme 2.67b).

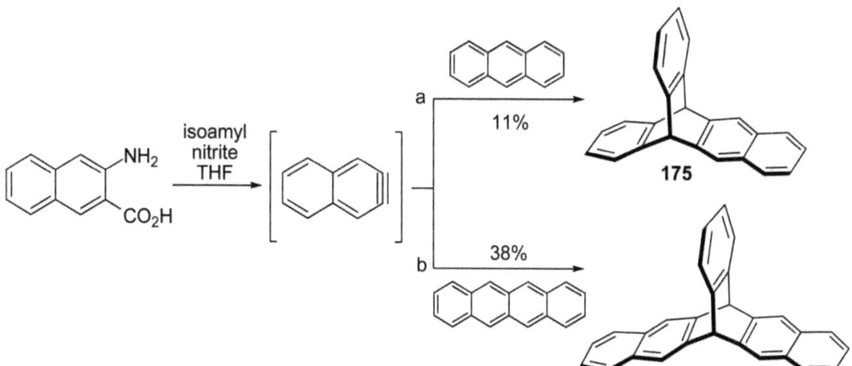

175

176

Scheme 2.64 Synthesis of **a** benzotriptycene **175** and **b** dibenzotriptycene **176**

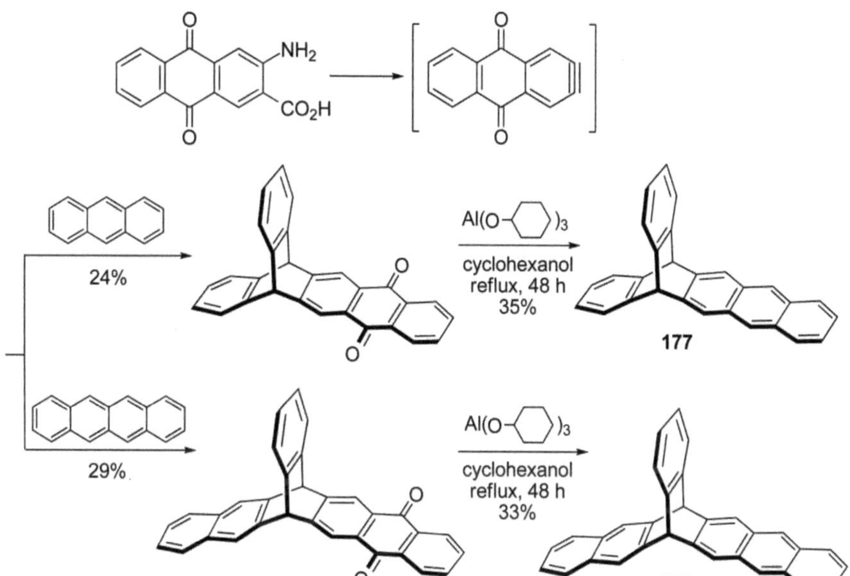

177

178

Scheme 2.65 Synthesis of naphthotriptycenes **177** and **178**

Although the triptycenes with fused ring could be generated by various routes as mentioned previously, the low yield and long process was the obstacle. In 1991, Patney [114] reported a general and facile route to the synthesis of fused ring derivatives of triptycene. As shown in Scheme 2.68, the addition of the quinone derivatives and anthracene, followed by the reduction of the carbonyl groups with lithium aluminum hydride and the further dehydration with *p*-toluenesulfonyl chloride in pyridine, gave the target extended triptycenes. However, the reaction time in this route was quite long; especially the last step reaction usually needed 3–7 days. Moreover, the target product was given in an instable yield (23–71 % based on the anthracene).

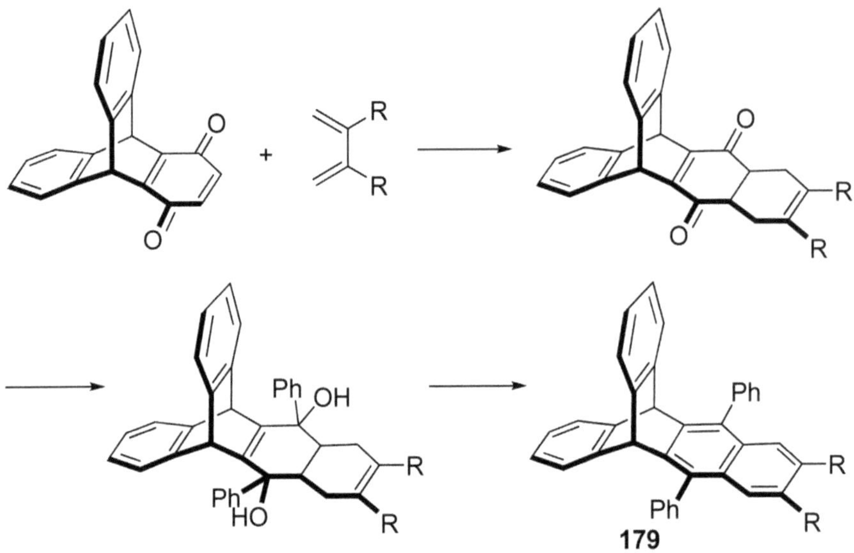

Scheme 2.66 Synthesis of the extended triptycene derivative **179**

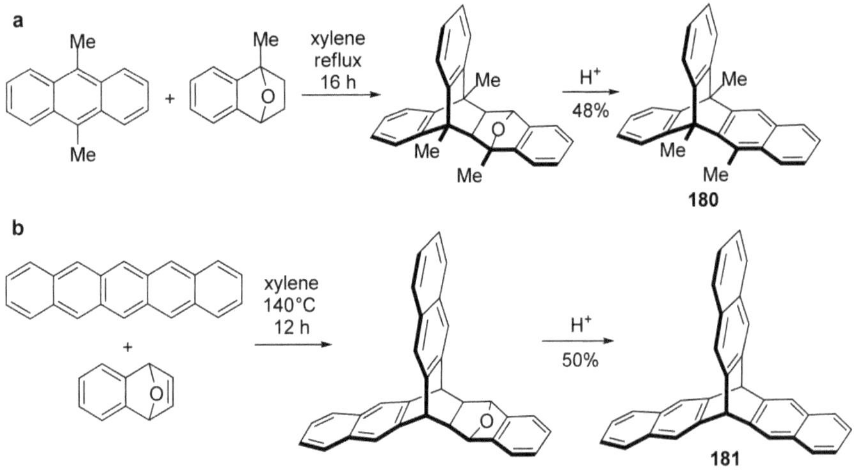

Scheme 2.67 Synthesis of **a** 2,3-benzotriptycene derivative **180** and **b** tribenzotriptycene **181**

Recently, Holý and co-workers [115] reported the synthesis of extended bifunctional triptycenes **182a, b** by two different methods. As shown in Scheme 2.69, the target products **182a, b** could be obtained by a one-pot reaction of aromatization with the primary adduct from dibromoanthracene with the dienophilic counterpart in xylene at 160 °C, and then followed by the acid hydrolysis. In contrast, the reaction of the dibromoanthracene and the aryne generated in situ from the amino acid gave the quinoid adduct **183** in a yield of 31 %, and the adduct **183** then proceeded

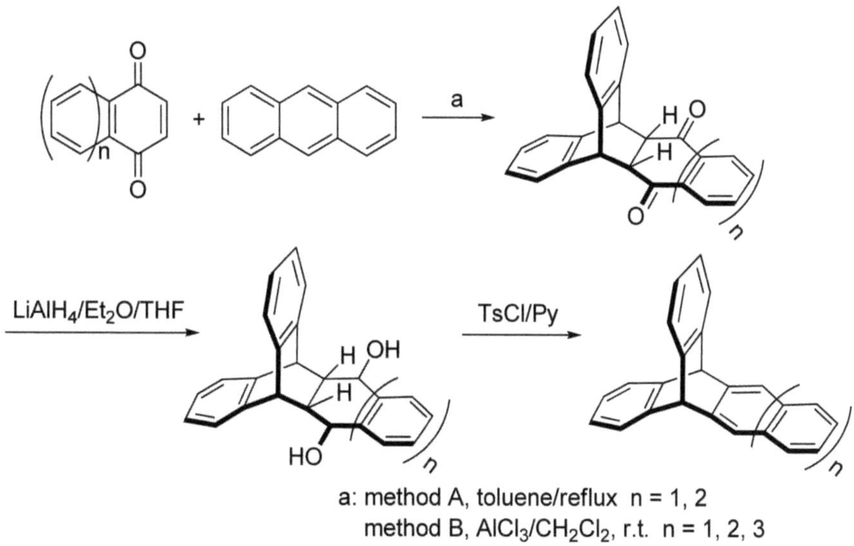

Scheme 2.68 Synthesis of extended triptycenes from quinone derivatives

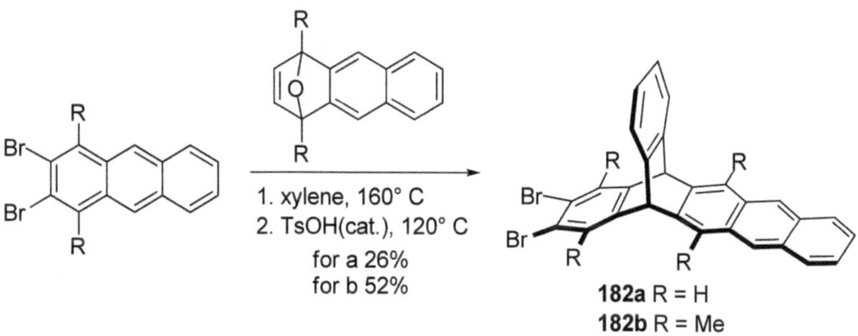

Scheme 2.69 Synthesis of extended bifunctional triptycenes **182a**, **b** by a one-pot reaction of aromatization

through a one-pot rearomatization to give the extended bifunctional triptycene **182a** in a yield of 34 % (Scheme 2.70).

For another extended bifunctional triptycenes **184a**, **b**, they could be achieved easily by the zinc-mediated reduction of the quinone carbonyl groups in **185**, and the conversion of the resulting air-unstable hydroquinone. As shown in Scheme 2.71a, treatment of the quinone **67** with NaOAc in the pressure of Zn in Ac_2O gave the product **184a** in a yield of 66 %. Although, the product **184b** could also be obtained by the similar zinc-mediated reduction of compound **67** with *t*-$BuPh_2SiCl$ in a mixture solution of THF and DMF (Scheme 2.71b). Moreover, the reaction of **67** with methyl lithium in a toluene solution at room temperature gave the intermediate compound **185**, which was then followed by the aromatization to give the extended bifunctional triptycene **184c** in a yield of 64 % (Scheme 2.72).

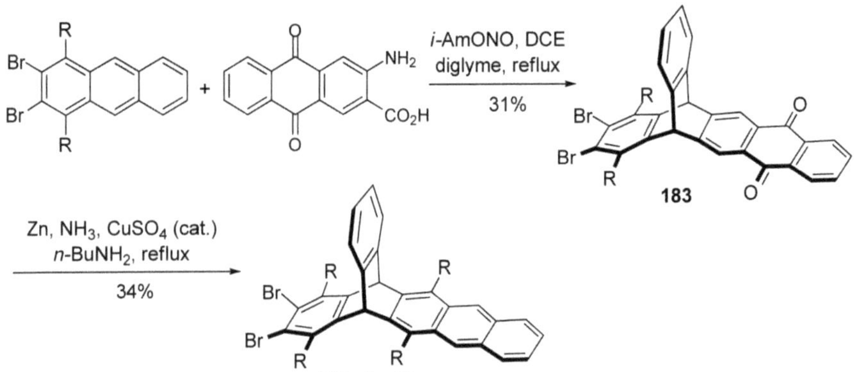

182a R = H

Scheme 2.70 Synthesis of extended bifunctional triptycenes **182a** by a one-pot rearomatization

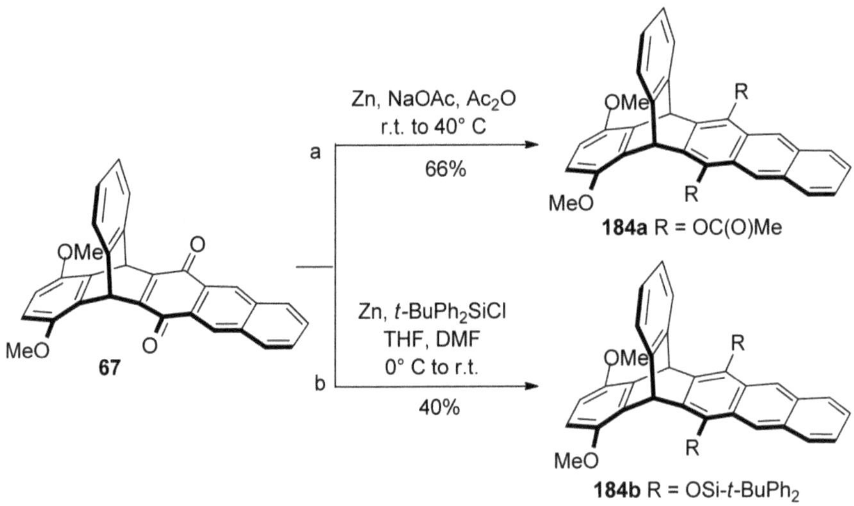

Scheme 2.71 Synthesis of extended bifunctional triptycenes **184a, b**

In 2010, Lee and co-workers [116] reported the synthesis of oxadisilole fused triptycene **187** and the extended oxadisilole fused triptycene **188**. As shown in Scheme 2.73, by the optimization of the reaction conditions, the triptycene **187** could be obtained in 82 % yield by the treatment of benzobisoxadisilole **189** with $PhI(OAc)_2$ in the presence of triflic acid, then with KF and a catalytic amount of TBAF, and further reaction with anthracene. Similarly, the extended triptycene derivative **188** was obtained by the addition of the aryne generated in situ with furan.

Oxadisilole fused triptycenes **187** and **188** could also be served as the precursors for the synthesis of other extended triptycene derivatives. Consequently, the reaction of **187** with phenyliodonium diacetate and triflic acid in CH_2Cl_2, followed by the

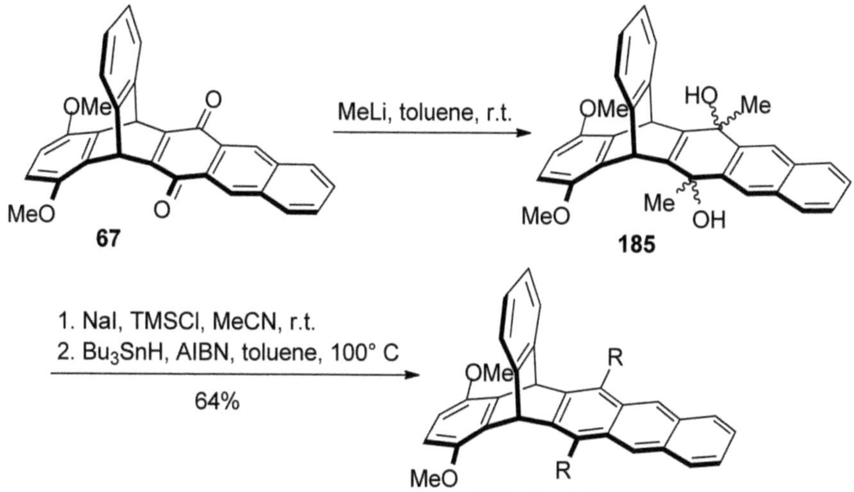

Scheme 2.72 Synthesis of extended bifunctional triptycene **184c**

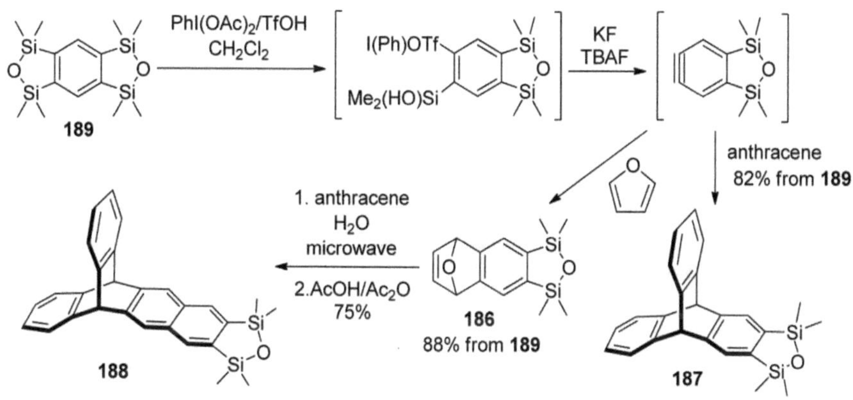

Scheme 2.73 Synthesis of oxadisilole fused triptycenes **187** and **188**

treatment with KF in the presence of TBAF gave the corresponding triptycynes (**190**, **191**), which further reacted with various dienes to give a series of extended triptycene derivatives (**192–196**, Scheme 2.74). Similarly, starting from the oxadisilole fused triptycene **188**, the extended triptycenes **197** and **198** could be synthesized via the aryne intermediate **191** as well (Scheme 2.75).

Soon after, Pei and Lee [117] further developed a highly efficient microwave-mediated Diels–Alder reaction between anthracene and various endoxides in water, which was then followed the dehydration in AcOH and Ac_2O to give the corresponding extended triptycenes (**175**, **177**, **188**, and **199a**, **b**) in 72–83 % yields (Scheme 2.76).

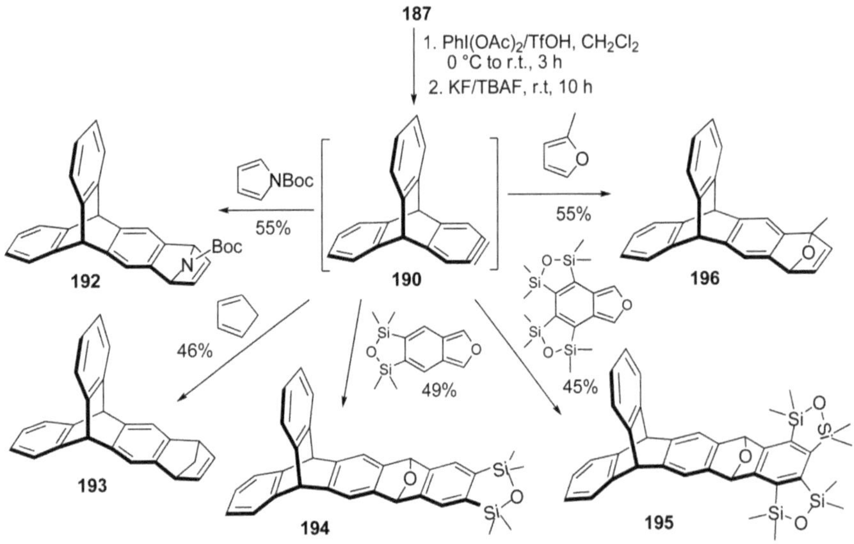

Scheme 2.74 Synthesis of extended triptycene derivatives **192–196**

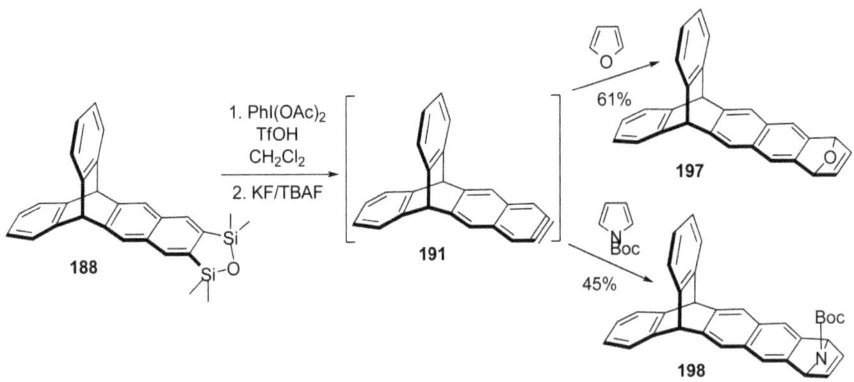

Scheme 2.75 Synthesis of extended triptycenes **197** and **198**

More recently, they [118] also synthesized a new synthon **200**, 5,6-(9,10-dihydroanthracen-9,10-yl)isobenzofuran for the preparation of extended triptycenes (**201**). The target extended triptycenes could be obtained in high yields (81–88 %) by the deoxygenation of the corresponding endoxide intermediates (**202**) with $TiCl_4/LiAlH_4/Et_3N$ (Scheme 2.77). Thus, the formation of endoxide intermediates (**202**) was the key to the preparation of extended triptycenes (**201**). As shown in Table 2.10, the endoxide intermediates (**202**) could be obtained in good isolated yields by the reaction of various dienophiles (**203**) with the freshly generated synthon **200**. In addition, they attempted to further obtain the extended triptycene quinones (**204**)

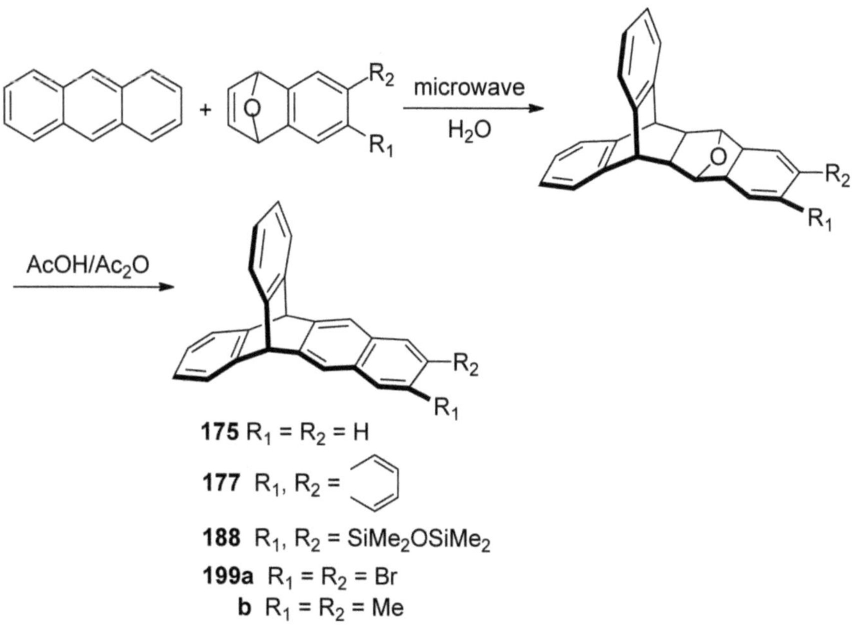

Scheme 2.76 Synthesis of extended triptycenes by highly efficient microwave-mediated Diels–Alder reaction

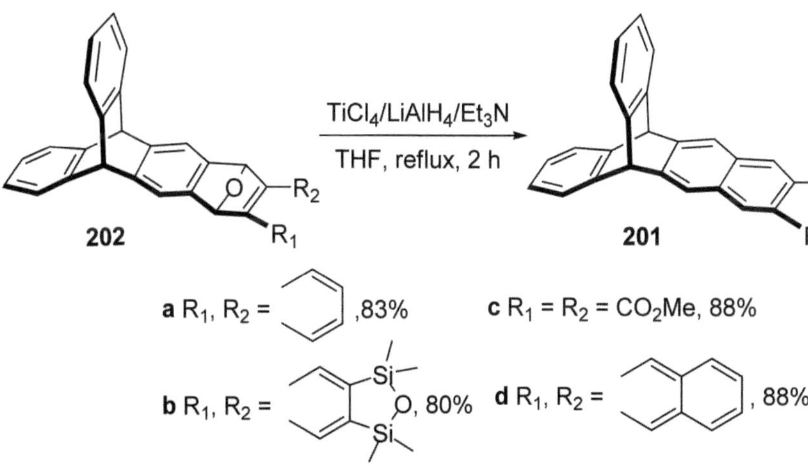

Scheme 2.77 Synthesis of extended triptycenes **201a, d** via endoxide intermediates **202a, d**

by the conversion of endoxides **202**. However, according to the method in literature [119], the endoxides **202** underwent the oxidation by O_2/K_2CO_3 only to give the trace amounts of the desired product (**204**). Finally, they found that the triptycene quinones could directly be converted from the corresponding endoxides in satisfactory yields (71–74 %) under $PhI(OAc)_2$/TfOH condition (Scheme 2.78).

Table 2.10 Synthesis of endoxides **202**

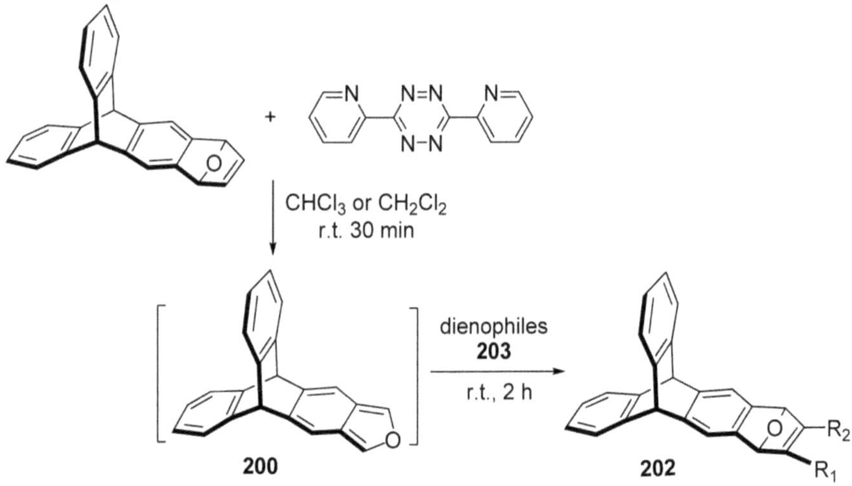

Entry	Dienophile (**203**)	Endoxide (**202**)	Yield (%)
1	$MeO_2C-\equiv-CO_2Me$	$R_1=R_2=CO_2Me$	80
2		$R_1, R_2=$ 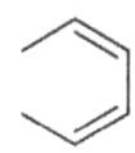	70
3	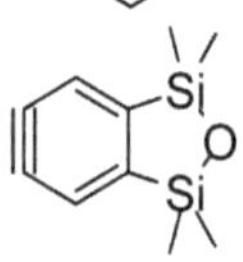	$R_1, R_2=$ 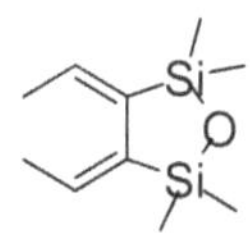	67
4	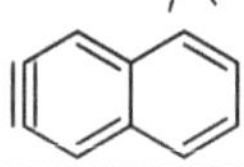	$R_1, R_2=$ 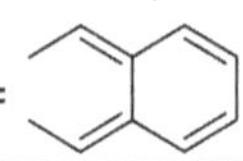	65

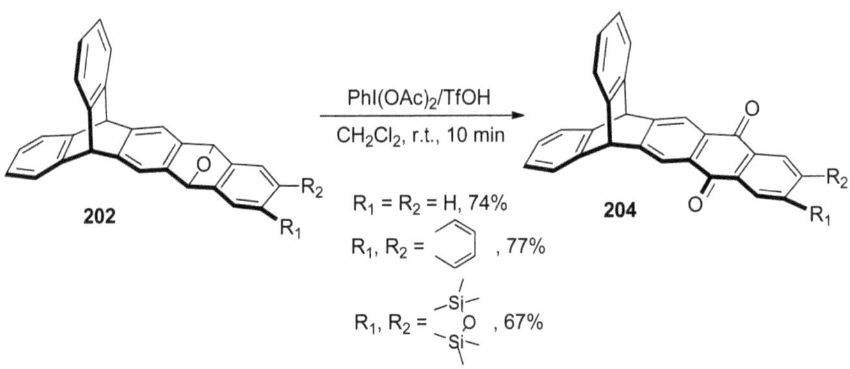

Scheme 2.78 Synthesis of extended triptycene quinone **204** via endoxide intermediate **202**

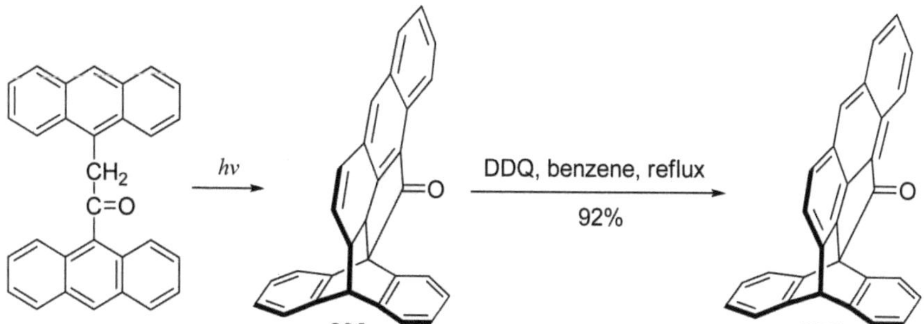

206

205

Scheme 2.79 Synthesis of anthracene-containing triptycene derivative **205**

Additionally, Kurata and co-workers [120] reported the synthesis of a new anthracene-containing triptycene derivative **205** by the dehydration of the photochemical [4 + 2] cycloadduct **206** of di(9-anthryl)methanone in a yield of 92 % (Scheme 2.79).

2.5 Synthesis and Reactions of Homotriptycenes

In 1970, Cristol and Pennelle [121] first synthesized the compound **207** called homotriptycene for the next higher homolog of triptycene. Compound **207** could be obtained by the ring expansion of 1-aminomethyltriptycene, or by the reduction of the chloride **208** with sodium in *t*-butyl alcohol. However, it was noteworthy that the generation of the isomer **209** as a side-product increased the difficulty of separation for compound **208** (Scheme 2.80). In 1985, Szeimies and co-workers [122] reported a new strategy for the synthesis of homotriptycene via the thermal dehydrogenation of anellated dibenzohomobarrelene, but the tedious preparation of the precursor propellane limited its general applicability as well. Soon after, Saito et al. [123] also synthesized the homotriptycene through the cycloaddition between strained benzocyclopropene and anthracenes.

Recently, Cao and co-workers [124, 125] reported a novel and facile route to synthesize homotriptycene **210** based on the acid-catalyzed cyclization of anthracenol derivatives. As shown in Scheme 2.81, starting from 10-benzyl-9,10-dihydroanthracen-9-ol **211** containing high electron density, the corresponding homotriptycenes could be quantitatively obtained by the transannular ring closure in the presence of acid. Furthermore, Cao and co-workers [126] obtained a series of new homotriptycenes (**212a**, **b**) containing heterocycles, like furan, thiophene, and pyridine rings in reasonable yields via the similar intramolecular electrophilic dehydration of the 10,10-dihetarylmethyl-9,10-di-hydroanthrancen-9-ols (**213a**, **b**).

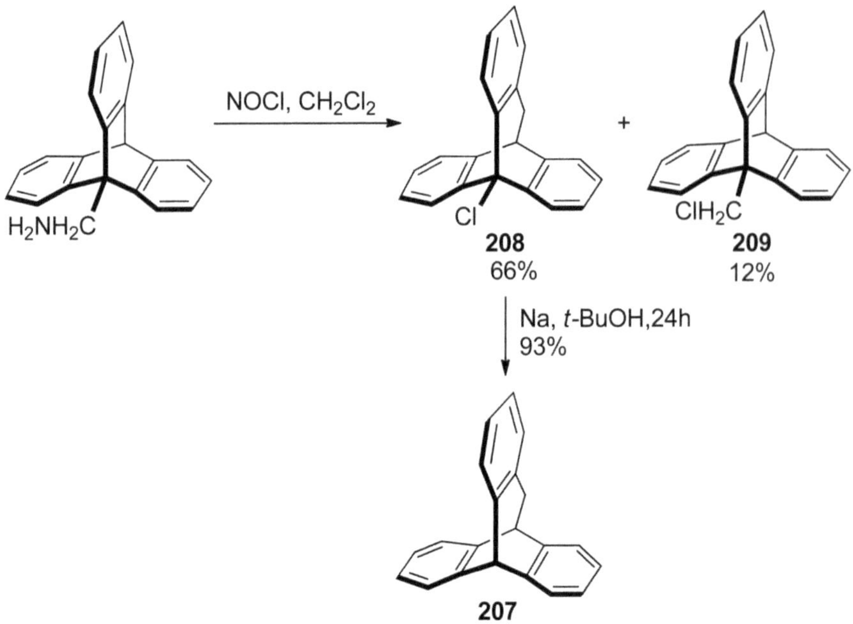

Scheme 2.80 Synthesis of homotriptycene **207**

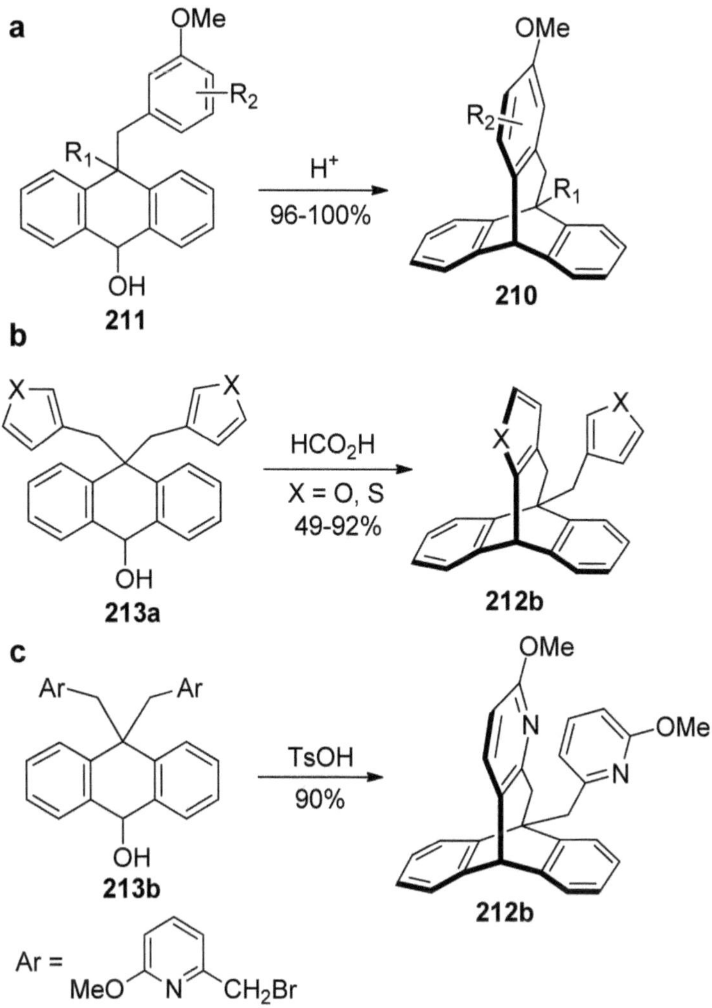

Scheme 2.81 Synthesis of homotriptycenes **210** and **212**

References

1. Bartlett PD, Ryan MJ, Cohen SG (1942) Triptycene (9,10-*o*-benzenoanthracene). J Am Chem Soc 64:2649–2653
2. Bartlett PD, Cohen SG, Cotman JD, Kornblum N, Landry JR, Lewis ES (1950) Bicyclic structures prohibiting the walden inversion. The synthesis of 1-bromotriptycene. J Am Chem Soc 72(2):1003–1004
3. Craig AC, Wilcox CF (1959) A new synthesis of triptycene. J Org Chem 24(10):1619–1619
4. Wittig G, Ludwig R (1956) Triptycen aus anthracen und dehydrobenzol. Angew Chem 68(1):40–40
5. Wittig G, Benz E (1958) Über lithium-natrium-organische komplexe, III. Über anionisierungsreaktionen mit diphenyl-lithium-natrium. Chem Ber 91(4):873–882
6. Wittig G (1959) Triptycene. Org Synth 39:75–77
7. Wittig G, Stilz W, Knauss E (1958) Anthracen-derivate mit *N*- und *O*-brückenatom. Angew Chem 70(6):166–166
8. Wittig G, Benz E (1960) Triptycen aus dehydrobenzol und anthracen. Tetrahedron 10(1–2):37–40
9. Cadogan JIG, Hall JKA, Sharp JT (1967) The formation of arynes by reaction of potassium *t*-butoxide with aryl halides. J Chem Soc C:1860–1862
10. Cadogan JIG, Harger MJP, Sharp JT (1971) Acylarylnitrosamines. Part III. Decomposition of 2,5-di-(*N*-nitrosoacetamido)-1,4-di-*t*-butylbenzene and related compounds. J Chem Soc B:602–607
11. Beyeler A, Belser P (2002) Synthesis of a novel rigid molecule family for the investigation of electron and energy transfer. Coord Chem Rev 230(1–2):29–39
12. Brewer JPN, Eckhard IF, Heaney H, Marples BA (1968) Aryne chemistry. Part V. Some addition reactions of tetrafluorobenzyne. J Chem Soc C 664–676
13. Heaney H, Jablonski JM (1968) Aryne chemistry. Part XII. Some cycloaddition reactions of tetrachlorobenzyne. J Chem Soc C 1895–1898
14. Cadogan JIG, Hibbert PG (1964) The anomalous decomposition of o-t-butyl-N-nitrosoacetanilide: evidence for the participation of an aryne. Proc Chem Soc 338–339
15. Cadogan JIG, Cook J, Harger MJP, Hibbert PG, Sharp JT (1971) Acylarylnitrosamines. Part II. The formation of arynes in the anomalous decompositions of o-t-butyl- and 2,5-di-t-butyl-N-nitrosoacetanilide. J Chem Soc B 595–601
16. Cadogan JIG, Mitchell JR, Sharp JT (1971) A simple, one-step, conversion of aniline into benzyne. J Chem Soc Chem Commun (1):1–2
17. Klanderman BH, Maier DP, Clark GW, Kampmeier JA (1971) Formation of benzyne by the decomposition of *N*-nitrosoacetanilide and *N*-(2-iodophenyl)-*N*-nitrosobenzamide. J Chem Soc Chem Commun (17):1003–1004
18. Stevens TE (1968) 2-azoxybenzoic acids as benzyne percursors. J Org Chem 33(2):855–856
19. Wittig G, Hoffmann RW (1961) Neuer zugang zu dehydrobenzol-reaktionen. Angew Chem 73(12):435–436
20. Wittig G, Hoffmann RW (1962) Dehydrobenzol aus 1.2.3-benzothiadiazol-1.1-dioxyd. Chem Ber 95(11):2718–2728
21. Stiles M, Miller RG (1960) Decomposition of benzenediazonium-2-carboxylate. J Am Chem Soc 82(14):3802–3802
22. Günther H (1963) Reaktionen von *o*-dijodbenzol mit zink. Chem Ber 96(7):1801–1809
23. Kessar SV, Singh P, Singh KN, Bharatam PV, Sharma AK, Lata S, Kaur A (2008) A study of BF3-promoted *ortho*-lithiation of anilines and DFT calculations on the role of fluorine–lithium interactions. Angew Chem Int Edit 47(25):4703–4706
24. Le Goff E (1962) Aprotic generation of benzyne from diphenyliodonium-2-carboxylate. J Am Chem Soc 84(19):3786–3786
25. Friedman L, Logullo FM (1963) Benzynes via aprotic diazotization of anthranilic acids. Convenient synthesis of triptycene and derivatives. J Am Chem Soc 85(10):1549

26. Kitamura T, Yamane M, Inoue K, Todaka M, Fukatsu N, Meng Z, Fujiwara Y (1999) A new and efficient hypervalent iodine–benzyne precursor, (phenyl)[*o*-(trimethylsilyl)phenyl]iodonium triflate: generation, trapping reaction, and nature of benzyne. J Am Chem Soc 121(50):11674–11679
27. Kornfeld EC, Barney P, Blankley J, Faul W (1965) Triptycene derivatives as medicinal agents. J Med Chem 8(3):342–347
28. Wilcox CF, Roberts FD (1965) Synthesis and some spectral properties of diphenyltriptycene. J Org Chem 30(6):1959–1963
29. Klanderman BH (1965) Novel products from the reaction of benzyne with anthracenes. J Am Chem Soc 87(20):4649–4651
30. Klanderman BH, Criswell TR (1969) Identity of benzyne from various precursors. J Am Chem Soc 91(2):510–512
31. Klanderman BH, Criswell TR (1969) Reactivity of benzyne toward anthracene systems. J Org Chem 34(11):3426–3430
32. Mori I, Kadosaka T, Sakata Y, Misumi S (1971) Synthesis and spectral properties of chloro-substituted triptycenes. Bull Chem Soc Jpn 44(6):1649–1652
33. Rogers ME, Averill BA (1986) Symmetrically trisubstituted triptycenes. J Org Chem 51(17):3308–3314
34. Klanderman BH, Faber JWH (1968) Novel bridged anthracene derivatives and polyesters and copolyesters therefrom. J Polym Sci Part A1 Polym Chem 6:2955–2965
35. Hoffmeister E, Kropp JE, McDowell TL, Michel RH, Rippie WL (1969) Triptycene polymers. J Polym Sci Part A1 Polym Chem 7:55–72
36. Chmiel J, Heesemann I, Mix A, Neumann B, Stammler HG, Mitzel NW (2010) The effect of bulky substituents on the formation of symmetrically trisubstituted triptycenes. Eur J Org Chem (20):3897–3907
37. Kuritani M, Nakagawa M, Sakata Y, Ogura F (1972) Absolute configuration of 2,7-disubstituted triptycenes as determined by chemical correlation. Chimia 26(9):470–471
38. Kricka LJ, Vernon JM (1971) Reactions involving deamination of isoindole adducts with acetylenic dienophiles. J Chem Soc Chem Commun (16):942–943
39. Marks V, Nahmany M, Gottlieb HE, Biali SE (2002) Polyethylated triptycene derivatives. J Org Chem 67(22):7898–7901
40. Lu J, Zhang JJ, Shen XF, Ho DM, Pascal RA (2002) Octaphenylbiphenylene and dodecaphenyltriptycene. J Am Chem Soc 124(27):8035–8041
41. Zhu XZ, Chen CF (2005) A highly efficient approach to [4]pseudocatenanes by threefold metathesis reactions of a triptycene-based tris [2] pseudorotaxane. J Am Chem Soc 127(38):13158–13159
42. Zyryanov GV, Palacios MA, Anzenbacher P (2008) Simple molecule-based fluorescent sensors for vapor detection of TNT. Org Lett 10(17):3681–3684
43. Rybackova M, Belohradsky M, Holy P, Pohl R, Dekoj V, Zavada J (2007) Synthesis of highly symmetrical triptycene tetra- and hexacarboxylates. Synthesis (10):1554–1558
44. Walborsky HM, Bohnert T (1968) A new synthesis of triptycene systems. J Org Chem 33(10):3934–3935
45. Taylor MS, Swager TM (2007) Triptycenediols by rhodium-catalyzed [2 + 2 + 2] cycloaddition. Org Lett 9(18):3695–3697
46. Sato K, Menggenbateer, Kubota T, Asao N (2008) AuCl-catalyzed reaction of *ortho*-alkynyl(oxo)benzene with benzenediazonium 2-carboxylate as a synthetic method towards anthracene, triptycene, and phthalazine derivatives. Tetrahedron 64(5):787–796
47. Clar E (1931) Über die konstitution des anthracens (zur kenntnis mehrkerniger aromatischer kohlenwasserstoffe und ihrer abkömmlinge, IX. mitteil.). Ber 64(7):1676–1688
48. Bartlett PD, Greene FD (1954) Triptycene 1-carboxylic acid and related compounds. The decomposition of ditriptoyl peroxide. J Am Chem Soc 76(4):1088–1096
49. Sonoda A, Ogura F, Nakagawa M (1962) Synthesis of trisubstituted triptycenes and the optical resolution of 7-carboxy-2,5-diacetoxytriptycene. Bull Chem Soc Jpn 35(6):853–857

50. Iwamura H, Makino K (1978) 5,8-dihydroxy-9,10-dihydro-9,10-[1,2] benzenoanthracene-1,4-dione. Intra-molecular triptycene quinhydrone. J Chem Soc Chem Commun (16):720–721
51. Hua DH, Tamura M, Huang XD, Stephany HA, Helfrich BA, Perchellet EM, Sperfslage BJ, Perchellet JP, Jiang SP, Kyle DE, Chiang PK (2002) Syntheses and bioactivities of substituted 9,10-dihydro-9,10-[1,2] benzenoanthracene-1,4,5,8-tetrones. Unusual reactivities with amines. J Org Chem 67(9):2907–2912
52. Wiehe A, Senge MO, Kurreck H (1997) One-step synthesis of functionalized triptycene-quinones as acceptors for electron-transfer compounds. Liebigs Ann 1997(9):1951–1963
53. Wiehe A, Senge MO, Schafer A, Speck M, Tannert S, Kurreck H, Roder B (2001) Electron donor-acceptor compounds: exploiting the triptycene geometry for the synthesis of porphyrin quinone diads, triads, and a tetrad. Tetrahedron 57(51):10089–10110
54. Fukuzumi S, Kochi JK (1982) Electron-transfer activation of the Diels–Alder reaction. Quantitative relationship to charge–transfer excited-states. Tetrahedron 38(8):1035–1049
55. Fukuzumi S, Kochi JK (1983) Importance of work terms in the free-energy relationship for electron-transfer. Bull Chem Soc Jpn 56(4):969–979
56. Dern M, Korth HG, Kopp G, Sustmann R (1985) On the role of radical ion-pairs in [4 + 2] cyclo-additions. Angew Chem Int Ed 24(4):337–339
57. Kochi JK (1988) Electron transfer and charge transfer: twin themes in unifying the mechanisms of organic and organometallic reactions. Angew Chem Int Ed 27(10):1227–1266
58. Maier G (1988) Tetrahedrane and cyclobutadiene. Angew Chem Int Ed 27(3):309–333
59. Sustmann R, Lucking K, Kopp G, Rese M (1989) [4 + 2]cycloadditions with 1,4-bis(*N, N*-dimethyl-amino)-1,3-dienes: stereochemical studies and observation of radical ions. Angew Chem Int Ed 28(12):1713–1715
60. Fukuzumi S, Okamoto T (1993) Magnesium perchlorate-catalyzed Diels–Alder reactions of anthracenes with *p*-benzoquinone derivatives: catalysis on the electron transfer step. J Am Chem Soc 115(24):11600–11601
61. Klanderman BH, Perkins WC (1969) Nitration of triptycene. J Org Chem 34(3):630–633
62. Hartshorn SR, Moodie RB, Schofield K (1971) Electrophilic aromatic substitution. Part VII. A critical re-examination of the reactivity of toluene towards nitration with acetyl nitrate in acetic anhydride, as determined by the kinetic and competition methods. J Chem Soc B 1256–1261
63. Rees JH (1975) Nitration of triptycene in acetic-anhydride. J Chem Soc Perk T 2(9):945–947
64. Taylor R, Wright GJ, Homes AJ (1967) Rate factors for the detritiation of triptycene, 9,10-dihydroanthracene and *o*-xylene. J Chem Soc B 780–782
65. Shigeru T, Ryusei K (1973) Electron spin resonance studies of bicyclo[2.2.1]heptanes and bicyclo[2.2.2]octanes spin labeled with nitrobenzene anion radicals. J Am Chem Soc 95(15):4976–4986
66. Chong JH, MacLachlan MJ (2006) Robust non-interpenetrating coordination frameworks from new shape-persistent building blocks. Inorg Chem 45(4):1442–1444
67. Chong JH, MacLachlan MJ (2007) Synthesis and structural investigation of new triptycene-based ligands: en route to shape-persistent dendrimers and macrocycles with large free volume. J Org Chem 72(23):8683–8690
68. Zhang C, Chen CF (2006) Synthesis and structure of 2,6,14-and 2,7,14-trisubstituted triptycene derivatives. J Org Chem 71(17):6626–6629
69. Chen Z, Swager TM (2008) Synthesis and characterization of poly(2,6-triptycene). Macromolecules 41(19):6880–6885
70. Mastalerz M, Sieste S, Cenicì' M, Oppel IM (2011) Two-step synthesis of hexaammonium triptycene: an air-stable building block for condensation reactions to extended triptycene derivatives. J Org Chem 76(15):6389–6393
71. Dahms K, Senge MO (2008) Triptycene as a rigid, 120 orienting, three-pronged, covalent scaffold for porphyrin arrays. Tetrahedron Lett 49(37):5397–5399
72. Paget CJ, Burger A (1965) Acetylation of triptycene. J Org Chem 30(4):1329–1331
73. Skvarchenko VR, Brunovle, II, Novikov AM, Levina RY (1970) Aromatic hydrocarbons. 40. Triptycene acylation. Zhur Org Khim 6(7):1501

74. Skvarchenko VR, Shalaev VK, Klabunovsk EI (1974) Advances in the chemistry of triptycene. Russ Chem Rev 43(11):951–966
75. Skvarchenko VR, Kha NB, Kokin VN, Levina RY (1971) Aromatic hydrocarbons. 42. 2-formyltriptycene. Zhur Org Khim 7(9):1951
76. Nguen Bik HA, Skvarchenko VR (1974). Vestnik Moskov Univ:74
77. Ballester M, Riera-Figueras J, Castaner J, Badfa C, Monso JM (1971) Inert carbon free radicals. I. Perchlorodiphenylmethyl and perchlorotriphenyl-methyl radical series. J Am Chem Soc 93(9):2215–2225
78. Shalaev VK, Getmanov EV, Skvarchenko VR (1973) Aromatic-hydrocarbons—trinitrotriptycene. Vestnik Moskov Univ 14(6):740–741
79. Shimizu Y, Tatemits H, Ogura F, Nakagawa M (1973) Absolute configuration of 1,5-disubstituted 9,10-dihydro-9,10-ethenoanthracenes as revealed by chemical correlation. J Chem Soc Chem Commun (1):22–23
80. Rybackova M, Belohradsky M, Holy P, Pohl R, Zavada J (2006) Versatile synthesis of triptycene di- and tetracarboxylic acids. Synthesis (12):2039–2042
81. Zonta C, De Lucchi O, Linden A, Lutz M (2010) Aynthesis and structure of D_{3h}-symmetric triptycene trimaleimide. Molecules 15(1):226–232
82. Theilacker W, Möllhoff E (1962) Spaltung von arylmethanen durch kalium. Angew Chem 74(20):781–781
83. Walsh TD, Ross RT (1968) Reductive cleavage of triptycene derivatives. Tetrahedron Lett (27):3123–3126
84. Bartlett PD, Lewis ES (1950) Bicyclic structures probhibiting the walden inversion. Further studies on triptycene and its derivatives, including 1-bromotriptycene. J Am Chem Soc 72(2):1005–1009
85. Bohm H, Kalo J, Yarnitzk C, Ginsburg D (1974) Attempted synthesis of a substituted [2.2.2]propellane derivative from triptycene derivatives. Tetrahedron 30(1):217–219
86. Akiyama S, Nakagawa M, Ogura F (1971) Preparation of 9-ethylnylanthracene and 9-ethynyltriptycene and their oxidative coupling. Bull Chem Soc Jpn 44(12):3443–3445
87. Skvarchenko VR, Kha NB, Frenkel EE (1974) Aromatic-hydrocarbons. 52. 2-ethynyltriptycene. Zhur Org Khim 10(7):1493–1495
88. Skvarchenko VR, Kha NB (1974) Aromatic-hydrocarbons. 50. isomeric vinyltriptycenes. Zhur Org Khim 10(6):1252–1256
89. Rabideau PW, Jessup DW, Ponder JW, Beekman GF (1979) Metal-ammonia reduction of triptycene and related benzobarrelene derivatives. J Org Chem 44(25):4593–4597
90. Walsh TD (1969) Photochemical isomerization of triptycene. J Am Chem Soc 91(2):515–516
91. Turro NJ, Tobin M, Friedman L, Hamilton JB (1969) Photochemistry of triptycene. J Am Chem Soc 91(2):516
92. Iwamura H (1974) Excited-state reactions of triptycenes. 2. carbene mechanism for photoisomerization of triptycene. Chem Lett (1):5–8
93. Iwamura H (1974) Excited-state reactions of triptycenes. 4. photolytic one-step synthesis and stereochemistry of 1,4-dimethyl-9-arylfluorenes. Chem Lett (10):1205–1208
94. Iwamura H, Yoshimur K (1974) Trapping of carbene intermediates in photolysis of triptycenes. J Am Chem Soc 96(8):2652–2654
95. Day RO, Day VW, Fuerniss SJ, Wheeler DMS (1975) Photorearrangement of dimethoxytriptycene to methoxybenz alpha aceanthrylene. J Chem Soc Chem Commun (8):296–297
96. Iwamura H, Tukada H (1978) Wittig rearrangement of alpha-alkoxycarbenes formed by photorearrangement of 1-alkoxytriptycenes. Tetrahedron Lett (37):3451–3454
97. Day RO, Day VW, Fuerniss SJ, Hohman JR, Wheeler DMS (1976) Photorearrangement of dihydroxytriptycene. J Chem Soc Chem Commun (21):853–854
98. Iwamura M, Tukada H, Iwamura H (1980) Contrasting photochemical bridging regioselectivity in bridgehead-substituted 9,10-ethenoanthracenes vs 9,10-(*ortho*-benzeno)-9,10-dihydroanthracenes. Tetrahedron Lett 21(50):4865–4868
99. Hemetsberger H, Neustern FU (1982) Photochemical deuterium effect on the rearrangement of triptycene. Tetrahedron 38(9):1175–1182

100. Kawada Y, Tukada H, Iwamura H (1980) Novel route from triptycenes to a dibenzo(hafner hydrocarbon), benz[a]indeno[1,2,3-*cd*]azulene. Tetrahedron Lett 21(2):181–182
101. Kitaguchi N (1989) Photochemical hydrogen abstraction reactions of 9,10-dihydro-9,10-*ortho*-benzenoanthracene-1,4-dione derivatives with xanthene. intramolecular charge–transfer quenching. Bull Chem Soc Jpn 62(11):3542–3548
102. Fu TY, Gamlin JN, Olovsson G, Scheffer JR, Trotter J, Young DT (1995) The novel photochemical behavior of triptycene-1,4-quinone. Tetrahedron Lett 36(12):2025–2028
103. Fu TY, Gamlin JN, Olovsson G, Scheffer JR, Trotter J, Young DT (1998) Photochemistry of triptycene-1,4-quinone. Acta Crystallogr C 54(1):116–119
104. Borecka B, Gamlin JN, Gudmundsdottir AD, Olovsson G, Scheffer JR, Trotter J (1996) Unusual photorearrangements of 9,10-disubstituted triptycene-1,4-quinone derivatives. Tetrahedron Lett 37(13):2121–2124
105. Schöllkopf U (1960) Substitutions-reaktionen am brückenkopf bicyclischer verbindungen. Angew Chem 72(5):147–159
106. Wittig G, Schöllkopf U (1958) Zum chemismus der Halogen–Lithium-austauschreaktion. Tetrahedron 3(1):91–93
107. Wittig G, Tochtermann W (1962) Reaktionen am bruckenkopf des triptycens. Liebip Ann Chem 660:23–33
108. Izumi G, Hatakeyama S, Nakamura N, Ōki M (1981) Facile intramolecular cyclization reactions of aromatic ethers with cationoids in triptycene systems. Bull Chem Soc Jpn 54(1):258–260
109. Hurd CD, Juel LH (1955) The reaction of aromatic nitro compounds with polynuclear hydrocarbons at elevated temperatures. J Am Chem Soc 77(3):601–606
110. Taylor GW (1957) Gas phase reactions of phenyl radicals. Can J Chem 35(7):739–741
111. Sugihash M, Sakata Y, Misumi S, Kawagita R, Otsubo T (1972) Benzo-derivatives and naphtho-derivatives of triptycene: synthesis and properties. Bull Chem Soc Jpn 45(9): 2836–2841
112. Regan TH, Miller JB (1967) Nuclear magnetic resonance spectra of some substituted benzotriptycenes: the effect of steric compression. J Org Chem 32(9):2789–2794
113. Wittig G, Harle H, Knauss E, Niethammer K (1960) Synthese von derivaten des triptycens und iso-aza-triptycens. Chem Ber-Recl 93(4):951–962
114. Patney HK (1991) A general and simple route to the synthesis of triptycenes. Synthesis (9):694–696
115. Rybacek J, Zavada J, Holy P (2008) Synthesis of extended bifunctional triptycenes. Synthesis (22):3615–3618
116. Pei BJ, Chan WH, Lee AWM (2010) Oxadisilole fused triptycene and extended triptycene: precursors of triptycyne and extended triptycyne. J Org Chem 75(21):7332–7337
117. Pei BJ, Lee AWM (2010) Highly efficient synthesis of extended triptycenes via Diels–Alder cycloaddition in water under microwave radiation. Tetrahedron Lett 51(34):4519–4522
118. Pei BJ, Chan WH, Lee AWM (2011) Anthracene capped isobenzofuran: a synthon for the preparations of iptycenes and iptycene quinones. Org Lett 13(7):1774–1777
119. Barluenga J, Martinez S, Suarez-Sobrino AL, Tomas M (2008) One-pot, two-step synthesis of substituted anthraquinones from chromium(0) alkynyl carbenes and isobenzofurans. Org Lett 10(4):677–679
120. Kurata H, Kyusho M, Nishimae Y, Matsumoto K, Kawase T, Oda M (2007) 5*H*,13*H*-5,13a-[1′,2′]benzenocyclopenta[*rst*]pentaphen-13-one: a new tripty-cene derivative with a strained structure. Chem Lett 36(4):540–541
121. Cristol SJ, Pennelle DK (1970) Bridged polycyclic compounds. LXI. Synthesis and some properties of tribenzobicyclo[3.2.2]nonatriene (homotriptycene) and derivatives. J Org Chem 35(7):2357–2361
122. Baumgart KD, Harnisch H, Szimies-Seebach U, Szeimies G (1985) Zur Chemie einiger [4.1.1]- und [3.1.1]Propellane. Chem Ber 118(7):2883–2916
123. Saito K, Ito K, Takahashi K, Kagabu S (1991) Facile synthesis of homotriptycenes via addition of benzocyclopropene and anthracenes. Org Prep Proced Int 23(2):196–197

124. Gao CM, Cao DR, Xu SY, Meier H (2006) Acid-catalyzed cyclization of anthracenol derivatives to homotriptycenes. J Org Chem 71(8):3071–3076
125. Cao DR, Gao CM, Meier H (2005) A facile synthesis of homotriptycenes from anthranol derivatives. Synlett (20):3166–3168
126. Zhang H, Cao D, Liu W, Jiang H, Meier H (2011) Synthesis of heterocyclic homotriptycenes. J Org Chem 76(14):5531–5538

Chapter 3
Synthesis and Reactions of Pentiptycenes and Their Derivatives

3.1 Synthesis of Pentiptycenes and Their Derivatives

Compared with the first synthesis of triptycene in 1942 [1], it was not until 32 years later that Skvarche and Shalaev [2] first accomplished the synthesis of pentiptycene (**2**). As shown in Scheme 3.1, pentiptycene **2** was synthesized in about 10 % total yield by a three-step route starting from 2-aminotriptycene **1**. Similar to the way for the synthesis of triptycene, the Diels–Alder reaction between anthracene and benzyne was also the key process to pentiptycene.

In 1981, on the basis of the strategy that tetra-halobenzenes were served as diaryne equivalents [3], Hart et al. [4] reported a one-pot synthesis of pentiptycene **2** in 94 % yield by the Diels–Alder addition reaction between anthracene and 1,2,4,5-tetrabromobenzene (**3**) in toluene at room temperature in the presence of *n*-butyl lithium in hexane solution (Scheme 3.2). Synthesis of the *ortho*-pentiptycene (**4**) took the similar approach. As shown in Scheme 3.3, 4,5-dibromo-3,6-diiodo-*o*-xylene **5** was first prepared from the corresponding dibromo compound (**6**) in a yield of 81 %. Then, the similar Diels–Alder reaction between compound **5** and anthracene at −23 °C in a toluene solution in the presence of *n*-butyl lithium gave pentiptycene **5** in 14 % yield. The *ortho*-pentiptycene **4** has a C_{2v} symmetry, thus there are only 12 kinds of "different" carbons shown in its^{13}C NMR spectrum.

Afterward, Hart et al. [5] further developed a new route for the synthesis of pentiptycene **2** by using 2,3-naphtho[b]triptycene **7**, which could be considered as a triptycene containing one anthracene moiety, as a synthon (Scheme 3.4). This route was complicated, but still provided a new strategy for the synthesis of pentiptycene and even other higher iptycenes. It was apparent that 2,3-naphtho[b]triptycene **7** was an important intermediate in this process.

In 2010, Lee and co-workers [6] reported another route to the synthesis of pentiptycene **2** by the addition reaction between triptycyne, which was generated from oxadisilole-fused triptycene (**8**) and anthracene. As shown in Scheme 3.5, the treatment of **8** with phenyliodonium diacetate and trifluoromethanesulfonic acid (TfOH) in a CH_2Cl_2 solution at room temperature gave the precursor (**9**) for triptycyne. Then without separation, the whole system was treated with KF in the presence of a catalytic amount of TBAF for 10 h to obtain the intermediate triptycyne, which was further directly reacted with anthracene to produce pentiptycene **2** in a moderate yield.

C.-F. Chen, Y.-X. Ma, *Iptycenes Chemistry*,
DOI 10.1007/978-3-642-32888-6_3, © Springer-Verlag Berlin Heidelberg 2013

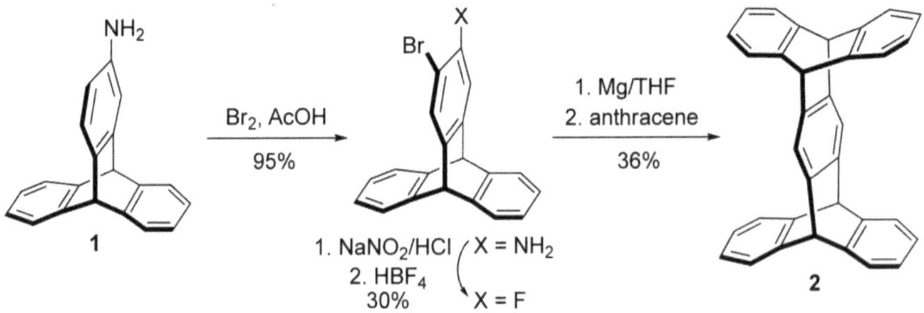

Scheme 3.1 Synthesis of pentiptycene **2** from 2-aminotriptycene **1**

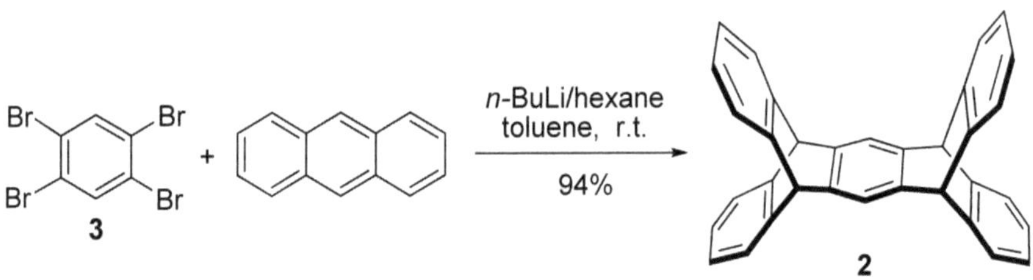

Scheme 3.2 Synthesis of pentiptycene **2** by the Diels–Alder addition reaction between anthracene and 1,2,4,5-tetrabromobenzene

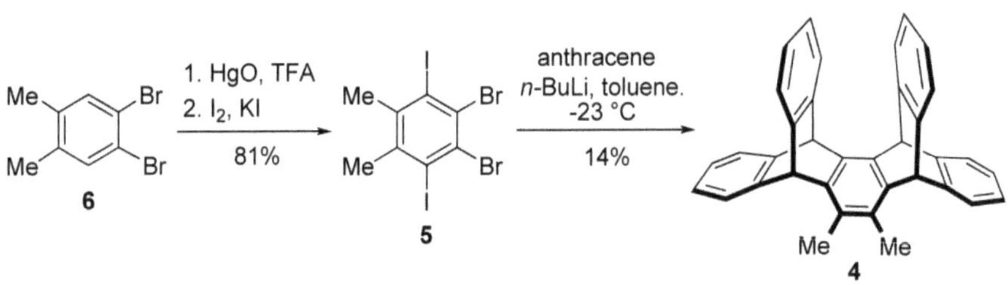

Scheme 3.3 Synthesis of *ortho*-pentiptycene **4**

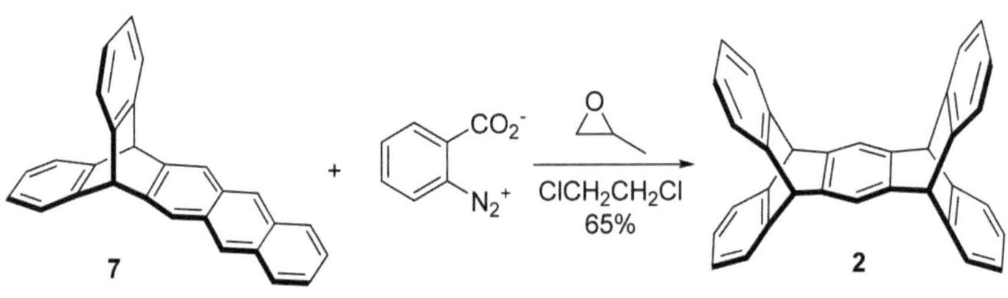

Scheme 3.4 Synthesis of pentiptycene **2** by using 2,3-naphtho[b]triptycene **7**

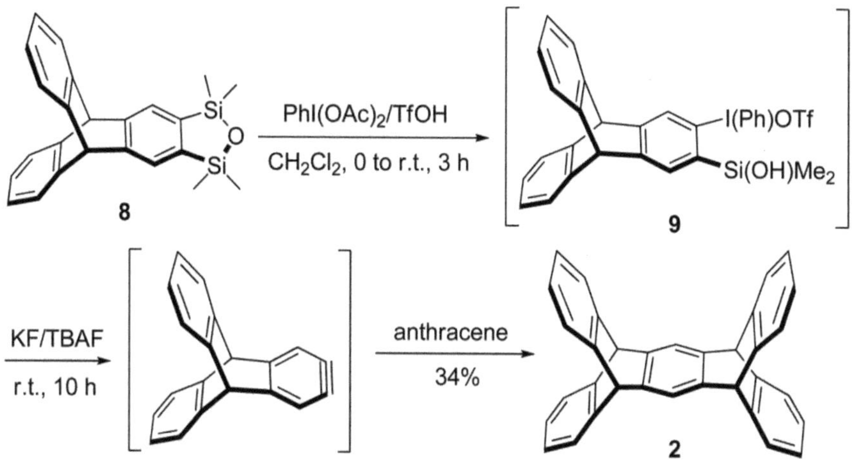

Scheme 3.5 Synthesis of pentiptycene **2** by the addition reaction between triptycyne and anthracene

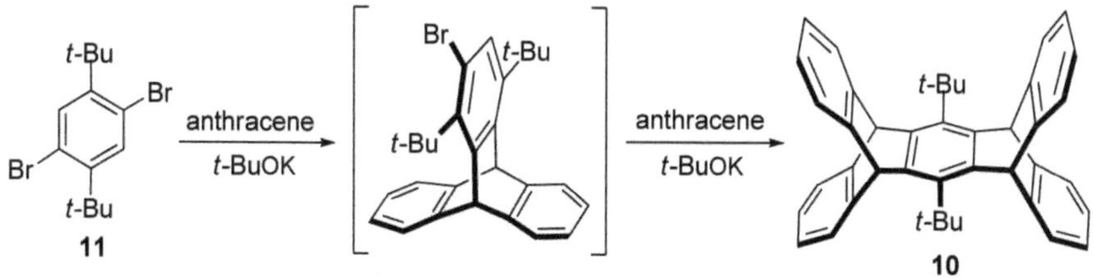

Scheme 3.6 Synthesis of di-*t*-butyl pentiptycene **10**

As pentiptycene contains one central benzene ring and four-phenyl side-chains, it has much more derivatives compared with triptycene. Such a structural feature also makes the pentiptycene to catch much attention to the synthesis of its derivatives. Generally, there are two strategies for the synthesis: (1) the Diels–Alder reaction between anthracene and arynes formed in situ from various halogenated benzenes, and (2) the similar addition reaction between anthracene and quinone afforded the pentiptycene quinones, which were then followed by the reduction of carbonyl groups to give the corresponding pentiptycene derivatives.

In 1971, Cadogan et al. [7] synthesized di-*t*-butyl pentiptycene (**10**) in very low yield (about 2 % after purification) by the reaction between aryne formed from **11** and anthracene in the presence of strong base (Scheme 3.6).

Ten years later, Hart et al. [4] reported the synthesis of substituted pentiptycenes by a one-pot reaction of anthracene and tetrabromo-*p*-xylene (**12**, R = Me) and tetrabromo-hydroquinone dimethyl ether (**12**, R = OMe), respectively, in the presence of *n*-butyl lithium (Scheme 3.7a). Moreover, the treatment of tetrabromobenzene (**12**, R = H) with 9,10-dimethoxyanthracene and *n*-butyl lithium gave the first example of a bridgehead substituted pentiptycene **14** (Scheme 3.7b). This method was straightforward, but the scope of application was limited to the poor solubility of substituted anthracenes and the too low yield of the reaction.

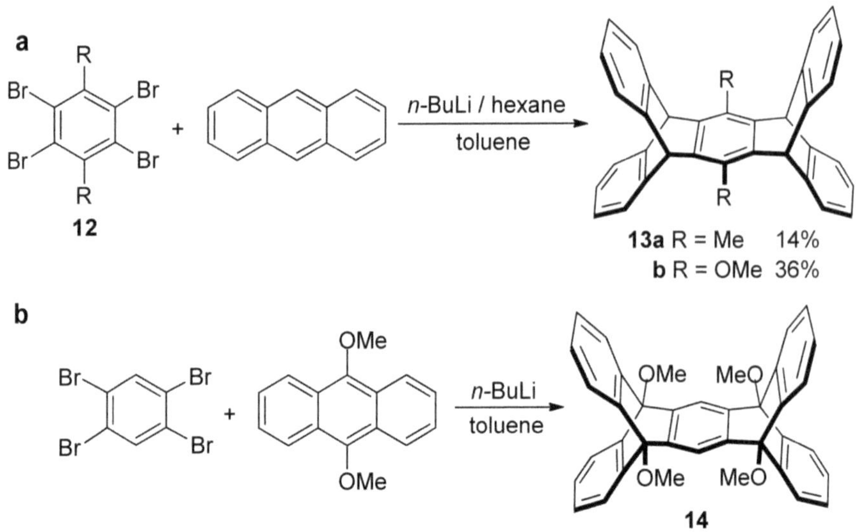

Scheme 3.7 Synthesis of substituted pentiptycenes **13** and **14**

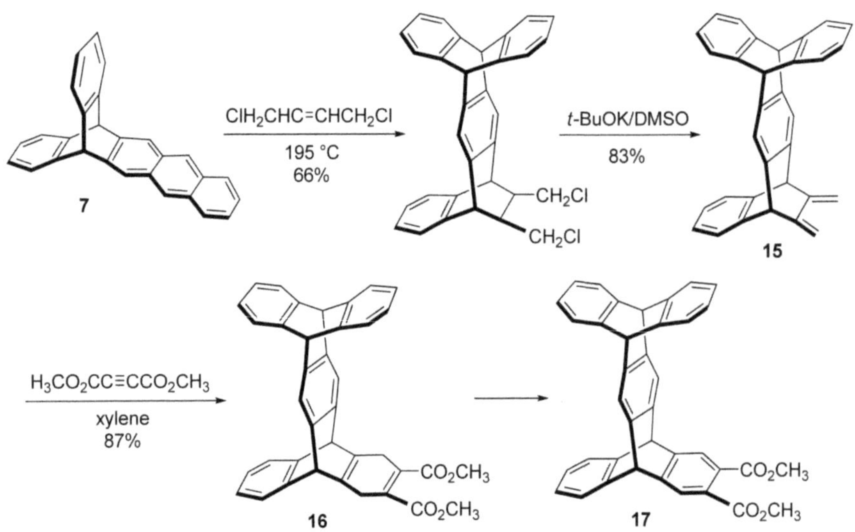

Scheme 3.8 Synthesis of pentiptycene dicarboxylate **17**

After that, Luo and Hart [8] successfully synthesized the diene (**15**) in about 56 % yield by the cycloaddition reaction of 1,4-dichloro-2-butene with the extended triptycene containing one anthracene moiety (**7**) under 195 °C for 3 days, and then followed by the treatment with potassium *tert*-butoxide (*t*-BuOK) in DMSO. Diene **15** reacted with dimethyl acetylenedicarboxylate in xylene under refluxing condition for 56 h to produce the adduct **16** in 87 % yield, which finally proceeded the isomerization to give the pentiptycene diester **17** (Scheme 3.8). However, this route had little practicability due to its too long reaction time and harsh reaction conditions.

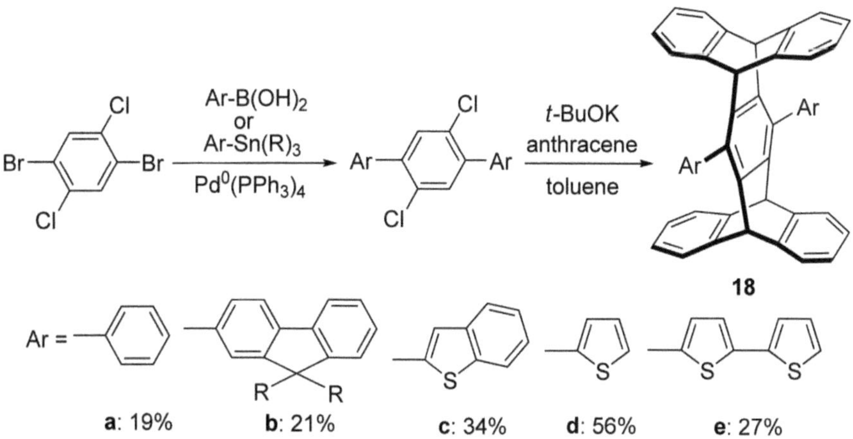

Scheme 3.9 Synthesis of 1,4-diarylpentiptycenes **18a–e**

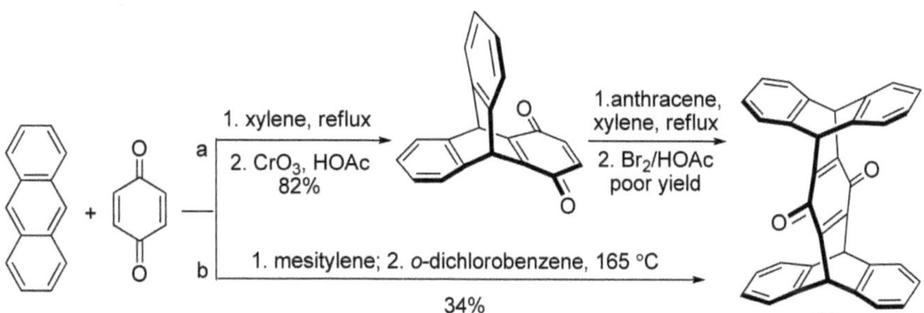

Scheme 3.10 Synthesis of pentiptycene quinone **19** from benzoquinone and anthracene

In 2008, Anzenbacher and co-workers [9] reported that the *t*-BuOK promoted the dehydrohalogenation/cycloaddition of 1,4-diaryl-2,5-dichlorobenzenes with an excess amount of anthracene to give the corresponding pentiptycenes **18a–e** in moderate yields (Scheme 3.9). Although this method could easily afford a series of 1,4-diarylpentiptycenes, approximately five equivalents consumption of anthracene and the mediocre yield led to the scope of the applications within certain limits.

Compared with the complicated synthesis of pentiptycenes, the pentiptycene quinone **19** [10, 11] seemed to be prepared easily from the inexpensive precursors (benzoquinone and anthracene). As shown in Scheme 3.10a, Clar [10] initially reported the synthesis of pentiptycene quinone **19** via a multistep method in 1931. Thirty years later, Theilacker et al. [11] modified the reaction conditions to synthesize the quinone **19** in 34 % yield by a two-step route starting from the addition of benzoquinone and anthracene (Scheme 3.10b).

With the development of pentiptycene chemistry, quinone **19** was gradually developed into a potential precursor for the synthesis of pentiptycene derivatives. Thus,

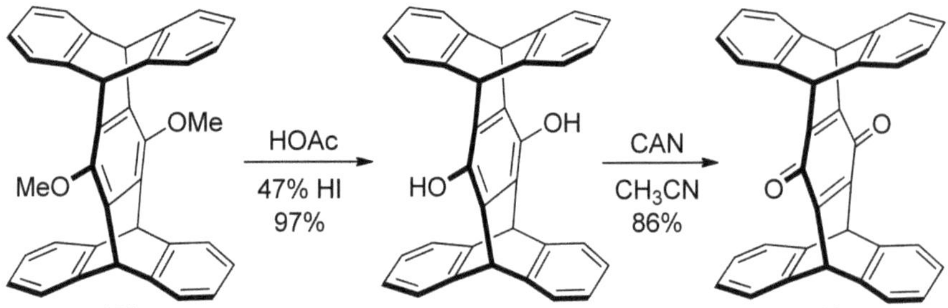

13b 20 19

Scheme 3.11 Synthesis of pentiptycene quinone **19** from dimethoxypentiptycene **13b**

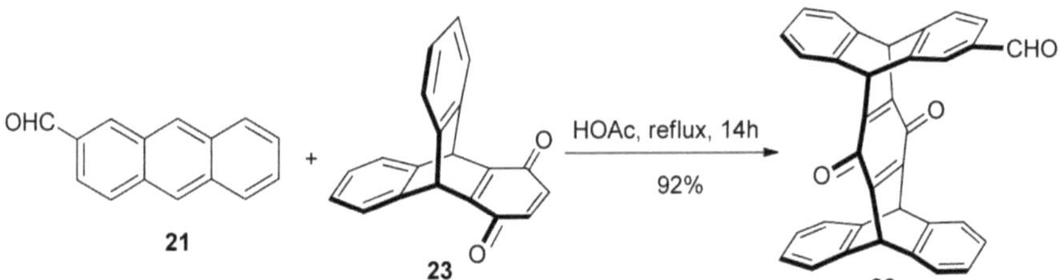

21 23 22

Scheme 3.12 Synthesis of pentiptycene quinone derivative **22**

the method and reaction conditions for the synthesis of **19** began to draw much attention on the improvement and optimization. In 1981, Hart et al. [4, 5, 12] obtained the pentiptycene quinone **19** via a two-step route from dimethoxy pentiptycene (**13b**). As shown in Scheme 3.11, compound **13b** was treated with hydrogen iodide in acetic acid to give the hydroquinone **20**, as a result of the departure of the methoxyl group; and then compound **20** was oxidized by ceric ammonium nitrate (CAN) in situ to give the target quinone **19** in an 83 % overall yield.

In 1997, Senge and co-workers [13] reported the synthesis of monoquinone **22** in a reasonable yield by the reaction of anthracene containing an aldehyde group (**21**) with excessive triptycene monoquinone **23** in acetic acid under reflux temperature (Scheme 3.12). In this process, the excess **23** served as an oxidation agent to oxidize in situ, thus, the large consumption of **23** limited its practicability.

Afterward, Yang and Swager [14, 15] obtained the quinone **19** through a two-step synthesis route starting from anthracene. Thus, two equivalents of anthracene reacted with benzoquinone, then followed by the oxidization of potassium bromate in acetic acid to give the target **19** in a 39 % overall yield (Scheme 3.13). Later, Williams and Swager [16] optimized the reaction conditions, and greatly improved the yield up to 84 %. This method was suitable for the large dose synthesis with an acceptable yield; although in the process, a mixture of triptycene quinone and pentiptycene quinone would actually be obtained, which needed to be purified by the silica–gel column chromatography.

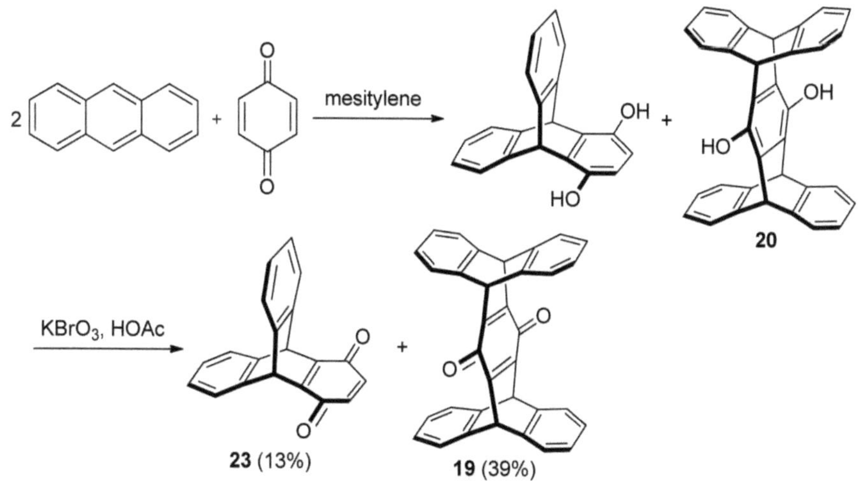

Scheme 3.13 Synthesis of **19** from anthracene through a two-step synthetic route

In 2003, Spyroudis and Xanthopoulou [17] unexpectedly isolated pentiptycene quinone **19** via the reaction of triptycene haloquinones **24** with sodium alkoxides in the refluxing alcohol solution (Scheme 3.14). The reaction associated with a certain amount of subsidiary triptycene derivatives (**23**; **25a**, **b**) as by-products, the yields of alkoxyquinones **25a** and **25b** even could reach up to 13–20 %. It was apparent that the existence of subsidiary products increased the separation difficulty, which also limited the practicability of this approach.

Inspired by the synthesis of iptycene quinones described by Senge and co-workers [13], Zhu and Chen [18] reported a new practical and efficient method for the synthesis of pentiptycene quinone **19** with a yield of 78 % in 2005. The quinone could be prepared by the reaction of anthracene with one equivalent of triptycene quinone in an acetic acid solution with the presence of *p*-chloranil (Scheme 3.15a). In this process, *p*-chloranil, a commercially available chemical, was served as an oxidant instead of the excess of triptycene quinone, thus, the practicability of this method immensely heightened. Similarly, one-pot reaction of triptycene quinone with 1,4-dimethoxyanthracene **26** gave pentiptycene quinone **27** in 82 % yield, which was then demethylated by CAN to give the diquinone **28** in a quantitative yield (Scheme 3.15b). When dimethoxytriptycene monoquinone **29** reacted with dimethoxyanthracene in refluxing acetic acid in the presence of *p*-chloranil, a mixture of two regioisomers **30a** and **30b** was obtained in an 82 % overall yield. It was hard to separate the isomers **31a** and **31b** by column chromatography but the separation of their CAN oxidation products of pentiptycene triquinones **31a** and **31b** by column chromatography could be easily achieved. Thus, after the oxidation, the pentiptycene triquinones **31a** and **31b** could be obtained in 45 and 28 % yield, respectively, by column chromatography (Scheme 3.16). According to the similar method, a series of other pentiptycene quinones (such as **32**, **33**, Fig. 3.1), even heptiptycene quinones, and noniptycene quinones could be easily synthesized in good yields. These iptycene quinones could

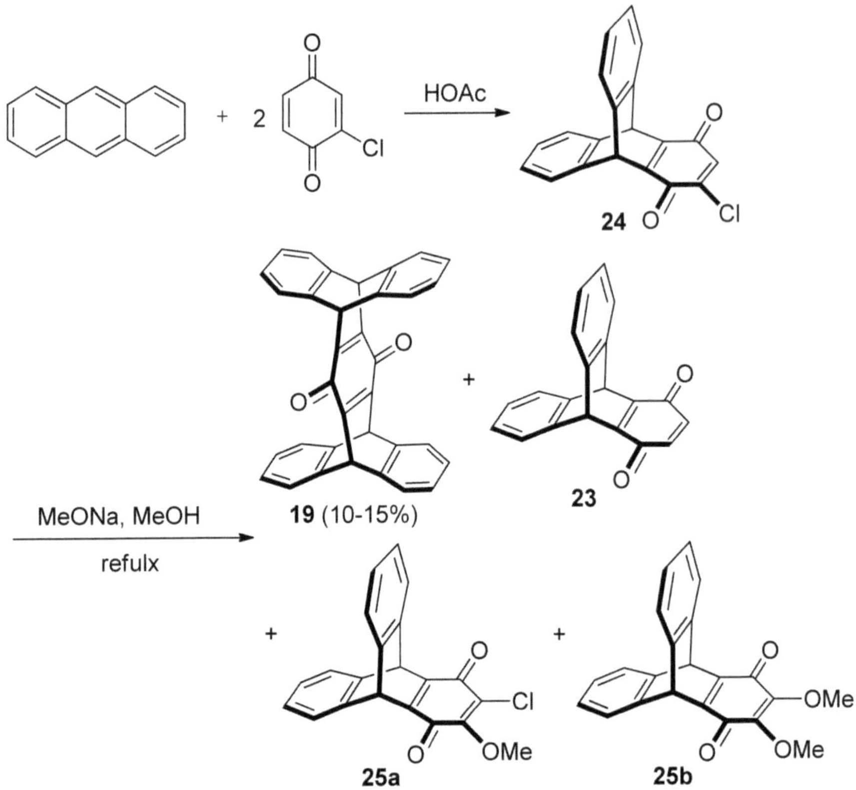

Scheme 3.14 Reaction of triptycene haloquinone **24** with sodium alkoxides

further serve as the principal precursors for the preparation of other higher iptycene quinones.

Following the similar reaction conditions as above, Chen and co workers [19] further synthesized a series of peripheral *o*-dimethoxy-substituted pentiptycene quinones. Consequently, treatment of the quinone **23** with 2,3-dimethoxyanthracene or treatment of anthracene with the triptycene quinone **34** in acetic acid in the presence of *p*-chloranil gave pentiptycene quinone **35** containing a *o*-dimethoxy-benzene unit in 75 and 70 % yield, respectively (Scheme 3.17). Similarly, a series of peripherally *o*-dimethoxy-substituted pentiptycene quinones (**36**–**40**) had been prepared in good yields (Fig. 3.2).

Furthermore, by the treatment of pentiptycene quinones with CAN in aqueous acetonitrile, their *o*-quinone derivatives could be easily prepared in high yields. Especially, it was found that if two *o*-dimethoxybenzene moieties were situated at the same side of the pentiptycene quinone, such as **36**, one of them was only oxidized by excess CAN in aqueous acetonitrile to give compound **41** (Scheme 3.18). According to a similar method, a series of pentiptycene *o*-quinone derivatives **42**–**47** were prepared in good yields (Fig. 3.3).

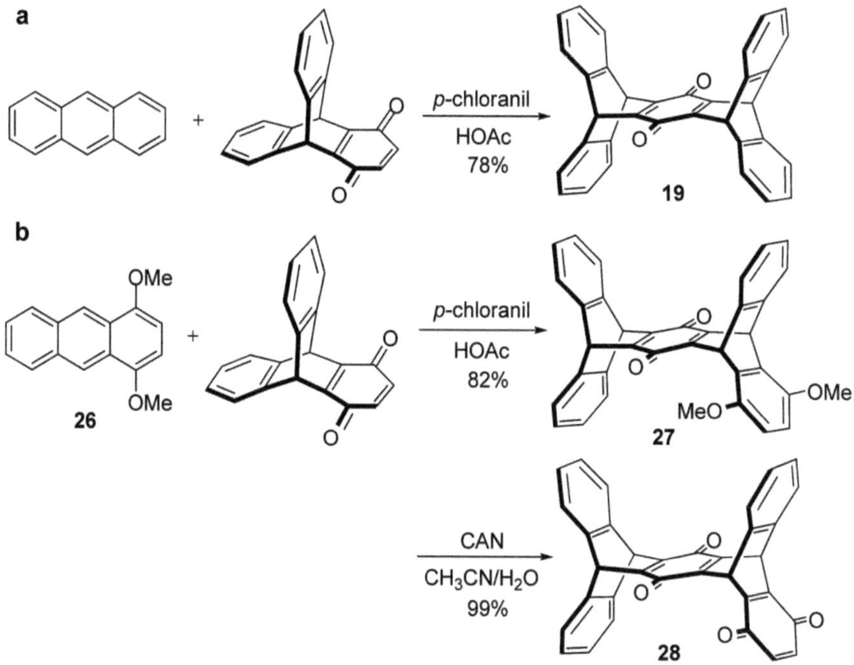

Scheme 3.15 Synthesis of pentiptycene quinones **19** and **28**

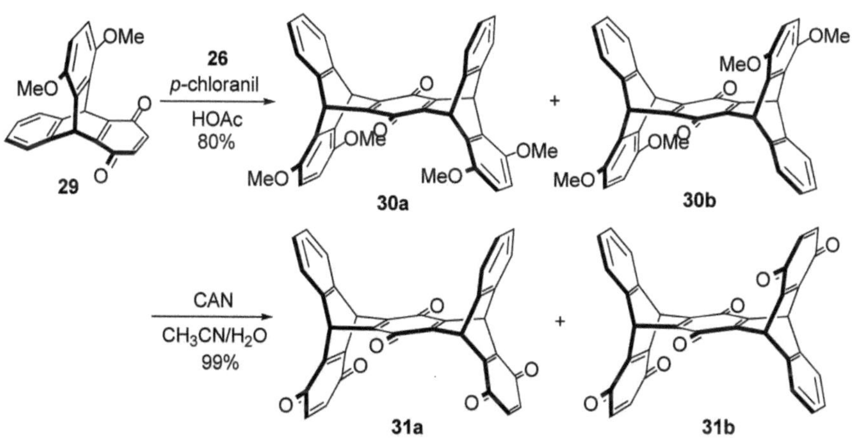

Scheme 3.16 Synthesis of pentiptycene triquinones **31a**, **b**

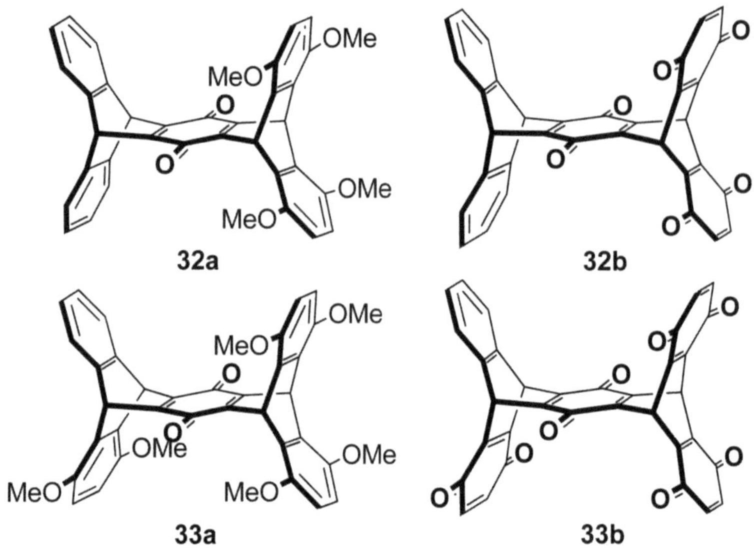

Fig. 3.1 Structures of pentiptycene quinones **32** and **33**

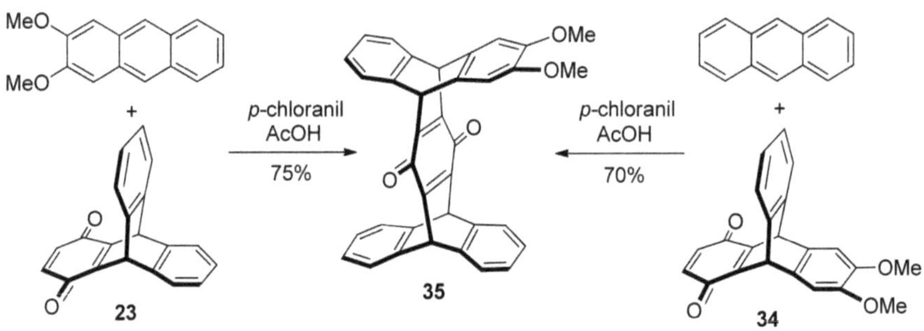

Scheme 3.17 Synthesis of *o*-dimethoxy-substituted pentiptycene quinone **35**

3.2 Reactions of Pentiptycenes and Their Derivatives

As mentioned in the previous section, pentiptycene quinones could easily undergo the reduction of carbonyl groups to give the corresponding pentiptycene derivatives. This feature makes the pentiptycene quinones potential precursors for the other central-ring substituted pentiptycene derivatives. In 1998, Yang and Swager [14, 15] reported the synthesis of the diethynyl-pentiptycene (**50**). As shown in Scheme 3.19, compound **48** could be prepared by the nucleophilic addition reaction of lithium (trimethylsilyl)acetylide to the quinone carbonyl groups of **19** under an atmosphere of argon in a THF solution. After overnight, the resulting compound **48** reacted with tin chloride in a 50 % acetic acid solution for 24 h at room temperature to give the TMS-protected diethynyl-pentiptycene **49** in a yield of 85 %, as the result of reduction

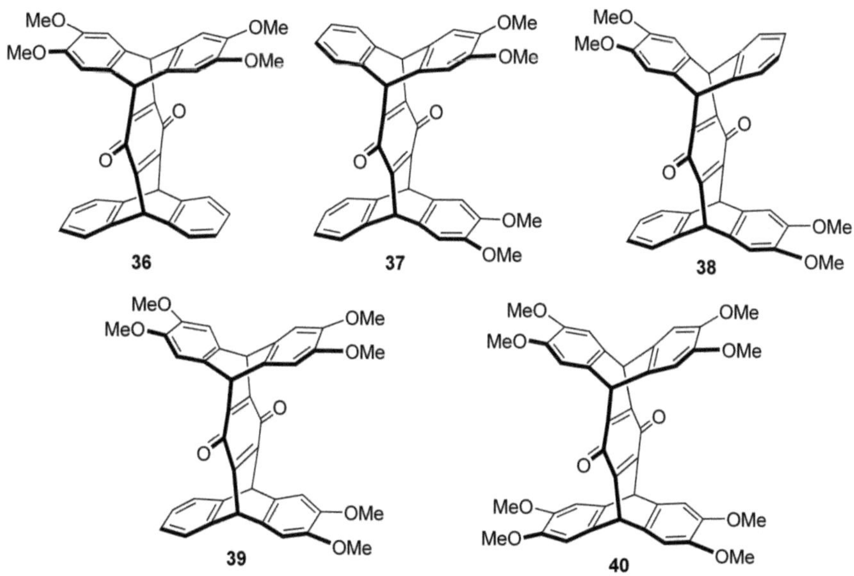

Fig. 3.2 Structures of pentiptycene quinones **36**–**40**

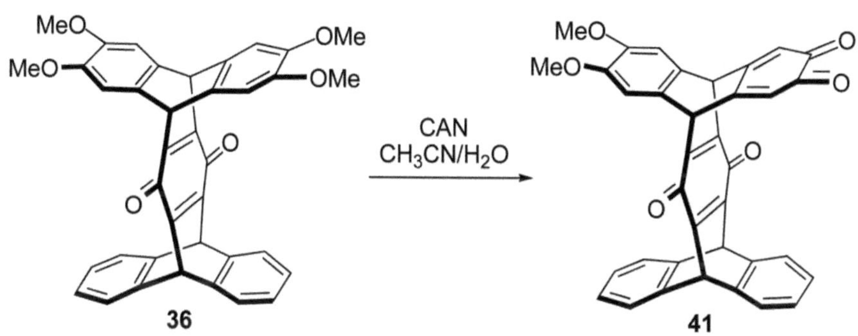

Scheme 3.18 Synthesis of compound **41**

and aromatization of compound **48**. Then, the departure of the protected groups in **48** with sodium hydroxide in a mixture solution of KOH, THF, and methanol (MeOH) at room temperature for another 5 h gave the pentiptycene derivative **50**. The compound **50** showed a poor solubility in common organic solvent. It was noteworthy that the pentiptycene derivative **51** with one alkynyl group could be obtained, if only one equivalent of lithium (trimethylsilyl)acetylide was added. In other words, the protecting group was crucially important for the synthesis of pentiptycene derivatives with one or two central-ring ethynyl groups. Thus, in this process for the incorporation of TMS protected groups, the amount of lithium (trimethylsilyl)acetylide was carefully concerned.

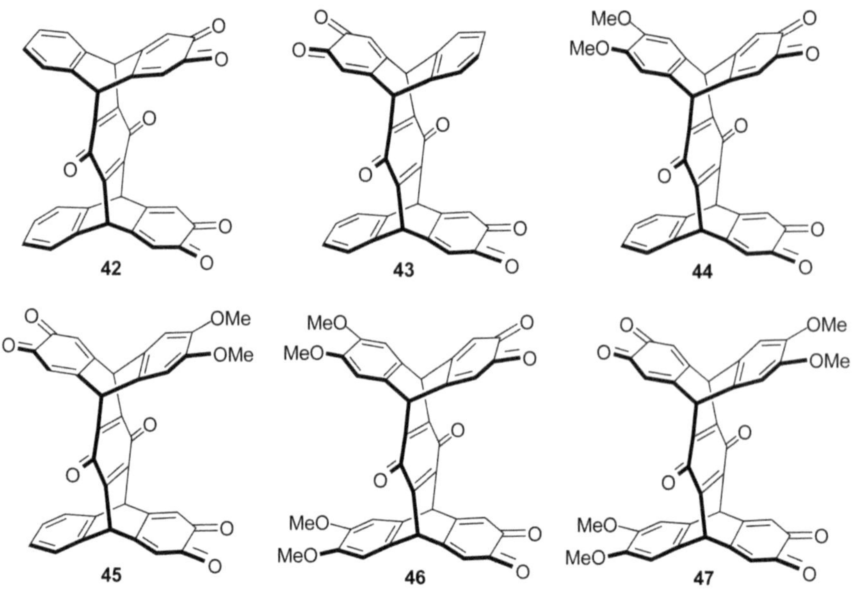

Fig. 3.3 Structures of pentiptycene *o*-quinone derivatives **42**–**47**

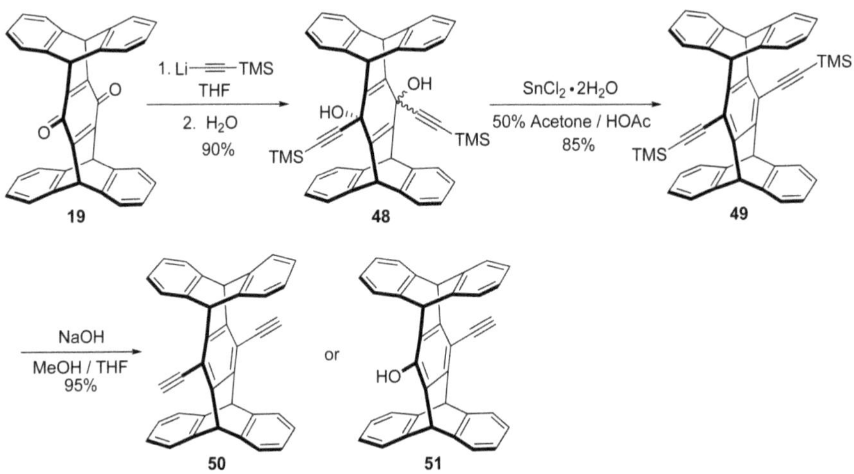

Scheme 3.19 Synthesis of pentiptycene derivatives **50** and **51**

In 2000, Williams and Swager [16] synthesized diiodotriptycene **52** from triptycene quinone **23** via the intermediate **53**, as shown in Scheme 3.20. They further attempted to extend this approach for triptycene to pentiptycene quinone **19**. Unfortunately, they failed to obtain diiodopentiptycene. They considered that the reason of the failure reaction was the lower reactivity of pentiptycene. In 2006, Yang and Ko [20] retried this directed iodination reaction of pentiptycene quinone **19**. They

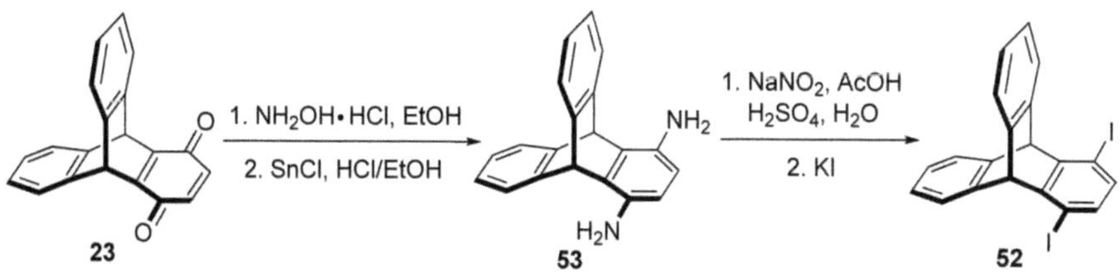

Scheme 3.20 Synthesis of diiodotriptycene **52**

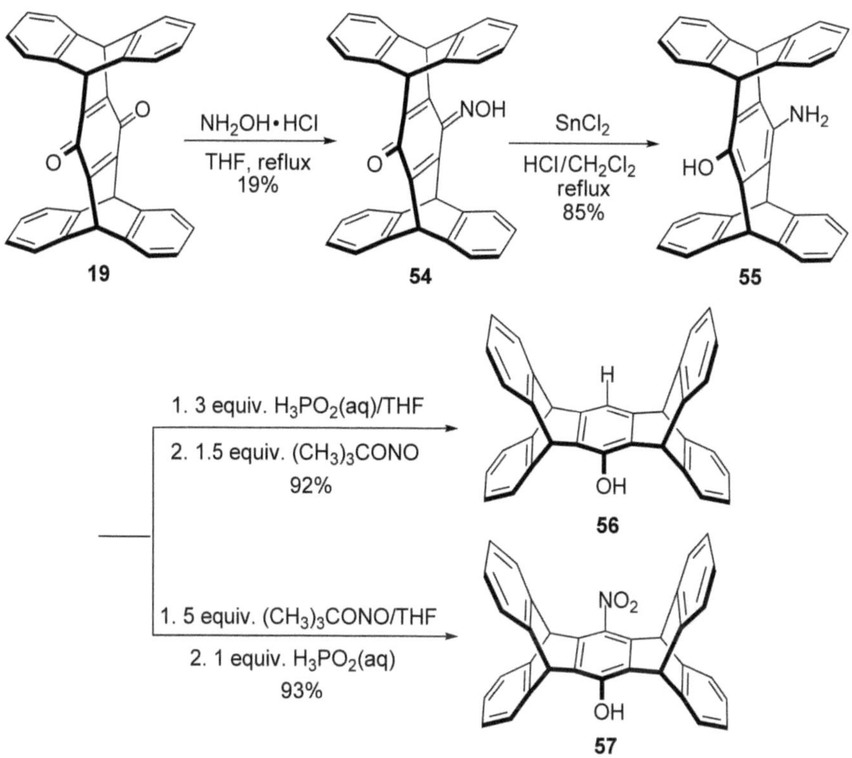

Scheme 3.21 Synthesis of pentiptycene derivatives **56** and **57**

deemed that the poorer solubility of monoquinone **19** rather than its lower reactivity might result in the failure of the reaction. Thus, they carried out the reaction between **19** and hydroxylamines to occur in THF instead of alcohol in the presence of about two equivalents of hydrochloric acid. As a result, they found that the monooxime **54** rather than the dioxime could be prepared in high yield (Scheme 3.21). Even though with a large excess of hydroxylamine and a prolonged reaction time, the product was still monooxime. It could not detect the presence of the corresponding dioxime, as the second carbonyl group was too inert to convert into oxime. Then, by the treatment of monooxime **54** with stannous chloride in dichloromethane, the reduction product **55** was obtained (Scheme 3.21). Furthermore, the amino group in **55** could be removed to form the pentiptycene derivative **56** with phenolic hydroxyl group by the simple treatment with *t*-butyl nitrite and H_3PO_2. According to the similar method, the amino group could also convert into a nitro group, and compound **57** could be obtained in almost quantitative yield (Scheme 3.21). Actually, it was interesting that only by changing the concentration and the order of reagents added to the THF solution of **55**, different products **56** or **57** could be obtained, respectively. It was noteworthy that both pentiptycene derivatives **56** and **57**, even triptycene monoquinone **19** could be obtained in a varied and instable yield of 40–96 % at the same time, if the above two reactions took place in air. Thus, keeping the reactions in a deaerated condition was very important for selectively producing the target product in a satisfactory and stable yield, along without the problem of separation.

As mentioned before, pentiptycenes **55–57** are all the important precursors for the synthesis of asymmetric substituted pentiptycene derivatives. Consequently, Yang et al. [20] attempted to directly synthesize the iodopentiptycene under the iodination conditions of I_2/Ag_2SO_4 starting from **56**. Unfortunately, the reaction did not occur smoothly as expected; the directed iodination seemed not to give the iodopentiptycene (**59**, R = OH). However, they found that the iodopentiptycene (**59**, R = OC_8H_{17}) could be readily prepared with a yield of 94 % from the alkoxy derivative **58** under the iodination conditions of I_2/Ag_2SO_4 in EtOH solution (Scheme 3.22). Similarly, the treatment of **58** with NBS in DMF at 80 °C could immediately give bromopentiptycene (**60**) in 90 % yield (Scheme 3.22). In addition, followed by the process in the literatures [21, 22], this bromopentiptycene (**60**) could convert to the pentiptycene derivatives **61** and **62** with the cyano, and formyl group, which was located at the *para*-position of the alkoxy group (Scheme 3.23). Moreover, they further synthesized a series of pentiptycene derivatives **63–65** with extended π-conjugated backbones in good-to-excellent yields (72–90 %) from **59** under the standard conditions of the Sonogashira, Heck, and Suzuki reactions, respectively (Scheme 3.24) [23].

On the basis of the previous work, Yang et al. [24] successfully prepared the center-ring halogenated pentiptycene phenols **66** in good yields, either in one-pot fashion (Scheme 3.25, route a) or by a two-step method (Scheme 3.25, route b) starting from the pentiptycene derivative **55**. The reactions all processed via the formation of the diazonium salt. Then, the treatment of the pentiptycene bromide under the Heck, Suzuki, and Sonogashira reaction conditions could give the pentiptycene derivatives **67–69**, which incorporated π-conjugated systems in the yields of 68–84 % (Scheme 3.26). In general, many other aryl triflates could be used as alternative substrates for these Pd-catalyzed coupling reactions. However, these coupling reactions for either pentiptycene monotriflates or ditriflates could not occur smoothly under the Heck, Suzuki, and Sonogashira reaction conditions. The reduction of **19** with sodium thiosulfate, followed by the reaction with triflic anhydride in pyridine at 0 °C gave 1,4-triflate, which then underwent the Pd^0-catalyzed Sonogashira, Heck, and Suzuki reactions to produce the target compounds **70** in a low yield (<9 %, Scheme 3.27), due to the little reactivity of pentiptycene triflates.

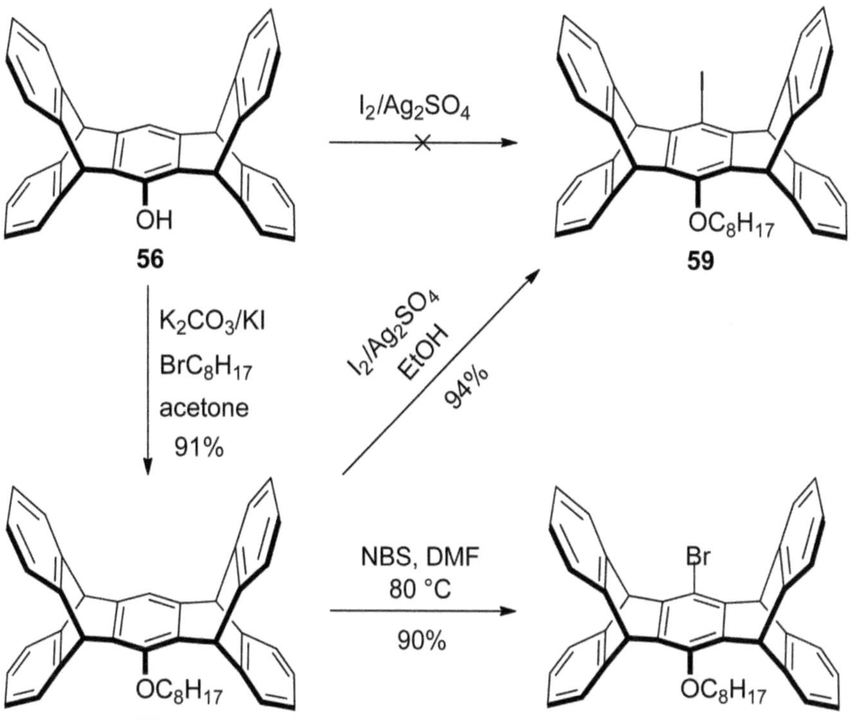

Scheme 3.22 Synthesis of iodopentiptycene **59** and bromopentiptycene **60**

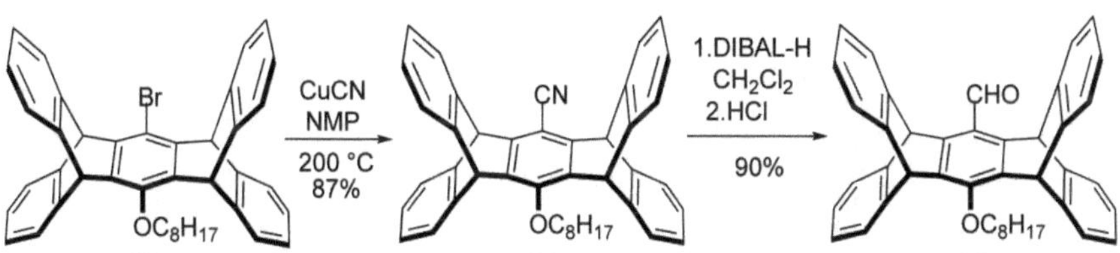

Scheme 3.23 Conversion of bromopentiptycene to pentiptycene derivatives **61** and **62**

Soon afterward, Yang et al. [24] successfully synthesized the dihalopentiptycenes **74–76** (Scheme 3.28 and Scheme 3.29) as well. As shown in Scheme 3.28, the key step for the synthesis of the dihalopentiptycenes was the Pd-catalyzed reduction of the triflate group in **71** to give the nitropentiptycene **72**, which could be efficiently achieved by triethylsilane in the presence of the more reactive Pd catalyst $Pd(PPh_3)_4$, and this reaction appeared to be more efficient for aryl triflates containing electron-withdrawing substituents. Based on the dibromo-pentiptycenes, a variety of 1,4-diaryl-pentiptycenes **77–79** could be further synthesized by the Heck, Suzuki, and Sonogashira reactions in the presence of palladium catalysts (Scheme 3.30).

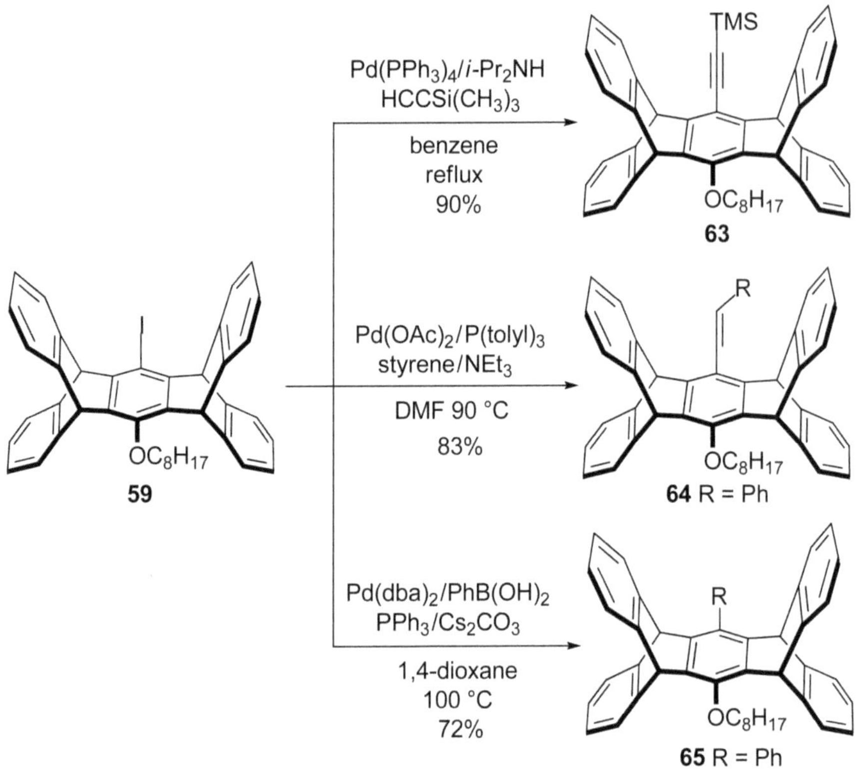

Scheme 3.24 Synthesis of pentiptycene derivatives **63–65** with extended π-conjugated backbones

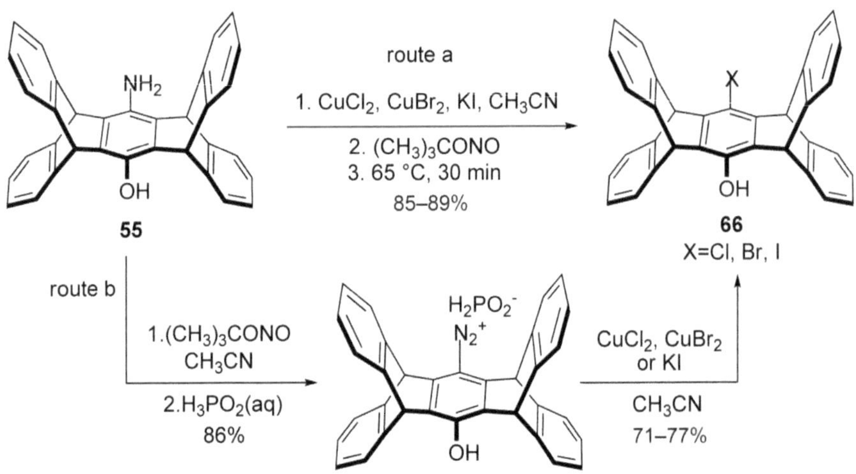

Scheme 3.25 Synthesis of center-ring halogenated pentiptycene phenols **66**

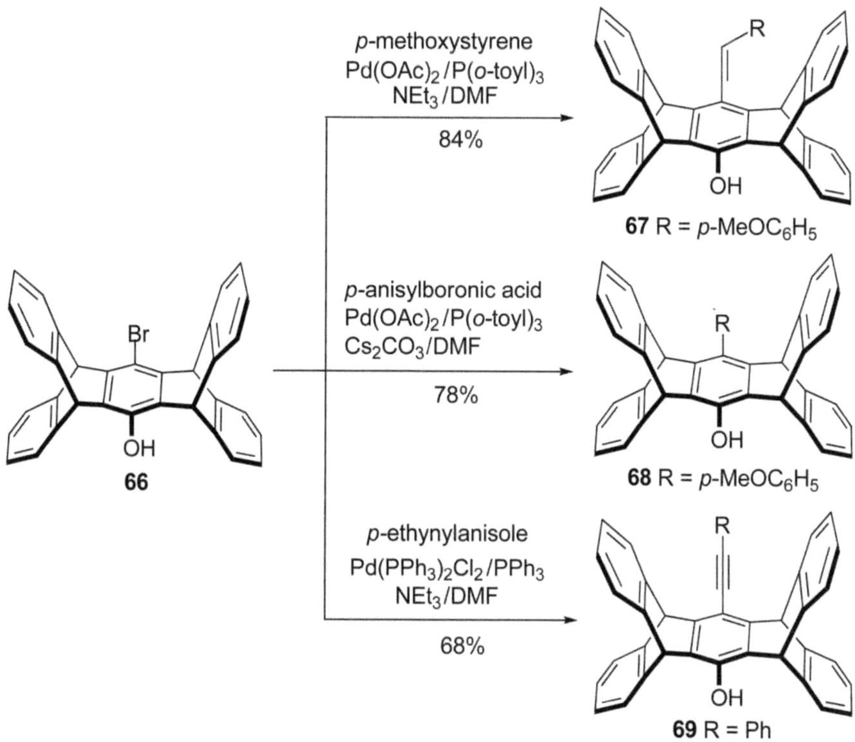

Scheme 3.26 Synthesis of pentiptycene derivatives **67–69** incorporated π-conjugated systems

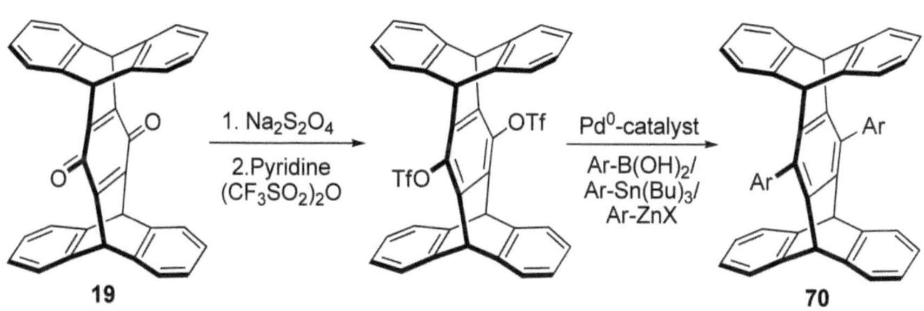

Scheme 3.27 Synthesis of pentiptycene derivative **70** via Sonogashira, Heck, and Suzuki reactions

In 2010, Yang and co-workers [25] also reported an efficient route to the synthesis of pentiptycene halides and other new pentiptycene building blocks via the nucleophilic aromatic substitution (S_NAr) reactions of the nitrosubstituted pentiptycene triflate **81** with LiBr and LiI and the resulting halides with N_3^-, CN^-, and ArS^- in DMF. As shown in Scheme 3.31, the halopentiptycenes **80a, b** were obtained through the C–O bond cleavage of route *a* from the reaction of **81** with LiBr and LiI in DMF, respectively. If the reaction processed via the S–O bond cleavage (route *b*), compound **57** would be obtained.

The halides **80a, b** could serve as key precursors for the synthesis of other pentiptycene derivatives. Consequently, it was found that the nucleophilic aromatic

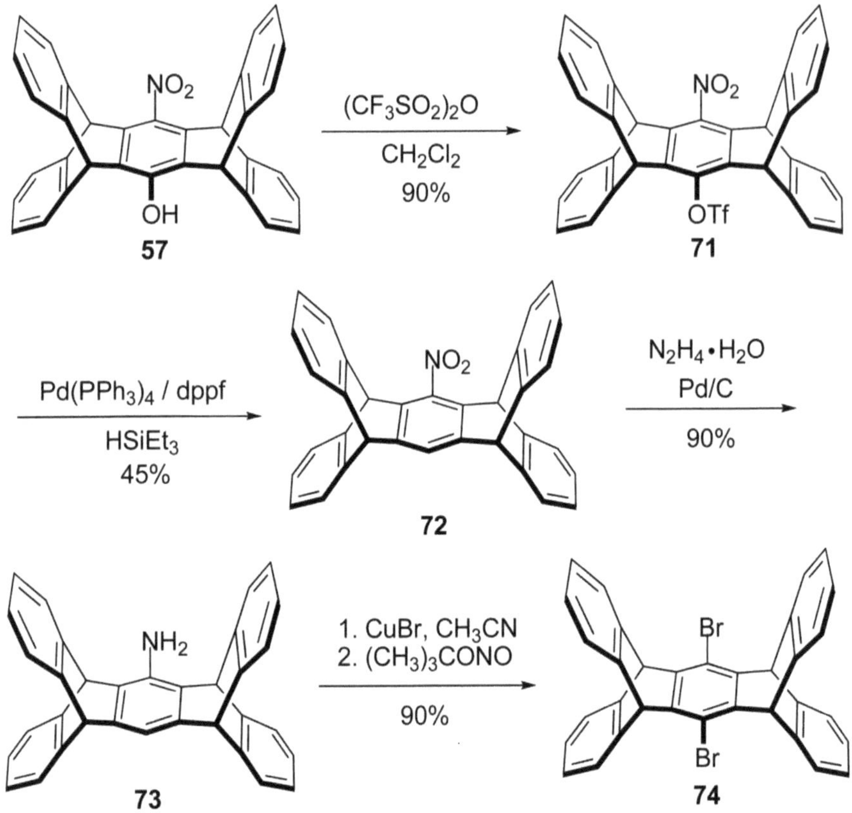

Scheme 3.28 Synthesis of dibromo-pentiptycene **74**

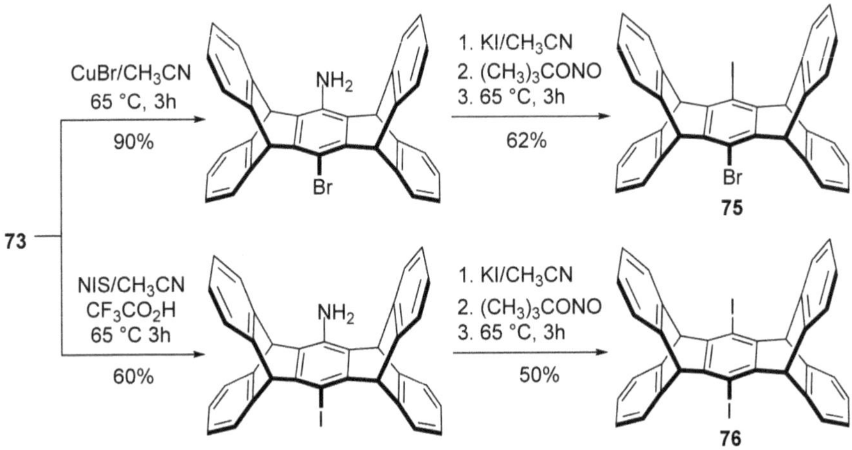

Scheme 3.29 Synthesis of dihalopentiptycenes **75** and **76**

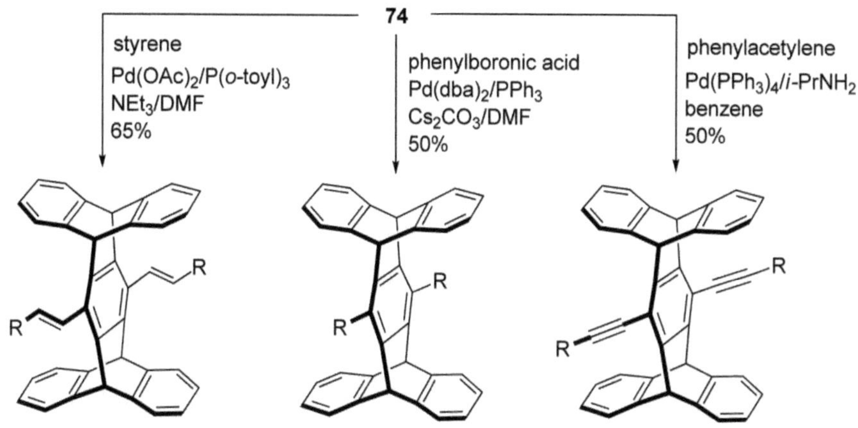

77 R = Ph
78 R = Ph
79 R = Ph

Scheme 3.30 Synthesis of 1,4-diarylpentiptycenes **77–79** from dibromo-pentiptycene **74**

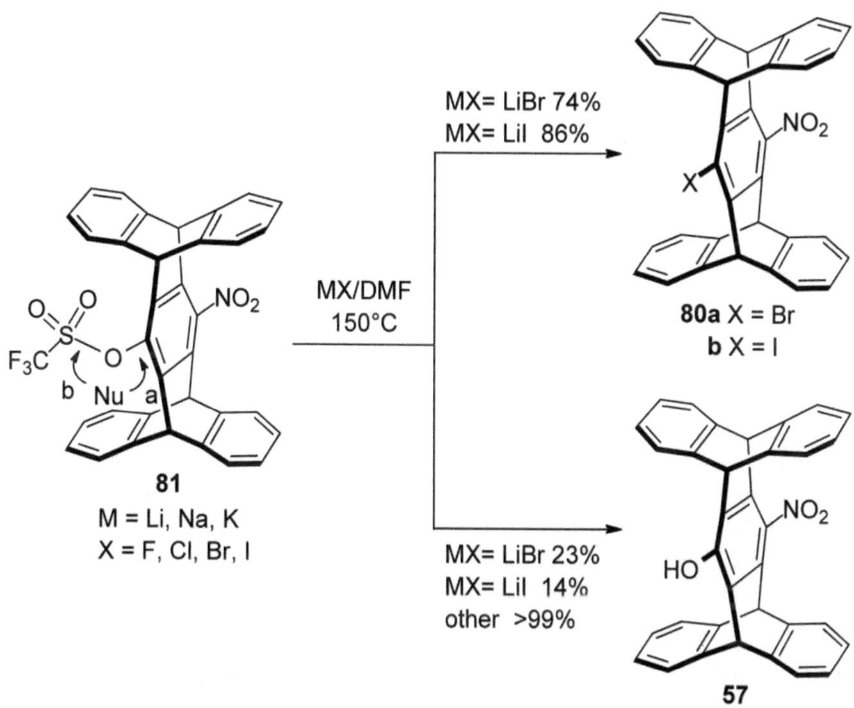

Scheme 3.31 Nucleophilic aromatic substitution of nitrosubstituted pentiptycene triflate **81**

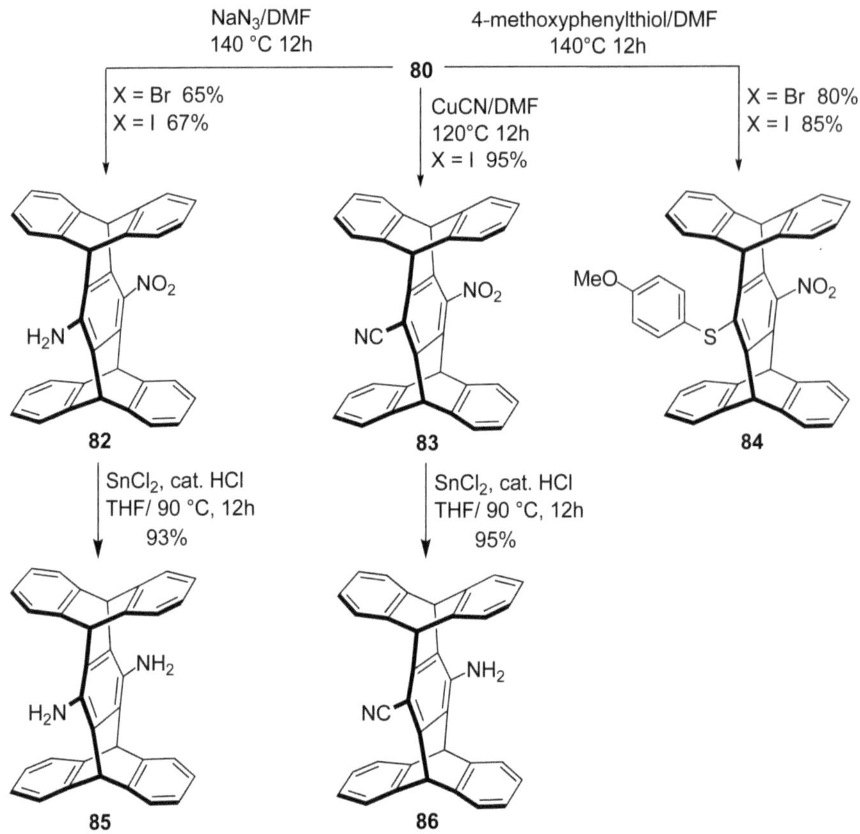

Scheme 3.32 Nucleophilic aromatic substitution reactions of halide **80**

substitution reactions of compound **80** with N_3^-, CN^-, and ArS^- in DMF gave pentiptycenes **82–84**, respectively. Then, the reduction of compounds **82** and **83** by tin chloride in a THF solution could produce the corresponding aminopentiptycenes **85** and **86**, respectively, in excellent yields (Scheme 3.32). It was noteworthy that the pentiptycene dibromide **76** and diiodide **74** could also be obtained in higher yields via the halide **80** rather than that with compound **72** as intermediate. In addition, the Pd-catalyzed Buchwald–Hartwig C–N coupling reaction of aminopentiptycene **85** with bromobenzene **87** could afford the pentiptycene **88** in 87 % yield (Scheme 3.33).

3.3 Synthesis of Extended Pentiptycenes Derivatives

In 1983, Hart et al. [26] prepared the synthons **89a** and **89b** by the addition reaction between tetrabromobenzene and furan, followed by the method depicted by Takehira [4] in 1981. This biadduct served as the useful synthon for the synthesis

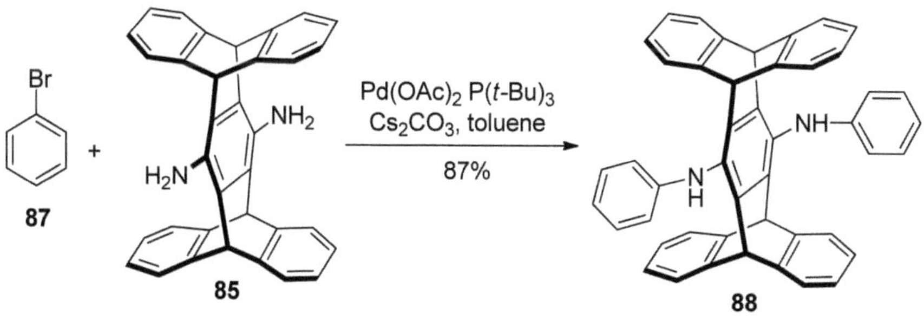

Scheme 3.33 Pd-catalyzed Buchwald–Hartwig C–N coupling reaction of aminopentiptycene **85** with bromobenzene **87**

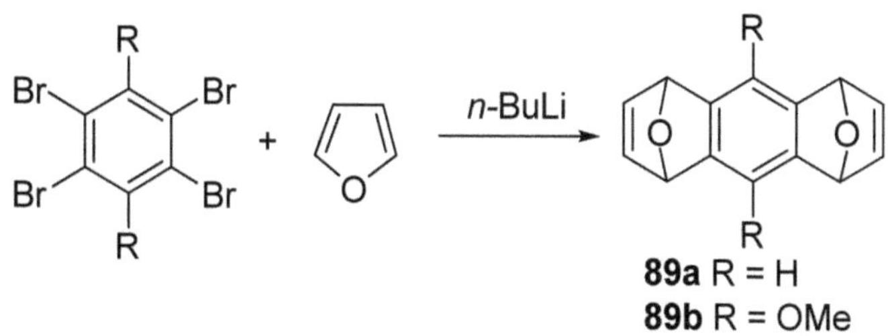

Scheme 3.34 Synthesis of synthons **89a, b**

of extended pentiptycene derivatives. As shown in Scheme 3.34, the treatment of 1,2,4,5-tetrabromobenzene and its derivative with *n*-butyl lithium at −23 °C with an argon atmosphere gave the isomers **89** (*syn/anti*), which could be separated by the different solubility in MeOH. With this synthon in hand, the biadduct **90** was further synthesized easily. Thus, the reaction of **89a, b** with two equivalents of anthracene in refluxing xylene for 48 h gave the biadduct **90**, which was then concentrated by hydrochloric acid in acetic anhydride to give the final target fused-ring pentiptycenes (**91a, b**) containing a central anthracene moiety in a yield of 33–41 % (Scheme 3.35). It was noteworthy that both *syn*-**89** and *anti*-**89** could afford the same product **91a, b** via the corresponding stereoisomers **90**.

Soon afterward, Hart et al. [5] found that the bis-methylene **92** could serve as the useful synthon for the synthesis of extended pentiptycene as well. Thus, they easily synthesized the extended pentiptycene **91a** from 1,2,4,5-tetrabromobenzene. As shown in Scheme 3.36, by the reaction of the tetrabromobenzene with two equivalents of the bis-methylene (**92**) in a toluene solution in the presence of butyl lithium, then followed the dehydrogenation of the resulting compound **93** by Pd/C, the target compound **91a** was obtained in about 84 % overall yield, which was much higher than the above described approach.

After that, they further synthesized the extended pentiptycene **94** (the region-isomer of extended pentiptycene **91a**) containing one "outer" anthracene moiety at side-chain ring. As shown in Scheme 3.37, the addition reaction of synthon **95** and

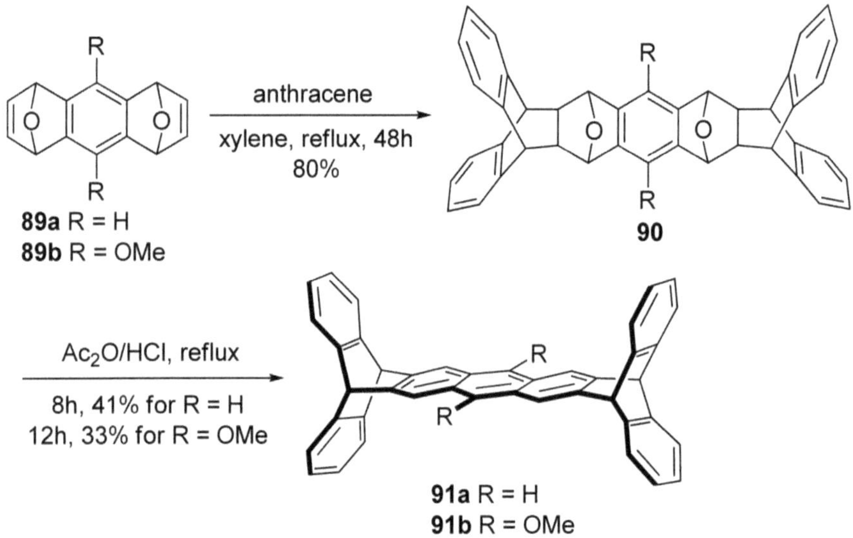

Scheme 3.35 Synthesis of fused-ring pentiptycenes **91** from synthons **89a**, **b**

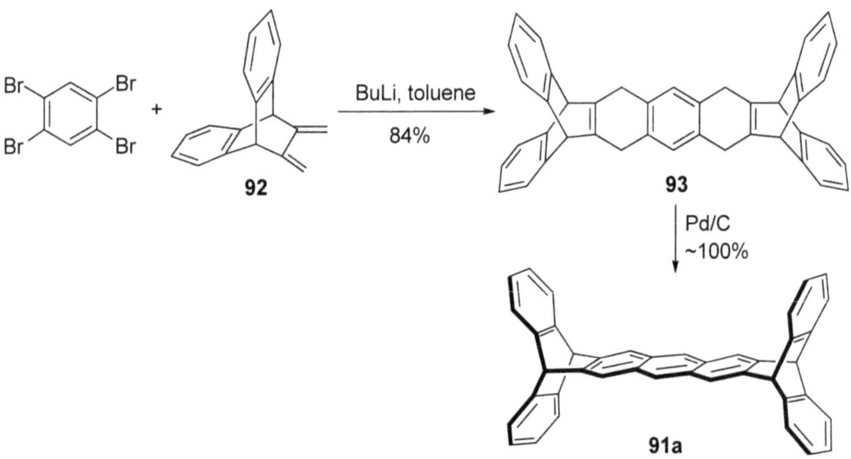

Scheme 3.36 Synthesis of fused-ring pentiptycene **91a** from bis-methylene **92**

extended triptycene **7** in reflux xylene gave a pair of region-isomeric adducts **96a** and **96b** in 91 % overall yield. Among them, **96a** with the less crowded structure accounted for 59 % of the total yield. Furthermore, after the mixture of compounds **96a**, **b** without separation directly treated with perchloric acid in ethanol (EtOH), the target **94** as the result of the dehydrogenation could be obtained in 41 % yield.

It was also desirable to synthesize more extended pentiptycenes with different fused ring moieties. Thus, they attempted to synthesize extended pentiptycenes **97**

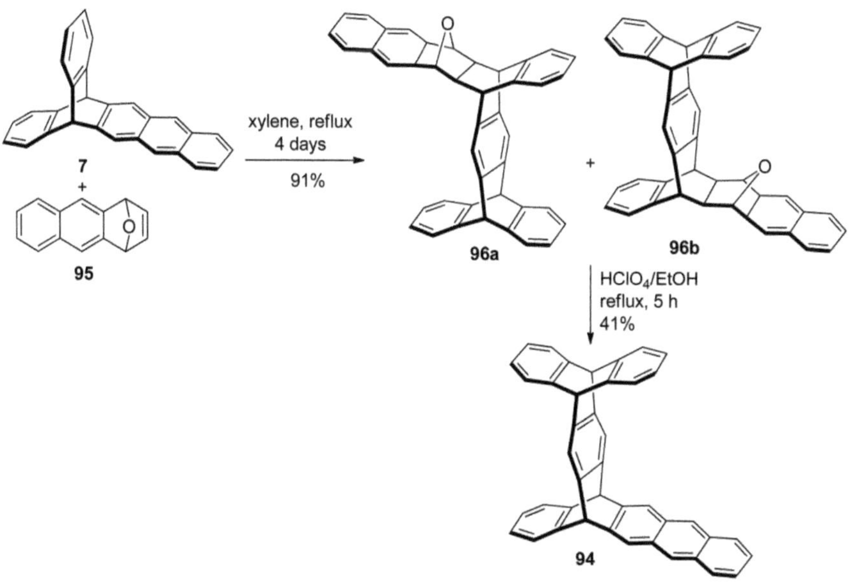

Scheme 3.37 Synthesis of extended pentiptycene **94**

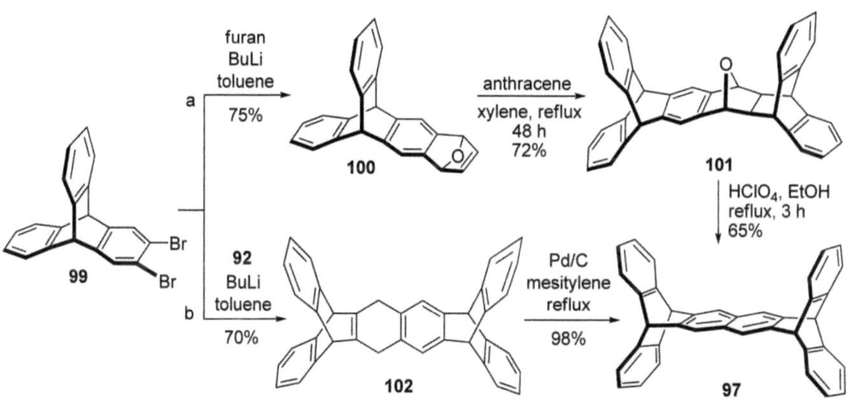

Scheme 3.38 Synthesis of extended pentiptycene **97** from synthon **99**

and **98** from the synthons **99** and **103**, respectively. As shown in Scheme 3.38, for pentiptycene **97** with central naphthalene moiety, it could be obtained from synthon **99** via two different methods. For route a, the treatment of tripytcene derivative **99** and furan in a toluene solution with butyl lithium gave the adduct **100** in 75 % yield. Then, compound **100** underwent the cycloaddition with anthracene in a reflux xylene for 48 h to produce the adduct **101**, which was dehydrated by perchloric acid in a reflux EtOH for another 3 h to afford the pentiptycene **97**. For another route, the reaction between **99** and **92** in the presence of butyl lithium gave the adduct

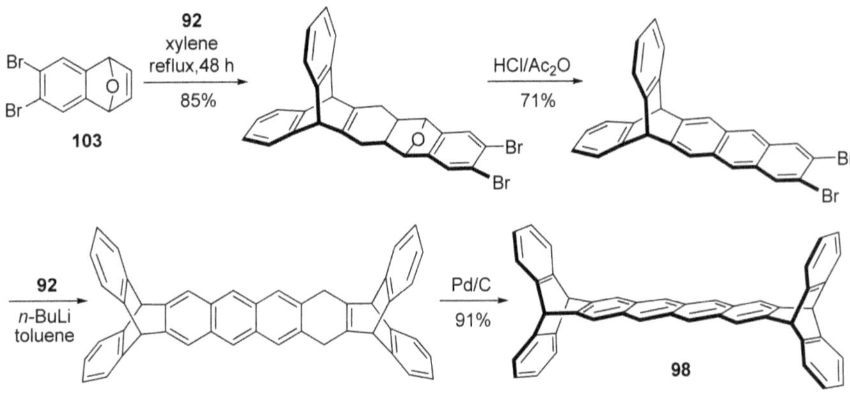

Scheme 3.39 Synthesis of extended pentiptycene **98** from synthon **103**

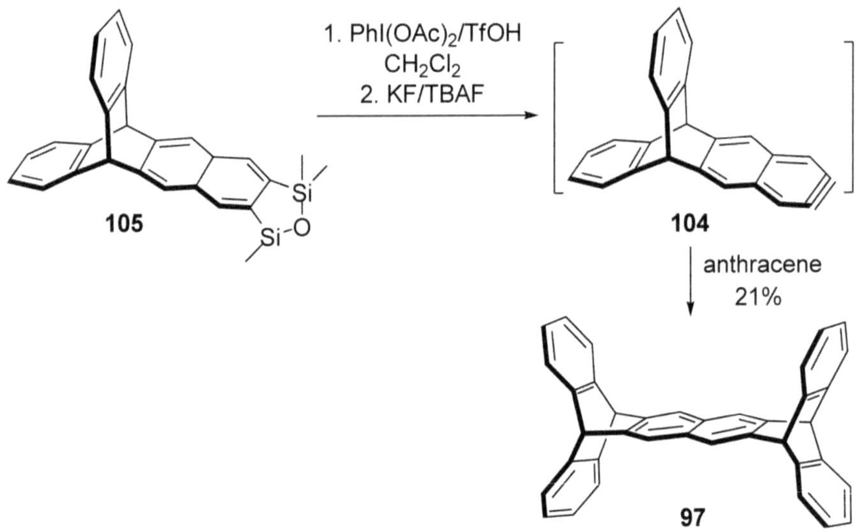

Scheme 3.40 Synthesis of extended pentiptycene **97** from oxadisilole fused extended triptycene **105**

102, which then underwent the dehydrogenation to give the target **97** in 70 % yield (Scheme 3.38, route b). For pentiptycene **98** with a tetracene moiety, it could be obtained in 52 % overall yield via a four-step route starting from synthons **103** and **92**, which was similar to the route b in Scheme 3.38 (Scheme 3.39).

In 2010, Lee and co-workers [6] synthesized extended pentiptycyne **97** in 27 % yield by the addition between anthracene and the extended triptycyne **104** (Scheme 3.40). It was noteworthy that the key intermediate **104** was generated from oxadisilole fused extended triptycene **105** in situ under the similar procedure for triptycyne generation, which is depicted in Chap. 2. Moreover, similar to the extended triptycene depicted in Chap. 2, extended pentiptycenes **91a** and **97** could also be obtained via the Diels–Alder cycloaddition in water under microwave radiation [27] as well (Scheme 3.41).

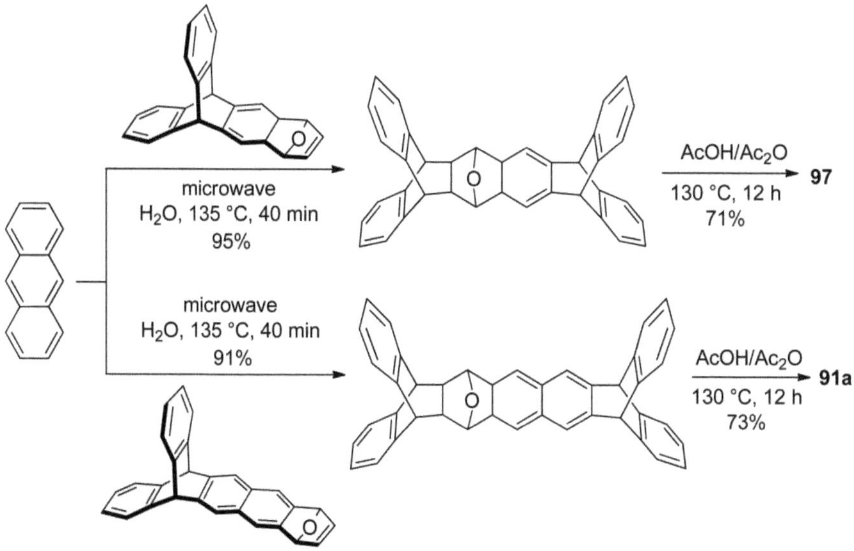

Scheme 3.41 Synthesis of extended pentiptycenes **91a** and **97** by highly efficient microwave-mediated Diels–Alder reaction

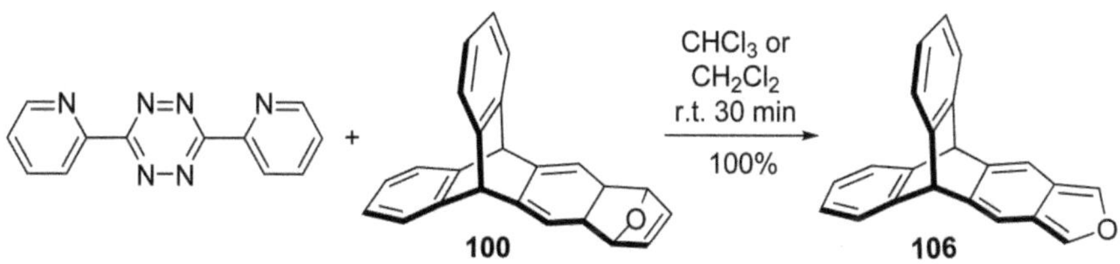

Scheme 3.42 Synthesis of extended triptycene **106** containing an isobenzofuran moiety

The H-shaped extended pentiptycenes **91a** and **98** could also be synthesized by the use of the extended triptycene **106** containing an isobenzofuran moiety as the precursor. As shown in Scheme 3.42, the extended triptycene **106** was prepared according to the Warrener's protocol [28, 29]. Consequently, the reaction of endoxide **100** with 2,6-bis-2-pyridyl-1,2,4,5-tetrazine in a mixture solution of chloroform/dichloromethane at room temperature under nitrogen atmosphere for 30 min gave compound **106** in a quantitative yield. Then, the freshly generated extended triptycene **106** immediately reacted with the dienophiles (**104**, **107**) under a nitrogen atmosphere to afford the corresponding iptycene cycloadducts (**108**, **109**) in good yields, which further underwent the deoxygenation with $TiCl_4/LiAlH_4$ in the presence of Et_3N to give the target extended pentiptycenes **91a** and **98** in high yield (Scheme 3.43). In addition, the endoxides **108**, **109** could also easily convert to the quinones **110, 111** by $PhI(OAc)_2$ and TfOH in CH_2Cl_2 at room temperature for 10 min in 78 and 81 % yield, respectively (Scheme 3.44).

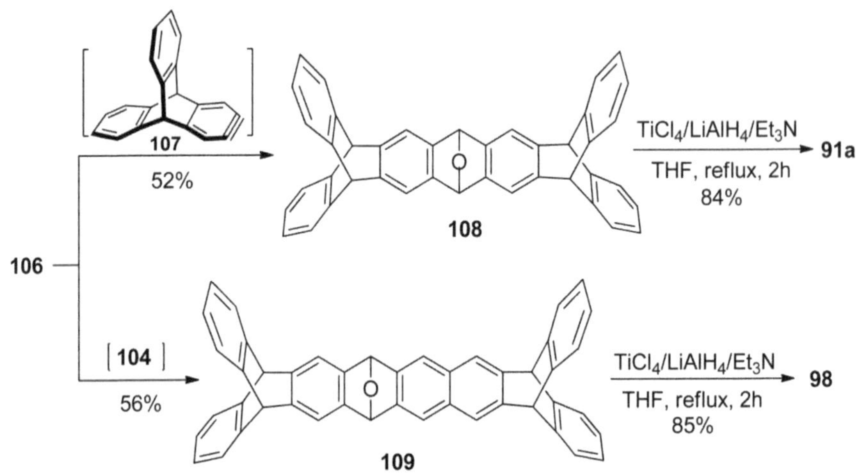

Scheme 3.43 Synthesis of extended pentiptycenes **91a** and **98** from synthon **106**

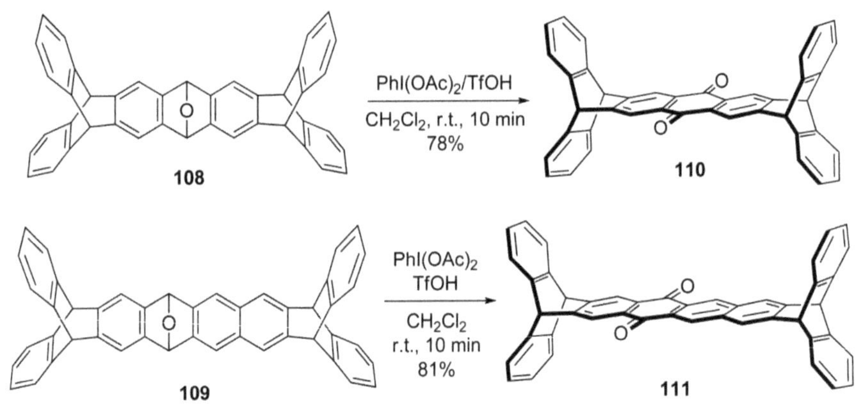

Scheme 3.44 Synthesis of quinones **110, 111** from endoxides **108, 109**

Lee and co-workers [30] also reported that the reaction of the dialdehyde, which was prepared from the oxidation of **106** by $PhI(OAc)_2$ in dichloromethane in the presence of TfOH, with 1,4-cyclohexane-dione in hot EtOH gave the H-shaped centrally extended pentiptycene quinone **112** (bianthracene capped pentacene quinone) in 95 % yield (Scheme 3.45). It was noteworthy that this pentiptycene quinone **112** could not be directly afforded by the reaction of **106** with cyclohexanedione. In addition, it was found that the reduction of the diester **113** with $LiAlH_4$, then followed by the Swern oxidation of the resulting diol **114** successfully gave the dialdehyde **115** in 93 % yield, which then treated with the 1,4-cyclohexanedione to give the extended heptaiptycene quinone **116** in 78 % yield (Scheme 3.46).

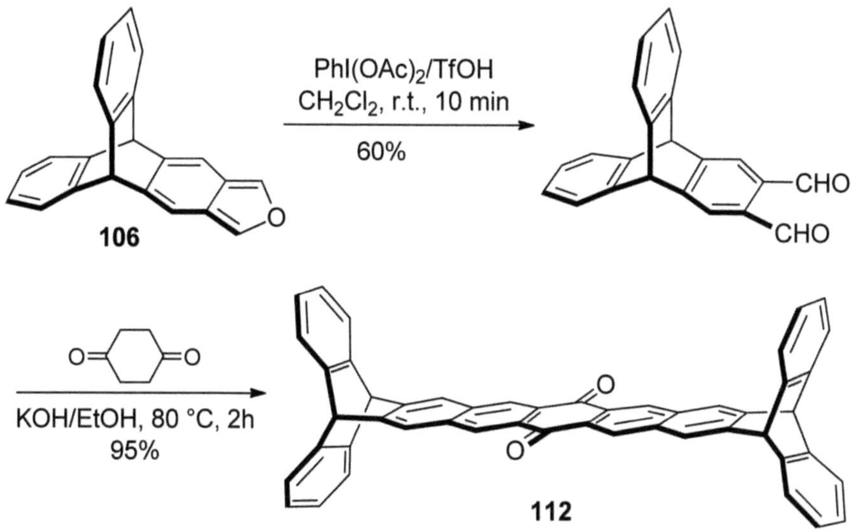

Scheme 3.45 Synthesis of extended pentiptycene quinone **112** from synthon **106**

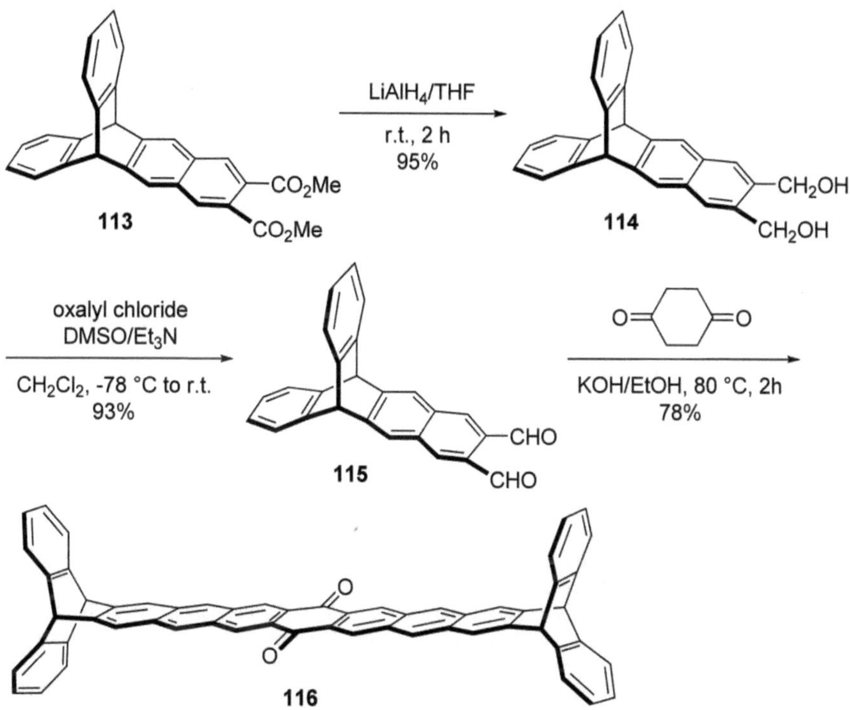

Scheme 3.46 Synthesis of extended heptaiptycene quinone **116** from synthon **113**

References

1. Bartlett PD, Ryan MJ, Cohen SG (1942) Triptycene (9,10-*o*-benzenoanthracene). J Am Chem Soc 64(11):2649–2653
2. Skvarche VR, Shalaev VK (1974) Aromatic-hydrocarbons: new aryn-2,3-dehydrotriptycene in reaction with anthracene. Dokl Akad Nauk SSSR 216(1):110–112
3. Hart H, Lai C, Nwokogu G, Shamouilian S, Teuerstein A, Zlotogorski C (1980) Bis-annelation of arenes with bis-aryne equivalents. J Am Chem Soc 102(21):6649–6651
4. Hart H, Shamouilian S, Takehira Y (1981) Generalization of the triptycene concept. Use of diaryne equivalents in the synthesis of iptycenes. J Org Chem 46(22):4427–4432
5. Hart H, Bashirhashemi A, Luo J, Meador MA (1986) Iptycenes-extended triptycenes. Tetrahedron 42(6):1641–1654
6. Pei BJ, Chan WH, Lee AWM (2010) Oxadisilole fused triptycene and extended triptycene: precursors of triptycyne and extended triptycyne. J Org Chem 75(21):7332–7337
7. Cadogan JIG, Harger MJP, Sharp JT (1971) Acylarylnitrosamines. Part III. Decomposition of 2,5-di-(*N*-nitrosoacetamido)-1,4-di-*t*-butylbenzene and related compounds. J Chem Soc B 602–607
8. Luo JM, Hart H (1987) New iptycenes using 2,3-naphtho[*b*]triptycene as a synthon. J Org Chem 52(16):3631–3636
9. Zyryanov GV, Palacios MA, Anzenbacher P (2008) Simple molecule-based fluorescent sensors for vapor detection of TNT. Org Lett 10(17):3681–3684
10. Clar E (1931) Über die konstitution des anthracens (zur kenntnis mehrkerniger aromatischer kohlenwasserstoffe und ihrer abkömmlinge, IX. Mitteil.). Ber 64(7):1676–1688
11. Theilacker W, Berger-Brose U, Beyer KH (1960) Untersuchungen in der triptycen-reihe, I. Synthese des triptycens und seiner 9- und 9,10-derivate. Chem Ber 93(7):1658–1681
12. Hart H (1993) Iptycenes, cuppedophanes and cappedophanes. Pure Appl Chem 65(1):27–34
13. Wiehe A, Senge MO, Kurreck H (1997) One-step synthesis of functionalized triptycene-quinones as acceptors for electron−transfer compounds. Liebigs Ann 1997(9):1951–1963
14. Yang JS, Swager TM (1998) Porous shape persistent fluorescent polymer films: an approach to TNT sensory materials. J Am Chem Soc 120(21):5321–5322
15. Yang JS, Swager TM (1998) Fluorescent porous polymer films as TNT chemosensors: electronic and structural effects. J Am Chem Soc 120(46):11864–11873
16. Williams VE, Swager TM (2000) Iptycene-containing poly(arylene-ethynylene)s. Macromolecules 33(11):4069–4073
17. Spyroudis S, Xanthopoulou N (2003) The reaction of triptycene haloquinones with alkoxides. An unusual route to pentiptycene quinones. Tetrahedron Lett 44(19):3767–3770
18. Zhu XZ, Chen CF (2005) Iptycene quinones: synthesis and structure. J Org Chem 70(3):917–924
19. Cao J, Lu HY, Chen CF (2009) Synthesis, structures, and properties of peripheral *o*-dimethoxy-substituted pentiptycene quinones and their *o*-quinone derivatives. Tetrahedron 65(39):8104–8112
20. Yang JS, Ko CW (2006) Pentiptycene chemistry: new pentiptycene building blocks derived from pentiptycene quinones. J Org Chem 71(2):844–847
21. Newman M, Boden H (1961) Notes: *N*-methylpyrrolidone as solvent for reaction of aryl halides with cuprous cyanide. J Org Chem 26(7):2525–2525
22. Raap J, Nieuwenhuis S, Creemers A, Hexspoor S, Kragl U, Lugtenburg J (1999) Synthesis of isotopically labelled L-phenylalanine and L-tyrosine. Eur J Org Chem 1999(10):2609–2621
23. Hegedus LS (1994) Organometallics in synthesis. Wiley, New York
24. Yang JS, Yan JL, Jin YX, Sun WT, Yang MC (2009) Synthesis of new halogenated pentiptycene building blocks. Org Lett 11(6):1429–1432
25. Kundu SK, Tan WS, Yan JL, Yang JS (2010) Pentiptycene building blocks derived from nucleophilic aromatic substitution of pentiptycene triflates and halides. J Org Chem 75(13):4640–4643

26. Hart H, Raju N, Meador MA, Ward DL (1983) Synthesis of heptiptycenes with face-to-face arene rings via a 2,3:6,7-anthradiyne equivalent. J Org Chem 48(23):4357–4360
27. Pei BJ, Lee AWM (2010) Highly efficient synthesis of extended triptycenes via Diels−Alder cycloaddition in water under microwave radiation. Tetrahedron Lett 51(34):4519–4522
28. Warrener RN (1971) Isolation of isobenzofuran, a stable but highly reactive molecule. J Am Chem Soc 93(9):2346–2348
29. Warrener RN, Hammer BC, Russell RA (1981) Regiospecific cycloaddition of 1-substituted isobenzofurans to quinone acetals. J Chem Soc Chem Commun (18):942–943
30. Pei BJ, Chan WH, Lee AWM (2011) Anthracene capped isobenzofuran: a synthon for the preparations of iptycenes and iptycene quinones. Org Lett 13(7):1774–1777

Chapter 4
Synthesis and Reactions of Other Iptycenes and Their Derivatives

4.1 Heptiptycene and Noniptycene

In contrast to triptycene or pentiptycene, the researches on heptiptycene and noniptycene are very few, probably due to the difficulties of synthesis. In the early 1970s, Huebner et al. [1] synthesized the first heptiptycene **1** with very low and instable yields by the reaction of 11-chloro-9,10-di-hydro-9,10-etheno anthrace [2] with *n*-butyl lithium at 25 °C. With the help of high-resolution mass spectrum, ultraviolet spectrum, and the space group of crystalline 1:1 chlorobenzene complex, the structure of compound **1** could be determined. Moreover, molecule **1** showed a remarkable thermal stability with high melting point. Even when melted to 580 °C in a sealed tube, it would not decompose with some minor sublimation.

A decade later, Hart et al. [2] modified the Huebner's method to obtain heptiptycene **1** with a reasonable yield. The reaction of compound **2** with *n*-butyl lithium first at −23 °C in THF gave the α-lithio derivative **3**, which was then refluxed in THF for 2 h to give the heptiptycene **1** in 20 % yield (Scheme 4.1). Besides compound **1**, the reaction also gave the coupling product **4** in 39 % yield. Under similar conditions, the methyl-substituted heptiptycene **5** could be obtained in 23 % yield along with the coupling product **5′**, by the reaction of compound **6** with 1.1 equivalents of butyl lithium in THF, and then quenching with methanol (Scheme 4.2) [3].

In 1988, Shahlai and Hart [4] investigated the formation mechanism of heptiptycene **1** in details. As shown in Scheme 4.3, it was found that the losing of one LiCl molecule from the α-lithio derivative **3** gave the intermediate of bicycloalkyne **7**, which was then reacted with compound **3** to give the "dimeric" organolithium **8**. Compound **8** was further reacted with another equivalent of bicycloalkyne **7** to afford hexatriene derivative **9**, which could produce cyclohexadiene **10** via the cyclization. The aromatic "cyclotrimer" **1** was finally obtained by the losing of one LiCl molecule from **10**. To verify the reliability of this mechanism scheme, they needed to prove the existence of bicycloalkyne intermediate **7**, "dimeric" and "trimeric" vinyllithiums **8** and **9**, respectively. They found that the adduct **11** could be first prepared in a good yield from the treatment of either **12** or **13** with BuLi under −78 °C in THF, and then followed by the Diels–Alder addition with 1,3-diphenylisobenzofuran in refluxed condition (Scheme 4.4). This result provided an evidence for the formation

C.-F. Chen, Y.-X. Ma, *Iptycenes Chemistry*,
DOI 10.1007/978-3-642-32888-6_4, © Springer-Verlag Berlin Heidelberg 2013

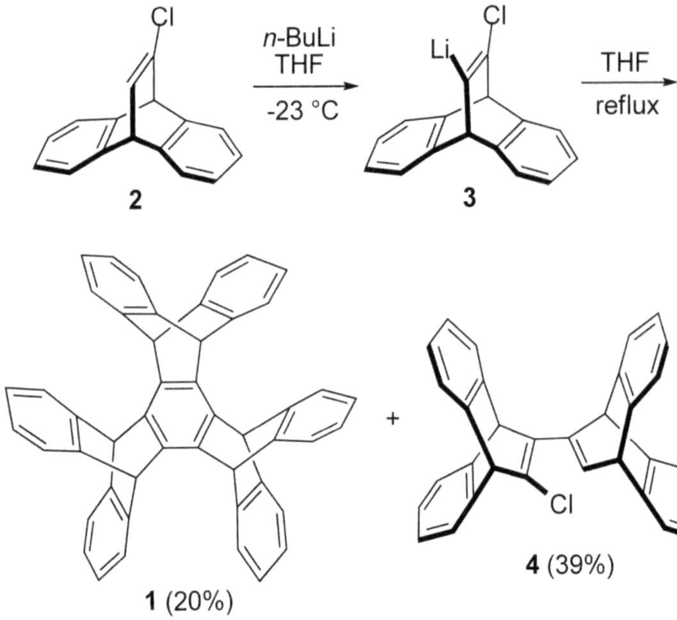

Scheme 4.1 Synthesis of heptiptycene **1** from compound **2**

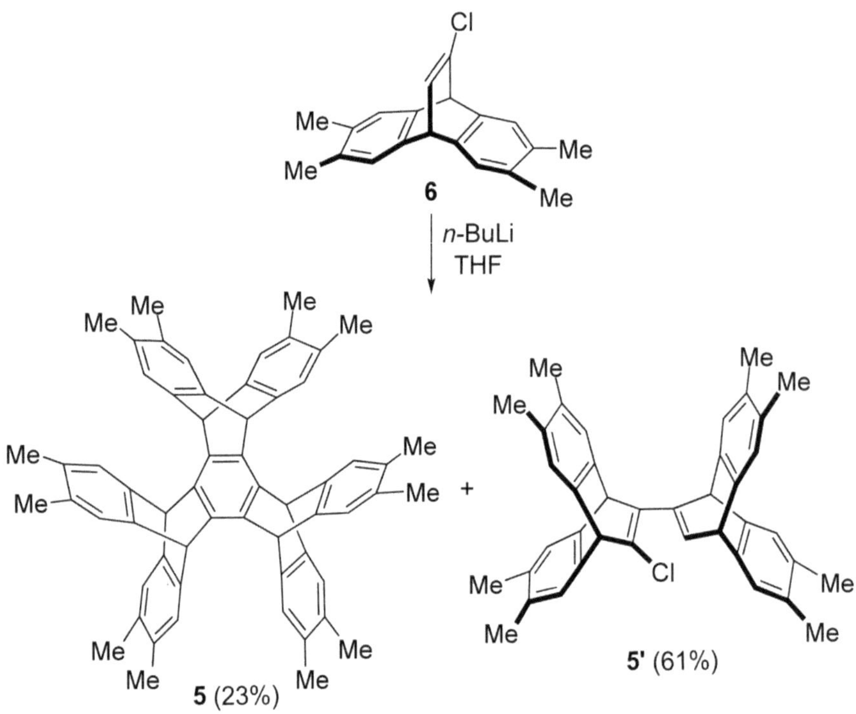

Scheme 4.2 Synthesis of methyl-substituted heptiptycene **5**

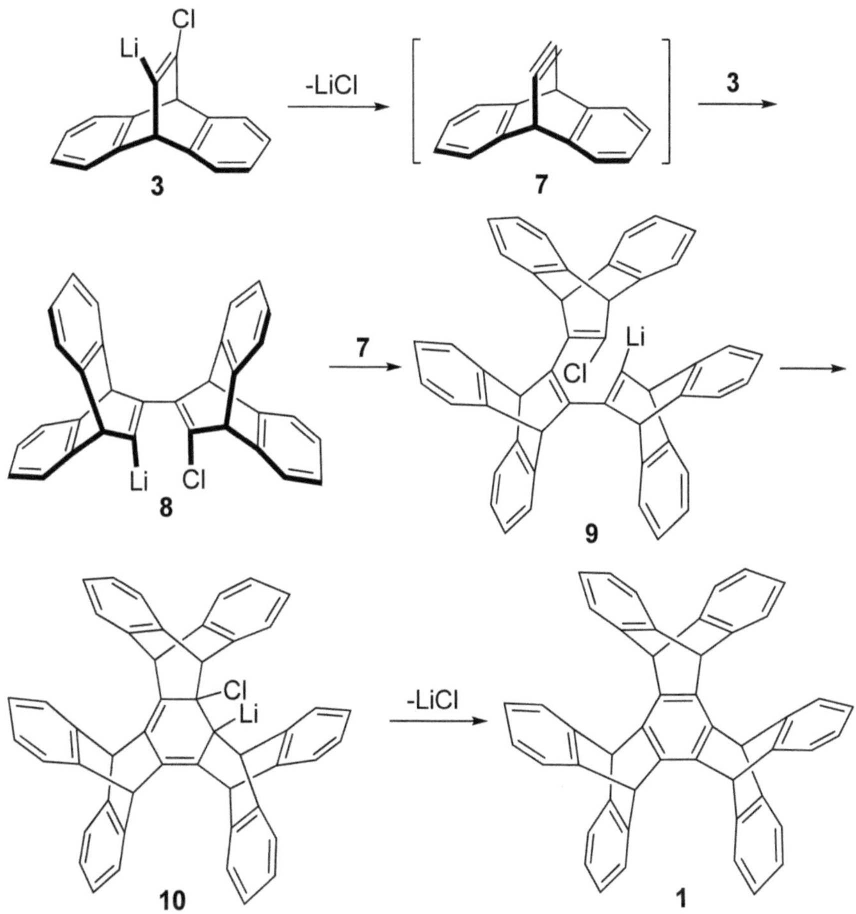

Scheme 4.3 Synthesis of heptiptycene **1** from α-lithio derivative **3**

of bicycloalkyne intermediate **7**. Similarly, the formation of "dimer" **14** and the C_{3h} "trimer" **15** from the vinyl chloride **12** and butyl lithium could also give the proof for the formation of molecule **1** followed by the Scheme 4.3 via the type intermediate of **8**. Likewise, the formation of "dimer" **16** and the C_s "trimer" **17** could be obtained from compound **13** through the same type of intermediate. The formation of the butadiene **18** and hexatriene **19** could seemingly prove the existence of type intermediate **9** as well. Compared to compound **9**, the hexatriene derivative **19** had a larger steric hindrance, thus, it would not finally achieve cyclization to form a closed ring molecule (Fig. 4.1).

In 1986, Hart et al. [5] reported the synthesis of the extended heptiptycene **20** containing an anthracene moiety in a two-step route. First, the reaction of the extended triptycene **21** with one anthracene moiety and compound **22** in refluxed xylene gave the intermediate **23** in 70 % yield, which then dehydrogenated in ethanol in the presence of perchloric acid to afford the target product **20** in 88 % yield (Scheme 4.5).

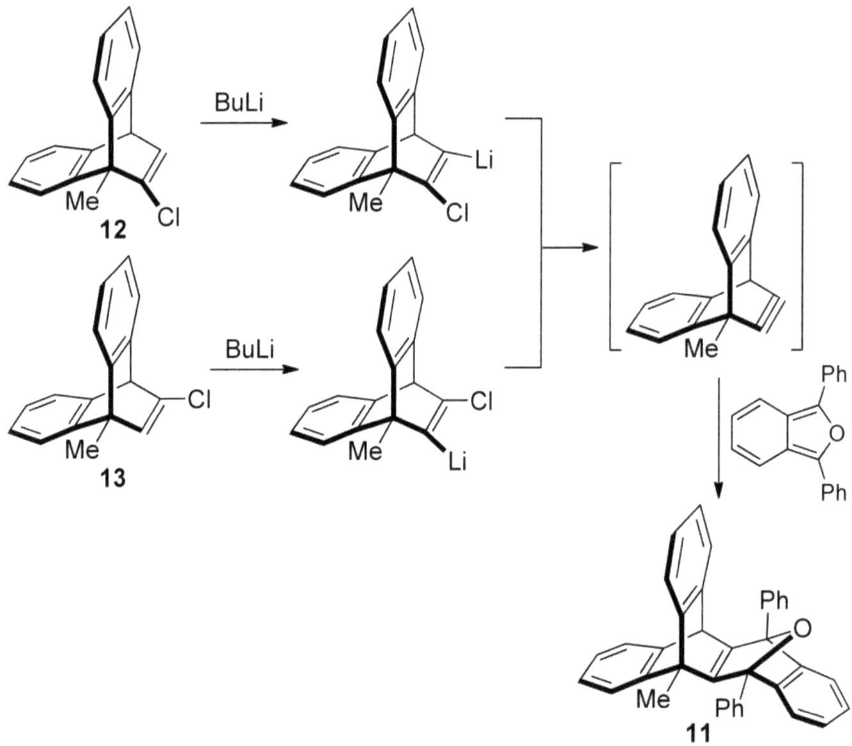

Scheme 4.4 Synthesis of compound **11**

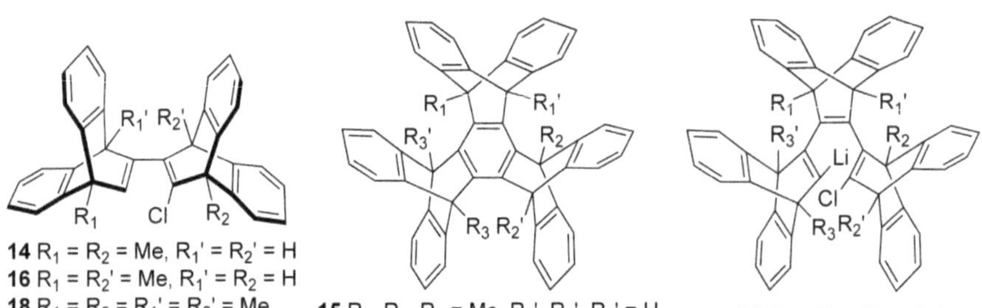

14 $R_1 = R_2 = Me$, $R_1' = R_2' = H$
16 $R_1 = R_2' = Me$, $R_1' = R_2 = H$
18 $R_1 = R_2 = R_1' = R_2' = Me$

15 R_1, R_2, $R_3 = Me$, R_1', R_2', $R_3' = H$
17 R_1, R_2, $R_3' = Me$, R_1', R_2', $R_3 = H$

19 $R_1 = R_2 = R_3 = R_1' = R_2' = R_3' = Me$

Fig. 4.1 Structures of the compounds **14–19**

By the Diels–Alder addition between the H-type pentiptycene **24** containing an anthracene moiety and benzyne in a 1,2-dichloroethane solution under reflux for 6 h, Hart et al. [6] successfully synthesized heptiptycene **25** in a U-shaped structure (Scheme 4.6), and also provided an efficient route for the synthesis of high order iptycenes from the extended triptycenes and/or pentiptycenes.

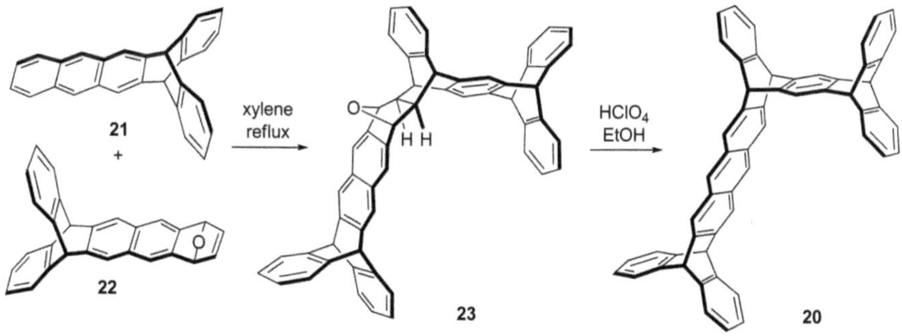

Scheme 4.5 Synthesis of extended heptiptycene **20** from extended triptycene **21**

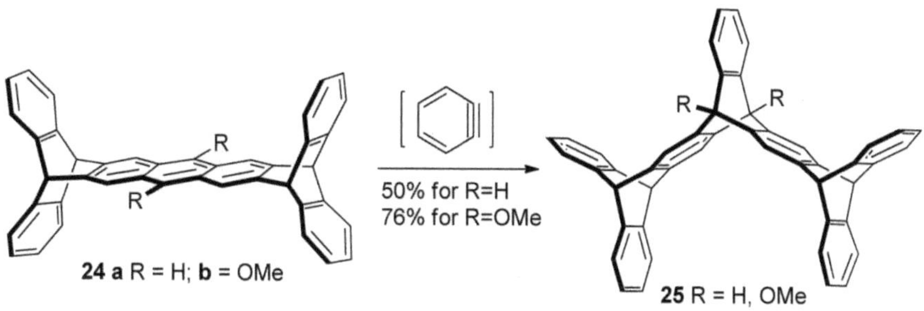

Scheme 4.6 Synthesis of heptiptycene **25** in a U-shaped structure

According to the similar strategy, Hart et al. [7, 8] succeeded in the synthesis of noniptycene **26** (also called tritriptycene) starting from the extended pentiptycene **24a** with one anthracene moiety via a multistep process, which is shown in Scheme 4.7. The key step in this process was the [4 + 2] addition between the diene moiety of **27** and 1,4-dihydronaphthalene 1,4-endoxide in a refluxed xylene solution, which gave the adduct **28** in 71 % yield. It was noteworthy that noniptycene **26** had an extraordinary thermal stability, and could not melt or decompose rapidly below 550°C. Moreover, it showed a moderate solubility in chlorinated hydrocarbons. In addition, noniptycene **26** with three equivalent U-shaped cavities could also make it to be the inclusion-forming substance with small guest molecules.

In 1998, Yang and Swager [9] reported the synthesis of noniptycene quinone **33** with a tweezer-like structure. As shown in Scheme 4.8, the Diels–Alder reaction between the triptycene quinone **29** and pentacene **30** in a refluxed toluene solution for 3 days gave a mixture of three semiquinone adducts. The mixture was then treated with glacial acetic acid in the presence of HBr till the color of the solution faded, which was further treated with potassium bromate under reflux condition for several minutes, and then followed by tautomerization and oxidation to give noniptycene diquinone **33** in 13 % yield, along with another two extended pentiptycenes **31** and **32**.

Several years later, Zhu and Chen [10] reported a practical and efficient method for the synthesis of iptycene quinones. Consequently, a series of heptiptycene quinones

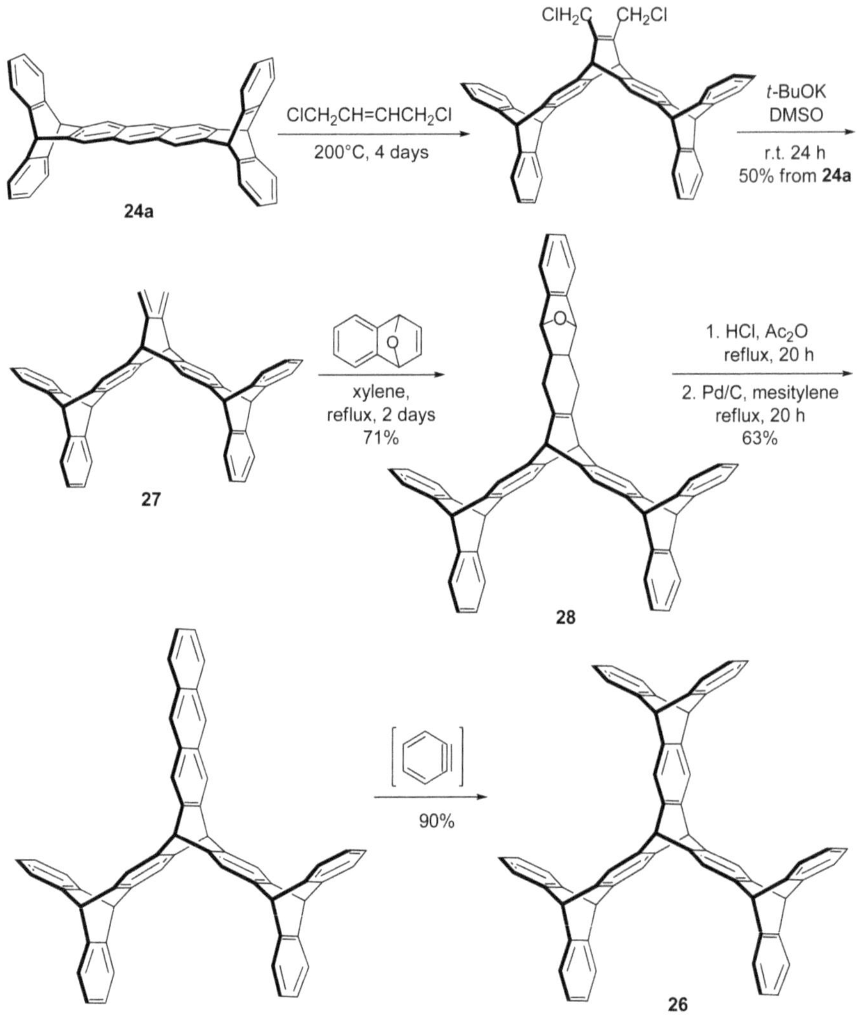

Scheme 4.7 Synthesis of noniptycene **26** from extended pentiptycene **24a**

with U-shaped cavities could be conveniently synthesized. As shown in Scheme 4.9, heptiptycene diquinone **35** was synthesized either by one-pot reaction of pentiptycene quinone **34** with anthracene in refluxing acetic acid with the presence of *p*-chloranil or by the reaction of triptycene diquinone **36** with two equivalents of anthracenes at the same conditions. In the latter case, heptiptycene semiquinone **35′** as an *endo*-adduct was also obtained, and the yield of **35** could be improved by prolonging the reaction time. When pentiptycene quinone **34** reacted with 1,4-dimethoxyanthracene in acetic acid in the presence of *p*-chloranil, a mixture of three adducts **37a** and **37b** were obtained. They could not be separated by conventional column chromatography method, but their CAN oxidative products **38a** and **38b** could be separated with column chromatography (Scheme 4.10). The structures of the two isomers were

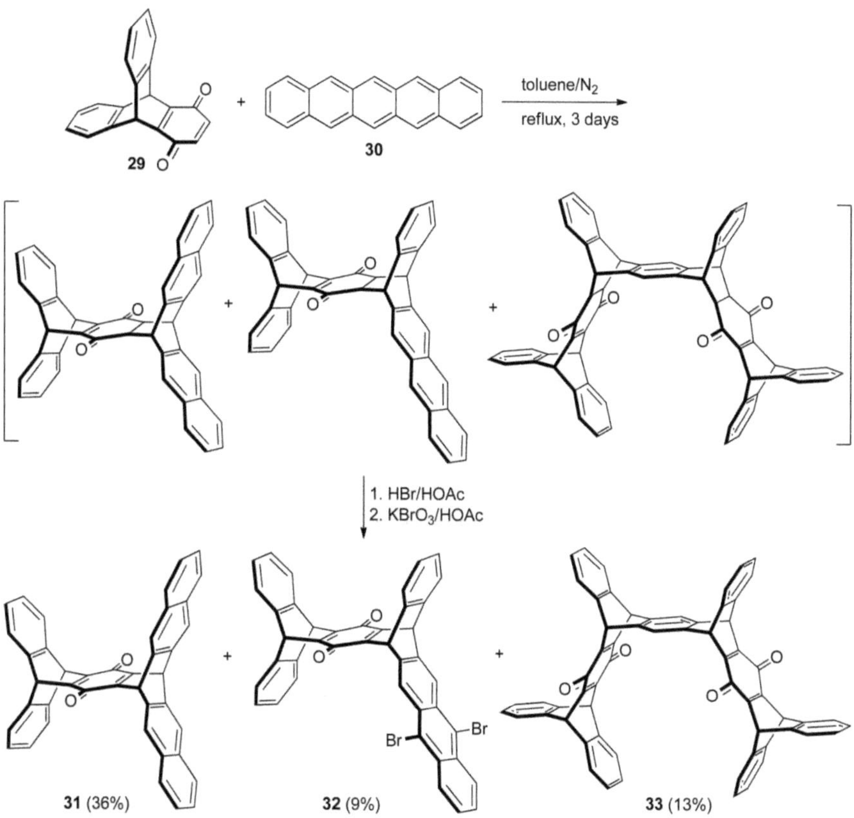

Scheme 4.8 Synthesis of noniptycene diquinone **33**

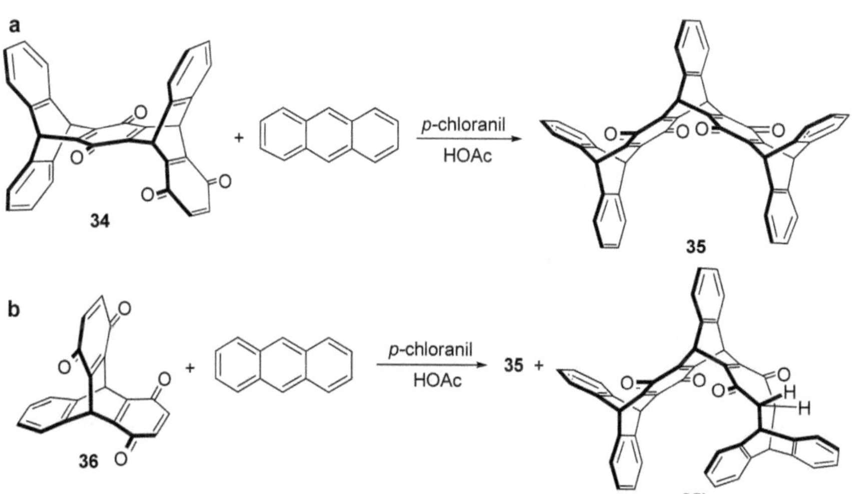

Scheme 4.9 Synthesis of heptiptycene diquinone **35**

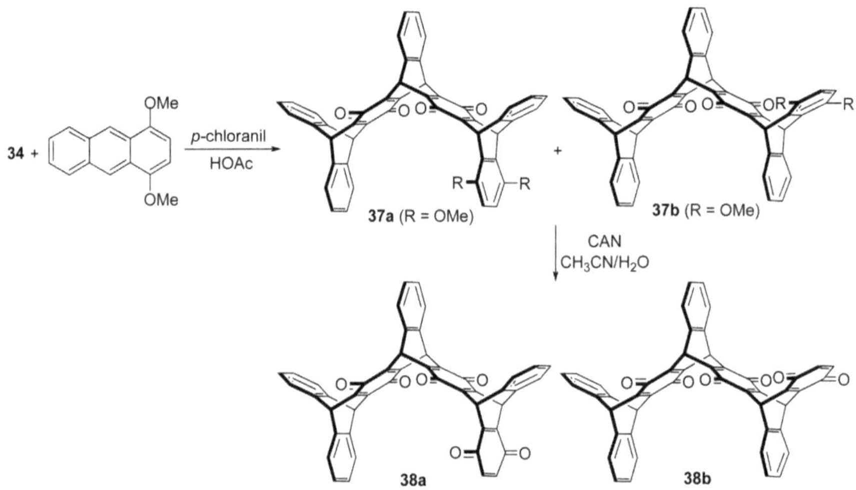

Scheme 4.10 Synthesis of heptiptycene triquinones **38a**, **38b** from pentiptycene quinone **34** with 1,4-dimethoxyanthracene

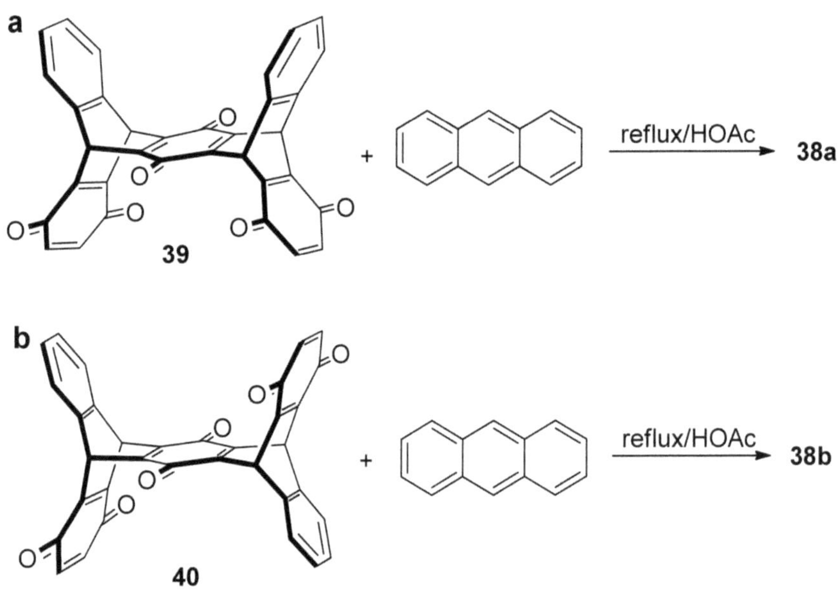

Scheme 4.11 Synthesis of heptiptycene triquinones **38a**, **38b** from pentiptycene triquinones **39** and **40** with anthracene

determined by the comparative reactions between the pentiptycene triquinones **39** and **40** with one equivalent of anthracene in acetic acid in the presence of *p*-chloranil, respectively (Scheme 4.11).

Similarly, by the reaction of pentiptycene quinones or triptycene quinones with anthracene or substituted anthracenes in acetic acid in the presence of *p*-chloranil,

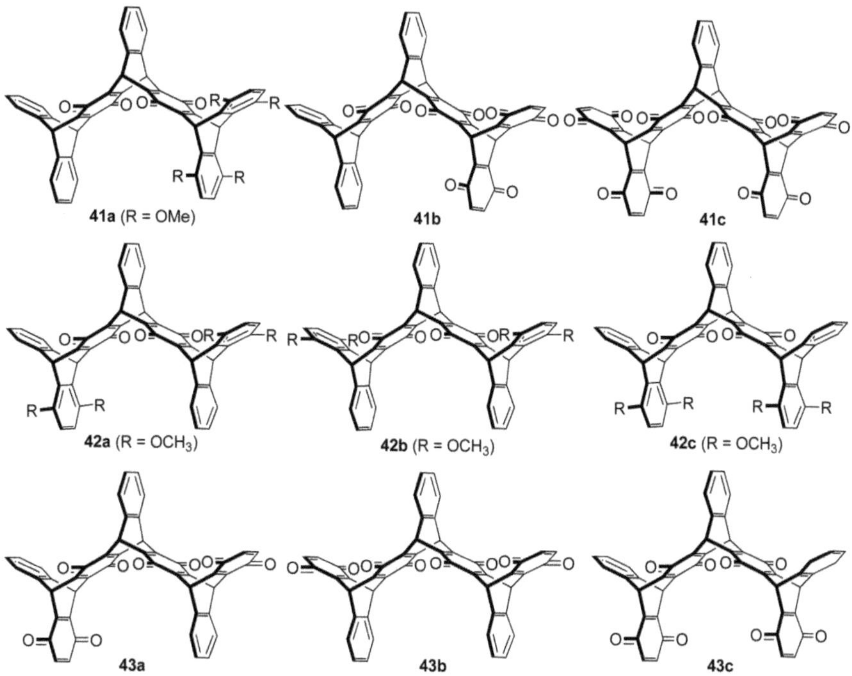

Fig. 4.2 Structures of heptiptycene quinones **41–43**

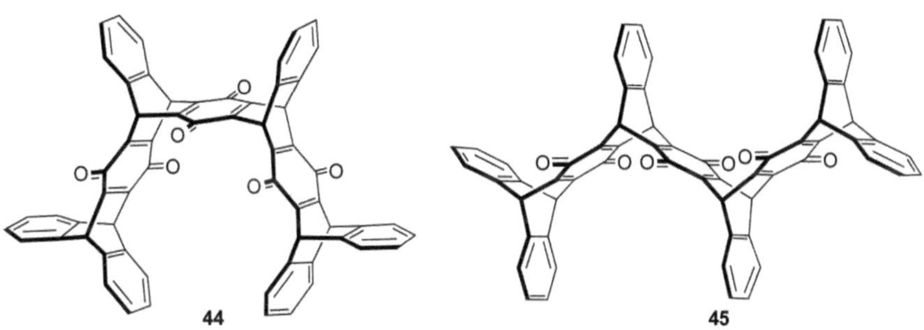

Fig. 4.3 Structures of noniptycene triquinones **44** and **45**

followed by oxidation with CAN, a series of other heptiptycene quinones **41–43** could be conveniently obtained as well (Fig. 4.2).

According to the similar method, Zhu and Chen further synthesized noniptycene triquinone **44** (Fig. 4.3) with a tweezer-shaped molecular cavity in 70 % yield by a one-pot reaction of pentiptycene triquinone **39** with two equivalents of anthracene in acetic acid with the presence of *p*-chloranil. Under the same conditions, the reaction of **40** with two equivalents of anthracene gave noniptycene triquinone **45** with two equivalent U-shaped cavities of heptiptycene triquinone. By the reaction

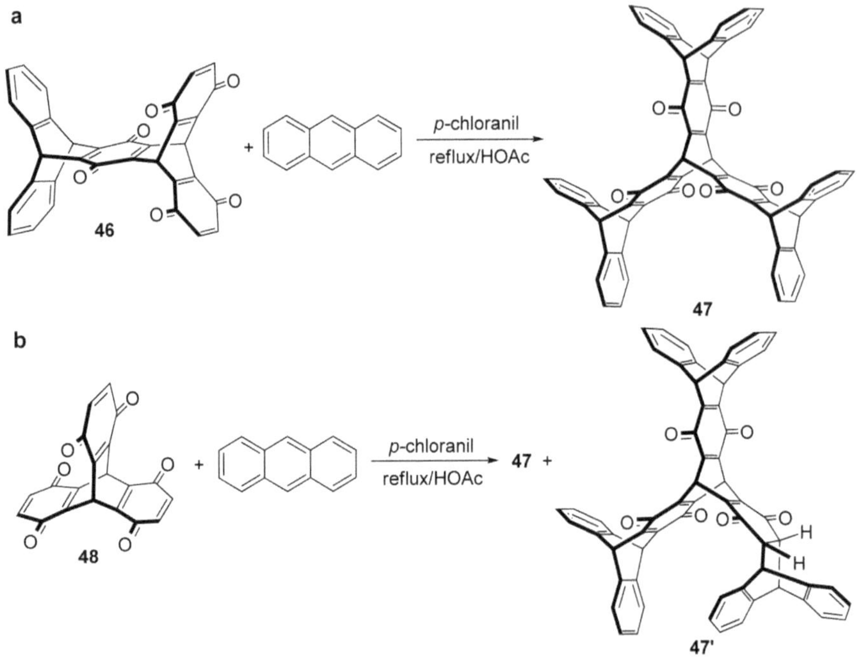

Scheme 4.12 Synthesis of heptiptycene triquinone **47**

of pentiptycene triquinone **46** with two equivalents of anthracene in refluxing acetic acid in the presence of *p*-chloranil, heptiptycene triquinone **47** with three equivalent U-shaped cavities was produced in 17.2 % yield. Similarly, it was found that the one-pot reaction of triptycene triquinone **48** with excess anthracene also gave **47** along with a semiquinone derivative **47′** (Scheme 4.12).

In the same year as the synthesis of iptycene quinones reported by Chen and co-workers, Zhao and Swager [11] also reported the synthesis of the noniptycene derivatives **49a** and **49b** containing both alkoxy and ethynyl substituents via the multistep reactions starting from 6,13-bis-(triisopropylsilylethynyl)pentacene **50** and triptycene quinone **29** (Scheme 4.13). The target iptycenes **49a** and **49b** with large and extended scaffolds showed high solubility in common solvents, which could also be the precursors for the further synthesis of soluble conjugated polymers with specific structures and properties.

4.2 Miscellaneous

In 1990, Hart and Singh [3] reported synthesis of a nonadecaiptycene **52** with a D_{3h} symmetric structure starting from compound **53**. As shown in Scheme 4.14, the treatment of compound **53** with butyl lithium in THF gave four products, including the C_{2v} symmetric molecule **54** in 14 % yield, “dimer” **55** in 25 % yield, and the

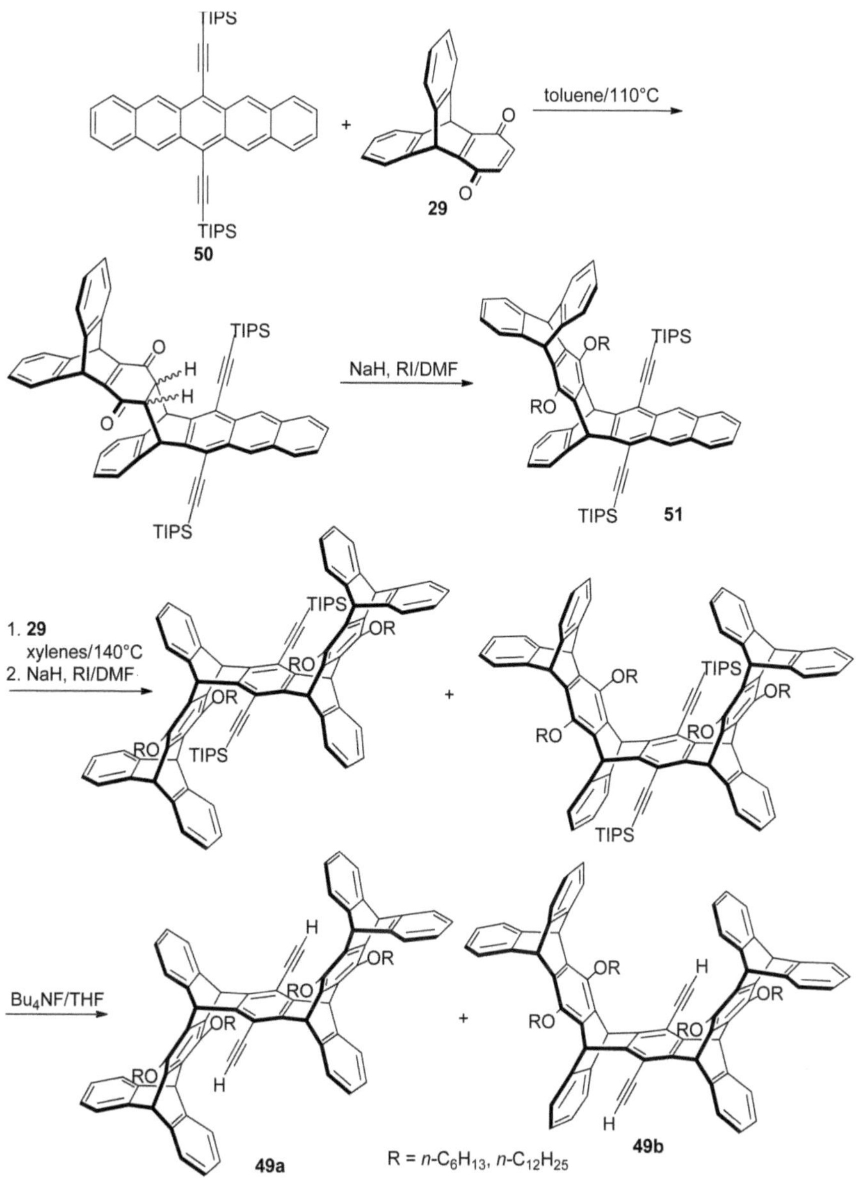

Scheme 4.13 Synthesis of noniptycene derivatives **49a, 49b**

target "trimer" **52** in only 2 % yield, along with the trace amount of pentiptycene **24a**. The nonadecaiptycene **52** (Fig. 4.4) contains six added triptycene moieties, compared with the heptiptycene **1**. In addition, there are two nearly enclosed sphere-like cavities, which are located at the top and bottom plane of the central benzene ring.

Soon after, Hart and co-workers [12] further synthesized more iptycenes (**56**–**58**, Fig. 4.5) with complicated helically chiral structure. The extended pentiptycene

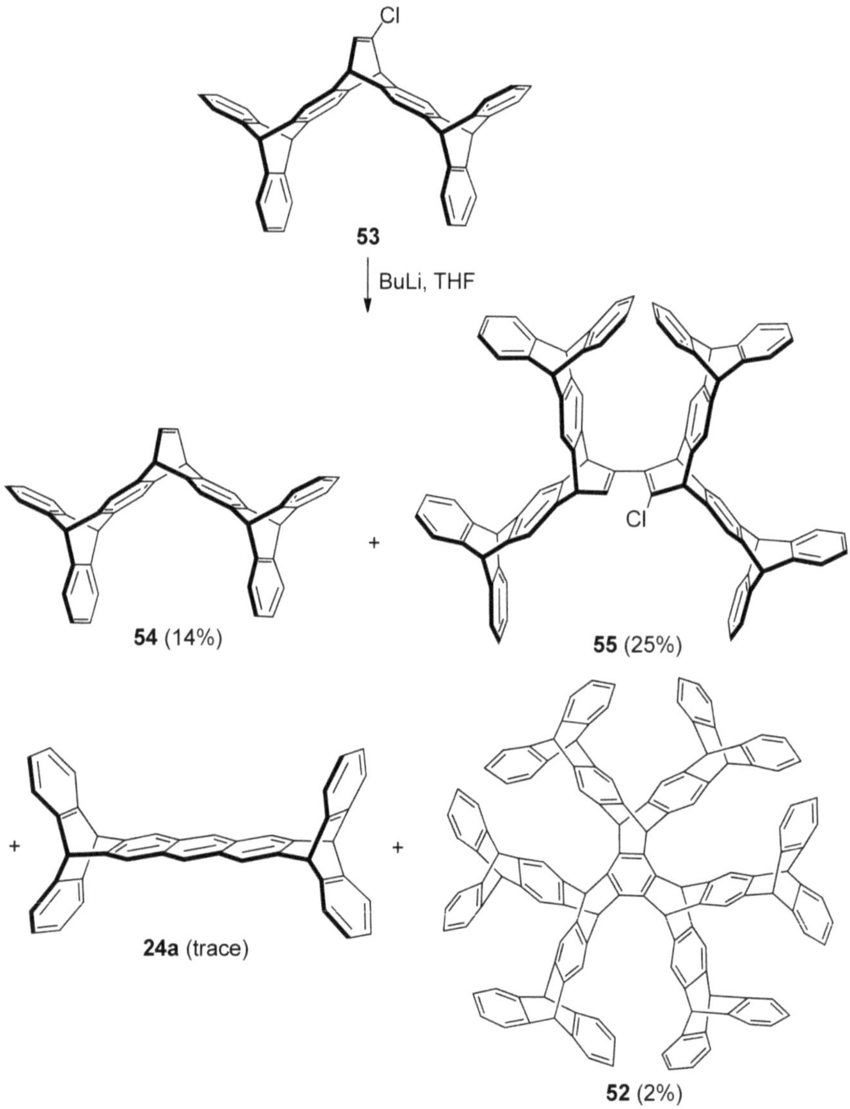

Scheme 4.14 Synthesis of nonadecaiptycene **52**

59 could serve as the key precursor for the formation of these three helically chiral iptycenes **56–58**. As shown in Scheme 4.15, the simply addition reaction of benzyne to the precursor **59** in 1,2-dichloroethane under reflux for 12 h afforded the target iptycene **56** with C_2 symmetry in 60 % yield. Although, precursor **59** reacted with *trans*-1,2-dichloroethene under 195–200 °C for 48 h to give the Diels–Alder adduct **60** in 79 % yield, which was eliminated a molecule of Cl_2 with lithium in THF, was treated with excess diene **62** in refluxing decalin, and then aromatized with DDQ in refluxing benzene to give the target iptycene **57** (Scheme 4.16). When the

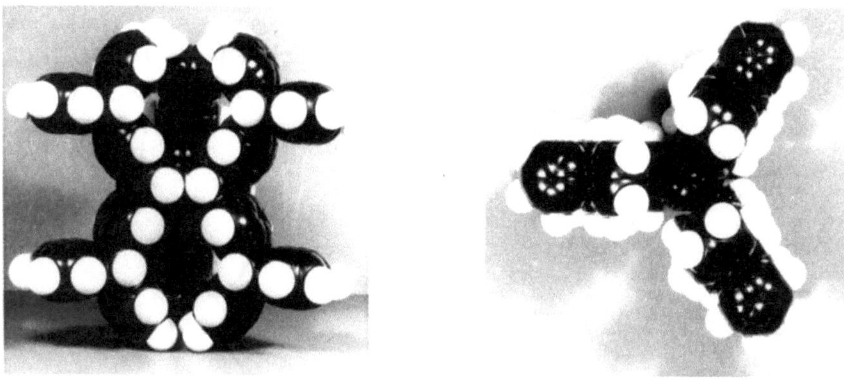

Fig. 4.4 CPK model of nonadecaiptycene **52**. (Reprinted with the permission from [3]. Copyright 1990 American Chemical Society)

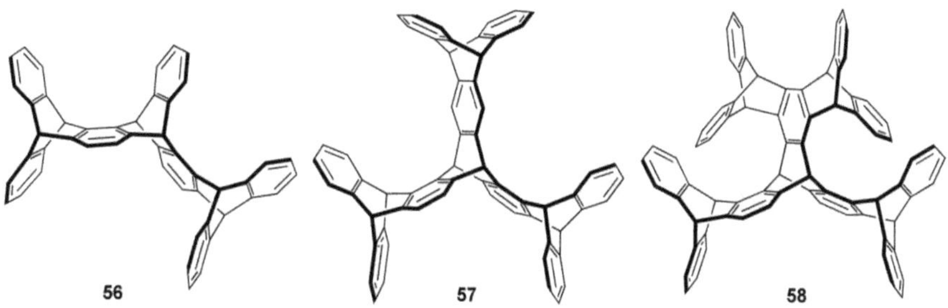

Fig. 4.5 Structures of iptycenes **56**–**58**

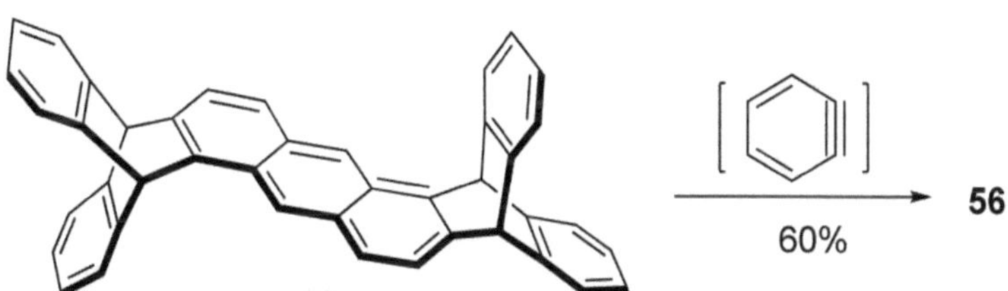

Scheme 4.15 Synthesis of helically chiral iptycene **56**

alkene **61** was treated with diene **64** instead of **62** in refluxing decalin, the adduct **65** could be obtained in 66 % yield, which could convert to the iptycene **58** by the dehydrogenation (Scheme 4.17).

On the basis of the similar route, Vinod and Hart [13, 14] further synthesized the iptycene (**70**) with the ultimate structure, which was also called supertriptycene in 33 % overall yield by the eight-step process starting from the reaction of diene **64** and benzoquinone. As shown in Scheme 4.18, the reaction of diene **64** and benzoquinone, followed by several steps of the tautomerization and reduction provided

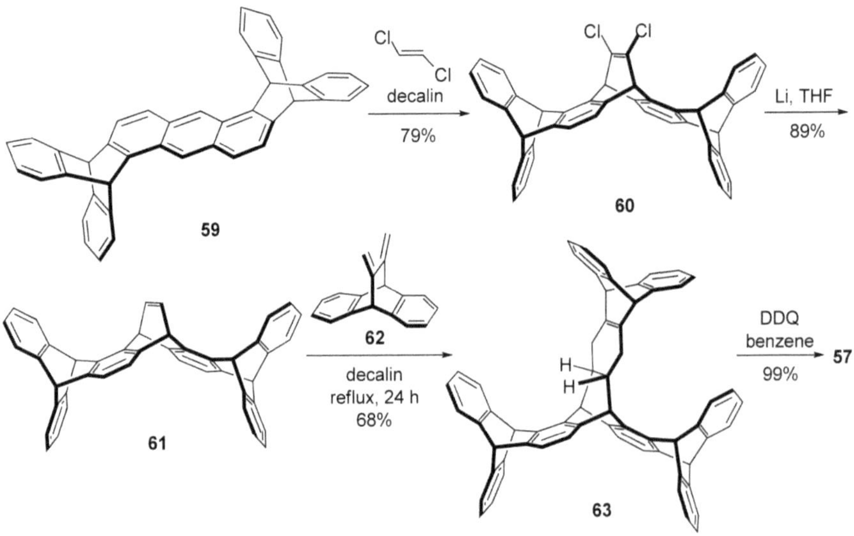

Scheme 4.16 Synthesis of helically chiral iptycene **57**

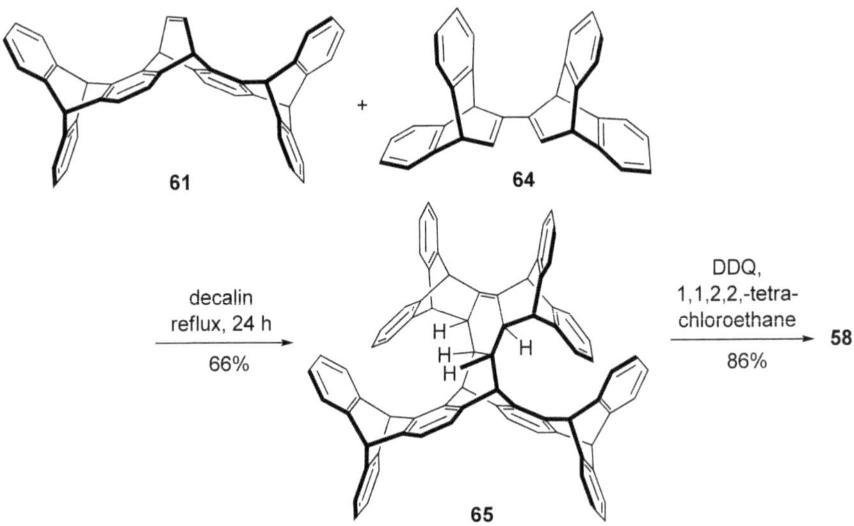

Scheme 4.17 Synthesis of helically chiral iptycene **58**

the alkene **68**, which could be considered as the key precursor for the synthesis of supertriptycene. Thus, the [2 + 4] addition of alkene **68** and diene **64** in refluxing decalin, and then followed by the dehydrogenation of compound **69** with Br_2 in a 1,1,2,2,-tetra-chloroethane gave the target supertriptycene **70**. Compared to triptycene, supertriptycene **70** has six added triptycene moieties, and contains three large cavities. In addition, it was found that **70** showed solubility in hot decalin, tetrachloroethene, and benzonitrile.

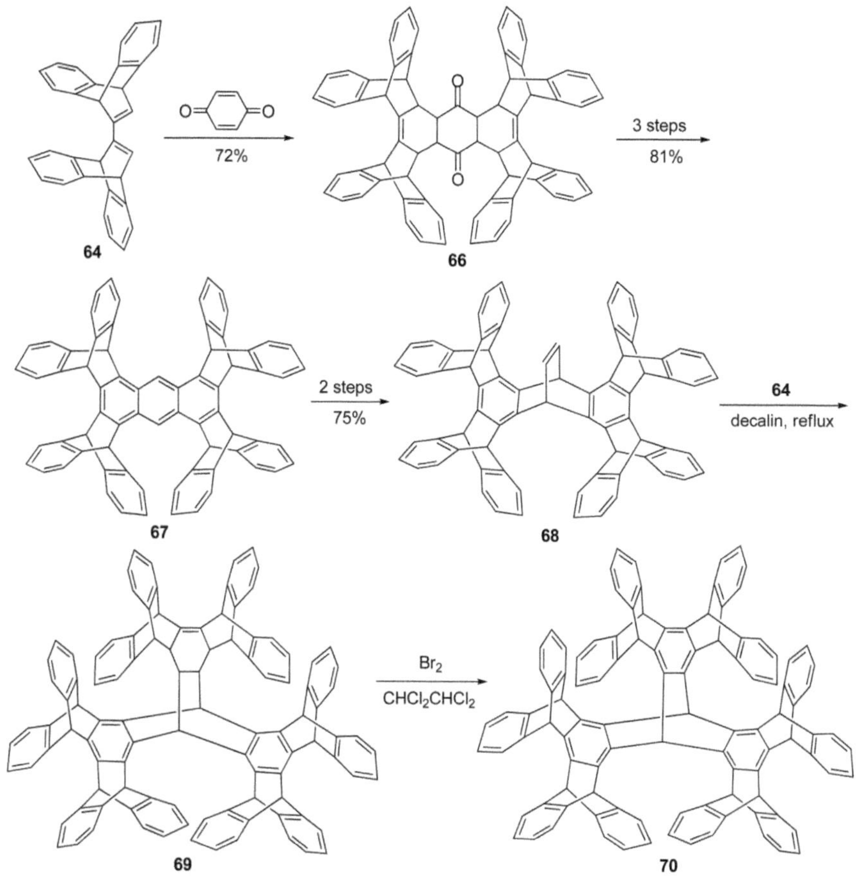

Scheme 4.18 Synthesis of supertriptycene **70**

On the basis of the synthesis of heptiptycene tetraquinone **43c** reported by Zhu and Chen [10], Hua and co-workers [15] recently achieved the syntheses of two cyclododeciptycene tetraquinones by a sequence of intermolecular and intramolecular Diels–Alder reactions. As shown in Scheme 4.19, the Diels–Alder reaction of heptiptycene tetraquinone **43c** and heptiptycene **71** in a refluxed 1,2-dichlorobenzene solution afforded the adduct **72** as major product, along with a small amount of other adducts. The mixture of these adducts was immediately reheated to reflux in an acetic acid solution, and the resulting product of enolization and cyclization was then followed by the addition of diacetoxyiodobenzene to give cyclododeciptycene **73** in 12 % overall yield. There were four methoxy and eight quinone groups in compound **73**, and its inner ring diameter was about 8.9 Å. The X-ray crystal structure of **73** further revealed that there were 16 cyclododeciptycene molecules in one unit cell, in which two molecules were parallel to each other and the other two intercalated between them with large void volume (Fig. 4.6). Moreover, it was found that cyclododeciptycene tetraquinone **73** (R = Me) could also self-assemble into a tube-like structure, and

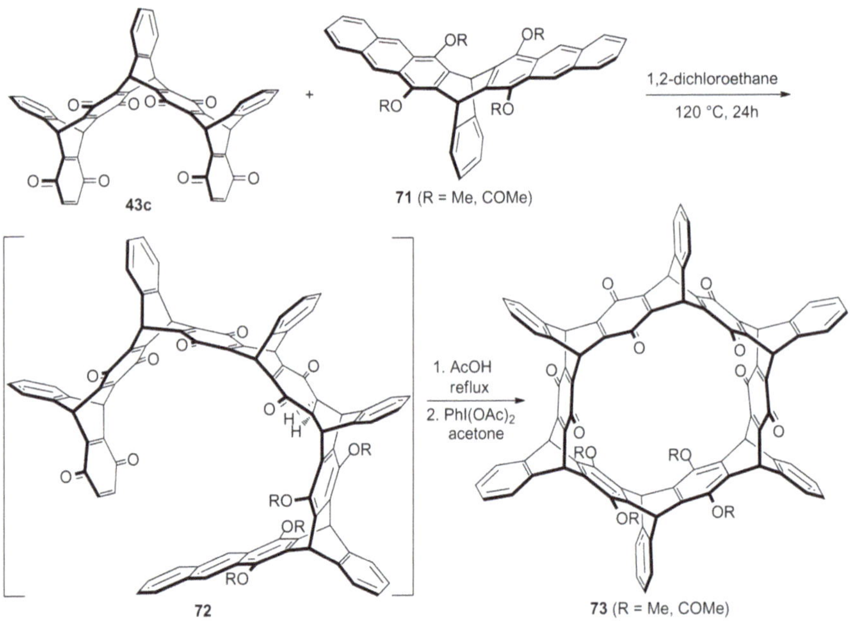

Scheme 4.19 Synthesis of cyclododeciptycene **73**

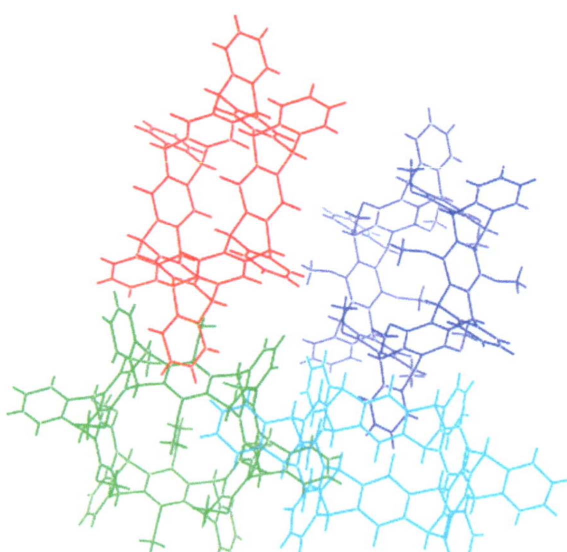

Fig. 4.6 Crystal packing of cyclododeciptycene **73** (R = Me). (Reprinted with the permission from [15]. Copyright 2010 American Chemical Society)

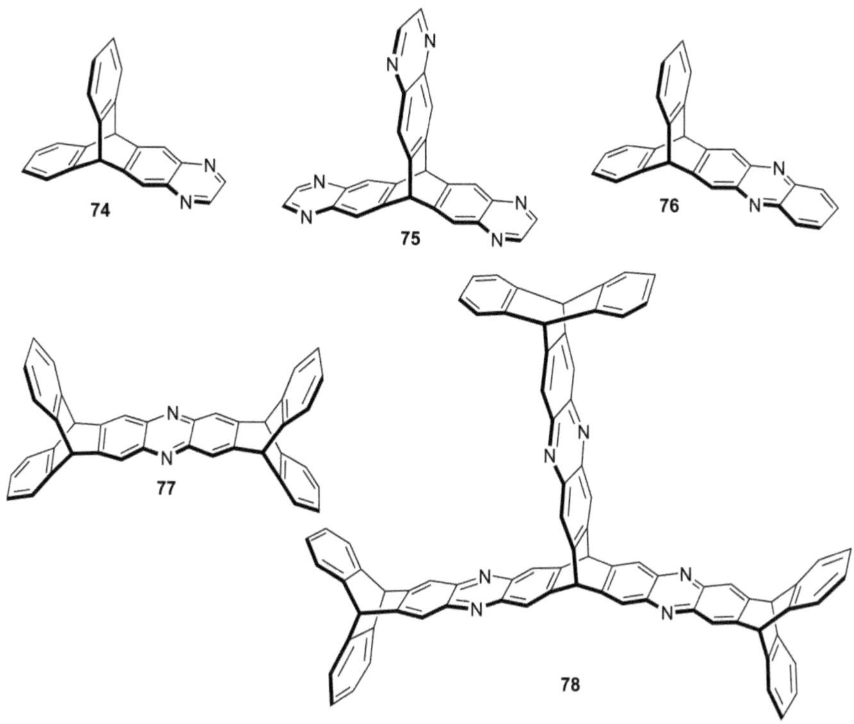

Fig. 4.7 Structures of iptycenes **74–78**

this feature made it a potential candidate as electrical conducting materials and ion channels.

In addition to the hydrocarbon iptycenes mentioned before, there are several kinds of higher iptycenes which contain hetero atoms, like N in their skeleton. For example, in 2006 Chong and MacLachlan [16] synthesized a new class of iptycene quinoxalines (**74** and **75**, Fig. 4.7) by the condensation of polyamino-triptycenes with 2,3-dihydroxy-1,4-dioxane. Furthermore, Chong and MacLachlan [17] also obtained a series of iptycenes (**73** and **74**, Fig. 4.7) containing pyrazine groups from the triptycene *o*-quinone in ethanol with the yields of 61 and 68 %. Although, the iptycene **75** was obtained in 35 % yield in the presence of a catalytic amount of piperidine, these iptycenes with the rigid "wings" were good building blocks for further generating the porous structures. The synthesis of these iptycene quinoxalines will be depicted with details in Chap. 5.

In 2010, Chen and co-workers [18] efficiently synthesized a series of pentiptycene-derived rigid tweezer-like molecules (**79–84**, Fig. 4.8) containing one or two pyrazine groups. The target molecules (**79–84**) were prepared conveniently in the reasonable yields by the condensation of *o*-diaminobenzene (or *o*-diaminotri-ptycene) with the corresponding *o*-quinones in refluxing ethanol for 3–6 h (or overnight). The single-crystal structure of **80** suggested that the molecule showed a little distortion

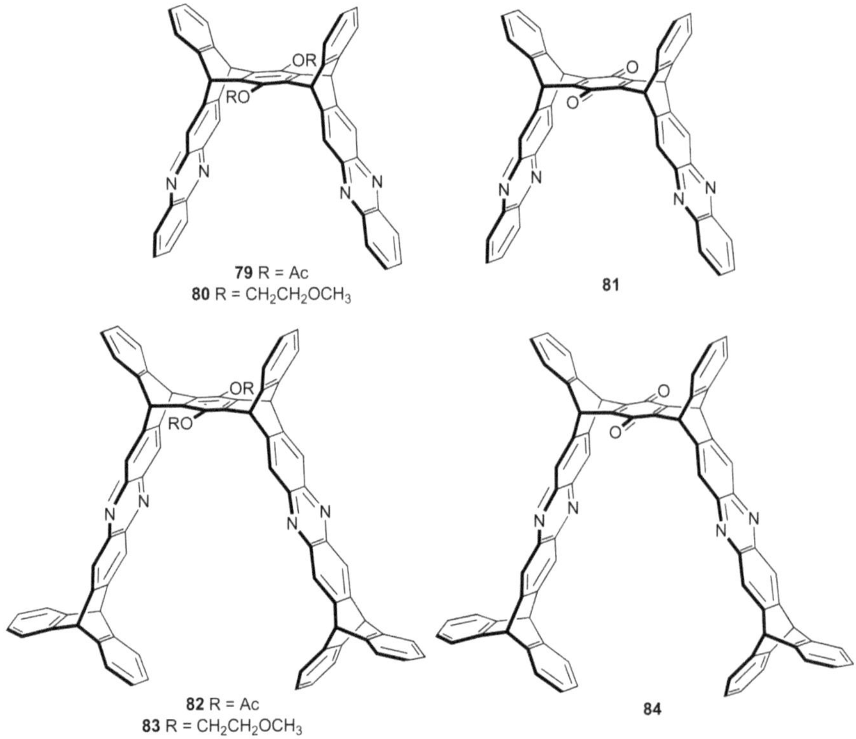

Fig. 4.8 Structures of pentiptycene-derived rigid tweezer-like molecules **79**–**84**

in the pyrazine rings, and the adjacent molecules could form a 1D structure with the alternate arrangement, and further assemble into a 2D-layered structure, and 3D open-framework in the solid state. The particular tweezer-like structures of these pentiptycene derivatives also made them potential applications in host–guest chemistry. For example, it was found that molecular tweezer **82** showed efficient binding ability toward C_{60} due to the suitable cavity size. The route to the synthesis of these iptycene tweezers, and their more structural characteristics will be depicted with details in Chap. 5 as well.

References

1. Huebner CF, Puckett RT, Schwartz SL, Brzechff M (1970) A trimeric $C_{48}H_{30}$ hydrocarbon of unusual structural interest derived from 9,10-dihydro-9,10-ethenoanthracene. Tetrahedron Lett (5):359–362
2. Hart H, Shamouilian S, Takehira Y (1981) Generalization of the triptycene concept. Use of diaryne equivalents in the synthesis of iptycenes. J Org Chem 46(22):4427–4432
3. Singh SB, Hart H (1990) Extensions of bicycloalkyne trimerizations. J Org Chem 55(10):3412–3415

4. Shahlai K, Hart H (1988) The mechanism of trimerization of bicyclo[2.2.2]alkynes. J Am Chem Soc 110(21):7136–7140
5. Hart H, Bashirhashemi A, Luo J, Meador MA (1986) Iptycenes-extended triptycenes. Tetrahedron 42(6):1641–1654
6. Hart H, Raju N, Meador MA, Ward DL (1983) Synthesis of heptiptycenes with face-to-face arene rings via a 2,3:6,7-anthradiyne equivalent. J Org Chem 48(23):4357–4360
7. Bashir-Hashemi A, Hart H, Ward DL (1986) Tritriptycene: a D_{3h} C_{62} hydrocarbon with three U-shaped cavities. J Am Chem Soc 108(21):6675–6679
8. Hart H (1993) Iptycenes, cuppedophanes and cappedophanes. Pure Appl Chem 65(1):27–34
9. Yang JS, Swager TM (1998) Fluorescent porous polymer films as TNT chemosensors: electronic and structural effects. J Am Chem Soc 120(46):11864–11873
10. Zhu XZ, Chen CF (2005) Iptycene quinones: synthesis and structure. J Org Chem 70(3):917–924
11. Zhao DH, Swager TM (2005) Conjugated polymers containing large soluble diethynyl iptycenes. Org Lett 7(20):4357–4360
12. Shahlai K, Hart H, Bashir-Hashemi A (1991) Synthesis of three helically chiral iptycenes. J Org Chem 56(24):6912–6916
13. Vinod TK, Hart H (1990) Synthesis of two noninterconvertible conformers of a single host. Self-filled and vaulted cappedophanes. J Am Chem Soc 112(8):3250–3252
14. Vinod TK, Hart H (1991) Synthesis of self-filled, vaulted, and intracavity functionalized cappedophanes. J Org Chem 56(19):5630–5640
15. Lou K, Prior AM, Wiredu B, Desper J, Hue DH (2010) Synthesis of cyclododeciptycene quinones. J Am Chem Soc 132(49):17635–17641
16. Chong JH, MacLachlan MJ (2006) Robust non-interpenetrating coordination frameworks from new shape-persistent building blocks. Inorg Chem 45(4):1442–1444
17. Chong JH, MacLachlan MJ (2007) Synthesis and structural investigation of new triptycene-based ligands: en route to shape-persistent dendrimers and macrocycles with large free volume. J Org Chem 72(23):8683–8690
18. Cao J, Zhu XZ, Chen CF (2010) Synthesis, structure, and binding property of pentiptycene-based rigid tweezer-like molecules. J Org Chem 75(21):7420–7423

Chapter 5
Synthesis and Reactions of Heterotriptycenes and Their Derivatives

5.1 The Bridgehead-Substituted Heterotriptycenes

5.1.1 *Derivatives of Nitrogen Group Elements*

As early as 1895, von Dechend and Wichelhaus [1] reported probably the oldest known member in iptycene family, but they only provided the molecular formula ($C_{18}H_{12}N_2$) rather than the constitutional formula, thus it could not eliminate that the molecule was in other conformation instead of the D_{3h} Y-shaped rigid structure. In 1927, McCleland and Whitworth [2] reported a new member of iptycene family, arsatriptycene **1**, in which the carbon atoms of triptycene at bridgehead position were replaced by two arsenic atoms. This target molecule **1** was afforded by the coupling reaction of diazotized *o*-aminodiphenyl-arsinic acid with phenylarsine oxide, and then followed by the reduction with phosphorus trichloride. Later, Bickelhaupt [3] also reported that the reaction of *o*-phenylenemagnesium with arsenic trichloride easily took place to form **1**, but the reaction could not stay at the tri-Grignard stage for its strain-free structure (Scheme 5.1).

In the early 1960s, Wittig and Steinhoff [4–6] succeeded in the synthesis of benzoazatriptycene (**2**) and azatriptycene (**4**). As shown in Scheme 5.2, the Diels–Alder reaction between acridine *N*-oxide and 1,4-epoxy-1,4-dihydro-naphthalene, followed by the departure of oxygen, gave the adduct **3**, which then went through a dehydration process to afford the final product **2**. Soon after, they [5, 6] provided another route for the synthesis of azatriptycene (**4**) in 54 % yield by the treatment of acridine **5** with KNH_2 in liquid NH_3 via the ring-closing process (Scheme 5.3).

In 1976, Kreil and Sandel [7] reported the pK_a value of azatriptycene **4** to be 2.1, which was determined by the potentiometric titration with $HClO_4$ in anhydrous acetic acid at 25 °C. This method was established by Wegmann and Simon [8], and they utilized the relationship between half-neutralization data and aqueous pK_a values to get the pK_a value at 25 °C in anhydrous acetic acid. Several years later, Wepster and co-workers [9] obtained the pK_a value of **4**, which was 1.12 based on spectroscopic and potentiometric measurements in highly aqueous solution. This method needed only small and reliable corrections and a short extrapolation, which was different

C.-F. Chen, Y.-X. Ma, *Iptycenes Chemistry*,
DOI 10.1007/978-3-642-32888-6_5, © Springer-Verlag Berlin Heidelberg 2013

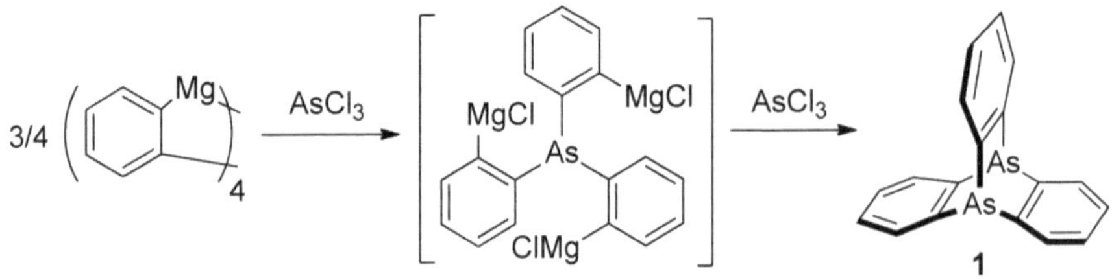

Scheme 5.1 Synthesis of arsatriptycene **1** from *o*-phenylenemagnesium

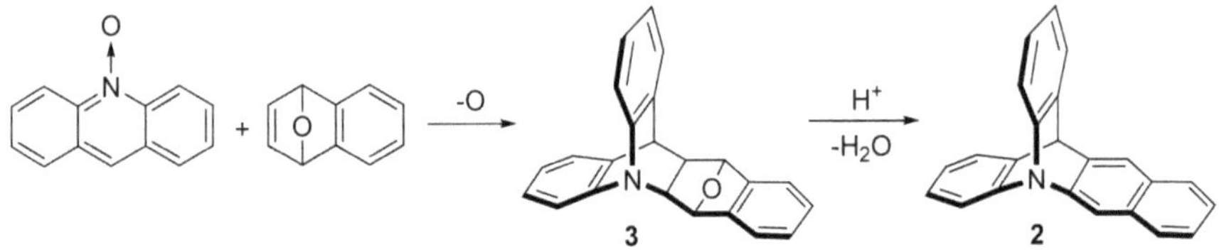

Scheme 5.2 Synthesis of benzoazatriptycene **2**

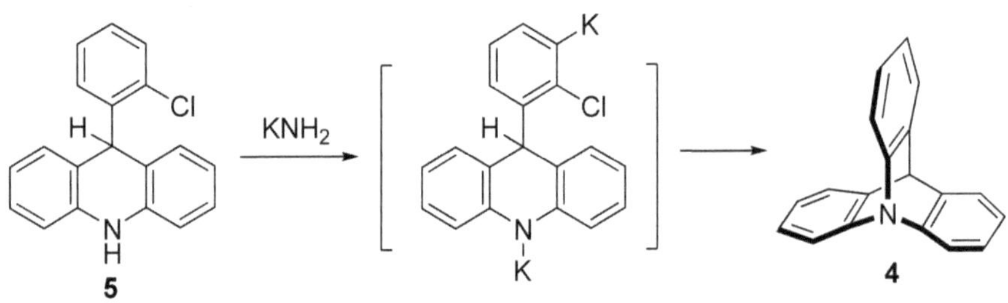

Scheme 5.3 Synthesis of azatriptycene **4**

from that of Wegmann and Simon. In addition, they also pointed out that the average value for the inductive effect of a phenyl group was 3.3 pK_a units, which was larger than the value (2.8 pK_a units) reported by Kreil and Sandel [7] because in the case of Kreil and Sandel a considerably lower pK_a value for quinuclidine was used [8].

The photochemical behavior of azatriptycene **4** was also investigated by Sugawara and Iwamura [10] in 1980. As shown in Scheme 5.4, the photoreaction of azatriptycene **4** via the nitrene intermediate gave the indenacridine **6** in 15 % yield.

In 1969, Hellwinkel and Schenk [11] first succeeded in the synthesis of azaphosphatriptycene **7** in 30 % yield by the treatment of tris(*o*-bromopheny)amine **8a** with *n*-butyl lithium to give the intermediate of tris(*o*-lithiopheny)amine **8b**, and then followed the reaction of **8b** with triphenyl phosphite in THF/ether (Scheme 5.5a). Several years later, they [12] also reported the synthesis of the trimethyl-substituted azaphosphatriptycene **12**. As shown in Scheme 5.5b, the reaction of phosphorus heterocycle **9** with the excess Grignard reagent **10** gave the intermediate **11**, which was then treated with potassium *t*-butoxide to afford the target azaphosphatriptycene **12**

Scheme 5.4 Photoreaction of azatriptycene **4**

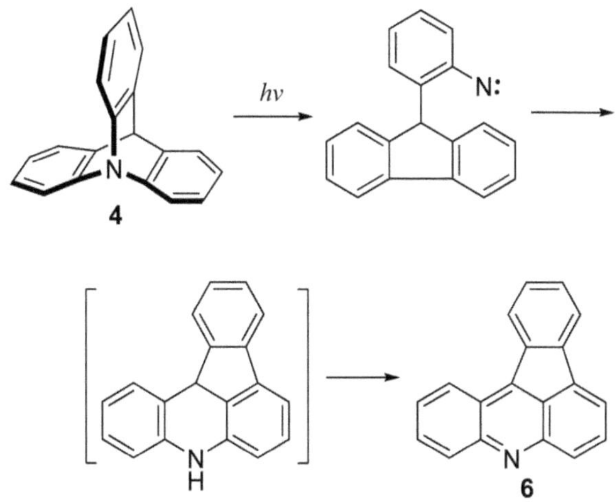

in 25 % yield. The further characterization of compound **7** revealed that its mass spectrum was similar to that of triptycene, and there was a particularly high proportion of doubly and triply charged ions in both the mass spectra of **7** and triptycene. Moreover, the UV spectrum of compound **7** had little resemblance to that of azatriptycene.

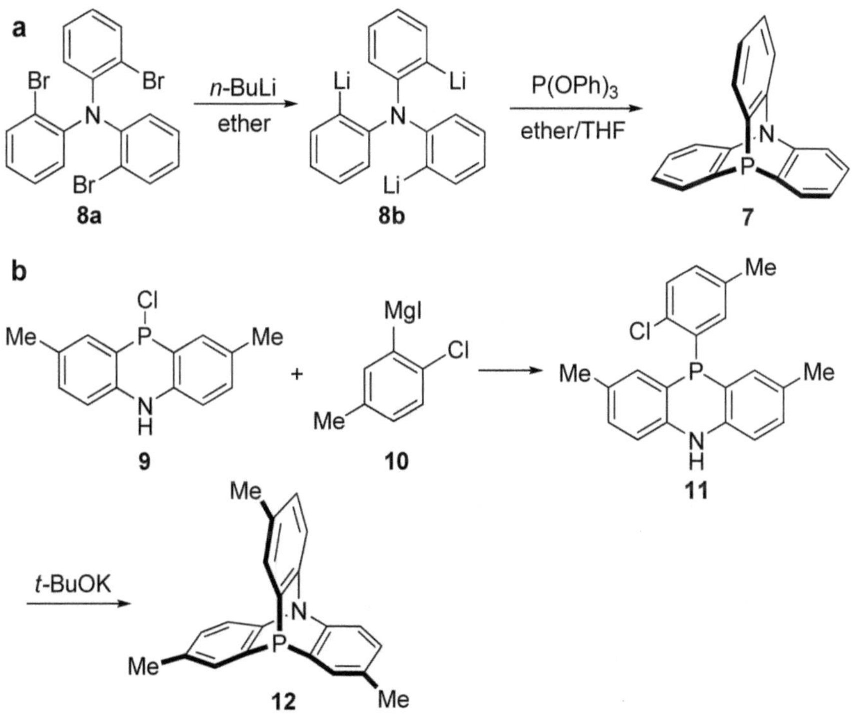

Scheme 5.5 Synthesis of **a** phosphatriptycene **7** and **b** trimethyl-substituted azaphosphatriptycene **12**

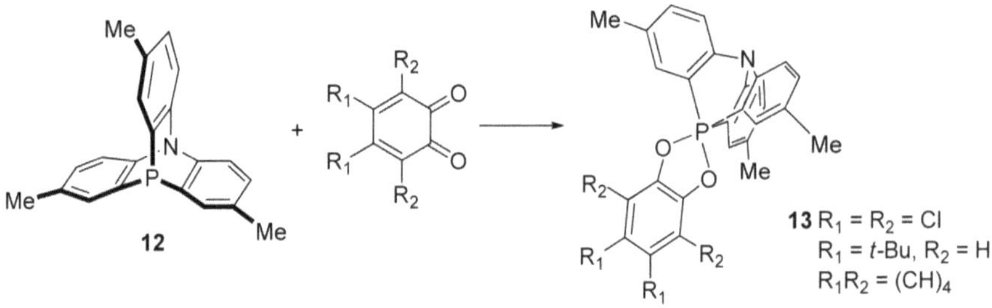

Scheme 5.6 Synthesis of spirophosphoranes **13**

Scheme 5.7 Synthesis of azarsatriptycene **14** from 2-chlorophenylarsine **15**

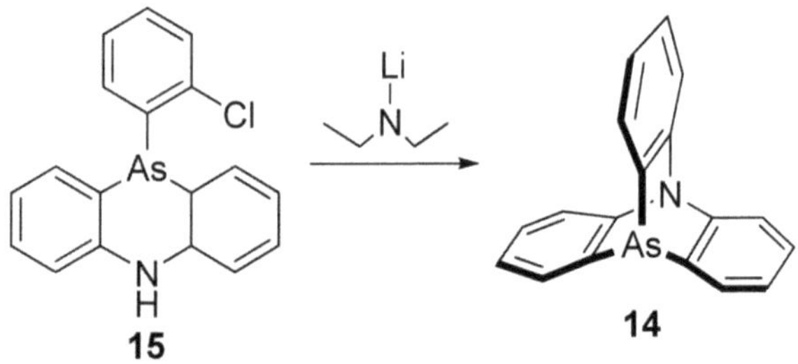

Furthermore, by the ring-closure reaction of trimethyl azaphosphatriptycene **12** with *o*-quinones at 170 °C without solvent, Hellwinkel et al. [13] and Osman and Samahy [14] obtained a series of spirophosphoranes (**13**, Scheme 5.6).

Similar to the Hellwinkel and Schenk's method for the synthesis of azaphosphatriptycene [13], Earley and Gallaghe [15] succeeded in the synthesis of the azarsatriptycene **14** in 48 % yield by the treatment of the 2-chlorophenylarsine **15** with lithium diethylamide in ether (Scheme 5.7). The UV spectrum of **14** was similar to that of azatriptycene and showed only weak absorption at above 220 nm; however, its nonresolved fine structure in the 265–280 nm region was differed from the weak band of azatriptycene **4** and triptycene.

Inspired by the work of azaphosphatriptycene, Weinberg and Whipple [16] reported a novel heterotriptycene system (**16**) containing two phosphorus atoms on the bridgehead positions, which was prepared by a one-step synthesis of *o*-dichlorobenzene with white phosphorus in a sealed glass tube at 280 °C in the presence of catalytic amount of ferric chloride (Scheme 5.8). After recrystallizing from tetrachloroethylene, the product **16** was obtained in 20 % yield. Furthermore, by treatment of diphosphatriptycene **16** with peracetic acid in ethyl acetate, the corresponding dioxide **17** was also obtained.

In 1972, Bickelha and co-workers [17] synthesized the oxide **18** in 68.7 % yield by heating compound **19** in poly-phosphoric acid up to 110 °C. The oxide **18** was further treated by SO_2/HCl to give arsatriptycene **20** with an arsenic atom at the bridgehead position (Scheme 5.9).

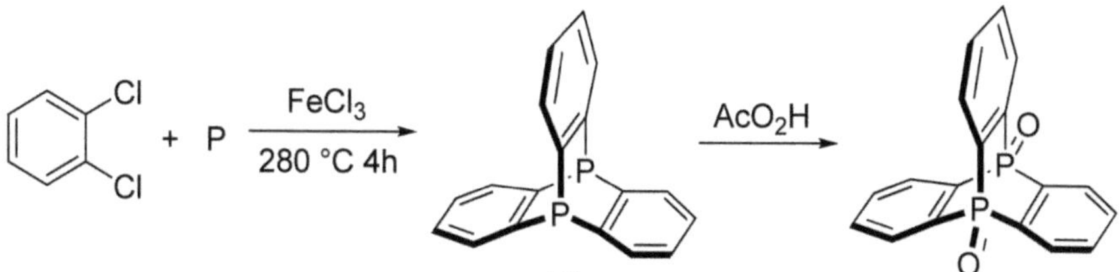

16 17

Scheme 5.8 Synthesis of dioxide **17**

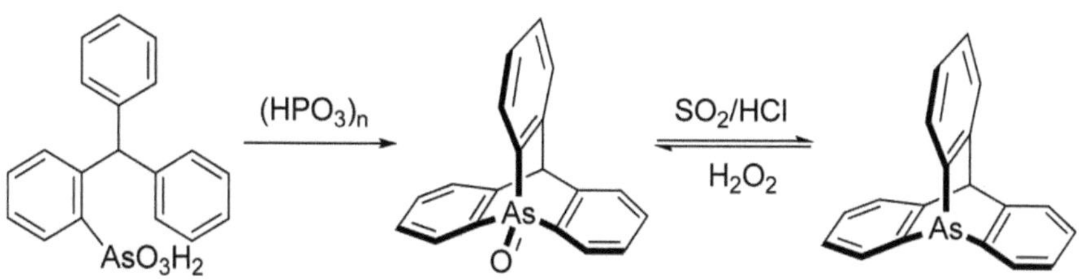

19 18 20

Scheme 5.9 Synthesis of arsatriptycene **20** from compound **19**

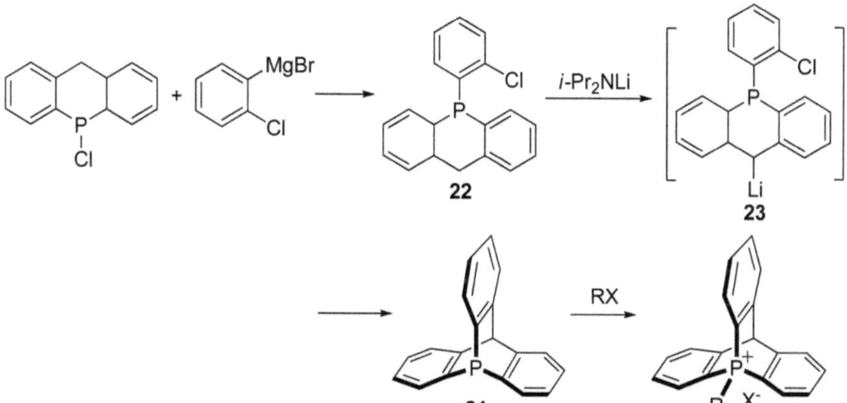

22 23 21

24a R = Me, X = I
24b R = CH_2Ph, X = Br

Scheme 5.10 Synthesis of phosphatriptycene **21** and its quaternary salts **24a, b**

After that, Jongsma et al. [18] obtained phosphatriptycene **21** in 35 % yield by the reaction of 9-(*o*-chlorophenyl)-9,10-dihydrophosphaanthracene (**22**) with the excess of lithium diisopropylamide in ether. In this progress, the intermediate **23** was probably involved (Scheme 5.10). Moreover, the reaction of compound **21** with an excess of methyl iodide or benzyl bromide gave the quaternary salts **24a** and

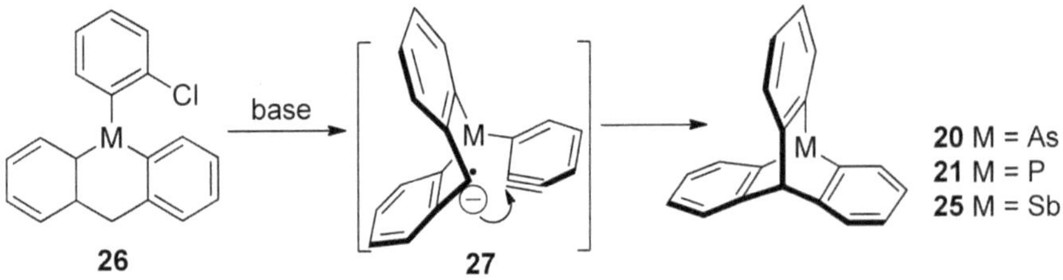

Scheme 5.11 Synthesis of heterotriptycenes **20**, **21**, and **25**

Fig. 5.1 Structure of 9,10-distibatriptycene **28**

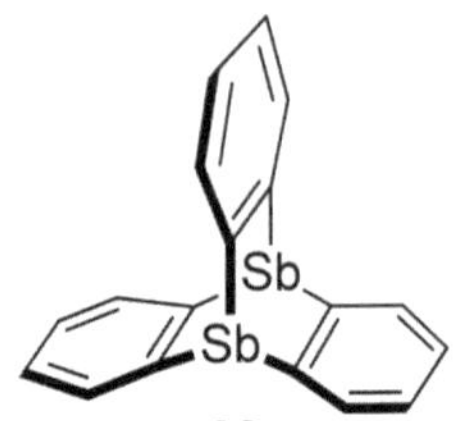

24b, respectively. It was noted that phosphatriptycene **21** could be oxidized by air in chloroform after 2 weeks, but its solid state was relatively stable under air.

To check the applicability of the above method for the synthesis of phosphatriptycene **21**, Jongsma et al. [19] further synthesized arsatriptycene **20** and stibatriptycene **25**. As shown in Scheme 5.11, the reaction of the corresponding arsenic compound **26** (M = As) with *o*-chlorophenyl magnesium bromide actually gave the product arsatriptycene (**20**). They deduced that the formation of the target compound **20** was the result of the attack of base to the carbanionic center in **27** on the benzyne. Thus, the treatment of compound **26** (M = As) with lithium piperidide instead of lithium diisopropylamide probably had higher tendency to form benzyne. As expected, the yield of **20** was up to 27 % with lithium piperidide as base. Similarly, stibatriptycene (**25**, M = Sb) could be prepared in 15 and 28 % yield, respectively, by the reaction of *o*-chloro-phenyl magnesium bromide or chlorophenyl trimethyltin with an excess of lithium piperidide. If lithium diisopropylamide was served as base, the yield of **25** (M = Sb) was only 8 %. These results confirmed that the rate of benzyne formation influenced the yield of the heterotriptycene to a certain extent.

Compared with stibatriptycene **25**, 9,10-distibatriptycene (**28**) had not been reported until 1985. Al-Jabar et al. [20] obtained the heterotriptycene **28** (Fig. 5.1) in 5 % yield by heating the mixture of *ortho*-phenylenemercury trimer and antimony powder. Actually, before compound **28** was reported, Massey and co-workers [21–23] had found that by simply heating the mixture of either arsenic or antimony powder and the corresponding 1,2-diiodotetra-halobenzene in sealed tubes, the perhalogenated diarsatriptycene and distibatriptycenes (**29**, Scheme 5.12) could be obtained, respectively.

In 1985, Al-Jabar and Massey [24] further reported a convenient and general method for synthesis of a series of fully fluorinated 9,10-disubstituted triptycenes

Scheme 5.12 Synthesis of perhalogenated diarsatriptycene and distibatriptycene

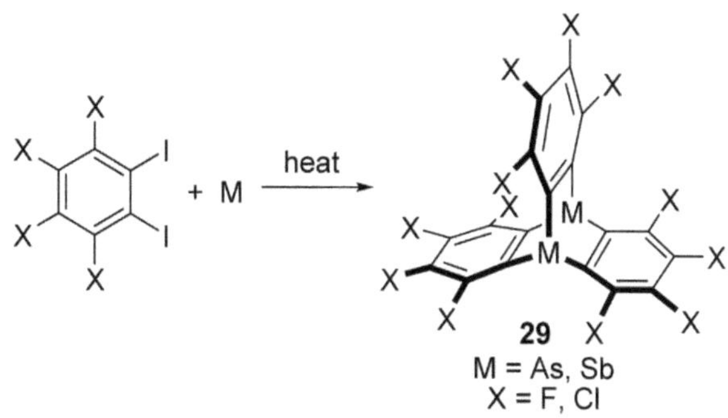

(**30**) containing the nitrogen family elements. Taking diarsatriptycene **30b** as an example, the reaction of tris(2-lithiotetrafluorophenyl)arsenic with arsenic trichloride gave the target product **30b** in only 4 % yield (Scheme 5.13). The authors considered that the low yield apparently arose from the formation of polymers via the reaction of tris(2–1ithiotetrafluorophenyl)arsenic with two, or three separated molecules of arsenic trichloride, instead of the single molecule required if the triple ring closure occurred [24]. Moreover, it was found that treatment of the tris(2-lithiotetrafluorophenyl)arsenic with antimony trichloride ($SbCl_3$) instead of $AsCl_3$ gave the mixed species of arsastibatriptycene **30g**. It was noteworthy that the intermediate $Bi(C_6F_4Br)_3$ was absolutely water-sensitive, it would readily lose all of the C_6F_4Br groups when it was exposed to air for a few hours. Thus, the product **30d** could be achieved in one-pot method by direct addition reaction of $BiCl_3$ to $Bi(C_6F_4Br)_3$ without prior isolation. In principle, this method was convenient and general, but the too low yield limited its practical applications.

Soon afterward, Al-Jabar and co-workers [25] also found that the treatment of *ortho*-phenylenemercury **31** with antimony powder gave the heat-sensitive compound **29** (M = Bi, X = Cl) in a very low yield. Similarly, the reaction of *ortho*-phenylene mercury with refined bismuth powder under 250 °C afforded 9,10-dibismutha-triptycene **32** and mercury. As triptycene would slowly decompose to tri- and hexa-phenylene at this temperature, the yield would reduce by prolonging

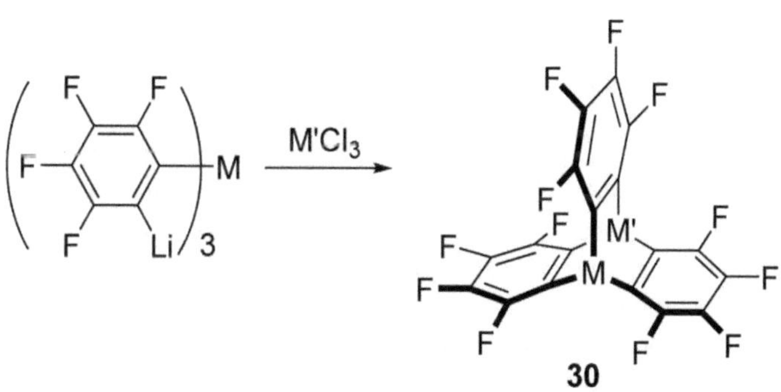

Entry	M	M'
a	P	P
b	As	As
c	Sb	Sb
d	Bi	Bi
e	P	As
f	P	Sb
g	As	Sb

Scheme 5.13 Synthesis of fully fluorinated 9,10-disubstituted triptycenes **30a–g**

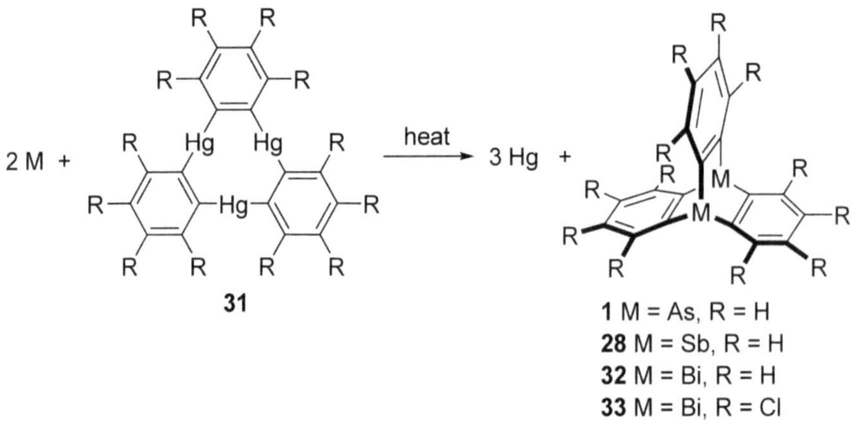

Scheme 5.14 Reaction of *ortho*-phenylenemercury **31** with metal powder

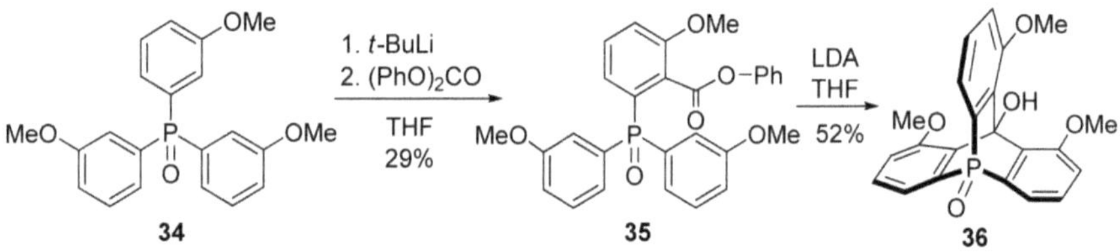

Scheme 5.15 Synthesis of phosphatriptycene oxide **36** via lithiation of compound **34**

the heating time. Heating arsenic instead, a bit of diarsatriptycene **1** could be detected. Both of these dimetallotriptycenes obtained by this method were rather unstable when associated with the decomposition, thus the yields were quite low for the synthesis themselves. However, the chloro-substituted dimetallotriptycenes (**33**, Scheme 5.14), which could be obtained by heating the mixture of metal and tetrachlorophenylenemercury trimer, seemed to be much more stable.

In 2004, Kobayashi et al. [26, 27] directly synthesized phosphatriptycene oxide **36** by the intramolecular cyclization of **35**, which was obtained by the reaction between the product of lithiation of **34** with diphenyl carbonate, as shown in Scheme 5.15.

In almost the same year, Agou et al. [28] reported a similar route to synthesize the 9-phosphatriptycene oxide **36** in an overall yield of 33 % via *ortho*-lithiation of a triaryl phosphine oxide, then followed by the introduction of an ester group and intramolecular ring-closing reaction in the presence of LDA (Scheme 5.16a). Similarly, the treatment of the lithio-derivative of **34** with $(PhO)_3P$ gave the compound **37**, which was then not isolated but directly reacted with an excess amount of LDA to give the heterotriptycene **38** in 6 % overall yield (Scheme 5.16b). The relatively low yield was probably caused by the side reaction from the nucleophilic attack of LDA to the phosphonite. Moreover, it was also found that treatment of compound **36** with a powerful methylation reagent, like Me_2SO_4, would afford the corresponding methyl derivative **39** in 94 % yield (Scheme 5.16c).

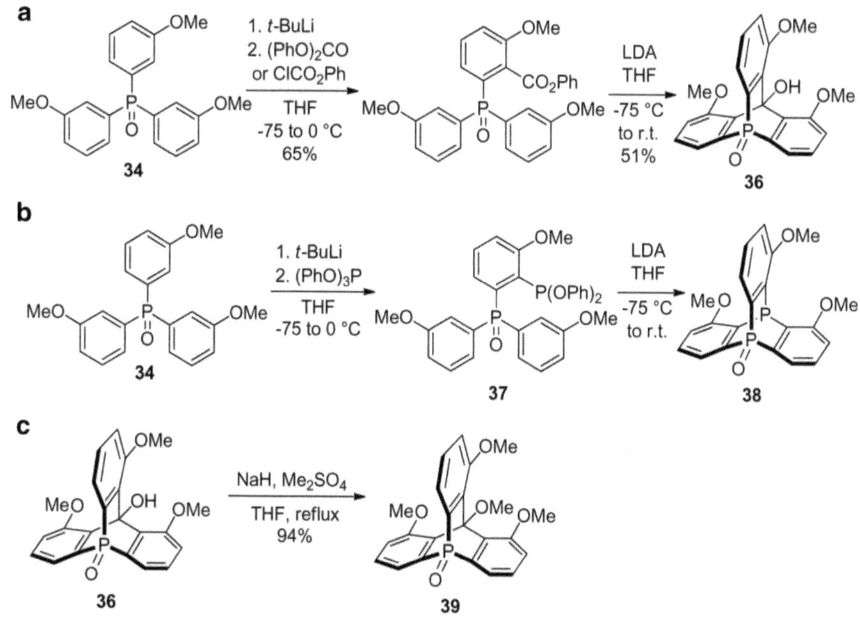

Scheme 5.16 Synthesis of compounds. **a 36**, **b 38**, and **c 39**

Recently, Mazaki and co-workers [29] attempted to introduce the alkyl groups into the skeleton of 9,10-diheteratriptycenes to enhance their solubility in common organic solvents. As a result, the intermediate tris(2-bromo-4,5-dimethyl-phenyl) phosphine **40** or tris(2-bromo-4,5-dimethylphenyl)-stibine **41** could be prepared in 38 and 59 % yield, respectively, by the reaction of 4,5-di-bromo-*o*-xylene with *n*-BuLi in a mixed solvent of THF/Et_2O under −110 °C, followed by the addition reaction with PCl_3 or $SbCl_3$ (Scheme 5.17a). Treatment of the intermediate **40** with *t*-BuLi and PCl_3 gave the product 9,10-diphosphatriptycene **42** in 7 % yield; whereas 9-phospha-10-stibatriptycene **43** was achieved in 13 % yield by the reaction of **41** with $SbCl_3$ in the presence of *t*-BuLi. Similarly, compound **41** reacted with *t*-BuLi and PCl_3 to give **43** in 10 % yield, whereas 9,10-distibatriptycene **44** were obtained in 9 % yield by the reaction of **41** and $SbBr_3$ in the presence of *t*-BuLi, which was shown in Scheme 5.17b.

In addition, the reaction of 9,10-diphosphatriptycene **42** with excess bromine in $CHCl_3$ gave the mono-(bromophosphonium) *P*-oxide salt **45**, which would easily be hydrolyzed by atmospheric moisture, and quantitatively converted to the quite stable compound *P*, *P*′-dioxide **46** (Scheme 5.18a) [29]. Similarly, the treatment of compounds **43** and **44** with bromine in CH_2Cl_2 or $CHCl_3$ gave the bromine adducts **47** and **48** [30], which were also very stable in the presence of air and moisture (Scheme 5.18b).

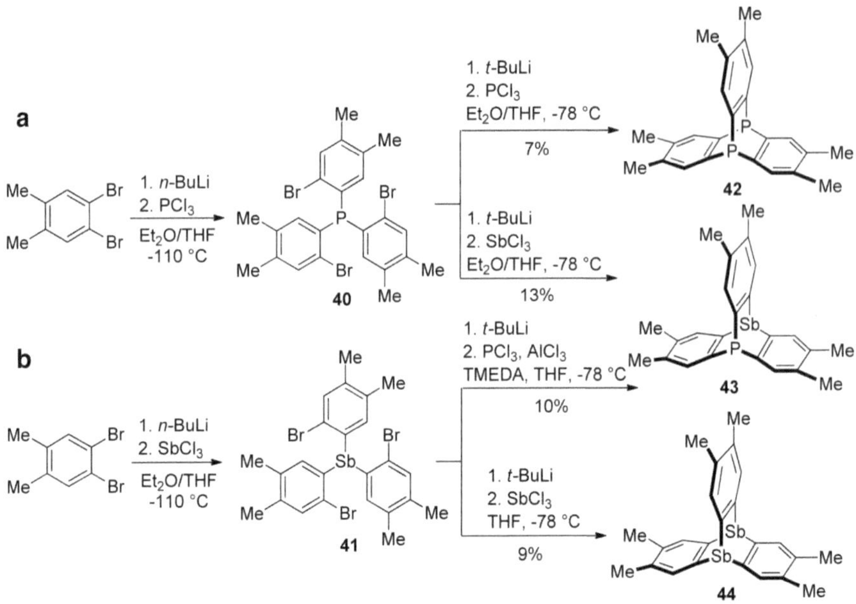

Scheme 5.17 Synthesis of 9,10-diheteratriptycenes **42**–**44**

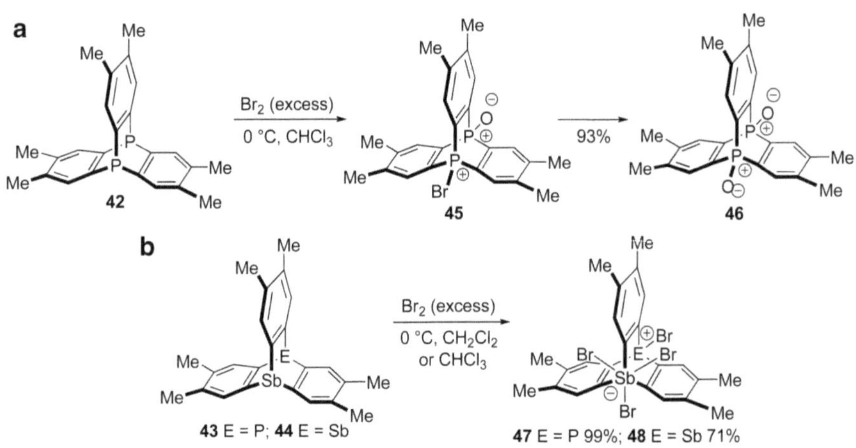

Scheme 5.18 Synthesis of **a** *P*, *P'*-dioxide **46** and **b** bromine compounds **47** and **48**

5.1.2 Derivatives of Carbon Group Elements

In the early stages, there were rare examples for the synthesis of the heterotriptycene with carbon group elements. Until 1985, Al-Jabar and Massey [24] reported the synthesis of disilatriptycene derivative **49** (Fig. 5.2) starting from tetrafluorinated

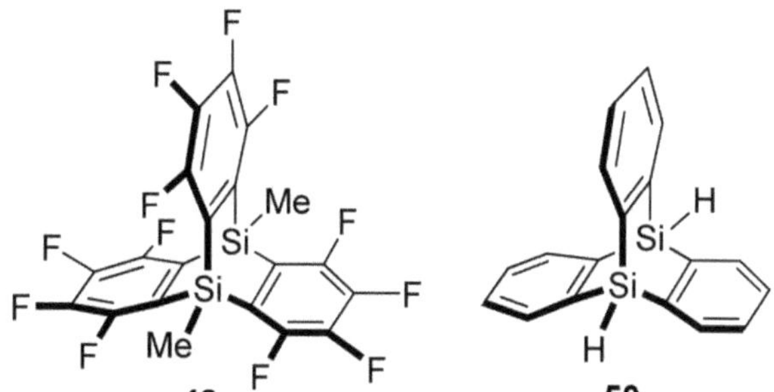

Fig. 5.2 Structures of disilatriptycenes **49** and **50**

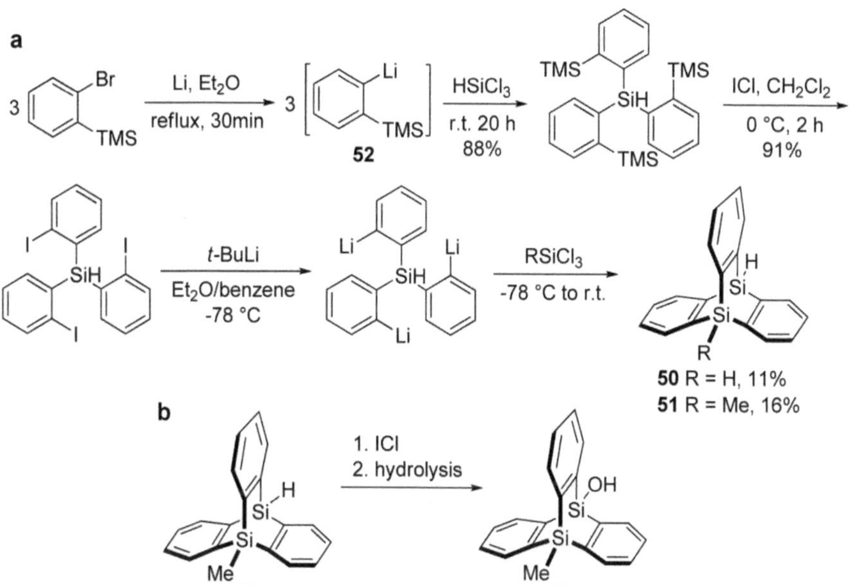

Scheme 5.19 Synthesis of 9,10-disilatriptycenes. **a 50**, **51** and **b 53**

o-phenylene. In the same year, Takahashi et al. [31] also synthesized disilatriptycene **50** (Fig. 5.2) by the co-pyrolysis of 5,10-dihydro-5,10-silaanthrene and *o*-dichlorobenzene.

In 1993, Takahashi et al. [31] reported a general and efficient route to synthesize 9,10-disilatriptycene **50** through the intermediate *o*-trimethylsilyl-phenyllithium **52** (Scheme 5.19a). This route successfully overcame the previous difficulty in preparing the suitable precursors. Moreover, the hydrolysis of the intermediate from **51** and ICl would afford compound **53** (Scheme 5.19b).

In 1996, Bickelhaupt and co-workers [32] reported a simple synthesis for a series of 9,10-dimetallotriptycenes of carbon group (like, M = Si, Ge, Sn) in reasonable yields. The one-pot reaction between *ortho*-phenylenemagnesium **54** and methylmetal trichlorides (**55**, $MeMCl_3$) in THF gave digermatriptycene **56** and

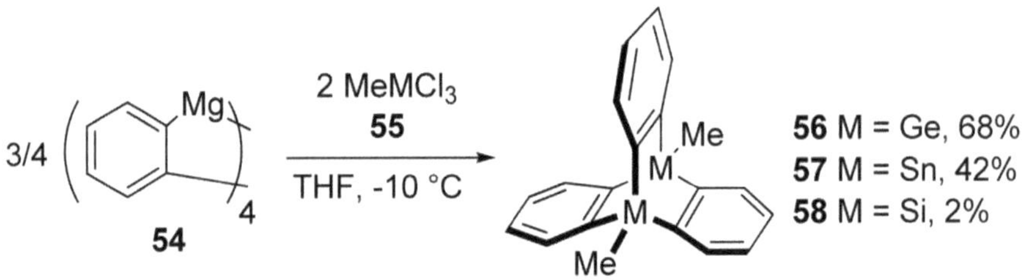

Scheme 5.20 Synthesis of 9,10-dimetallotriptycenes **56–58**

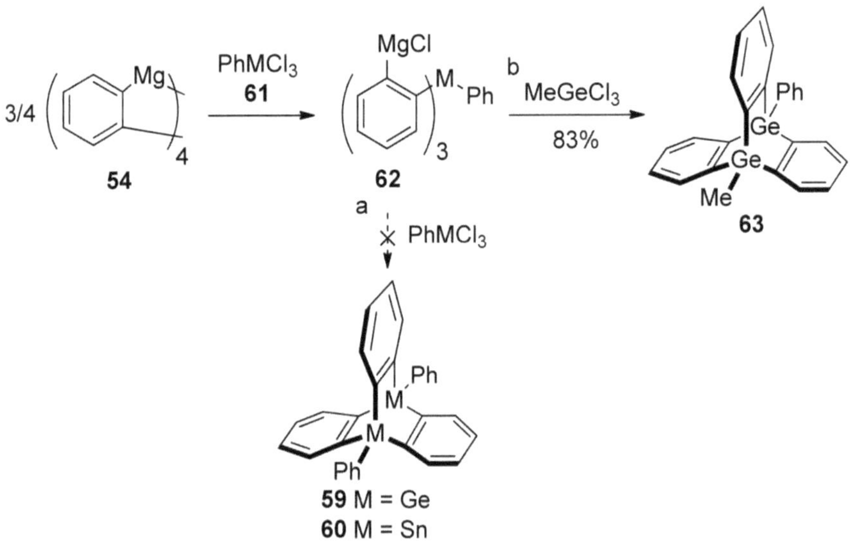

Scheme 5.21 Synthesis of digermatriptycene **63**

distannatriptycene **57** in 68 and 42 % yield, respectively; however, the yield of disilatriptycene **58** was only 2 % under the reaction conditions (Scheme 5.20). Followed by the similar synthetic strategy, they also attempted to obtain diphenyl-substituted dimetallotriptycenes (**59** and **60**) from the reaction of **54** with the phenyl-substituted metal trihalides (**61**, $PhMCl_3$). In consequence, the tri-Grignard reagent (**62**) could be easily obtained from **61**, but the subsequent reaction with another equivalent of **61** did not occur as expected (Scheme 5.21a). However, it was interestingly found that the subsequent reaction of tri-Grignard reagent (**62**, M = Ge) with methyl-substituted **55** (M = Ge) instead of phenyl-substituted **61** could give digermatriptycene **63** in 83 % yield (Scheme 5.21b).

On the basis of the results, it was further revealed that the process for formation of 9,10-dimetallotriptycenes was divided in two stages: firstly, the treatment of **54** with one equivalent of **55** (M = Ge, Sn) gave the corresponding tri-Grignard reagent; secondly, the dimetallotriptycene was obtained by the addition of another equivalent of **55** (M = Ge, Sn). Thus, a 1:1 mixture of $MeGeCl_3$ and $MeSnCl_3$ was treated with

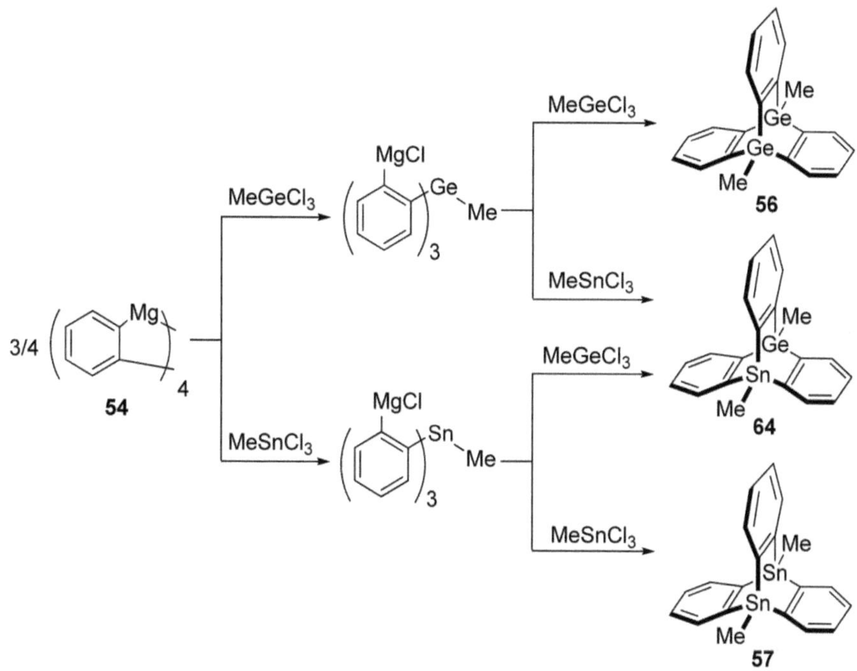

Scheme 5.22 Synthesis of 9,10-dimetallotriptycenes **56**, **57**, and **64**

Scheme 5.23 Synthesis of 9-germa-10-silatriptycene **65**

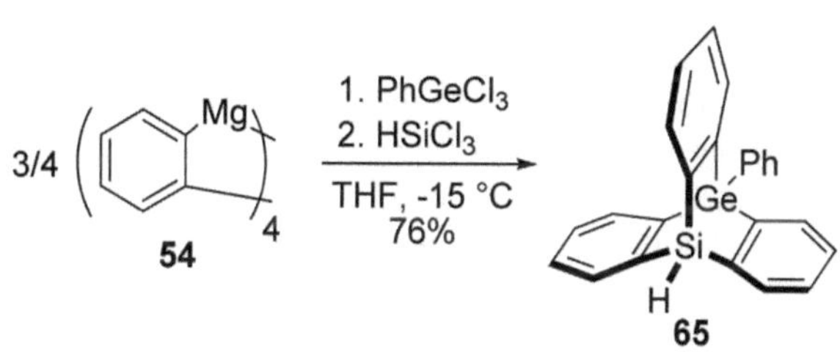

three equivalents of **54** to provide the products **56**, **57**, and **64** in a ratio of 5:25:70, as expected (Scheme 5.22).

Clearly, this way could extend to the synthesis of more mixed dimetallotriptycenes. In 1998, Bickelhaupt and co-workers [33] reported the synthesis of 9-phenyl-9-germa-10-silatriptycene **65** in 76 % yield by the stepwise one-pot approach from the reaction of trichlorophenylgermane with *ortho*-phenylenemagnesium (**54**) in THF under −15 °C, and then immediately treated with trichlorosilane without separation and purification, which is shown in Scheme 5.23.

In addition, various mixed dimetallotriptycenes (**66**) shown in Table 5.1 were also synthesized. For the diversity of yields, Bickelhaupt [3] deemed that metal and the

Table 5.1 Synthesis of mixed dimetallotriptycenes **66**

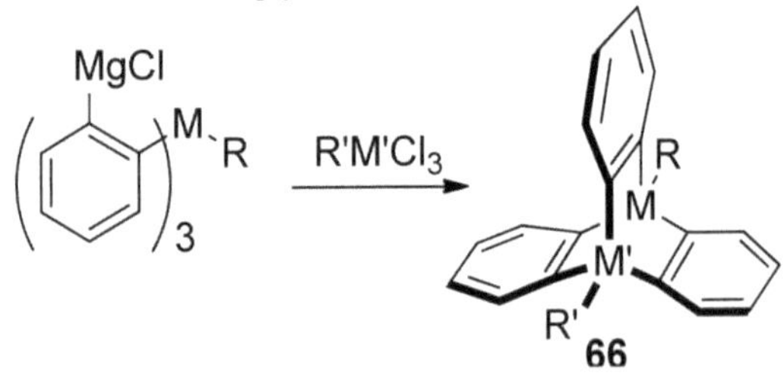

Entry	M	R	M′	R′	Yield (%)
1	Si	H	Si	H	20–30
2	Si	Me	Si	H	73
3	Si	Me	Si	Me	2
4	Ge	Me	Ge	Me	68
5	Ge	Ph	Ge	Me	83
6	Ge	Ph	Ge	Ph	0
7	Ge	Ph	Si	H	76
8	Ge	Ph	Si	Me	0
9	Sn	Me	Sn	Me	42
10	Sn	Ph	Sn	Ph	0

substituents in the dimetallotriptycenes had no influence on the formation of the tri-Grignard reagent at the first stage, but the steric factors were apparently decisive at the second stage to a great extent. Thus, the smaller silicon had only one methyl group, whereas, the larger germanium allowed ring closure with up to one methyl and one phenyl group, and two phenyl groups could not exist simultaneously. In the case of tin, it had the similar behavior to germanium. These results indicated that the second phenyl group in diphenyl-substituted dimetallotriptycenes (**66**, R = R′ = Ph) could not be introduced by this approach, which is not a consequence of these compounds being incapable of existence [3].

In addition, it was found that diphenyl-substituted silatriptycene **67** could be obtained by the reaction of disilatriptycene **50** with PhLi (Scheme 5.24a). Similarly, the mixed dimetallotriptycene **68** could also be achieved in 70 % yield by the reaction of **69** with an excess of phenyllithium in diethyl ether at room temperature via the intermediate **70** (Scheme 5.24b).

5.1.3 Other Bridging Atoms

The different elements from different groups could form the mixed dimetallotriptycenes via the route of the tri-Grignard stage as well [32]. Consequently, the mixed triptycenes **71a–c** containing one germanium atom and another atom of nitrogen group were obtained by the reaction of **72** with the corresponding trichloride [3], which is shown in Scheme 5.25.

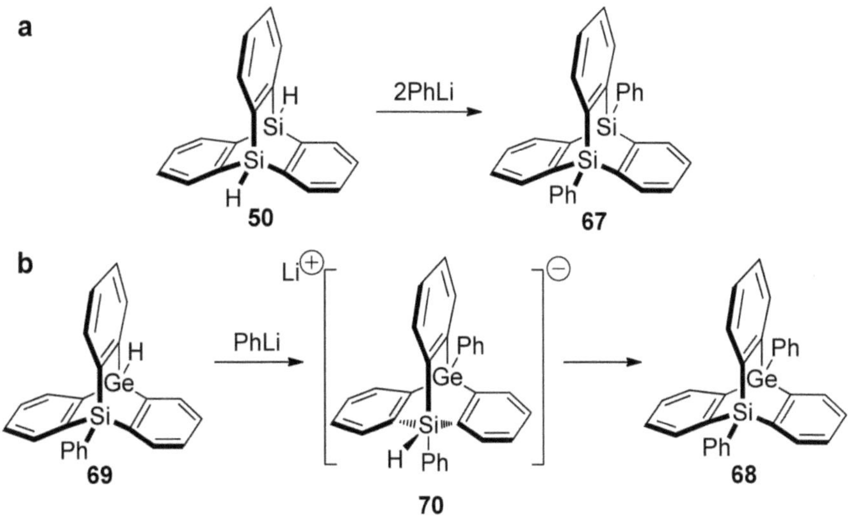

Scheme 5.24 Synthesis of **a** diphenyl-substituted silatriptycene **67**, and **b** mixed dimetallotriptycene **68**

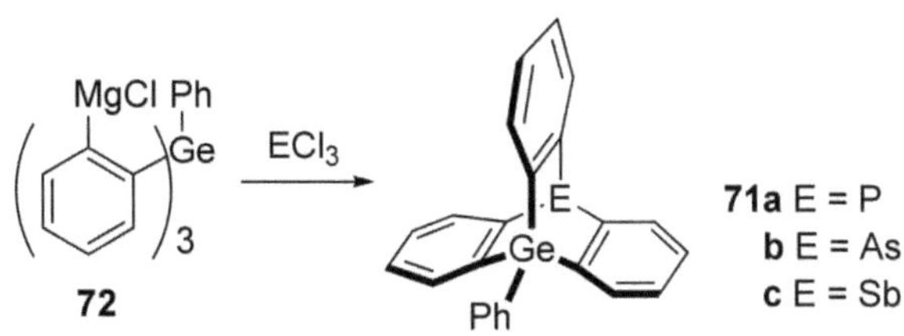

Scheme 5.25 Synthesis of mixed dimetallotriptycenes **71a–c**

Compared with the phosphatriptycene and phosphagermatriptycene, the synthesis of the phosphasilatriptycene seemed to face more problems. Until 2006, Tsuji et al. [34] reported the first synthesis of 9-phospha-10-sila-triptycenes **73** and their derivatives. The target products **73a, b** as moisture- and air-stable solids could be obtained in the reasonable yields by a three-step route starting from *o*-dibromobenzene. As shown in Scheme 5.26, the designed intermediate compound chloro-substituted phosphasilatriptycene **75a** was first achieved by the reaction of tris(2-bromo-phenyl) phosphine with *t*-butyl-lithium and then silicon tetrachloride, and then converted to fluorosilane **75b**. Compound **75b** was further treated with methylmagnesium bromide or *n*-dodecyl-magnesium bromide to give the corresponding **73a** and **73b** in 54 and 32 % yield, respectively. According to the similar route, a series of the phosphasilatriptycenes **76a–f** with various substituents on the bridgehead silicon atom could also be synthesized in 28–68 % yield starting from 1,2-dibromo-4,5-diisopropoxy-benzene (Scheme 5.27). In this process, the introduction of two isopropoxy groups into the benzene rings could enhance the solubility of the target phosphasilatriptycenes.

To clarify the structures and properties of the framework of phosphasilatriptycenes, a series of phosphine selenides (**77**, **78**) were also synthesized in good

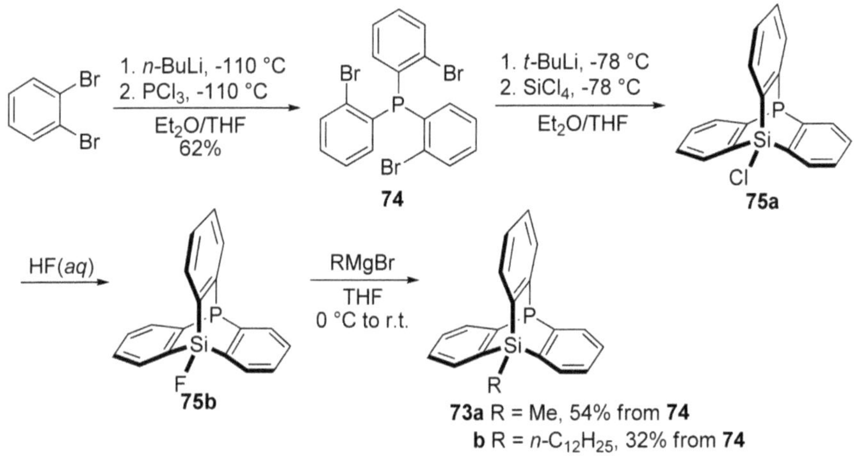

Scheme 5.26 Synthesis of phosphasilatriptycenes **73**

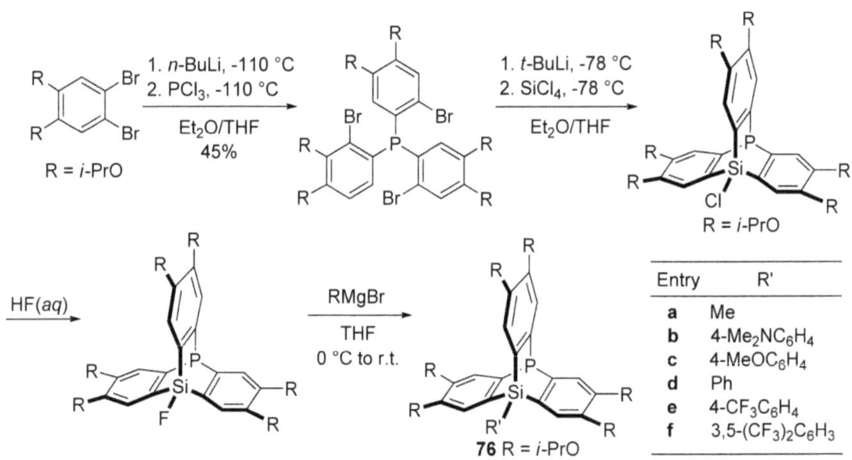

Scheme 5.27 Synthesis of phosphasilatriptycenes **76a–f**

yields by the reaction of the corresponding phosphasilatriptycenes (**79**, **80**) with selenium powder in $CHCl_3$ under reflux condition or in C_6D_6 at room temperature (Scheme 5.28).

A successful example of heterotriptycenes containing transition metals in the bridgehead positions was rarely reported. In 1999, Bickelhaupt [3] reported the synthesis of **81** in up to 50 % yield by the reaction of **54** with $CpTiCl_3$ (Scheme 5.29), which was determined with ^1H NMR spectroscopy. However, he failed to isolate compound **81** as pure form since the compound would decompose above −5 °C.

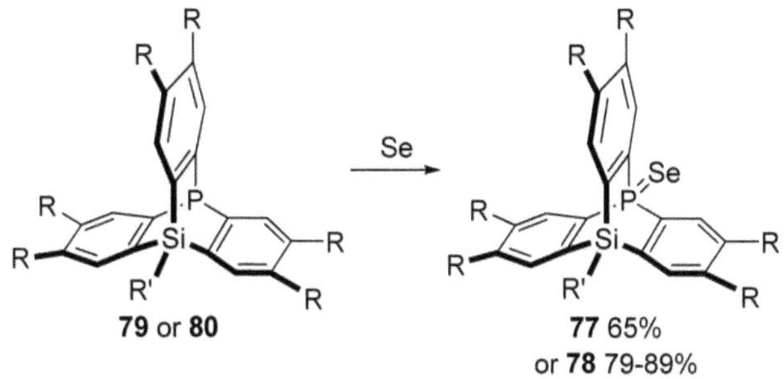

77, **79** R = H, R' = Me
78, **80** R = *i*-PrO

Entry	R'
a	Me
b	4-$Me_2NC_6H_4$
c	4-$MeOC_6H_4$
d	Ph
e	4-$CF_3C_6H_4$
f	3,5-$(CF_3)_2C_6H_3$

Scheme 5.28 Synthesis of phosphine selenides **77** and **78**

Scheme 5.29 Synthesis of compound **81**

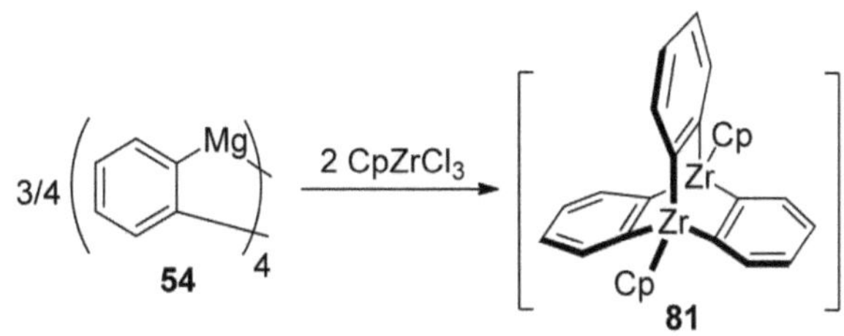

5.2 The Heterotriptycenes with Heterocycles

5.2.1 Derivatives of Nitrogen-Containing Heterocycles

In 1980s, Skvarchenko et al. [35, 36] first reported the synthesis of 2-azatriptycene. Afterward, Quast and Schon [37] extended this route to the synthesis of 1-azatriptycene from 9,10-bridged anthracene via the intermediate **83**, which was obtained in 55 % yield by the reaction between the enamine **84** and phenyl vinyl ketone (Scheme 5.30a). It was found that treatment of the intermediate **83** with NH_2OH in *n*-butanol could produce the oximes **86** and **87**, which then led to the 1-azatriptycene derivative **82** and the by-product **85**, respectively (Scheme 5.30b). Moreover, two derivatives of 1,8-diazatriptycene (**88**, R = *t*-Bu; R = Ph) were also synthesized by the reaction of diazanthracene and benzyne generated in situ (Scheme 5.30c). For 1,8-diazatriptycene (**88**, R = H), they failed to synthesize it. But it was noteworthy that the benzoazatriptycene derivative **89** could be obtained by the reaction of 9,10-bridged anthracene and *o*-aminoacetophenone as shown in Scheme 5.30d [37, 38].

In 1991, Markgraf et al. [39] synthesized the cycloadduct **90** from the 1-azaanthracene **91** which was obtained by the Skraup cyclization reaction of 1-chloro-2-aminonaphthalene. However, the reduction of the adduct **90** with neither magnesium nor chemically activated magnesium failed to obtain 1-azatriptycene **92**. Finally, it was found that 1-azatriptycene **92** could be achieved in 47 % yield only by the treatment of **90** with tri-*n*-butyl tin hydride (Scheme 5.31).

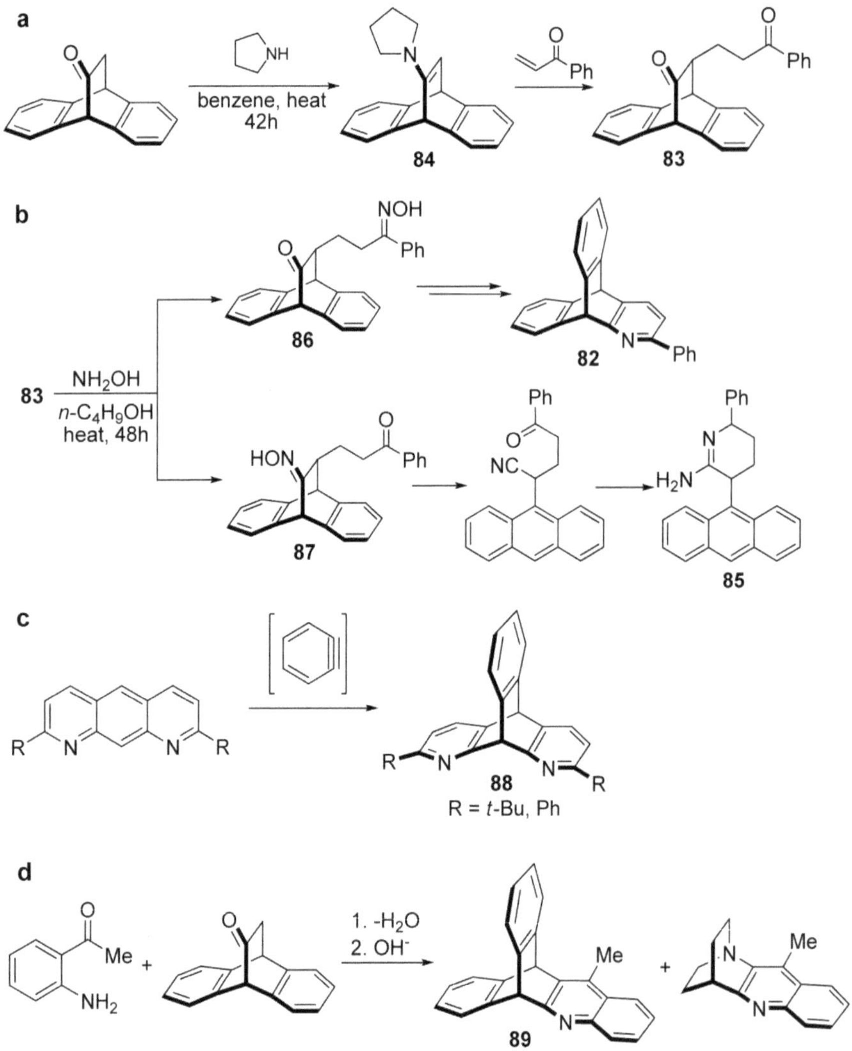

Scheme 5.30 Synthesis of 1-azatriptycene from 9,10-bridged anthracene via the intermediate **83**

Besides iptycene-derived pyridines, Gorgues and Le Coq [40] obtained the iptycene-derived pyridazine **94** in 1979 based on the common route to the synthesis of pyridazine rings. As shown in Scheme 5.32, the treatment of the starting materials **93** in hot ethanol with an alcoholic solution of hydrazine hydrate in few seconds gave the product **94** in 75 % yield.

In 2007, Swager and co-workers [41] reported a new route for the preparation of iptycene-derived pyridazines. As shown in Scheme 5.33, the saponification and dehydration of the adduct **95** (**a**, $R_1 = R_2 = H$) produced the bicyclic maleic anhydride **96** (**a**, $R_1 = R_2 = H$), which was then reacted with hydrazine monohydrochloride of

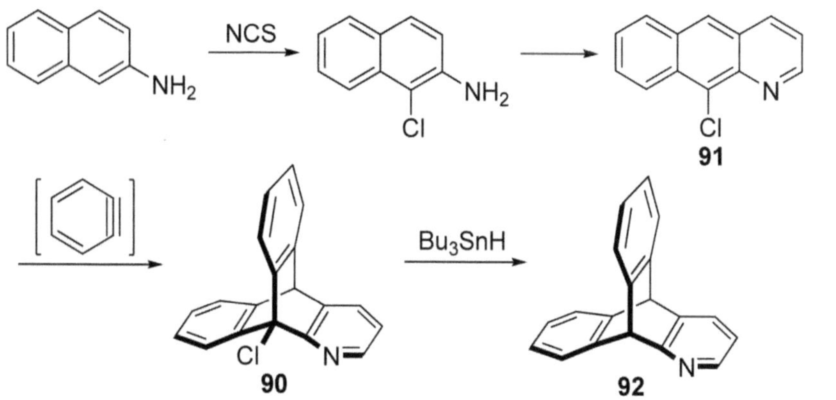

Scheme 5.31 Synthesis of 1-azatriptycene **92**

Scheme 5.32 Synthesis of iptycene-derived pyridazine **94**

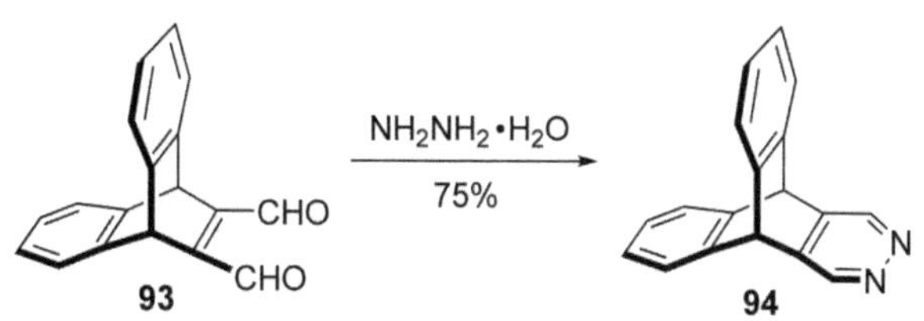

acetic acid under reflux condition to give the hydrazide in a high yield. The further treatment of the hydrazide with neat phosphorus oxychloride or molten phosphorus oxybromide gave the dihalides **97a** and **97d**. Finally, the selective S_NAr reaction of **97** with sodium methoxide afforded compound **98** in an excellent yield without chromatographic purification.

Following the synthetic route of iptycene-derived pyridazines, Swager and coworkers [41] further synthesized iptycene-derived phthalazines. Consequently, the addition of exocyclic diene **99** to DMAD, and then followed by the oxidative rearomatization with potassium permanganate in the presence of a phase-transfer catalyst afforded the dimethyl phthalate **100**. It was noteworthy that **100** could directly react with hydrazine to give the target phthalhydrazide **101**, which was then treated with sodium methoxide to produce the methoxy chloride **102** (Scheme 5.34).

Interestingly, it was found that the pyridazine ring showed sufficiently electron deficiency, and was activated for the oxidative addition with standard palladium catalysts even for the nominally less reactive chlorides [41]. Thus, they further synthesized a series of the symmetrically and unsymmetrically 3,6-disubstituted iptycene-derived pyridazines (**104a–d, 105a–d**) from the newly synthesized halopyridazines (**103**) by the cross-coupling reactions (Table 5.2).

Furthermore, it was found that the cross-coupling intermediate **106** of the 3-chloropyridazines **103** ($X_1 = OMe$, $X_2 = Cl$), which beared an electron-donating methoxy group could easily convert to the corresponding chloropyrizadine **107**

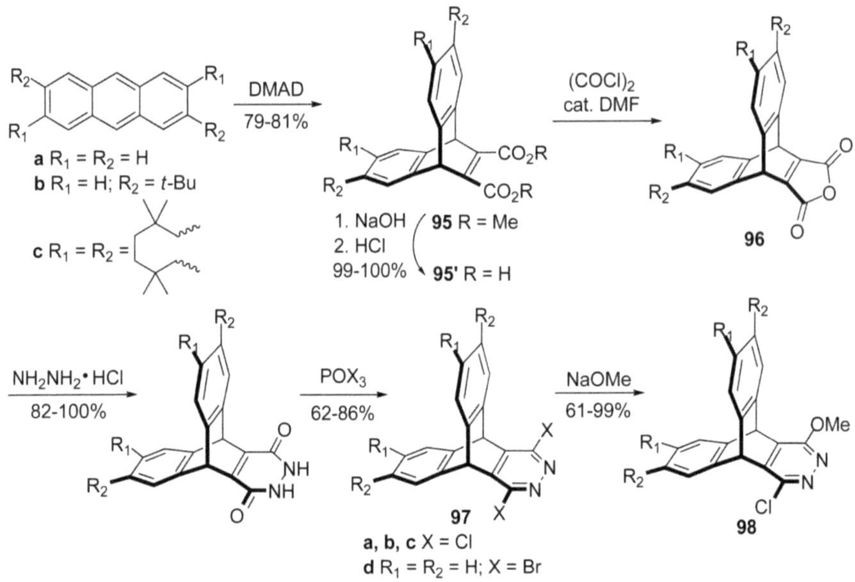

Scheme 5.33 Synthesis of iptycene-derived pyridazines **98**

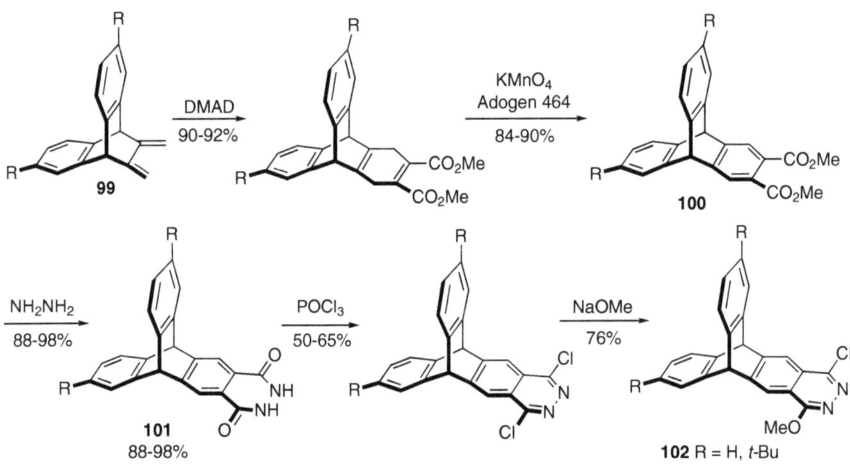

Scheme 5.34 Synthesis of methoxy chloride **102**

(Scheme 5.35a). In contrast, the hydrodechlorination of compound **103**, followed by the deprotection and chlorination could afford the monosubstituted pyridazine **108** (Scheme 5.35b).

Actually, as early as 1971, Fields et al. [42] reported the efficient addition of benzyne, which was generated from anthranilic acid, diazotized in situ, to azonianthracene **109** gave a series of azoniatriptycene-type adducts (**110**) in reasonable

Table 5.2 Synthesis of 3,6-disubstituted iptycene-derived pyridazines by the cross-coupling reaction

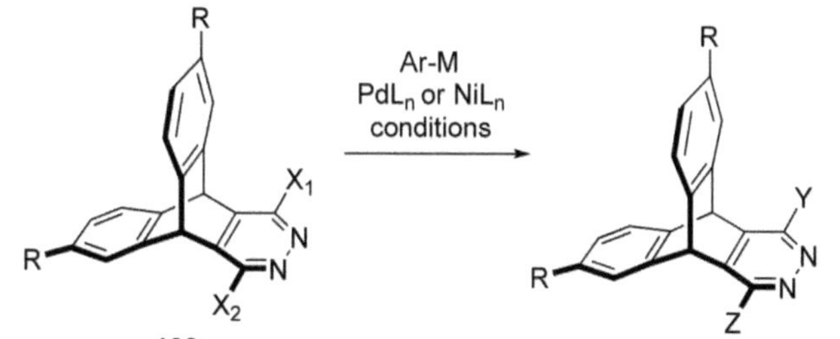

103 **104-105**

Entry	X_1, X_2 (**103**)	Ar-M	Y, Z	Yield (%)
1	$X_1 = X_2 = Cl$	$PhSnMe_3$	**104a** Y = Z = Ph	54
	$X_1 = X_2 = Cl$	$PhB(OH)_2$	**104a** Y = Z = Ph	83
2	$X_1 = X_2 = Cl$	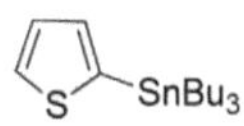	**104b** Y = Z = 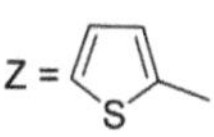	95
3	$X_1 = X_2 = Cl$	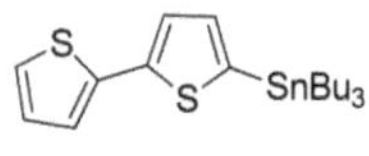	**104c** Y = Z = 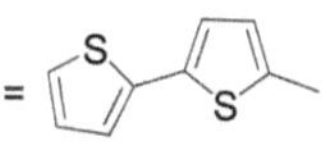	86
	$X_1 = X_2 = Cl$	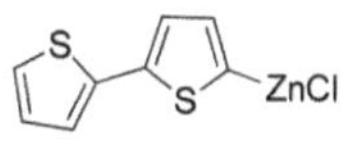	**104c** Y = Z = 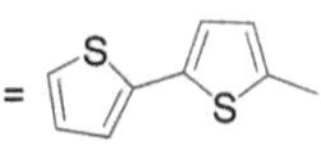	77
4	$X_1 = X_2 = Cl$	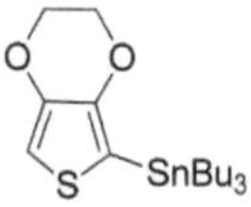	**104d** Y = Z = 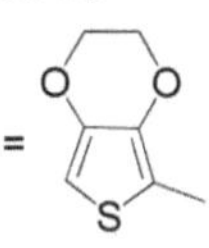	85
5	X_1 = OMe X_2 = Cl	$PhB(OH)_2$	**105a** Y = OMe Z = Ph	91
6	X_1 = OMe X_2 = Cl	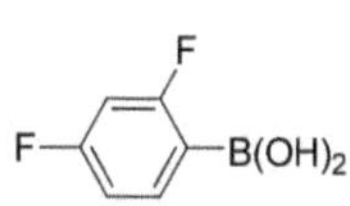	**105b** Y = OMe, Z = 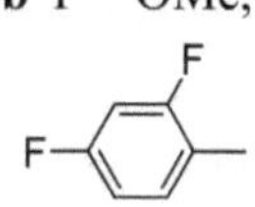	37
7	X_1 = H, X_2 = Cl	$PhB(OH)_2$	**105c** Y = H, Z = Ph	64
8	X_1 = H, X_2 = Cl	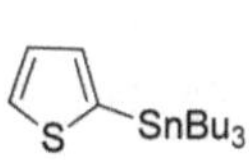	**105d** Y = H, Z = 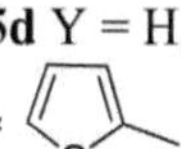	78

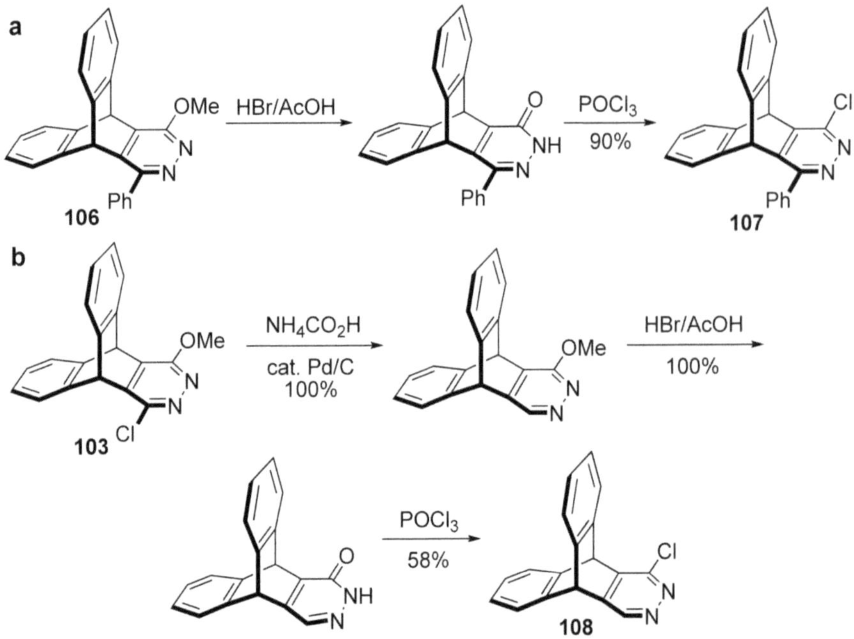

Scheme 5.35 Synthesis of **a** chloropyrizadine **107** and **b** monosubstituted pyridazine **108**

yields, and the results are summarized in Table 5.3. It was noteworthy that the pure target azoniatriptycene salt could be simply precipitated in ether without column chromatography and recrystallization.

Besides the six-membered nitrogen heterocyclic ring, Scheffer and Ihmels [43] reported the synthesis of the triptycene containing the pyrrole ring in 1997. As shown in Scheme 5.36, the pyrrole derivative **111** was unexpectedly obtained in 50 % yield by the irradiation of ground crystals of ethenoanthracene **112**, which might be the resulting compound of the oxidization of the pyrroline **113** in air.

In 2006, Chong and MacLachlan [44] synthesized a new class of triptycene derivatives **114** and **115** containing a quinoxaline moiety by the condensation of polyamino-triptycenes **116** and **117** with 2,3-dihydroxy-1,4-dioxane (Scheme 5.37). The compounds **114** and **115** with the rigid "wings" were good precursors for generating the porous structures.

Soon afterward, they [45] further made the use of triptycene to build a series of ligands containing pyrazine groups **118** and **119** from quinone **120** (Scheme 5.38). By the optimization study, the targets **118** and **119** could be obtained in ethanol instead of THF in 61 and 68 % yield, respectively. However, the yield of **121** could be improved to be 35 % in the presence of a catalytic amount of piperidine.

Recently, Chen and co-workers [46] reported a convenient and efficient method for the synthesis of a series of pentiptycene-derived rigid tweezer-like molecules incorporating nitrogen-containing heterocycles. As shown in Schemes 5.39 and 5.40,

Table 5.3 Synthesis of azoniatriptycene-type adducts **110**

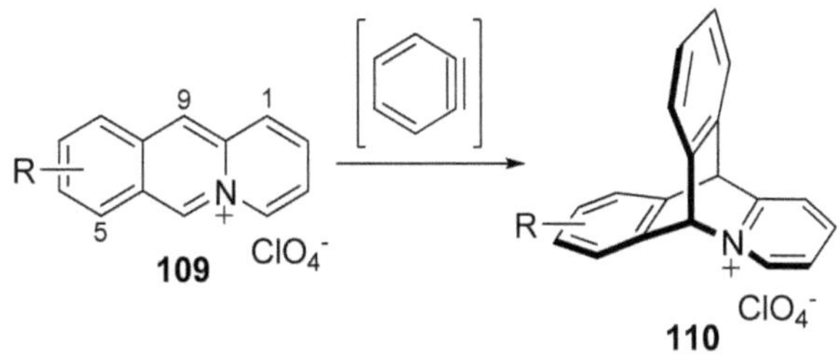

Entry	Substrate (**109**)	Yield (%)
1	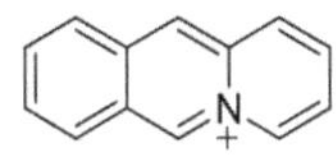	78
2	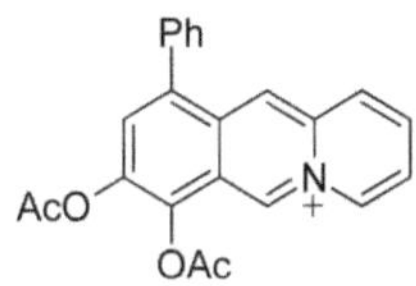	67
3	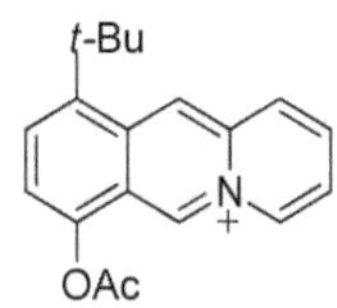	75
4	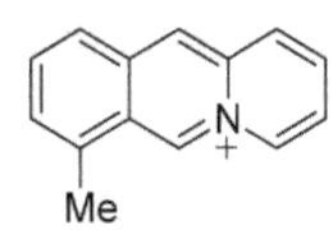	75
5	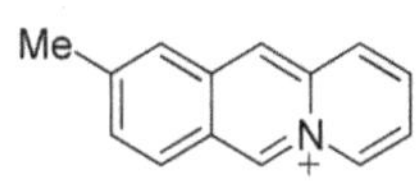	72
6	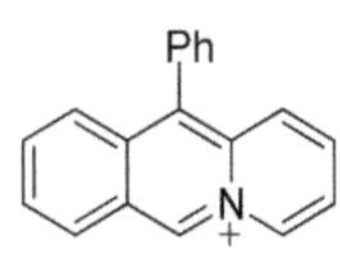	59
7	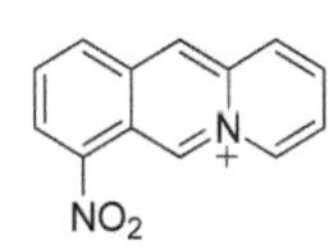	55

Scheme 5.36 Synthesis of pyrroline **113**

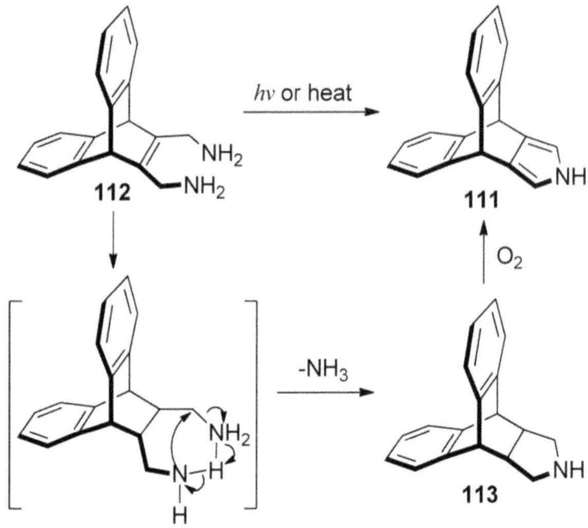

the condensation of *o*-diaminobenzene or *o*-diaminotriptycene with the corresponding *o*-quinones **122–125** in refluxing ethanol afforded the target molecules **128–139** in the reasonable yields (58–64 %). The particular topology structures made these tweezer-like molecules promising candidates for the applications in host–guest chemistry.

According to the similar approach, Jiang and Chen [47] also reported the synthesis of a series of 1,10-phenanthroline-based extended triptycene derivatives, which

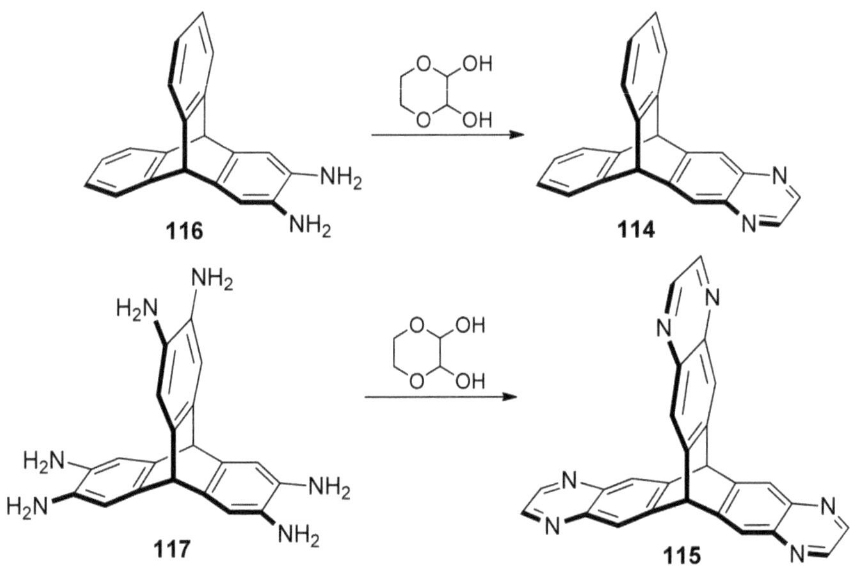

Scheme 5.37 Synthesis of triptycene derivatives **114** and **115** containing a quinoxaline moiety

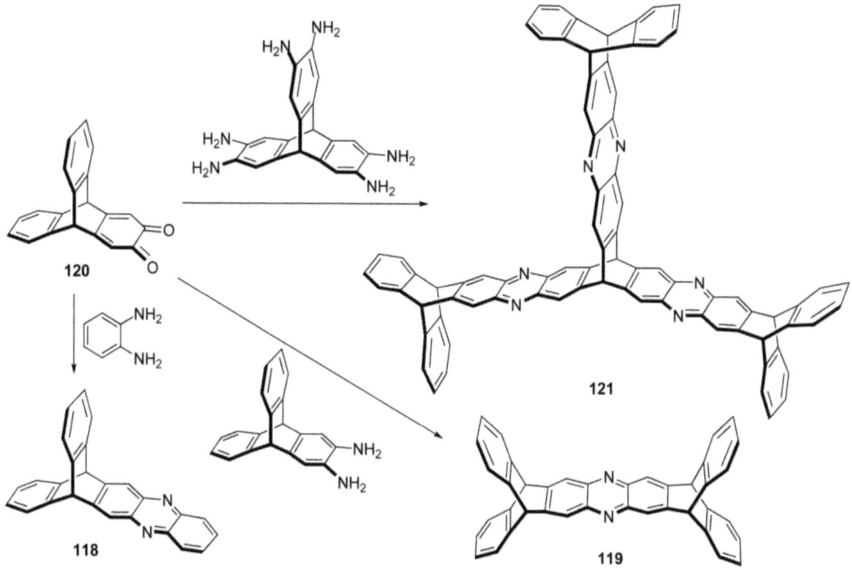

Scheme 5.38 Synthesis of ligands containing pyrazine groups **118**, **119**, and **121**

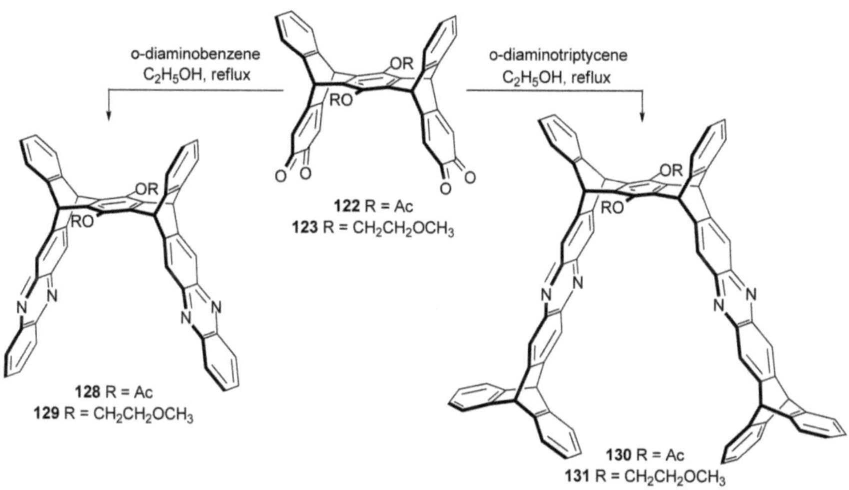

Scheme 5.39 Synthesis of pentiptycene-derived rigid tweezer-like molecules **128–131**

formed a new rigid scaffold geometry and generated large IFVs. As shown in Scheme 5.41, the extended triptycene derivatives (**140a–d**) with good solubility in chloroform and dichloromethane were obtained in good yields by the condensation of **141** and 2,3-diaminotriptycene **116** in methanol under refluxing. When compound **141c** or **141d** was reacted with 2,3,6,7,14,15-hexaaminotriptycene **117**, the corresponding

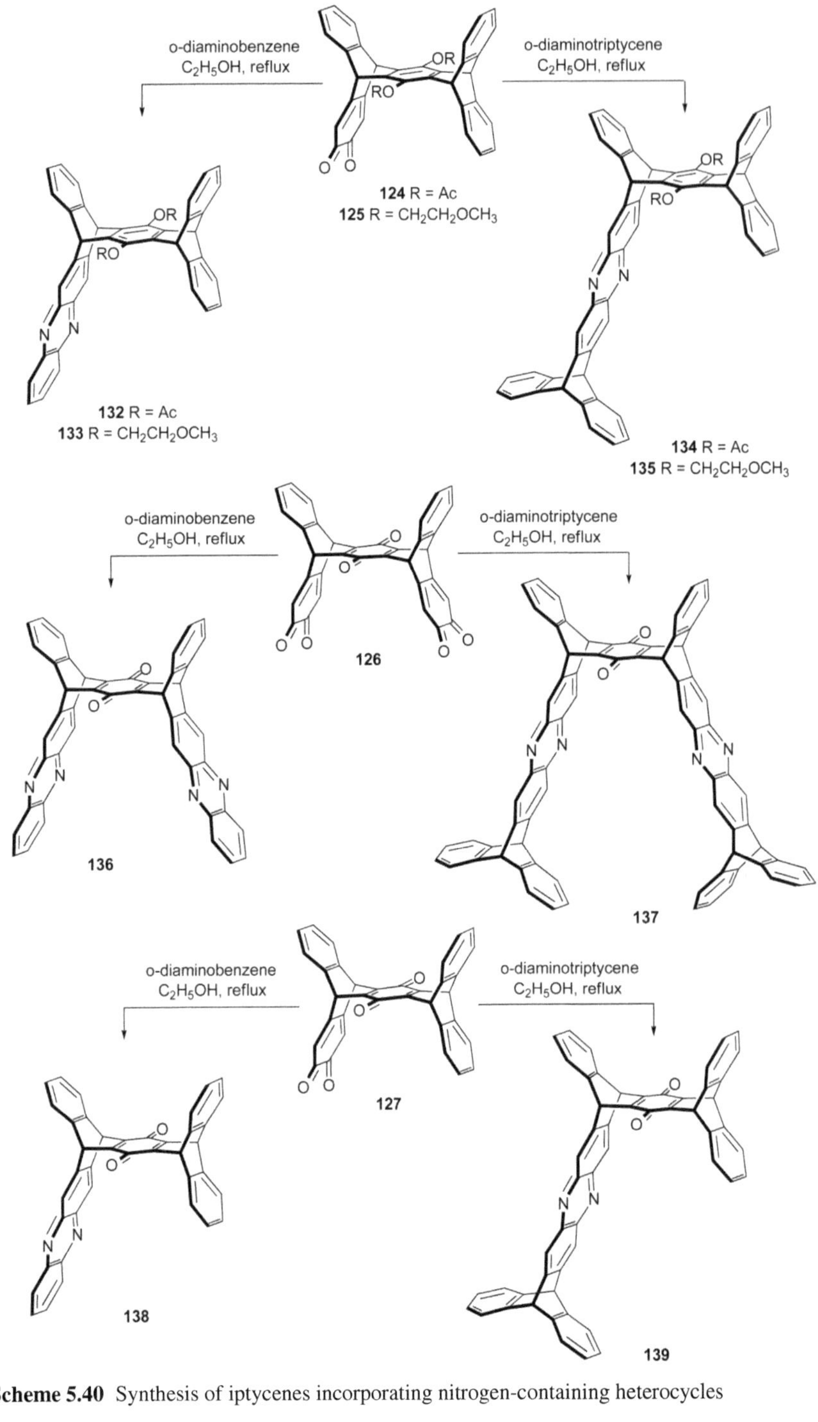

Scheme 5.40 Synthesis of iptycenes incorporating nitrogen-containing heterocycles

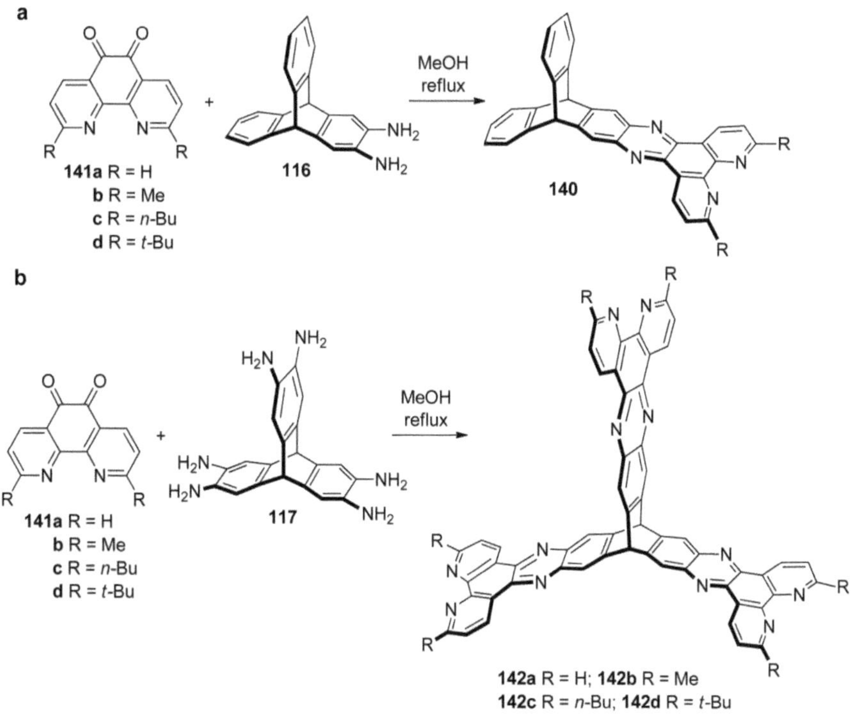

Scheme 5.41 Synthesis of extended triptycene derivatives. **a 140** and **b 142**

extended triptycene derivatives **142c** and **142d** were afforded in 28 and 32 % yield, respectively. However, the extended triptycene derivatives **142a** and **142b** showed poor solubility in common organic solvents; thus, it was difficult to confirm them by ^{1}H NMR and ^{13}C NMR spectra, but they could be detected by the MALDI-TOF MS spectra, which was different from those of **142c** and **142d**.

Recently, Mastalerz et al. [48] reported a facile method for the synthesis of an air-stable hexaammoniumtriptycene **143**. By the various condensation reactions based on molecule **143**, the corresponding benzimidazole, benzotriazole, and quinoxaline derivatives could be obtained in high yield, which is shown in Scheme 5.42.

5.2.2 Derivatives of Sulfur-Containing Heterocycles

To make the triptycene analogues with particular heterocyclic system's character, McKinnon and Wong [49] first introduced the thiophene ring to triptycene in 1971, and the results are summarized in Table 5.4. The Diels–Alder reaction of the *trans*-dibenzoylethylene with a variety of substituted anthracenes afforded the corresponding dibenzodibenzoylbicyclooctadienes (**148**), which could then be converted

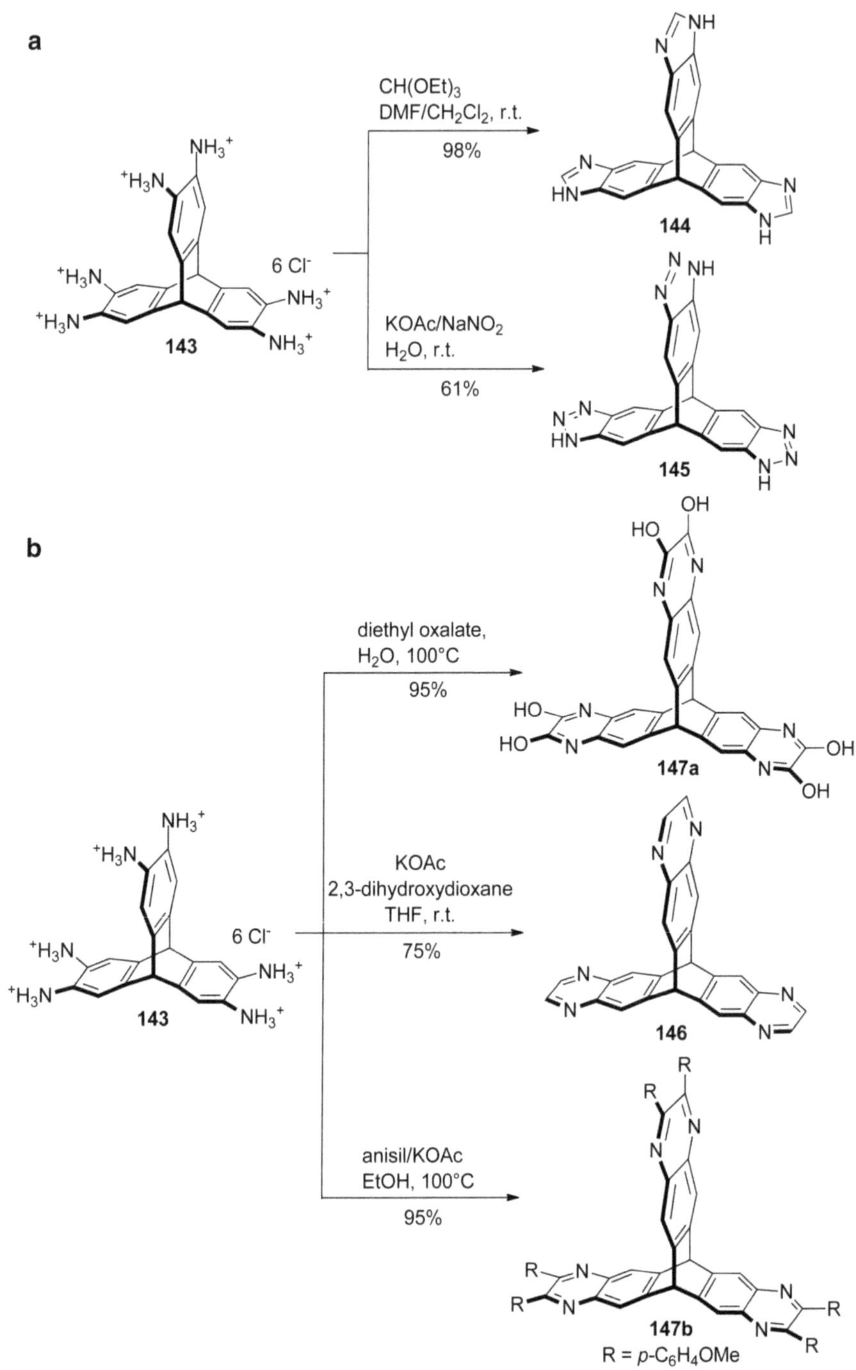

Scheme 5.42 Synthesis of iptycene-based benzimidazole, benzotriazole, and quinoxaline derivatives

Table 5.4 Synthesis of thiophene derivatives (**149**)

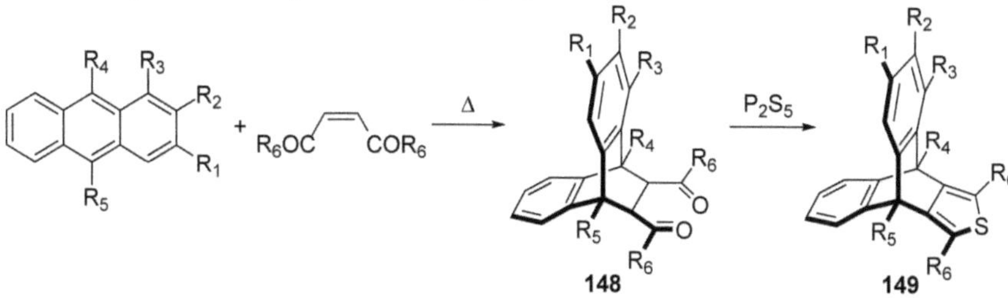

Entry	Yields (%)	R_1	R_2	R_3	R_4	R_5	R_6
1	56	Ph	Ph	Ph	Ph	Ph	Ph
2	88	H	H	H	H	Me	Ph
3	98	H	H	H	Me	Me	Ph
4	70	H	Me	H	H	H	Ph
5	43	H	H	H	H	OMe	Ph
6	12	H	H	H	H	Br	Ph
7	19	H	(–CH=CH–)$_2$	(–CH=CH–)$_2$	H	H	Ph
8	67	(–CH=CH–)$_2$	(–CH=CH–)$_2$	H	H	H	Ph
9	41	H	H	H	H	H	Me
10	55	H	H	H	H	Me	Me

to the target thiophene derivatives (**149**) in reasonable yields by its reaction with phosphorus pentasulfide in pyridine.

In 1973, De Wit and Wynberg [50] failed to synthesize the thiophene **150** directly by the reaction of disodium salt with P_2S_5. But they developed another route to prepare the desired product **150**. As shown in Scheme 5.43, the adducts **151** and **152** were firstly prepared by the reaction of anthracene with diacetylethylene. It was found that the reaction of anthracene and the *cis*-diacetylethylene gave a mixture of *cis*-adduct **152** and *trans*-adduct **151** in 82 % overall yield; whereas, the compound **151** could be obtained in 78 % yield from the *trans*-diacetylethylene. The dimethyl-substituted heterotriptycene **153** was then produced by the ring-closure reaction between **151** or the mixture of **150** and **151** with H_2S/HCl. Finally, following several steps of reactions, compound **153** could covert to the unsubstituted thiophene derivative **149**. According to the similar method, the extended triptycenes containing one naphthalene moiety and one thiophene ring, **154** and **155**, could also be synthesized in reasonable yields (Fig. 5.3).

At the same time, De Wit and Wynberg [50] pointed out that the triptycene containing a thiophene ring might also be obtained by the reaction of 3,4-didehydrothiophene and anthracene. Till 1996, Wong and co-workers [51] made that idea come true. Consequently, the reaction of 3,4-didehydrothiophene and anthracene provided the mixture of the 9,10-adduct **150** and the 1,4-adduct **156**, but the total yield was only 10 % (Scheme 5.44). It was noted that compounds **150** and **156** were chromatographically inseparable; however, through careful partial recrystallization of the mixture from methanol, the pure product **150** could be obtained.

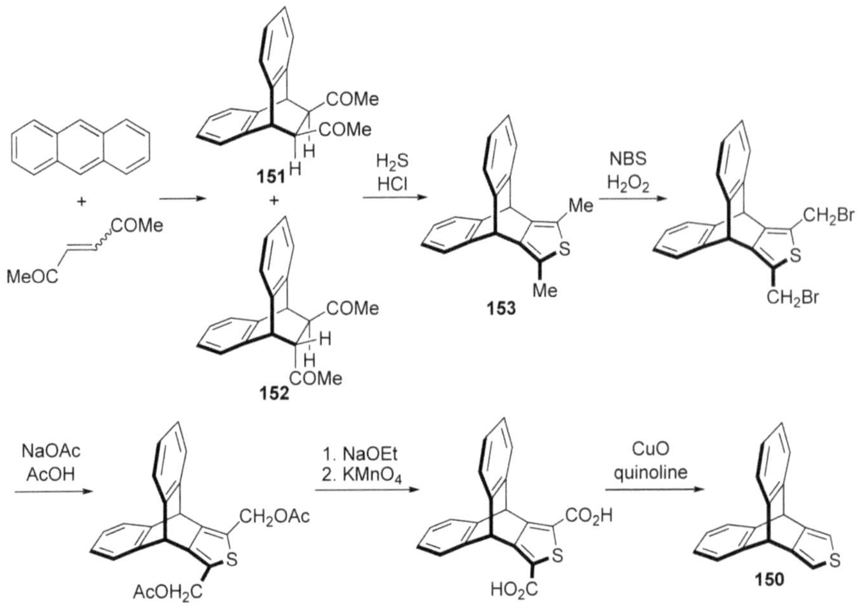

Scheme 5.43 Synthesis of thiophene **150**

Fig. 5.3 Structures of the compounds **154** and **155**

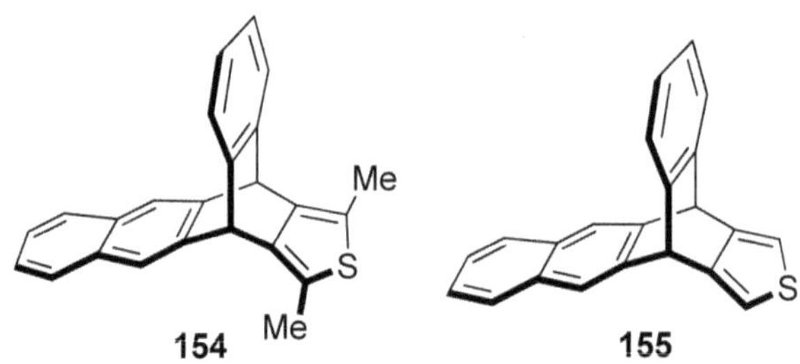

In 1970, Wynberg et al. [52] depicted the synthesis of the unsubstituted heterocyclic asymmetric triptycene **155** containing a 2,3-thiophene moiety by the reaction of naphtho[[2],[3]-*b*]thiophene **157** and 1,4-epoxy-1,4-dihydro-naphthalene **158**, then following the isomerization in the presence of hydrochloric acid and acetic acid (Scheme 5.45).

In 1986, Meth-Cohn and van Vuuren [53] reported a new versatile class of sulphimides: thiophene *S*,*N*-ylides. These *S*,*N*-ylides could be used as dienophiles to react with anthracene or 9,10-dimethylanthracene in benzene solution under reflux condition, which afforded the adducts **159a**, **b** in reasonable yields (Scheme 5.46). Under the similar reaction conditions, the adduct **159c** would be obtained by the addition of *S*,*C*-ylide with the corresponding anthracene derivative. In addition, the reduction of adducts **159a**, **b** with zinc dust in hot methanol would give the corresponding triptycene analogues (**160**) in essentially quantitative yields.

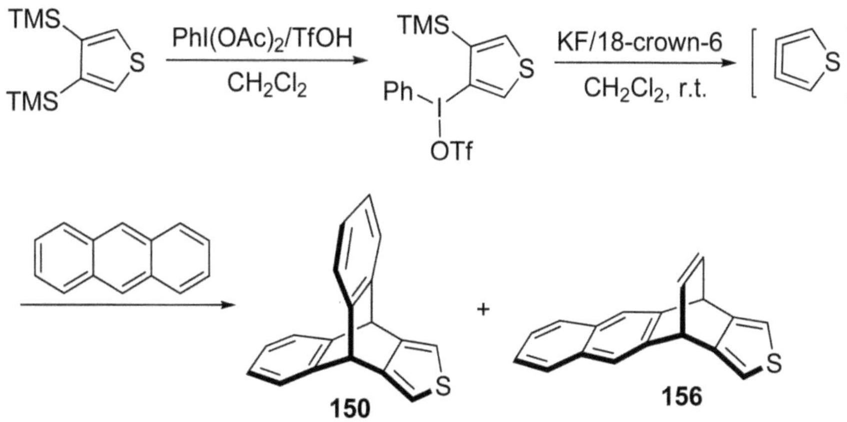

Scheme 5.44 Synthesis of thiophene **150** and its analogue **156**

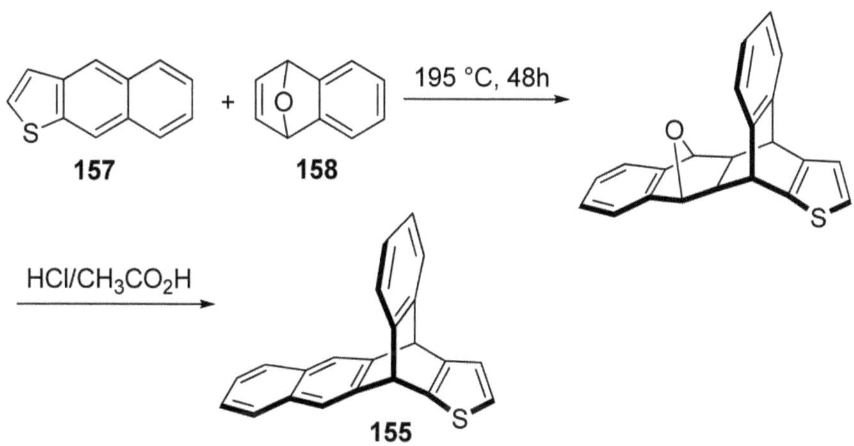

Scheme 5.45 Synthesis of unsubstituted heterocyclic asymmetric triptycene **155**

In 1991, Nakayama and co-workers [54] synthesized the hydroxyl tetramethyl-substituted heterotriptycene **161** containing three 2,3-thiophene rings orientated in the same direction via several steps of reactions starting from 4-bromo-2-methylthiophene. As outlined in Scheme 5.47, the 4-bromo-2-methylthiophene was treated by the lithiation with *n*-butyl lithium, and then reacted with diethyl carbonate to afford the alcohol **162** as an intermediate compound. The crude alcohol **162** without purification immediately reacted with 60 % perchloric acid in acetic anhydride, and then followed by the reaction of MeMgI to give compound **163** in 16 % yield from the alcohol intermediate compound **162**. The compound **163** was then brominated with bromine in a mixture of acetic acid and carbon tetrachloride solution to obtain the tribromide **164** in 96 % yield. Furthermore, after tribromide **164** was lithiated with *tert*-butyllithium in THF at −78 °C, and then followed by the reaction with diethyl carbonate, the target heterotriptycene **161** was afforded in 14 % yield

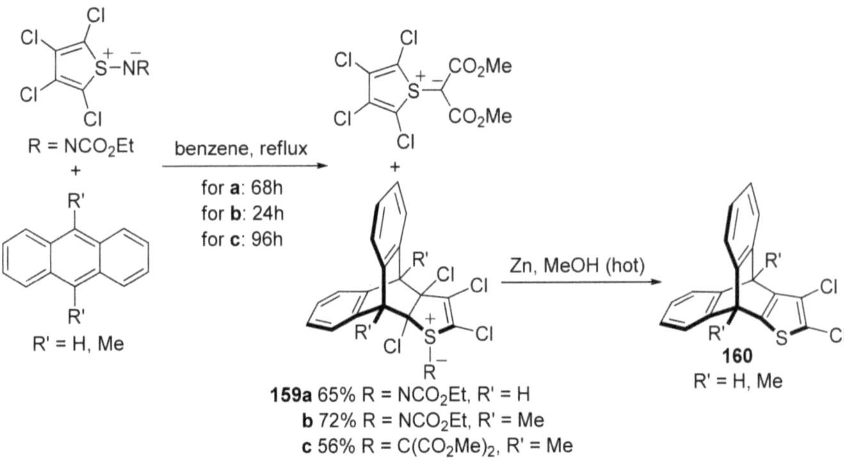

Scheme 5.46 Synthesis of compound **160**

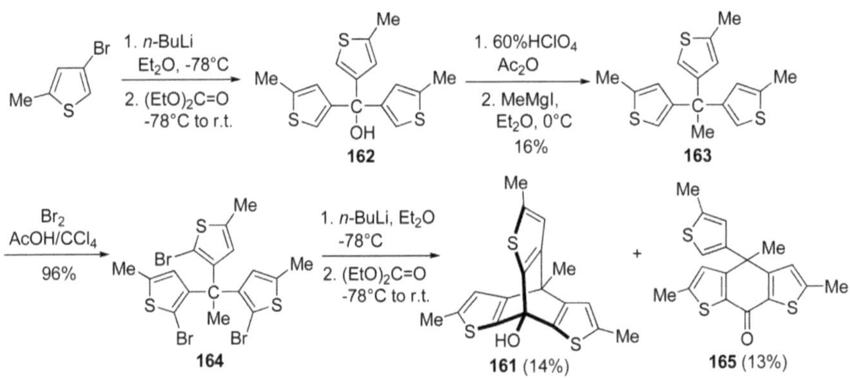

Scheme 5.47 Synthesis of hydroxyl tetramethyl substituted heterotriptycene **161**

after chromatographic purification along with the side product ketone **165** in 13 % yield.

Soon after, Nakayama and co-workers [55] further studied the synthesis of thiophene-containing triptycene derivatives. Consequently, under the similar conditions as above, they obtained 8-hydroxy-4-ethylthiophenetriptycene **166** with the yield of 19 %, accompanied with the by-product **165** in 23 % yield, from the tribromide **167** (Scheme 5.48a). In addition, the lithiation of 2,3-dibromo-5-methylthiophene **168**, followed by the treatment with diethyl carbonate, afforded the alcohol **169** in 90 % yield. The treatment of **170** with 60 % $HClO_4$ in Ac_2O gave the corresponding perchlorate, which then reacted with MeMgI to give the desired compound **170** in 65 % yield (based on alcohol **169**). Finally, the tribromide **170** reacted with *t*-BuLi and the trilithium salt in the presence of dimethyl carbonate to provide the target 4-hydroxy-8-methyl-thiophenetriptycene **171** in a 42 % yield (Scheme 5.48b).

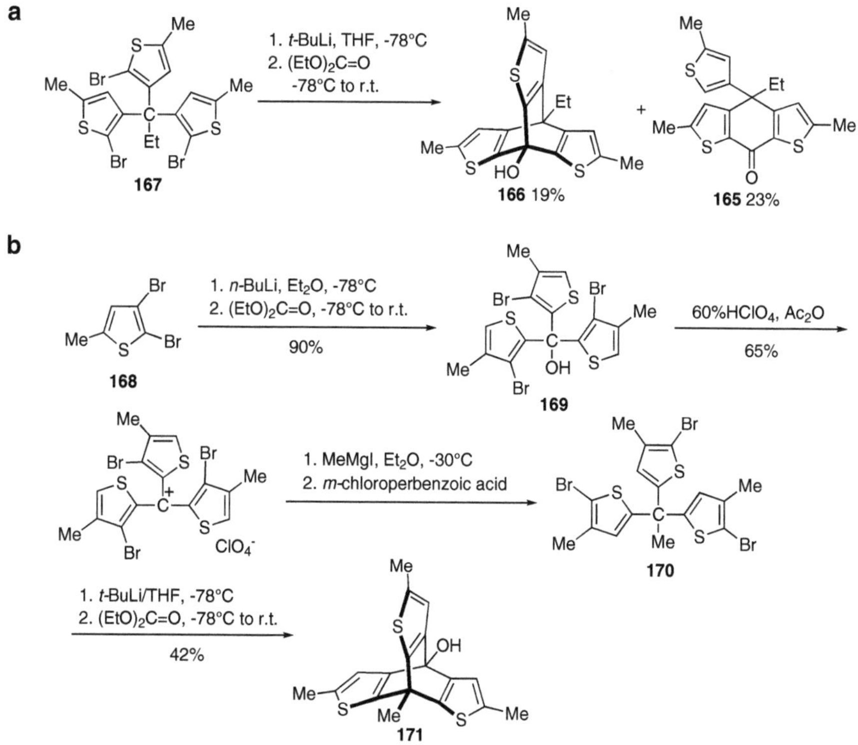

Scheme 5.48 Synthesis of **a** 8-hydroxy-4-ethylthiophenetriptycene **166** and **b** 4-hydroxy-8-methyl-thiophenetriptycene **171**

Furthermore, it was found that the thiophenetriptycene **161** would be directly deuterated in D_2SO_4 at room temperature to give the deuteriated ketone **172** (Scheme 5.49a). Moreover, the decomposition of compound **161** under the temperature near its melting point could provide the ketone **165**. Reaction of compound **161** with butyl lithium either at room temperature for 1.5 h or in refluxing *o*-dichlorobenzene gave the ketone **165** (Scheme 5.49b). This ring-opening of compound **161** to form ketone **165** would be mainly dependent on the large inherent ring strain. In addition, the treatment of the compound **161** with methane sulfonyl chloride in pyridine afforded the 8-methanesulfonate **173** (Scheme 5.49c), and reaction of **161** with acetic anhydride in the presence of triethylamine and DMAP gave the corresponding 8-acetoxy derivative **174** (Scheme 5.49d).

In 1996, Nakayama and co-workers [56] reported the synthesis of thiophenetriptycene-8-thiol **175**. As shown in Scheme 5.50, by the lithiation of 1,1,1-tris(2-bromo-5-methyl-3-thienyl)ethane **164** with *t*-BuLi, followed by the reaction with $S=(OEt)_2$, and quenched with aqueous ammonium chloride, the desired thiol **175** was afforded in 42 % yield. If the reaction was quenched by MeI or BuBr, the corresponding sulfides **176** (R = Me) and **177** (R = Bu) could be obtained in 51 and 40 % yield, respectively.

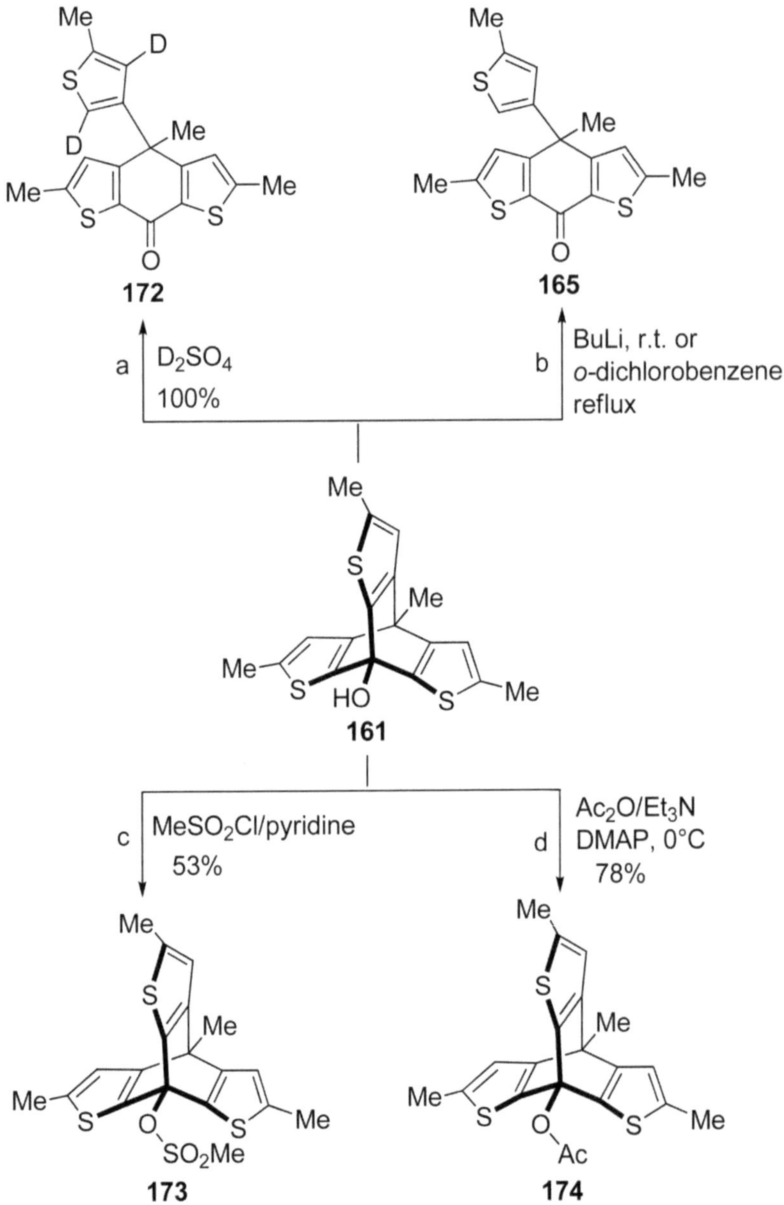

Scheme 5.49 Reactions of thiophenetriptycene **161**

Furthermore, the oxidization of sulfide **176** with MCPBA in CH_2Cl_2 gave the corresponding sulfoxide **178** in 91 % yield (Scheme 5.51a). And the isolable sulfenic acid **179** could be afforded by the oxidation of with a 1.3 molar amount of MCPBA. The intermediate product sodium salt was achieved by the treatment of the thiol **175** with 1.5 molar amount of NaH in THF at room temperature (Scheme 5.51b).

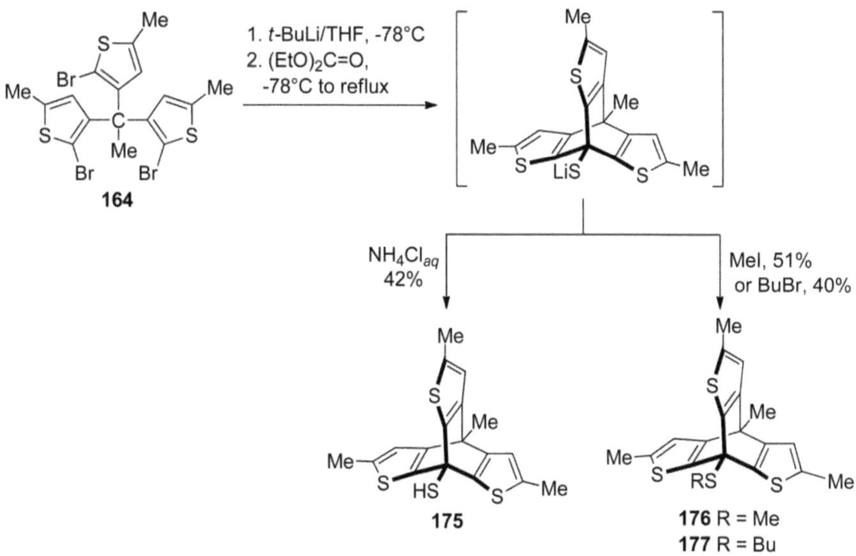

Scheme 5.50 Synthesis of thiophenetriptycene-8-thiol **175**, and sulfides **176** and **177**

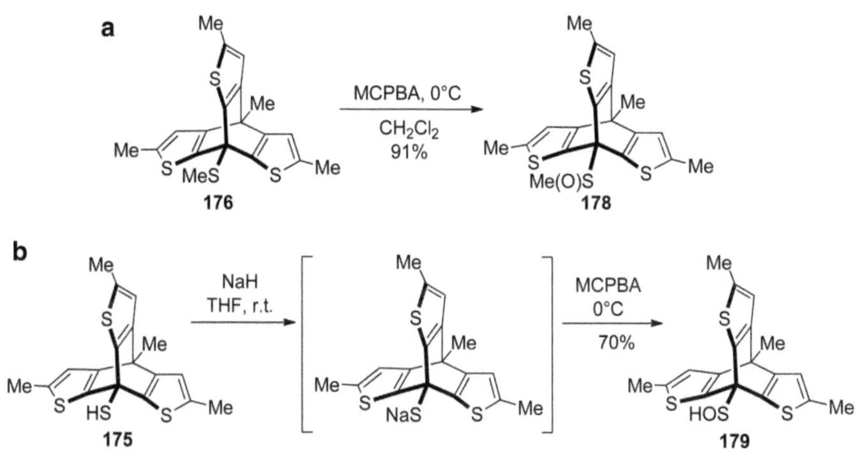

Scheme 5.51 Synthesis of **a** sulfoxide **178** and **b** sulfenic acid **179**

5.3 Miscellaneous Heterotriptycenes and Their Derivatives

In 1962, Wilcox and Stevens [57] reported the preparation of a new kind of triptycene analogs, in which the other aromatic nucleus took place of the benzene rings. As shown in Scheme 5.52, the unsaturated 1,4-diketone **180** was halogenated by the PBr_5 under 200 °C to give the furan **181** in 67 % yield. However, heating **180** to 140 °C

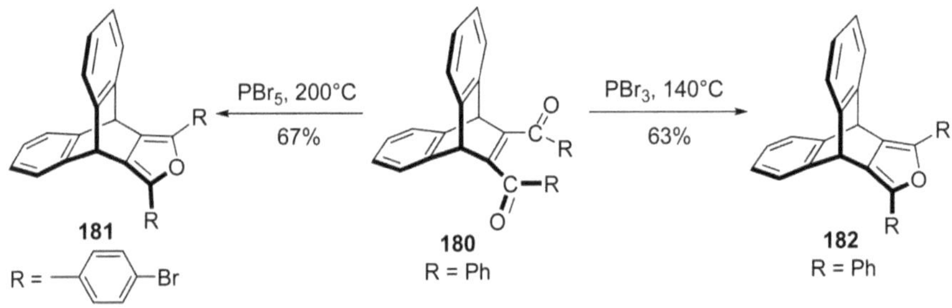

Scheme 5.52 Synthesis of unhalogenated furan **182**

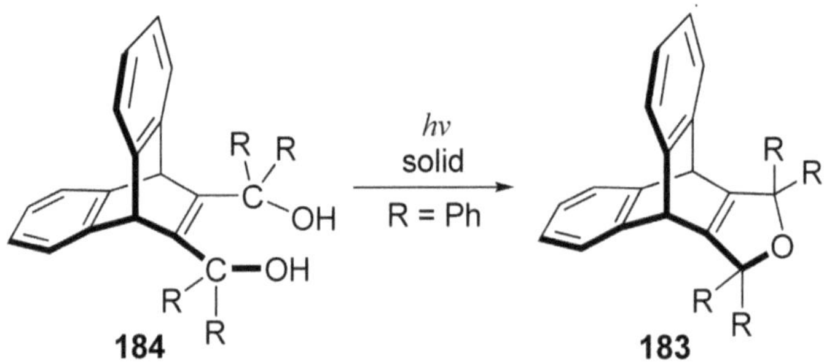

Scheme 5.53 Synthesis of 2,5-dihydrofuran **183**

in the presence of PBr_3 gave the unhalogenated furan **182** in 63 % yield. Afterward, Scheffer and co-workers [58] reported the synthesis of 2,5-dihydrofuran **183** by the solid state photolysis of 11,12-bis(hydroxyl-methyl)-substituted ethenoanthracene **184** (Scheme 5.53).

In 1993, Nakayama and co-workers [59–61] synthesized a series of novel heterotriptycenes containing three five-membered thiophene rings and one or two heteroatom bridgeheads. As shown in Scheme 5.54, the silane **185** and its brominated product (**186**) could serve as the precursors for the preparation of 4-silathiophenetriptycenes **161** and 8-phospha-4-silathiophenetriptycene **187**. It was noteworthy that the direct lithiation of the silane **185** with *n*-BuLi in a mixed solvent of ether and THF at 0 °C could give the compound **161** in only 6 % yield. In contrast, the target **161** could also be afforded in 5 % yield from the compound **186** by the treatment with *t*-BuLi and $(EtO)_2C{=}O$. However, the desired product **187** could be obtained by the reaction of **188** and $P(OPh)_3$ or directly from the silane (**185**). Moreover, it was also found that the reaction of compound **187** with MCPBA or the excess of cyclohexene sulfide in refluxing benzene could give the corresponding phosphineoxide **188a** and sulfide **188b** in 97 and 79 % yield, respectively.

Following the similar method, the 8-phospha-thiophenetriptycene **189** could be synthesized in 18 % yield by the treatment of the 1,1,1-tris(2-bromo-5-methyl-3-thienyl)ethane with *t*-BuLi in Et_2O, and followed by the reaction with an excess amount of $P(OPh)_3$ (Scheme 5.55a). Similarly, the reaction of compound **190** with

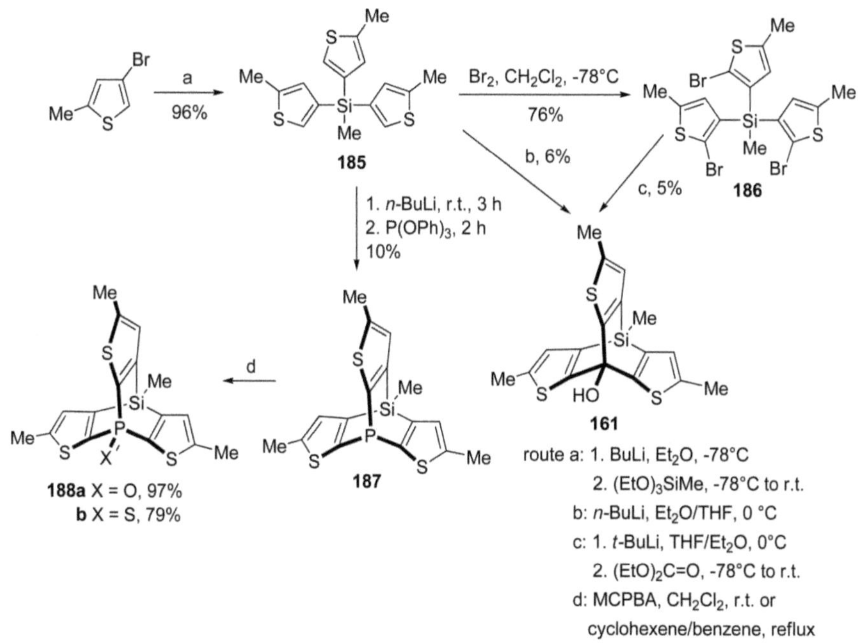

Scheme 5.54 Synthesis of heterotriptycenes containing thiophene rings and heteroatom bridgeheads

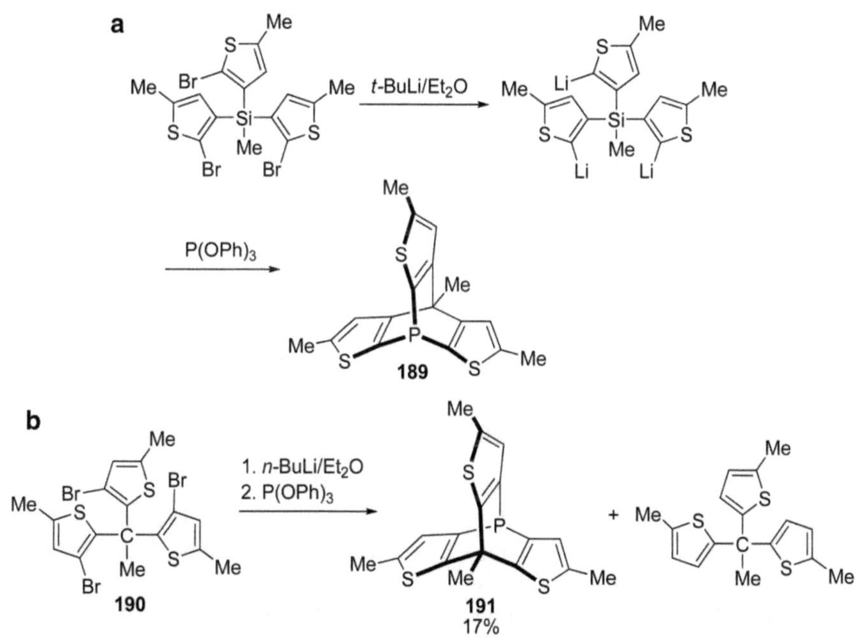

Scheme 5.55 Synthesis of **a** 8-phospha-thiophenetriptycene **189** and **b** 4-phosphathiophenetptycene **191**

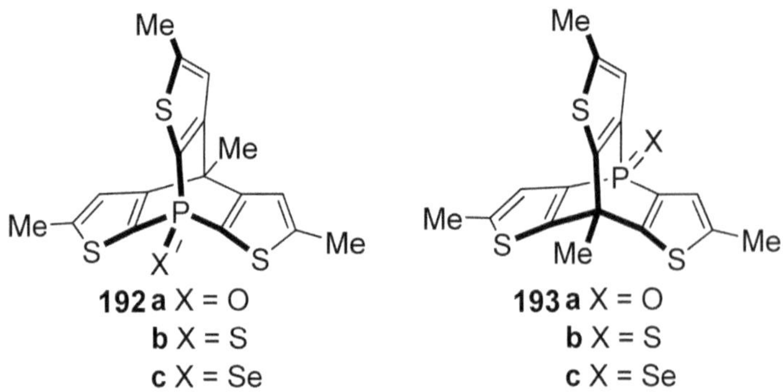

Fig. 5.4 Structures of the compounds **192** and **193**

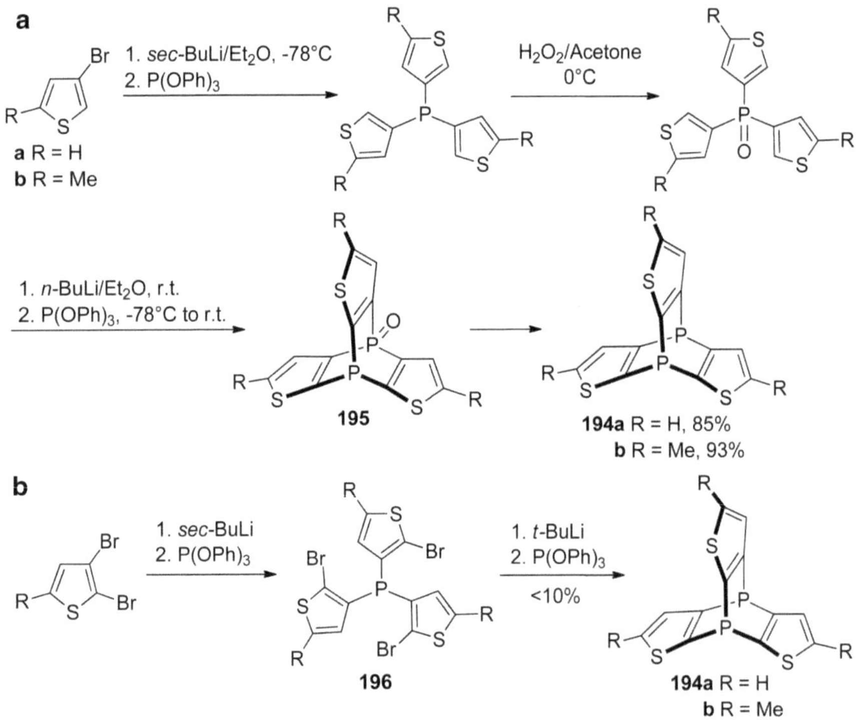

Scheme 5.56 Synthesis of 4,8-diphosphathiophenetriptycenes **194**

t-BuLi and then $P(OPh)_3$ afforded 4-phosphathiophenetptycene **191** in 17 % yield after recrystallization (Scheme 5.55b). Under the same conditions, the treatment of both compounds **189** and **191** with MCPBA gave the corresponding phosphine oxides **192a** and **193a**. Similarly, the compounds **189** and **191** could also be sulfurized and selenized with elemental sulfur and selenium to give **192b** and **193b**, and **192c** and **193c** (Fig. 5.4), respectively. On the whole, **191** seemed to be more reactive than

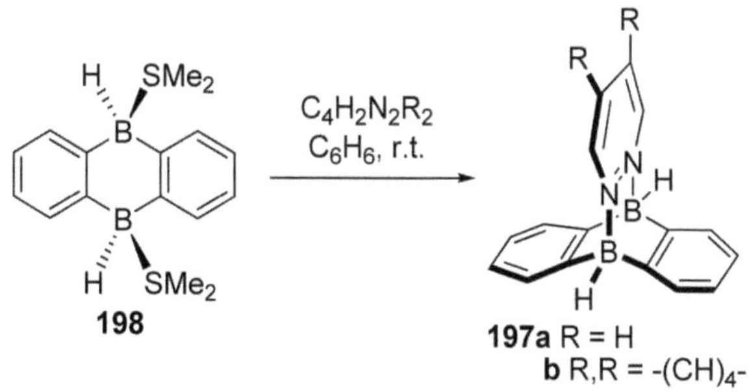

Scheme 5.57 Synthesis of B–N analogues of triptycene **197**

189, which could be considered due to the larger steric hindrance of the three sulfur atoms and the decreased nucleophilicity by higher *s*-orbital character of the lone pair electrons [59].

As shown in Scheme 5.56a, 4,8-diphosphathiophenetriptycenes **194a** could be synthesized by the reduction of **195b** with trichlorosilane in refluxing benzene. Under the same reaction condition, the reduction of **195a** with the excess tri-chlorosilane gave 4,8-diphosphathiophenetriptycenes **194b**. However, it was found that the mixture of **194a, b** could be obtained in very low yields by the direct cyclization of the trilithium salts from tris(3-bromo-2-thienyl)phosphines **196** (Scheme 5.56b).

In 2010, Wagner and co-workers [62] reported that 9,10-dihydro-9,10-dibora-anthracene could serve as the potential hydroboration reagent for terminal alkynes. Thus, the B–N analogues of triptycene **197** could be afforded by the reaction between the DMS adduct **198** and pyridazine, which is shown in Scheme 5.57.

In 1994, Kräutler and co-workers [63] reported a convenient and practical method for the synthesis of novel porphyrin **199** with 3D and D_4-symmetric biconcave structure. As shown in Scheme 5.58, the porphyrin **199** was synthesized by the reduction

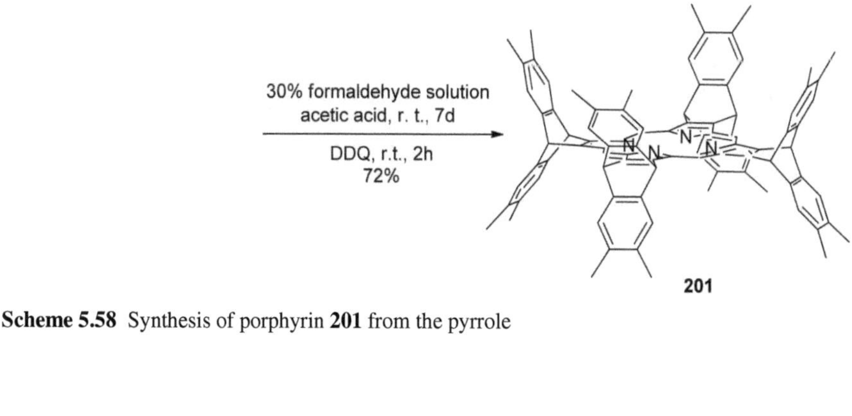

Scheme 5.58 Synthesis of porphyrin **201** from the pyrrole

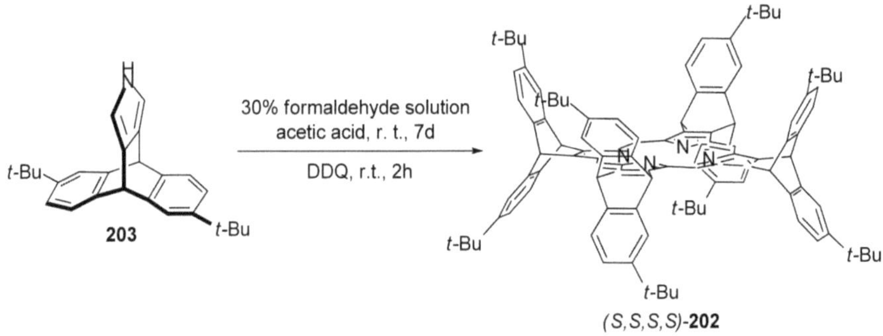

Scheme 5.59 Synthesis of biconcave porphyrin **202**

of the dinitrile **200** with DIBAH at −20 °C, and then followed by the common method for the generation of porphyrin from the pyrrole **201**.

Finally, they [64] further synthesized the enantiopure biconcave porphyrin **202** from the chiral pyrrole **203** via a sequence of reactions. As shown in Scheme 5.59, the acid-catalyzed condensation of the chiral pyrrole **203** with formaldehyde at room temperature for 7 days, then the resulting mixture further reacted with DDQ at room temperature for another 2 h to afford the chiral porphyrin **202** in 70 % yield after separation and purification. Taking advantage of this method, the enantiomeric purity of the chiral biconcave porphyrin **202** could be about 10^9:1 with an unambiguously assigned absolute configuration. Moreover, this biconcave porphyrin **202** could form the the D_4-symmetric biconcave Co^{II} porphyrinate with cobalt ion, which could be served as a potentially promising chiral shift reagent.

References

1. von Dechend F, Wichelhaus H (1875) Ueber einwirkung von anilin auf nitrobenzol. Ber 8(2):1609–1615
2. McCleland NP, Whitworth JB (1927) Organic compounds of arsenic. Part III. Tri-*o*-phenylenediarsine. J Chem Soc 2753–2757
3. Bickelhaupt F (1999) Travelling the organometallic road: a Wittig student's journey from lithium to magnesium and beyond. Chem Soc Rev 28(1):17–23
4. Wittig G, Steinhoff G (1962) Benzo-azatriptycen. Chem Ber 95(1):203–212
5. Wittig G, Steinhoff G (1963) Azatriptycene. Angew Chem Int Ed 2(7):396
6. Wittig G, Steinhoff G (1964) Azatriptycen. Liebigs Ann Chem 676:21–31
7. Kreil DJ, Sandel VR (1976) pK_a of azatriptycene. J Chem Eng Data 21(1):132–133
8. Wegmann D, Simon W (1962) Potentiometrische mikrobestimmung von scheinbaren dissoziationskonstanten in essigsäure. Helv Chim Acta 45(3):962–973
9. Hoefnagel AJ, Hoefnagel MA, Wepster BM (1981) Substituent effects. 8. Basic strength of azatriptycene, triphenylamine, and some related amines. J Org Chem 46(21):4209–4211
10. Sugawara T, Iwamura H (1980) Formation of *ortho*-(9-fluorenyl) phenylnitrene in the photoisomerization of 1-azatriptycene. J Am Chem Soc 102(23):7134–7136
11. Hellwinkel D, Schenk W (1969) Azaphosphatriptycene. Angew Chem Int Ed 8(12):987–988

12. Hellwinkel D, Schenk W, Blaicher W (1978) Heterotriptycenes-structural calculations and NMR relations. Chem Ber-Recl 111(5):1798–1814
13. Hellwinkel D, Blaicher W, Krapp W, Sheldrick WS (1980) Penta-coordinated phosphoranes with azaphosphatriptycene skeleton. Chem Ber-Recl 113(4):1406–1413
14. Osman FH, El-Samahy FA (2002) Reactions of α-diketones and *o*-quinones with phosphorus compounds. Chem Rev 102(3):629–678
15. Earley RA, Gallaghe MJ (1970) Synthesis of some nitrogen–arsenic heterocycles: 5,10-dihydro-5,10-*ortho*-benzenophenarsazine (azarsatriptycene), 10,11-dihydro-5-phenyl-5*H*-dibenzo[*b, f*][1,4]azarsepine, and 2,3-dihydro-1,2-diphenyl-1*H*-benz[*c*]-azarsole. J Chem Soc C (1):158–163
16. Weinberg KG, Whipple EB (1971) 1,6-Diphosphatriptycene. J Am Chem Soc 93(7):1801–1802
17. Vermeer H, Bickelha F, Kevenaar PC (1972) Arsatriptycene. Liebigs Ann Chem 763(NSEP):155–161
18. Jongsma C, Dekleijn JP, Bickelha F (1974) Phosphatriptycene. Tetrahedron 30(18):3465–3469
19. Jongsma C, Dekok JJ, Weustink RJM, Vanderley M, Bulthuis J, Bickelhaupt F (1977) Stibatriptycene. Tetrahedron 33(2):205–209
20. Al-Jabar NAA, Bowen D, Massey AG (1985) The synthesis of 1,6-distibatriptycene (5,10-*o*-benzenostibanthrene), $Sb_2(C_6H_4)_3$. J Organomet Chem 295(1):29–32
21. Woodard CM, Hughes G, Massey AG (1976) Perfluorophenyl derivatives of elements. 29. Synthesis of some heterocycles of Hg, S, Se, Te, As and Sb. J Organomet Chem 112(1):9–19
22. Mistry TK, Massey AG (1981) Dodecafluoro-5,10-*ortho*-benzenoarsanthrene and dodecafluoro-5,10-*ortho*-benzenostibanthrene. J Organomet Chem 209(1):45–47
23. Al-Jabar NAA, Massey AG (1984) Dodecafluoro-5,10-*ortho*-benzenostibanthrene and dodecachloro-5,10-*o*-benzenostibanthrene. J Organomet Chem 276(3):331–340
24. Al-Jabar NAA, Massey AG (1985) The synthesis of perfluoro-1,6-disubstituted triptycenes containing group-IV and group-V elements. J Organomet Chem 287(1):57–64
25. Humphries RE, Al-Jabar NAA, Bowen D, Massey AG, Deacon GB (1987) Transmetallation reactions involving phenylenemercurials. J Organomet Chem 319(1):59–67
26. Kobayashi J, Agou T, Kawashima T (2003) A novel and convenient synthetic route to a 9-phosphatriptycene and systematic comparisons of 9-phosphatriptycene derivatives. Chem Lett 32(12):1144–1145
27. Kobayashi J, Agou T, Kawashima T (2004) Synthesis and structure of a novel symmetrically substituted phosphatriptycene oxide. Phosphorus Sulfur 179(4–5):959–960
28. Agou T, Kobayashi J, Kawashim T (2004) Synthesis, structure, and reactivity of a symmetrically substituted 9-phosphatriptycene oxide and its derivatives. Heteroat Chem 15(6):437–446
29. Uchiyama Y, Sugimoto J, Shibata M, Yamamoto G, Mazaki Y (2009) Bromine adducts of 9,10-diheteratriptycene derivatives. Bull Chem Soc Jpn 82(7):819–828
30. Uchiyama Y, Yamamoto G (2005) A bromine adduct of 2,3,6,7,14,15-hexamethyl-9,10-distibatriptycene. The first Stibonium–Stiborate zwitterion. Chem Lett 34(7):966–967
31. Takahashi M, Hatano K, Kawada Y, Koga G, Tokitoh N, Okazaki R (1993) Synthesis and crystal structure of 9,10-disilatriptycenes. J Chem Soc Chem Commun (24):1850–1852
32. Dam MA, Akkerman OS, De Kanter FJJ, Bickelhaupt F, Veldman N, Spek AL (1996) 9,10-Dimetallatriptycenes of group 14: a novel synthetic approach from *ortho*-phenylenemagnesium. Chem Eur J 2(9):1139–1142
33. Dam MA, de Kanter FJJ, Bickelhaupt F, Smeets WJJ, Spek AL, Fornies-Camer J, Cardin C (1998) 9,10-dimetallatriptycenes of group 14: a structural study. J Organomet Chem 550(1–2):347–353
34. Tsuji H, Inoue T, Kaneta Y, Sase S, Kawachi A, Tamao K (2006) Synthesis, structure, and properties of 9-phospha-10-silatriptycenes and their derivatives. Organometallics 25(26):6142–6148
35. Skvarchenko VR, Koshkina NP, Abramov AV (1981) Cationoidal forms of 2-azatrypticene in nucleophilic-substitution reaction. Zhur Org Khim 17(5):1018–1023
36. Skvarchenko VR, Lapteva VL, Gorbunova MA (1990) Convenient synthesis of 2-azatriptycene using methyleneimine hydrochloride. Zhur Org Khim 26(12):2588–2590

37. Quast H, Schon N (1984) 1-Aza-azatriptycene and 1,9-diazatriptycene. Liebigs Ann Chem (2):381–388
38. Kempter G, Heilmann D, Mühlstädt M (1972) Heterocyclen aus aminoketonen. XX. Synthesen von chinolinen, 1-aminonortricyclen- und -quadricyclenderivaten. J f Prakt Chem 314(3–4):543–556
39. Markgraf JH, Davis HA, Ernst PS, Hirsch KS, Leonard KJ, Morrison ME, Myers CR (1991) Strained heterocyclic-systems. 19. 1-azatriptycene and derivatives. Tetrahedron 47(2):183–188
40. Gorgues A, Le Coq A (1979) A useful synthon: acetylenedicarbaldehyde (but-2-ynedial). J Chem Soc Chem Commun (17):767–768
41. Bouffard J, Eaton RF, Mueller P, Swager TM (2007) Iptycene-derived pyridazines and phthalazines. J Org Chem 72(26):10166–10180
42. Fields DL, Regan TH, Graves RE (1971) Overcrowded molecules. 3. 13,14-bis(2-pyridyl)pentaphene and relatedcompounds. J Org Chem 36(20):2995–3001
43. Scheffer JR, Ihmels H (1997) Photochemistry of 11,12-bis(aminomethyl)-9,10-dihydro-9,10-ethenoanthracene in solution and in the solid state. Liebigs Ann (9):1925–1929
44. Chong JH, MacLachlan MJ (2006) Robust non-interpenetrating coordination frameworks from new shape-persistent building blocks. Inorg Chem 45(4):1442–1444
45. Chong JH, MacLachlan MJ (2007) Synthesis and structural investigation of new triptycene-based ligands: en route to shape-persistent dendrimers and macrocycles with large free volume. J Org Chem 72(23):8683–8690
46. Cao J, Zhu XZ, Chen CF (2010) Synthesis, structure, and binding property of pentiptycene-based rigid tweezer-like molecules. J Org Chem 75(21):7420–7423
47. Jiang Y, Chen CF (2010) Synthesis and structures of 1,10-phenanthroline-based extended triptycene derivatives. Synlett (11):1679–1681
48. Mastalerz M, Sieste S, Cenicǐ' M, Oppel IM (2011) Two-step synthesis of hexaammonium triptycene: an air-stable building block for condensation reactions to extended triptycene derivatives. J Org Chem 76(15):6389–6393
49. McKinnon DM, Wong JY (1971) Preparation of some heterocyclic analogues of triptycene. Can J Chem 49(19):3178–3184
50. De Wit J, Wynberg H (1973) Heterocyclic triptycenes: synthesis and ultraviolet spectroscopy. Tetrahedron 29(10):1379–1391
51. Ye XS, Li WK, Wong HNC (1996) 3,4-Didehydrothiophene: generation, trapping reactions, and *Ab initio* study. J Am Chem Soc 118(10):2511–2512
52. Wynberg H, Wit JD, Sinnige HJM (1970) Synthesis of an asymmetric heterotriptycene. J Org Chem 35(3):711–715
53. Meth-Cohn O, van Vuuren G (1986) Thiophene *S,N*-ylides: a new versatile class of sulphimides. J Chem Soc, Perk T 1:233–243
54. Ishii A, Kodachi M, Nakayama J, Hoshino M (1991) Synthesis and isomerization of the novel heterotriptycene, 8-hydroxy-2,4,5′,6-tetramethyl-4,8-dihydro-4,8[3′,2′]thiophenobenzo[1,2-*b*: 5,4-*b*′]dithiophene. J Chem Soc Chem Commun (11):751–752
55. Ishii A, Maeda K, Kodachi M, Aoyagi N, Kato K, Maruta T, Hoshino M, Nakayama J (1996) Syntheses and chemical and physical properties of thiophenetriptycenes. J Chem Soc Perk T 1(11):1277–1286
56. Ishii A, Komiya K, Nakayama J (1996) Synthesis of a stable sulfenic acid by oxidation of a sterically hindered thiol (thiophenetriptycene-8-thiol) and its characterization. J Am Chem Soc 118(50):12836–12837
57. Wilcox CF, Stevens MP (1962) The mechanism of reductive furan formation. J Am Chem Soc 84(7):1258–1262
58. Fu TY, Olovsson G, Scheffer JR, Trotter J (1995) Modification of host photobehavior by formation of crystalline host–guest assemblies. Tetrahedron Lett 36(25):4353–4356
59. Ishii A, Takaki I, Nakayama J, Hoshino M (1993) Syntheses and reactions of 4-phosphathiophenetriptycene and 8-phosphathiophenetriptycene. Tetrahedron Lett 34(51): 8255–8258

60. Ishii A, Tsuchiya T, Nakayama J, Hoshino M (1993) Syntheses and reactivities of 4-sila and 8-phospha-4-silathiophenetriptycenes. Tetrahedron Lett 34(14):2347–2350
61. Ishii A, Yoshioka R, Nakayama J, Hoshino M (1993) 4,8-Diphosphathiophenetriptycenes. Tetrahedron Lett 34(51):8259–8262
62. Lorbach A, Bolte M, Lerner HW, Wagner M (2010) Lewis-base adducts of 9,10-dihydro-9,10-diboraanthracene: ditopic hydroboration reagents and a B-N analogue of triptycene. Chem Commun 46(20):3592–3594
63. Ramondenc Y, Schwenninger R, Phan T, Gruber K, Kratky C, Kräutler B (1994) Porphyrins with biconcave frameworks. Angew Chem Int Ed 33(8):889–891
64. Schwenninger R, Ramondenc Y, Wurst K, Schlögl J, Kräutler B (2000) A highly enantiopure biconcave porphyrin with effective D_4-structure. Chem Eur J 6(7):1214–1223

Chapter 6
Preparation of Iptycene-Containing Polymers and Oligomers

In this chapter, the methods for the preparation of various iptycene-containing polymers and oligomers will be discussed in details. The structural features and the properties of these polymers and oligomers will also be described in the following corresponding sections.

6.1 Triptycene-Containing Polymers

6.1.1 Triptycene-Containing Non-conjugated Polymers

When a 3D structural unit, especially, a rigid 3D skeleton was introduced into a polymer, some special physical properties could be shown. As early as 1968, Klanderm and Faber [1] attempted to prepare the polymers containing 9,10-bis(hydroxylmethyl)triptycene unit, which was expected to enhance the glass transition temperature (T_g). Consequently, they obtained the modified polyesters with 9,10-bis(hydroxylmethyl)triptycene instead of the common glycol components, and found that the target poly(ethylene terephthalate)s expectedly showed the high heat-distortion temperatures and high glass transition temperatures, but the crystallizability was simultaneously declining. Obviously, the rigid bulky triptycene moiety seemed to interfere with the segmental rotation of the polymer chain, leading to an influence on the thermal properties of the polyesters.

Soon after, Hoffmeister et al. [2] synthesized a series of bridgehead-substituted triptycene monomers **1** (Fig. 6.1), and then introduced them into the polymer segments to form various kinds of triptycene-containing polymers, including polyesters, polyamides, polyurethanes, and polyoxadiazoles. Moreover, it was also found that the polyesters and polyurethanes of 9,10-bis(hydroxylmethyl)-triptycene **1a** and the polyamides based on 9,10-triptycenedicarbonyl chloride **1d** could be obtained in the moderate-to-high molecular weights, and the polyesters and polyamides were thermally stable and colorless. As the limit of the thermal stability for preparation of thin films existed, only some of the colorless polymers with high melting points were suitable to be solution-cast into thin films; although, the films were generally weak or brittle for their high crystallinity.

C.-F. Chen, Y.-X. Ma, *Iptycenes Chemistry*,
DOI 10.1007/978-3-642-32888-6_6, © Springer-Verlag Berlin Heidelberg 2013

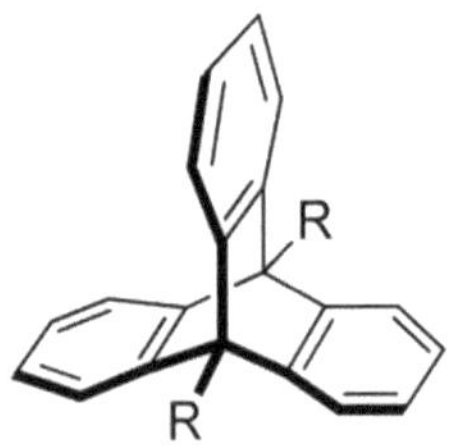

1a R = CH_2OH
1b R = CH_2CH_2OH
1c R = CH_2OCOCl
1d R = COCl
1e R = CO_2CH_3
1f R = $CONHNH_2$
1g R = CO_2H
1h R = NH_2
1i R = NCO
1j R = $CH_2CO_2CH_2CH_3$

Fig. 6.1 Structures of bridgehead-substituted triptycene monomers

In the early works on the preparation of polymers, the main focus was on the triptycene monomers with highly symmetrical structures. However, Akutsu and co-workers [3] attempted to introduce the unsymmetrical triptycenediamine **2** into the main chain of aromatic polyamide to enhance the solubility, along with keeping the high thermal stability. Consequently, in 1994 they reported the preparation of a series of aromatic polyamides **3** (Scheme 6.1a) via the low-temperature polycondensation of the diamine **2** and several aromatic diacyl dichlorides in NMP. They found that the desired polymers expectedly exhibited both satisfactory solubilities and high thermal stabilities. Soon after, Akutsu et al. [4] also incorporated the triptycenediol **4** into the backbone of polyarylates, and found that the polymers **5a**, **b** (Scheme 6.1b) showed good solubilities as well as high thermal stabilities. Moreover, it was found that introduction of the diol **4** into the poly(ether ether ketone)s could afford the increased solubilities of the polymers **5c**, **d** (Scheme 6.1c) without sacrificing their high thermal stabilities, except that the polymers **5d** showed semicrystalline property and low solubility.

In 2006, Swager and co-workers [5] realized that the triptycene with unique structure might not only serve as the interlocking subunit to weave the chains of polymer to enhance its strength, but also allow the polymeric chains to have some degree of freedom to slide by each other and thus show high ductility. Consequently, the authors designed and successfully prepared the polyester **6** (Scheme 6.2), and found that the triptycene units could simultaneously improve both stiffness and ductility of the polymer. Incorporation of triptycene units into the polymers enhanced not only the values for Young's modulus (about 3 times), but also the strength (about 3 times) and strain to failure (over 20 times) compared with the reference polyester **7**, which indicated that the rigid and pendant units might indeed enhance the mechanical properties of the polymers due to the steric effect. Concretely speaking, the molecular threading and molecular interlocking steric interactions could minimize the "internal molecular free volumes" (IMFVs) of the system (Fig. 6.2).

Soon after, Swager's and co-workers [6] reported a series of polyesters **8–10** (Fig. 6.3) with different compositions (*x* and *y* varied) based on different pendant triptycene units via the similar strategy. The presence of triptycene units actually brought less crystalline, lower melting points, and higher T_g, compared with the common polyester without iptycene moiety. More importantly, the introduction of triptycene moiety could also enhance the Young's modulus and strength (about 2 times),

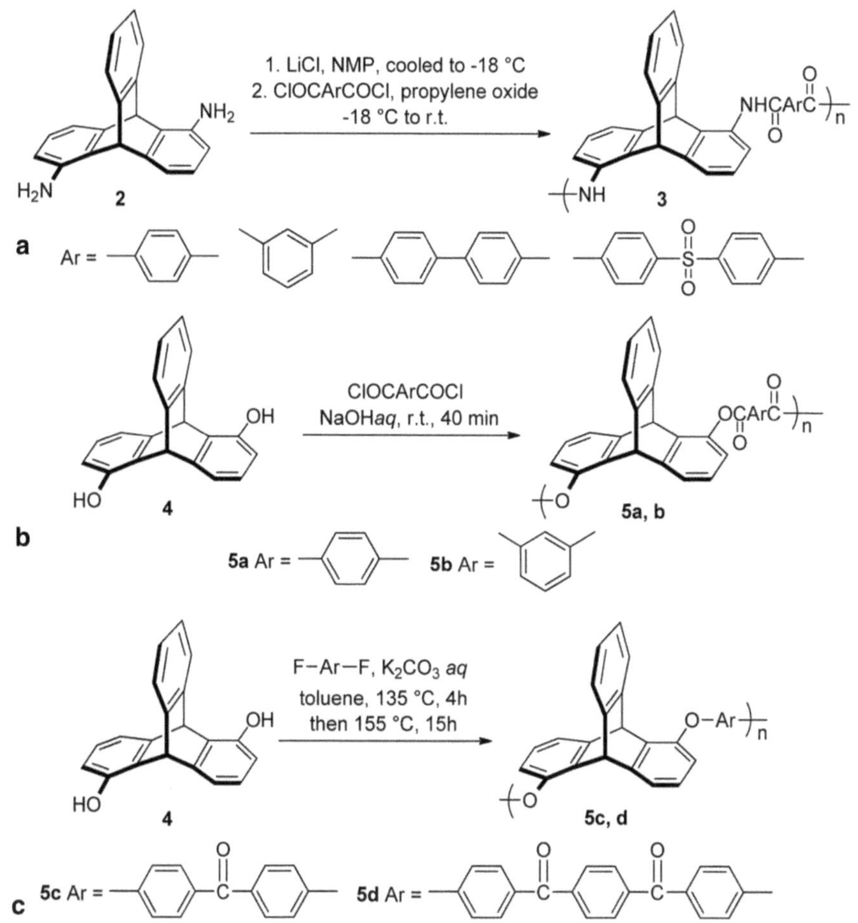

Scheme 6.1 Synthesis of polymers from triptycene monomers

along with an increase of 14–21 times in the stress required to induce mechanical failure compared with the reference polyester **7**. Additionally, the stress strain curves for the polymer **8** exhibited that the threading and interlocking of the polymer chains through molecular-level steric interactions could actually simultaneously improve all aspects of the properties of the polymers.

Furthermore, Swager's and co-workers [7] also synthesized the triptycene-based polymer **11** containing bisphenol-A PC fragment. As shown in Scheme 6.3, the mixture of triptycene hydroquinone and bisphenol-A was added into a diphenyl carbonate solution, and then the mixture was heated to 190 °C for 30 min under a slow flow of argon, which provided the target polymer **11**. Upon the addition of tetramethylammonium hydroxide and sodium hydroxide, the temperature and pressure were then changed in a gradient (Table 6.1) to ensure the polymerization to occur well. To study

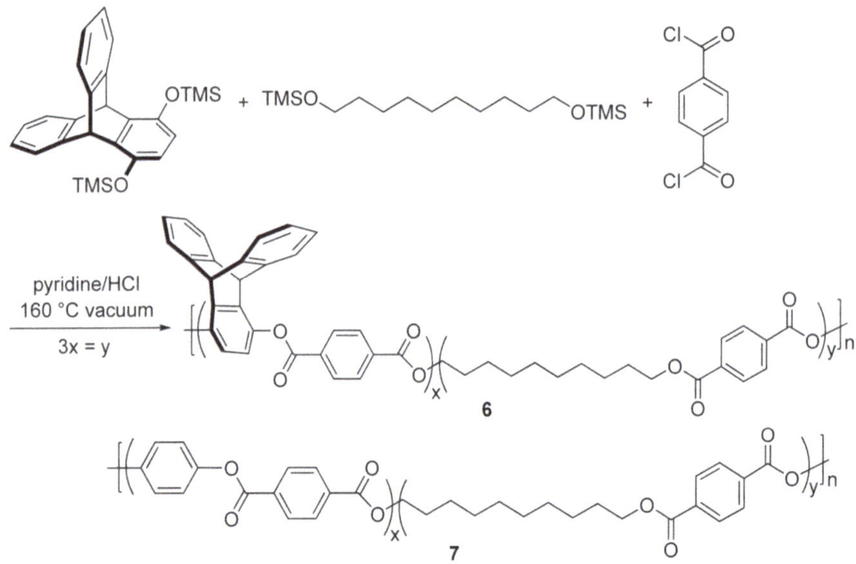

Scheme 6.2 Synthesis of polyester **6**

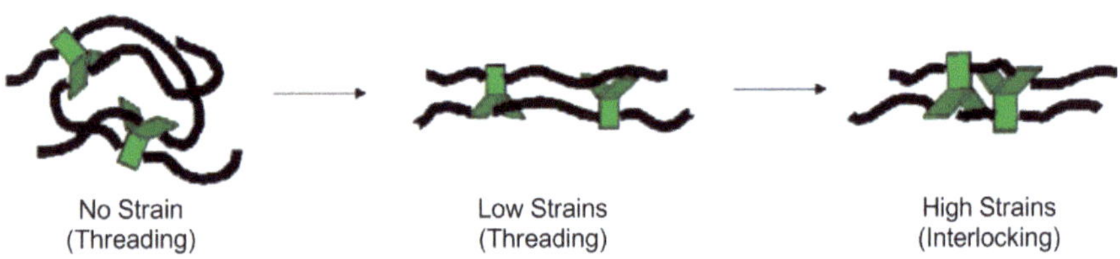

Fig. 6.2 Schematic of the steric interactions induced by the triptycene unit and the minimization of IMFVs. (Reprinted with the permission from [5]. Copyright 2006 American Chemical Society)

the mechanical behavior of the triptycene units incorporated with PC, the authors further prepared the polymer with low molecular weight and low triptycene content by blending triptycene-PC with a bisphenol-A PC to extend the scope of triptycene contents from 0 to 26 wt%. As a result, they found that the triptycene-containing PCs actually exhibited great improvements in various mechanical properties, including modulus, compressive yield stress, and tensile strength at all temperatures, compared with the Iupilon® PC. The results also showed that the mechanical properties increased obviously with the increase of amount of triptycene.

Recently, Turner and co-workers [8] reported a melt polycondensation method for the synthesis of triptycene-containing copolyesters. As shown in Scheme 6.4, by the copolymerization of the monomer triptycene diol **12**, comonomers ethylene glycol, 1,4-butanediol, and 1,6-hexanediol with cyclohexanedicarboxylate via the melt polycondensation reaction, a series of copolyesters **13** were afforded. The copolyesters showed good thermal stabilities and relatively high glass transition

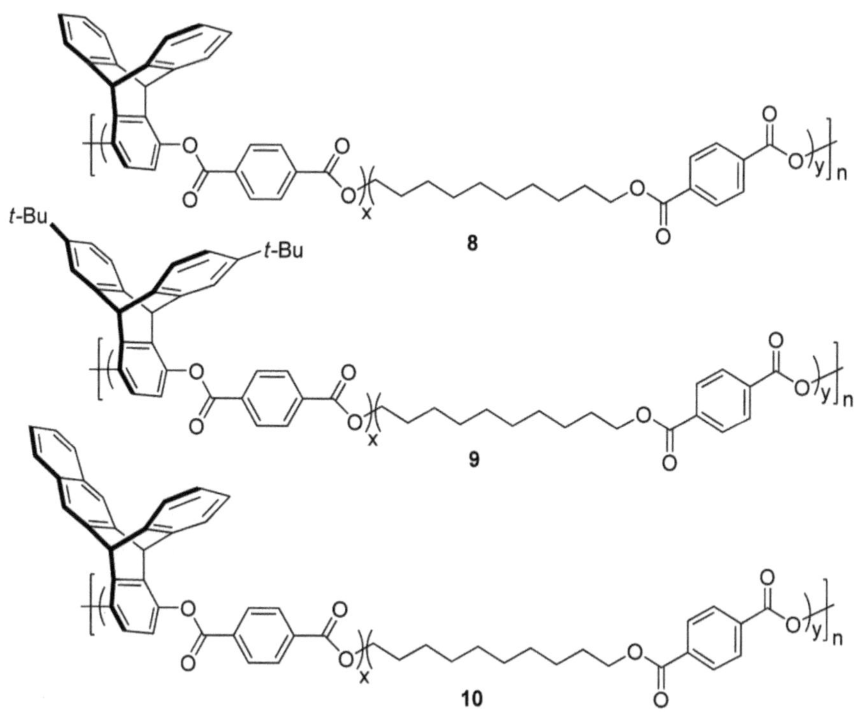

Fig. 6.3 Structures of polyesters **8–10**

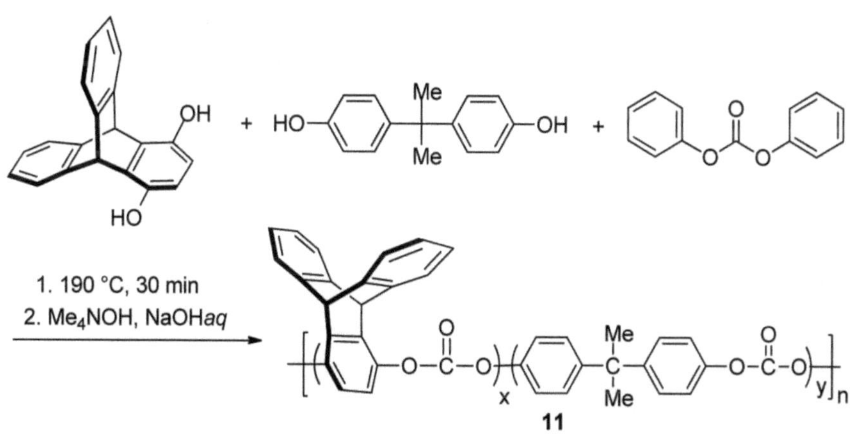

Scheme 6.3 Synthesis of triptycene-based polymer **11**

temperatures (T_g). Moreover, it was also found that the 1,4-butanediol-based triptycene copolyester exhibited the enhancement of modulus along with maintaining the high elongation at 23 °C.

In 2007, Zhang et al. [9] designed and synthesized a series of new polyimides **14a–f** by incorporating the triptycene moieties into the backbone of the polymers. As

Table 6.1 The temperature and pressure in reaction step

Reaction step	Temperature (°C)	Pressure (Torr[a])	Time (h)
1	190	760	0.5
2	220	200	1
3	250	200	0.5
4	250	150	0.5
5	250	100	0.5
6	250	15	0.5
7	280	2	0.5
8	280 to r.t.	~ 0.8 (full vacuum)	–

[a] 1 Torr (mmHg) = 0.01934 psi (lb/in^2) = 1.316×10^{-3} atm

shown in Scheme 6.5, the desired polyimides **14a–f** were obtained by the one-step polycondensation between the triptycene-derived dianhydride monomer and various aromatic diamines in *m*-cresol in the presence of a catalytic amount of isoquinoline under high temperature. As expected, the polyimides **14a–f** exhibited good solubilities in common organic solvents, along with high T_g and thermal stabilities. Moreover, the rigid triptycene moieties created the large free volumes, which made the polymers potential for membrane applications.

Recently, Hsiao et al. [10] also synthesized two kinds of new polyimides containing triptycene units **15a–f** and **16a–f** by a two-step thermal or chemical imidization method from the reaction of the triptycene-based bis(ether amine)s **17** or **18** and various aromatic tetracarboxylic dianhydrides (Scheme 6.6). It was noteworthy that the viscosities of the reaction system would become very high, resulted in the formation of high molecular weight poly(amic acid) precursors. Then, the poly(amic acid) precursors could be converted to the polyimides by the chemical dehydration with acetic

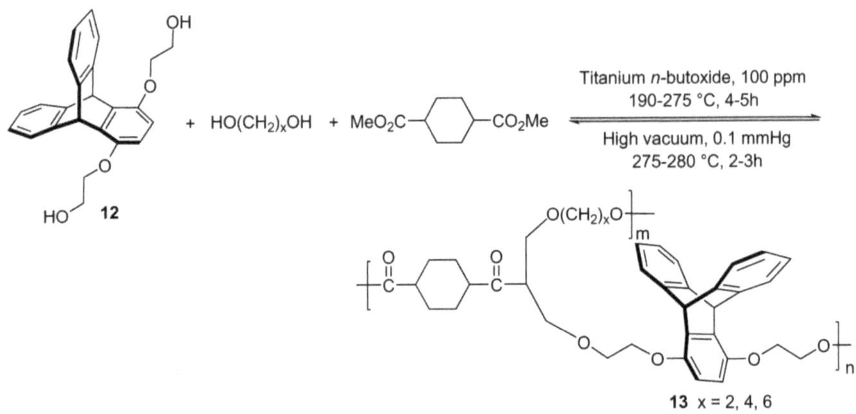

Scheme 6.4 Synthesis of triptycene-containing copolyesters **13**

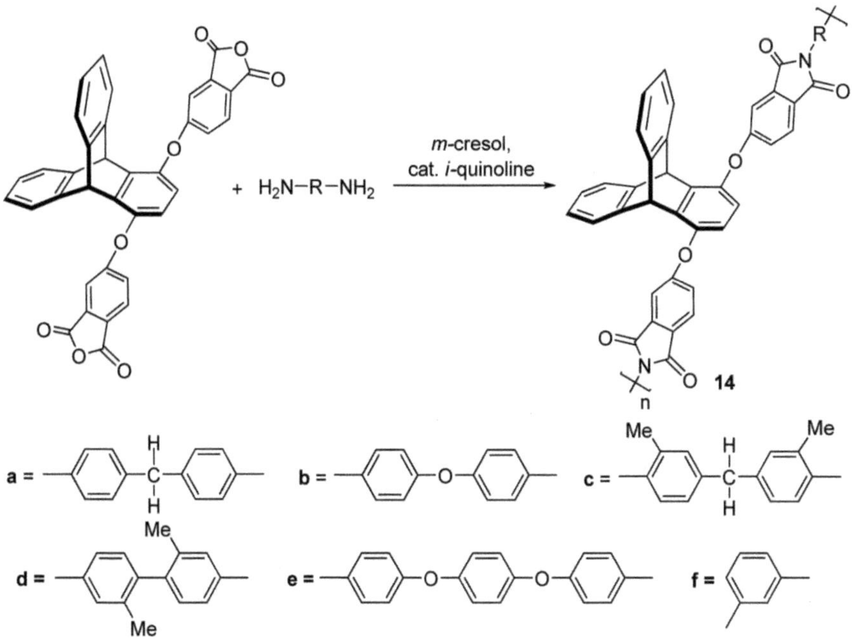

Scheme 6.5 Synthesis of polyimides **14**

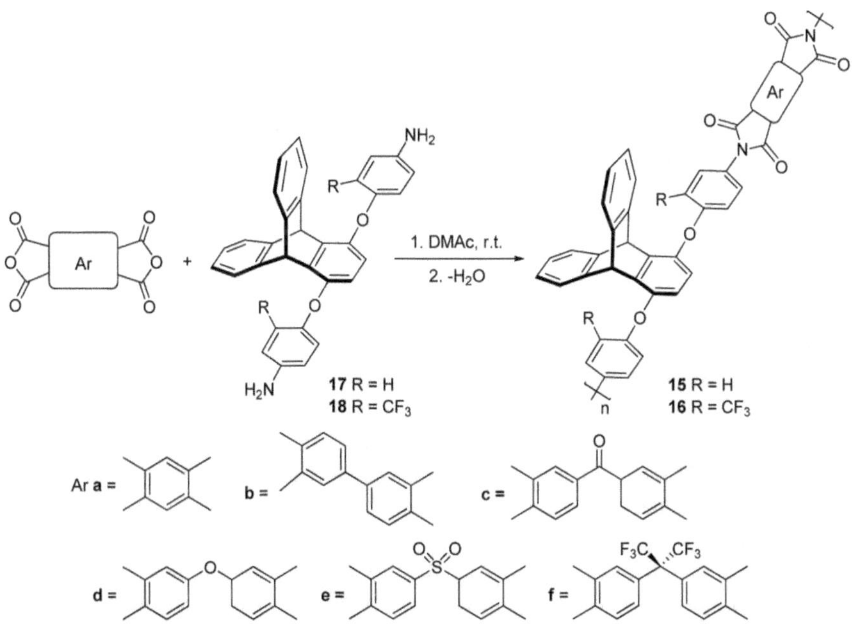

Scheme 6.6 Synthesis of polyimides **15** and **16** containing triptycene units

anhydride and pyridine or the thermal conversion at 250 °C. As a consequence, these desired polymers with high enough molecular weights could be suitable for casting of flexible and strong films due to their good mechanical properties. In particular, the fluorinated polyimides **16a–f** exhibited the good solubility in common organic solvents. Moreover, because of the decreased interchain interactions, the thin films made from these fluorinated polyimides also showed the light color. The combination of colorless, low dielectric constants, and good mechanical properties made the polyimides good candidates for micro-electronics and optoelectronics applications.

Soon after, Swager and co-workers [11] also reported the polymerization between the 2,6-diaminotriptycene derivatives and various five- and six-membered ring anhydrides to give the aromatic polyimides **19a–h** via two different routes (Scheme 6.7). For the five-membered ring imides, it was found that the ring-opening polyaddition reaction and then ring-closing dehydration provided the polyimide **19f** in the yields of 78–93 %. In contrast, for the polymer with hexatomic ring, the initial polymerization between the *m*-cresol and benzoic acid in the presence of an acid catalyst occurred, which then afforded the polymers **19g, h** by the addition of isoquinoline as the base to cease the reaction, after the end of imidization (Scheme 6.7a). Moreover, the five-membered ring polymers **19c**, **19d**, and **19f** with high molecular weights could be obtained by the similar route to six-membered ring, which is shown in Scheme 6.7b.

To afford good solubility and high mass exchange properties, Cho and Park [12] also synthesized a kind of polyimides based on 2,6-diamino-triptycene in 2011. As shown in Scheme 6.8, the target 6FDA–DATRI polyimide **20** was obtained by the ester–acid chemical imidization method. As expected, the target polyimide showed good solubility in common organic solvents, and high gas permeability due to the high internal free volumes, which could make it potential material for gas separation.

In the same year, Zhang and co-workers [13] reported the synthesis of novel poly(aryl ether sulfone) **21** containing triptycene groups with high molecular weight. As shown in Scheme 6.9, the polymers could be obtained by the nucleophilic aromatic substitution polycondensation between 2,5-dihydroxytriptycene, bis(4-hydroxyphenyl) sulfone, and 4,4′-difluorodiphenyl sulfone, followed by the sulfonation reaction. The degree of sulfonation could be controlled by changing the ratios of momomers. The target polymer **21** exhibited good thermal stability, good mechanical strength, and excellent water swelling resistance. These features made this polymer potential for proton exchange membranes in low humidity conditions.

Long and Swager [14, 15] considered that introducing the iptycene structure into polymers might have potential applications in electronics, as the rigid skeleton of iptycene created the molecular-scale pores. The increasing IFVs may be beneficial for lowering the dielectric constant of polymers. Therefore, Long and Swager [14] synthesized two different copolymers **22** and **23** from the strained cyclic olefinic monomers and iptycene-derived monomers by ROMP with a "second generation" Grubbs catalyst in CH_2Cl_2, which is shown in Scheme 6.10. For polymer **22**, the triptycene units could not freely rotate owing to the fixed orientation polymer backbone. On the contrary, the pendant triptycene units in the polymer **23** were attached to the backbone of polymer through a single bond, thus, these triptycene moieties

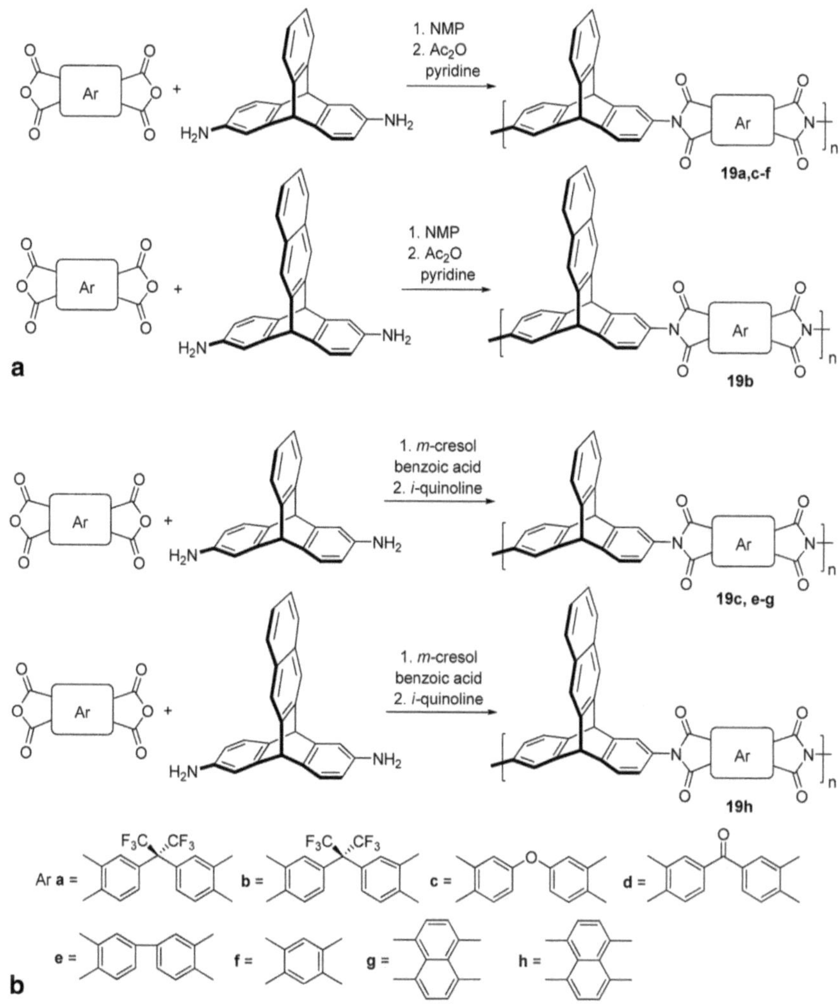

Scheme 6.7 Synthesis of aromatic polyimides **19a–h**

in polymer **23** could rotate freely. At the same time, Swager et al. also synthesized poly(aryl ether) **24** by the reaction of triptycene-1,4-hydroquinone and decafluoro biphenyl in the presence of K_2CO_3 (Scheme 6.11), and proved that the introduction of triptycenes into the polymer backbones increased thermal stability and resulted in lower dielectric constants.

Recently, McKeown and co-workers [16] reported a kind of insoluble tape-like polymers containing triptycene moieties. As shown in Scheme 6.12, the target polymer **26** (R = Et) could be synthesized by the condensation of 9,10-diethyl-2,3,6,7,12,13-hexahydroxytriptycene derivative **25** (R = Et) with 2,3,5,6-tetrafluoroterephthalonitrile at 80 °C in DMF in the presence of K_2CO_3. This new

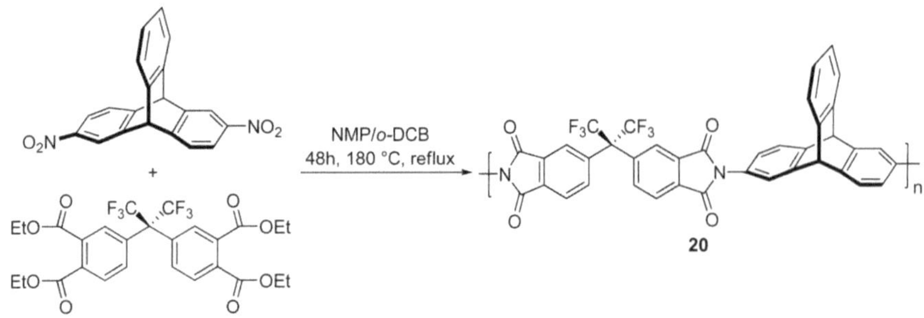

Scheme 6.8 Synthesis of polyimide **20** based on 2,6-diamino-triptycene

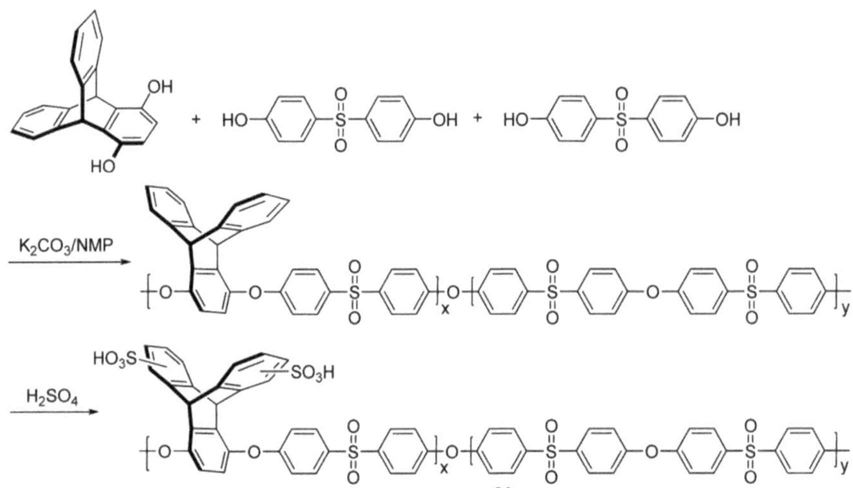

Scheme 6.9 Synthesis of poly(aryl ether sulfone) **21**

kind of triptycene-based polymers showed the enhanced surface area for the intrinsic microporosities, and thus had great potential applications in gas adsorption. Furthermore, they [17] obtained a series of network polymers **26** with intrinsic microporosity from the triptycene derivatives with different alkyl groups at 9,10-bridgehead positions. They also found that the shorter (e.g., methyl) or branched (e.g., isopropyl) alkyl chains were beneficial to the greatest microporosity, whereas the rigid organic framework of the longer alkyl chains seemed to hinder the formation of microporosity. These porous polymers could be regarded as the potential gas adsorption materials.

6.1.2 Triptycene-Containing Conjugated Polymers

In 2001, Zhu and Swager [18] first introduced the triptycene structure into the PPVs. As shown in Scheme 6.13a, starting from 1,4-dihydroxyl substituted triptycene, the

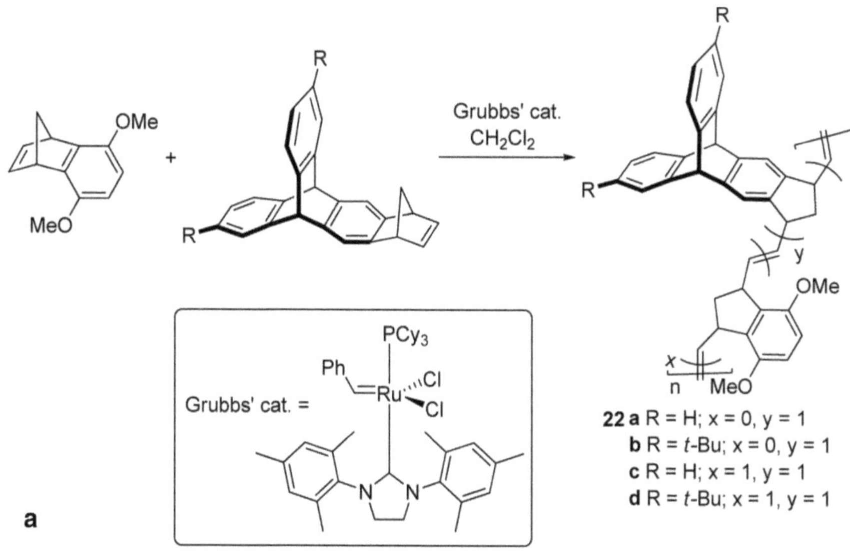

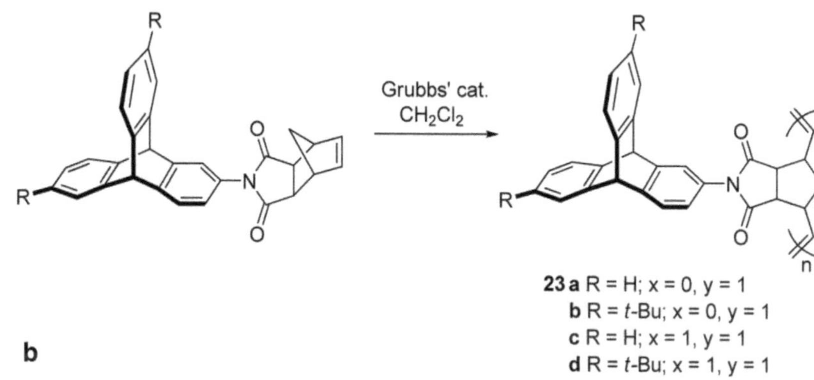

Scheme 6.10 Synthesis of copolymers **22** and **23**

triptycene derivative **27** was prepared by four-step reactions. Compound **27** could serve as a key precursor for the synthesis of triptycene-derived PPVs. Consequently, the target triptycene-derived PPVs **28a** could be obtained in 71 % yield by the Suzuki-type palladium-catalyzed cross-coupling reaction of **27** with 1,4-di(dodecyloxy)-2,5-diiodobenzene in dioxane in the presence of cesium carbonate. Similarly, the reaction of **27** with the 1,4-diiodo-2,3-di-(dodecyloxy)benzene could give the corresponding triptycene-containing PPVs **28b** in 79 % yield (Scheme 6.13b).

According to the similar approach as above, Zhu and Swager [19] further synthesized a series of triptycene-containing conjugated polymers **29a–d** by the Suzuki cross-coupling reaction of bisborolane **27** and 1,4-diiodobenzene derivatives (Scheme 6.14a). Moreover, they also synthesized two kinds of triptycene-based

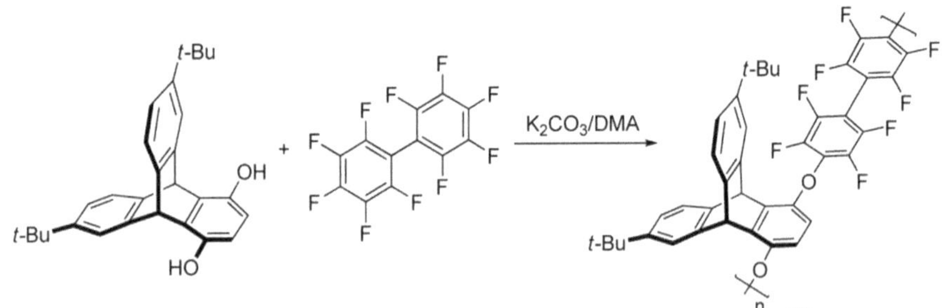

Scheme 6.11 Synthesis of poly(aryl ether) **24**

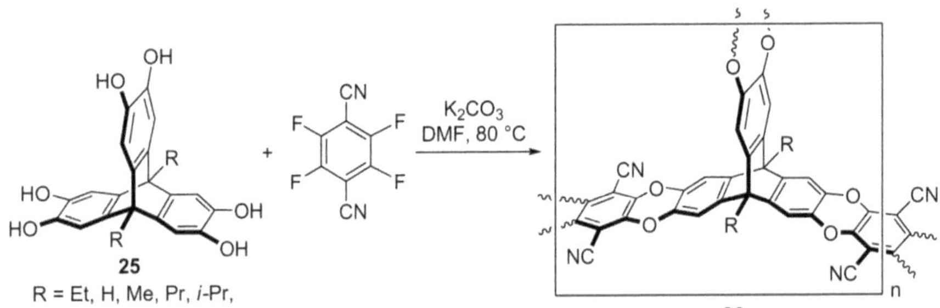

Scheme 6.12 Synthesis of network polymers **26**

PPEs **30a, b** via the Sonogashira protocols starting from the diethynyltriptycene **31** (Scheme 6.14b). Furthermore, they found that these conjugated copolymers **29** and **30** with high molecular weights exhibited high solubility and the controlled anisotropy in LC solutions.

Soon after, in order to achieve the improvements of alignment of polymers with high molecular weight, Hoogboom and Swager [20] further designed and synthesized the triptycene-containing polymer **32** with the hydrogen bonding end groups, Upy. As shown in Scheme 6.15, the polymer with hydrogen bonding donors of Upy end-capped groups was obtained by the Sonogashira cross-coupling reaction between diethynyltriptycene and di-2,5-(2-ethylhexyl)-1,6-diiodo-benzene, followed by the addition reaction of the end capping hydrogen bonding unit **33**. It was noteworthy that the reaction time for the polymer growth in this process could determine the properties of the target polymer **32**. Due to the Upy end-capped group as the source of hydrogen bonding, there were hydrogen bonding interactions between the polymer chains, which dramatically increased the dichroic ratio of this polymer system. Interestingly, the polymer **32** could self-assemble into ultrahigh molecular weight materials and gels via the hydrogen bonding interactions between the polymer chains.

In 2008, Swager and co-workers [21] synthesized a series of iptycene-derived conjugated copolymers containing fluorene subunit via the palladium-catalyzed coupling reaction. Consequently, the target copolymers **38–40** were obtained in high

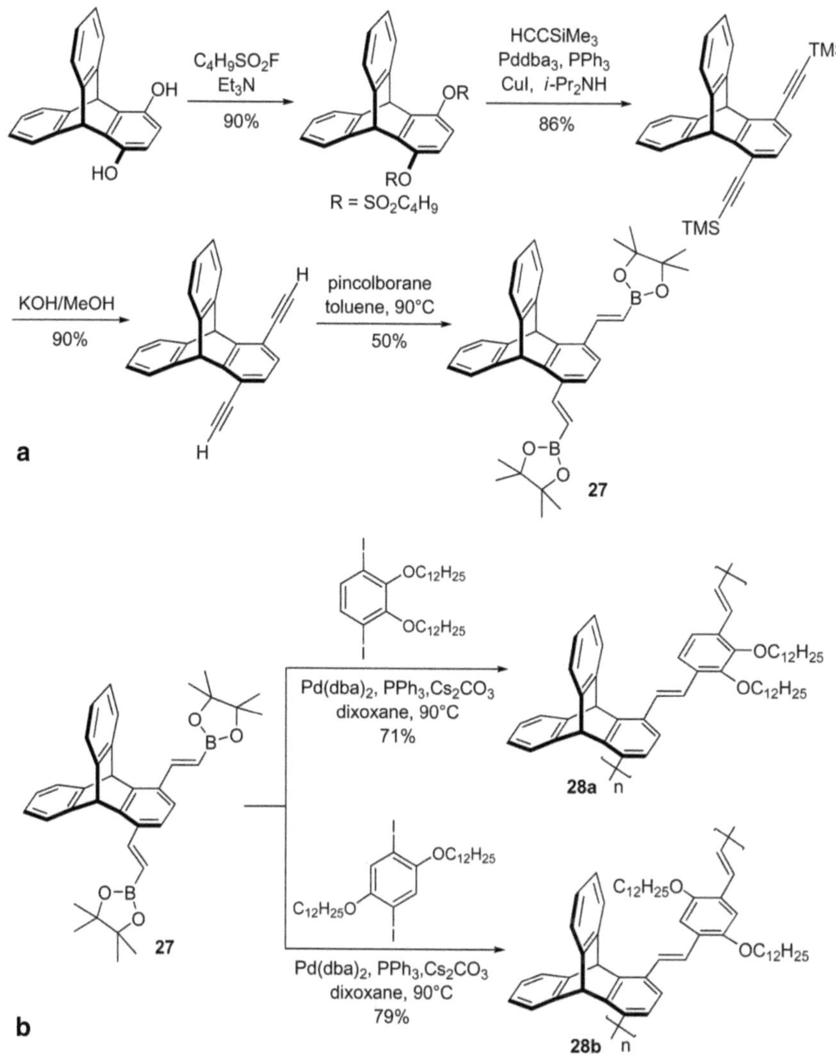

Scheme 6.13 Synthesis of PPVs **a 27** and **b 28**

yields by the palladium-catalyzed Suzuki reactions of triptycene-type monomers **34–36** with 9,9-dioctylfluorene-2,7-bis(trimethyleneboronate) **37** in the presence of Cs_2CO_3 (Scheme 6.16). The copolymers all exhibited good solubilities in appropriate organic solvents, and also emitted blue, greenish-blue, or red color in solution and the solid state for different repeating units. Moreover, the insulation effect of iptycene units clearly displayed in their spectroscopic data, because the rigid triptycene units blocked the intermolecular interactions between the chromophores, and prevented from the formation of the excimers of solid state.

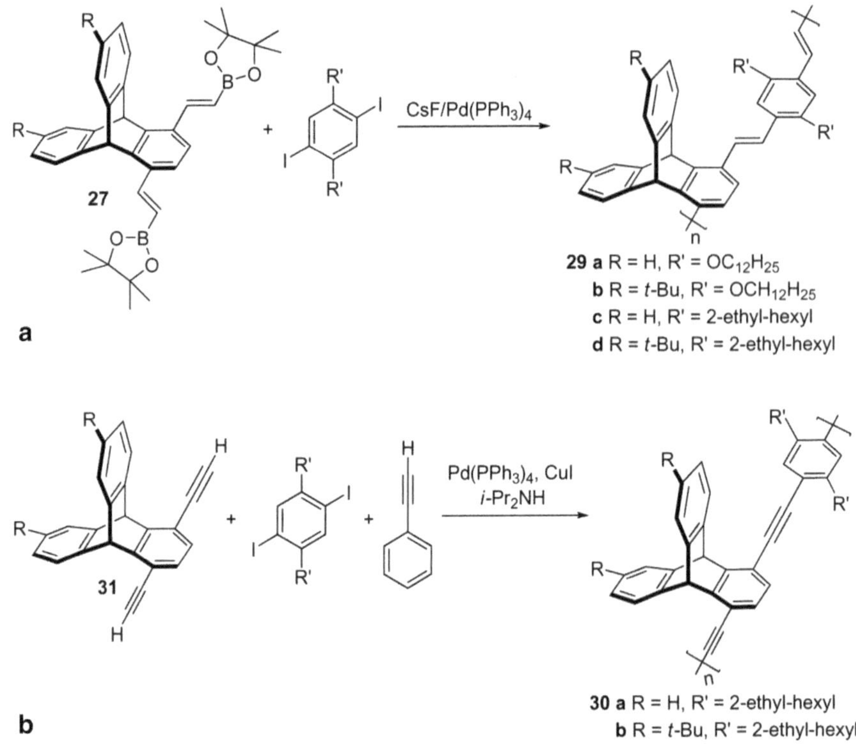

Scheme 6.14 Synthesis of triptycene-containing conjugated polymers **29** (**a**) and **30** (**b**)

6.2 Pentiptycene-Containing Polymers

As early as 1998, Yang and Swager [22, 23] first introduced the rigid 3D pentiptycene moieties, which are larger than the rigid triptycene ones, into the backbone of PPEs. As shown in Scheme 6.17a and b, the pentiptycene diacetylenes **44** and **45** reacted with disubstituted diiodobenzenes under the palladium-catalyzed cross-coupling reaction conditions to give the corresponding PPEs **41–43**. As the π–π stacking and the formation of excimer at the solid state of PPEs were prevented by the rigid pentiptycene structure, the pentiptycene-based polymers **41–43** could form the films from solutions without any significant red shifts in their absorption spectra. Moreover, the introduction of rigid frameworks of pentiptycene moieties also enhanced the solubility and the fluorescence quantum yields of the polymers **41–43**. It was also noteworthy that the pentiptycene-derived conjugated polymers **41–43** showed unprecedented high sensitivity for the recognition of electron-deficient unsaturated species including TNT, DNT, and BQ, which could be used as excellent fluorescent chemosensors for these species. These features for the sensor applications will be depicted in Chap. 12 in detail. According to the similar method, Zhu and Swager

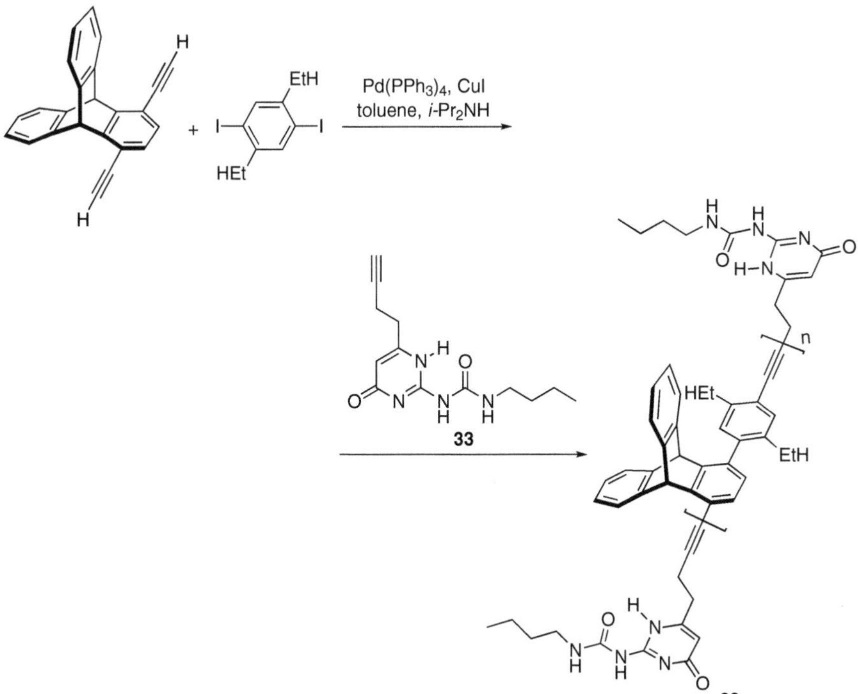

Scheme 6.15 Synthesis of triptycene-containing polymer **32**

[18] also synthesized another pentiptycene-based polymer **46** with high molecular weight by the palladium-catalyzed cross-coupling reaction between pentiptycene diacetylene and diiodidetriptycene under the modified Sonogashira conditions (Scheme 6.17c).

In a certain sense, the work of Swager and co-workers about the pentiptycene-based polymers broke the limitation of the previous work, and brought more possibilities to create iptycene-derived polymers via the palladium-catalyzed coupling reactions. In 2000, Williams and Swager [24] synthesized a broad range of iptycene-based conducting polymers via the cross-coupling reaction from the diiodo pentiptycene monomer **47**. With the diiodo monomer in hand, the random terpolymers **48** (Fig. 6.4) were easily obtained by the copolymerization between monomer **47**, 1,4-diiodotriptycene, and a number of diethynylphenyl monomers under a standard Sonogashira–Hagihara coupling reaction conditions. Due to the decreased number of chromophores associated with dialkoxybenzene groups, the principal absorption band of the terpolymers **48** in solid state showed a hypsochromic shift compared with the polymer **41**. Moreover, the absorption spectra of terpolymers **48** displayed only a small (< 2 nm) shift between the solution and thin film maxima; however, this shift was up to 14 nm for the reference triptycene polymer **49**. It was probably because the pentiptycene moiety could prevent the interactions between the

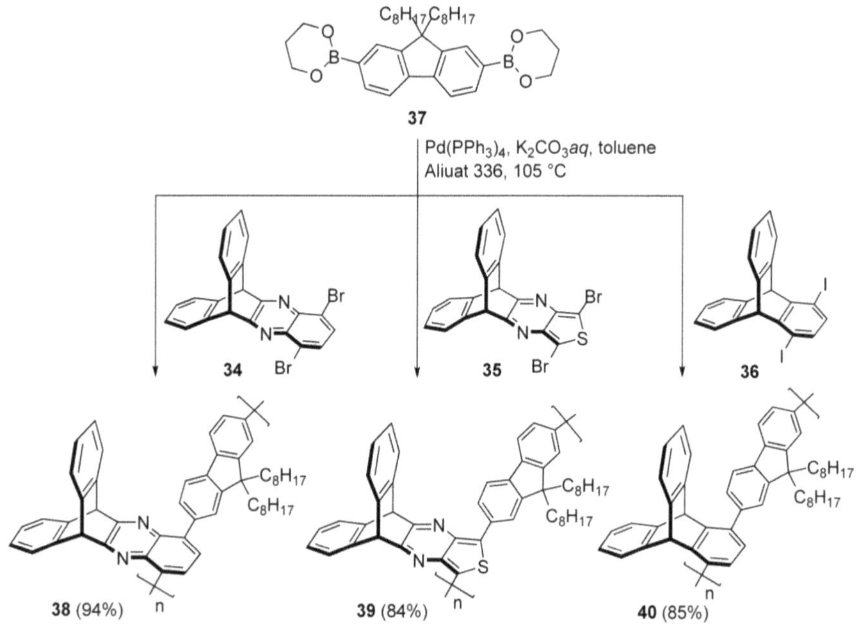

Scheme 6.16 Synthesis of iptycene-derived conjugated copolymers **38–40** containing fluorene subunit

polymer chains, which resulted in the slight perturbation from solution to the solid state. Moreover, the varying degrees of chromophore interactions in the solid state also influenced the thin-film fluorescence quantum yields of these polymers. As a result, the fluorescence quantum yields of terpolymers **48a** and **48b** were 0.22 ± 0.05 and 0.30 ± 0.04, respectively; whereas the thin-film fluorescence quantum yield was 0.09 ± 0.3 for the reference triptycene polymer **49**.

Amara and Swager [25] further obtained a series of pentiptycene-derived PPEs **50** in the yields of 75–87 % by the palladium-catalyzed Sonogashira–Hagihara cross-coupling polymerization of dialkynyl-substituted pentiptycene monomer **44** and different monomers with varying pendant HFIP groups (Scheme 6.18).

It was proved that π-conjugated frameworks with ring expansion reaction could effectively eliminate the conformational disorder. Therefore, Yamaguchi and Swager [26] attempted to introduce the dibenzochrysene into π-conjugated pentiptycene copolymer chain to afford the copolymers with excellent fluorescence properties, which would probably be suited for sensory applications. As shown in Scheme 6.19, the Pd/Cu-catalyzed cross-coupling reaction between dibenzochrysene monomer **51** and the monomer **44** gave the pentiptycene-based polymer **52** in 91 % yield, which exhibited a high quantum yield, small Stokes shift, and long excited-state lifetime.

In addition, Zahn and Swager [27] also introduced the chiral side chains into the pentiptycene framework to afford the chiral polymer **53** (Fig. 6.5a), which showed an excellent quantum yield for its 3D chiral grid-like aggregated structure. The further study discovered that this chiral polymer **53** could aggregate into a helical grid

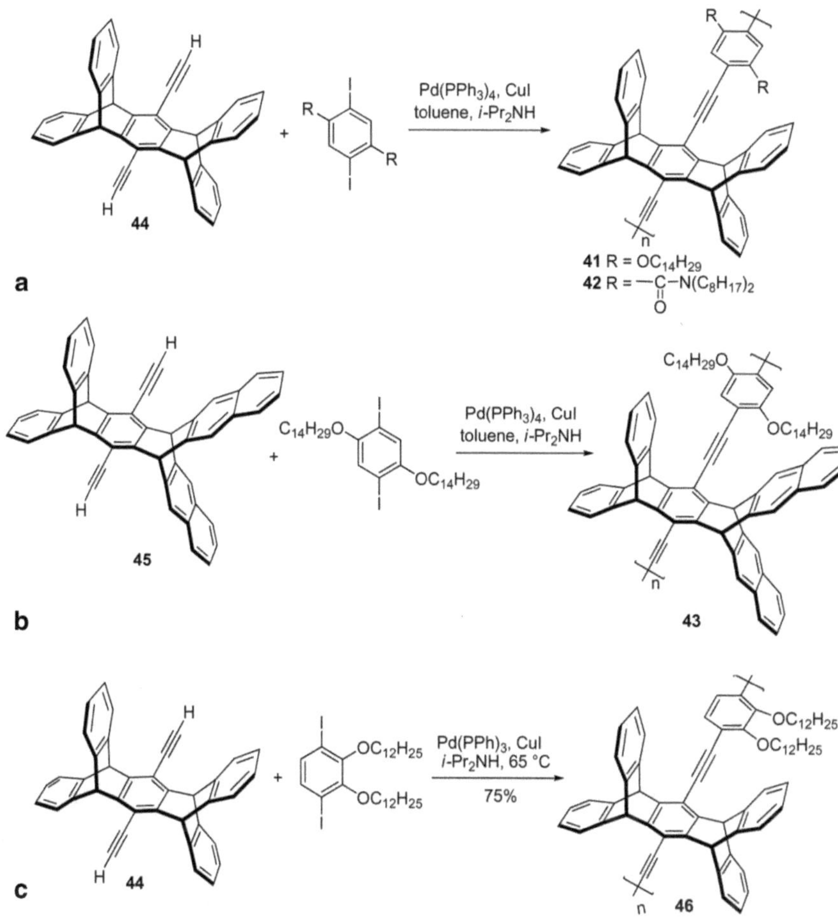

Scheme 6.17 Synthesis of pentiptycene-based PPEs

structure with a high quantum yield. This probably was the first example of chiral polymer with strong and systematical electronic coupling while maintaining high fluorescence efficiency. Soon after, Schanze and co-workers [28] reported the synthesis of Pt–acetylide-based polymer **54** (Fig. 6.5b) containing the bulky pentiptycene moiety by the reaction between *cis*-dichlorobis(tri-*n*-butyl-phosphine)-platinum(II) and diethynyl pentiptycene.

In 2006, Bailey and Swager [29] reported the synthesis of the PPEs containing masked maleimide groups which could participate in the thiol addition reactions or Diels–Alder additions. However, it was found that in the process of palladium-catalyzed cross-coupling polymerization, these maleimide groups were left unmasked, which resulted in the formation of short chain oligomers with number-average molecular weights less than 3,000. In order to overcome the problem of the incompatibility of maleimide functionality under cross-coupling-based

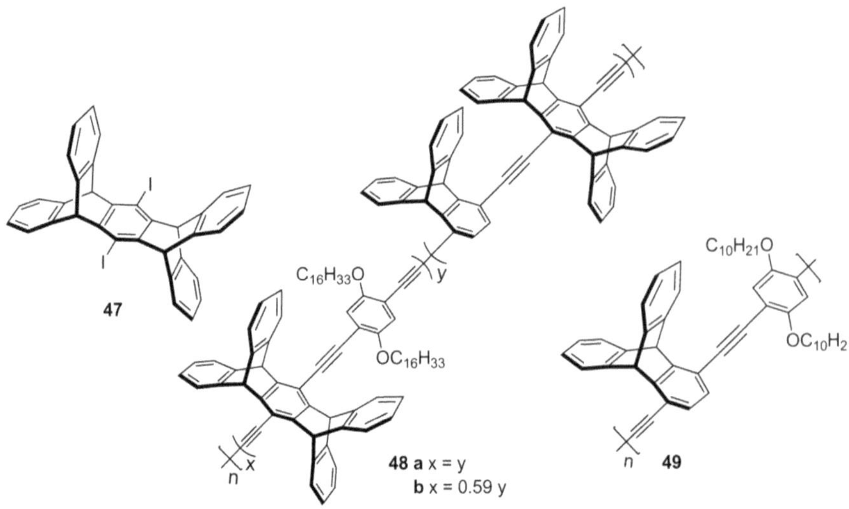

Fig. 6.4 Structures of compound **47**, and the polymers **48**, **49**

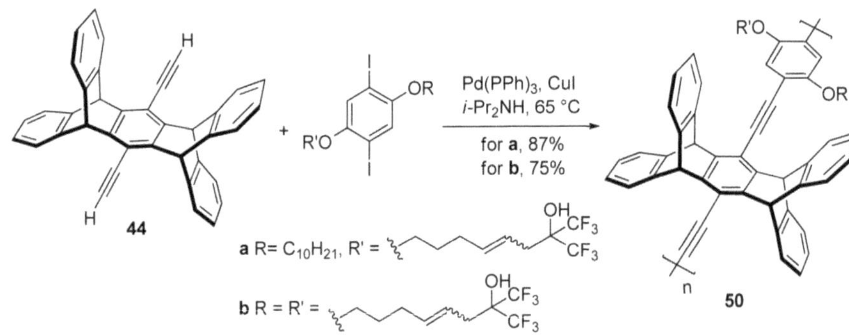

Scheme 6.18 Synthesis of pentiptycene-derived PPEs **50**

polymerization conditions, the blocking furan group was introduced to ensure the polymerization. As furan could be removed quantitatively under relatively mild thermal conditions via cycloreversion, the maleimide-functionalized PPEs with number-average molecular weights of 8,000–11,000 could be favorably obtained. As shown in Scheme 6.20, treatment of monomer **44** with suitable aryl diiodide monomers in a step-growth polymerization under the Sonogashira–Hagihara cross-coupling reaction conditions afforded the protected maleimide-containing polymers **55** and **56** with the number-average molecular weights of 11,000 and 8,000, respectively. The protected maleimide-containing polymers **55** and **56** then reacted with the ROX dye containing sulfhydryl group and DTT in refluxed THF solution to give the corresponding rhodamine-tethered conjugated polymers **57** and **58**, respectively (Scheme 6.21). The thiol-based biomolecule conjugation polymers **57** and **58** could be promising candidates for biosensory or chemosensory applications.

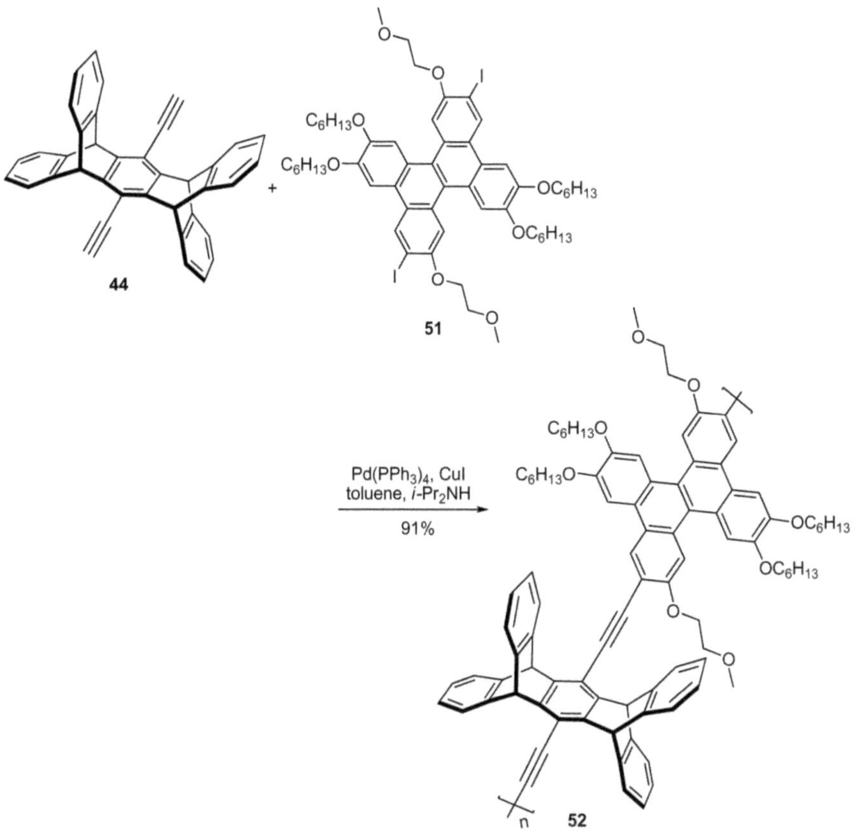

Scheme 6.19 Synthesis of pentiptycene-based polymer **52**

In 2005, Swager and co-workers [30] designed and synthesized a special iptycene-containing PPEs **59**, which was end-capped with 10-(phenyl-ethynyl)-anthracenyl groups. As shown in Scheme 6.22, the copolymerization between the pentiptycene-based monomer **60** and 10 % excess of diiodo monomer **61** via a Sonogashira coupling protocol afforded the copolymer with the chain ended with iodine groups, which was then treated with an excess of the acetylene reagent to give the target end-capped polymer **59**. As expected, the polymer **59** could dissolve in a nematic LC phase, enhance the quantum yields, and reduce Stokes shift probably because of its more rigid and planar conformation.

Furthermore, Ohira and Swager [31] prepared several novel PPEs **62–64** (Fig. 6.6) by the palladium-catalyzed cross-coupling reaction to introduce the larger and more complex iptycene skeleton into the chains. As expected, the more steric crowding iptycene moieties inhibited the anisotropic ordering and prevented the interactions between the polymer chains, and these molecules could control the orientation of the liquid crystalline. Thus, these polymers with high molecular weights (> 10^5 Da) could still be completely soluble in nematic LC solvents.

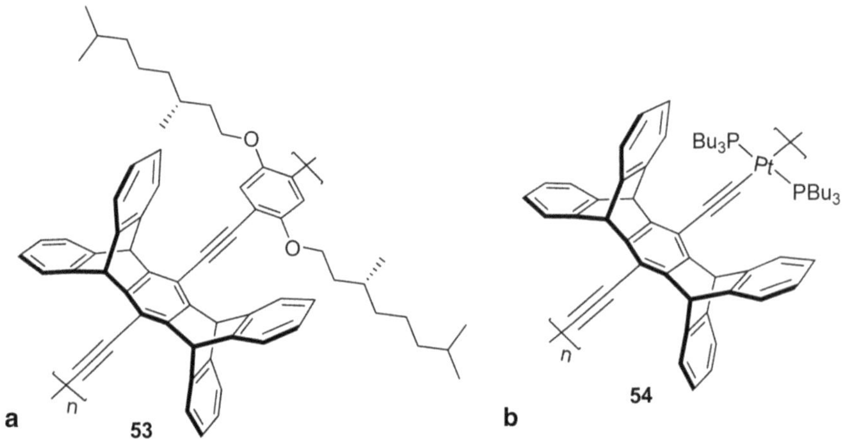

Fig. 6.5 Structures of polymers **53** and **54**

Besides the pentiptycene-containing conjugated polymers, Long and Swager [14] also synthesized a series of poly(aryl ether)s **65–67** with iptycene moieties incorporated into the polymer skeleton by the condensation of iptycene diols with decafluorobiphenyl in DMA in the presence of potassium carbonate (Scheme 6.23). These poly(aryl ether)s were proved to show increased thermal stabilities and lower dielectric constants.

In 1980s, Meador et al. [32, 33] reported the synthesis of the copolymers **68a**, **b** with anthracene end-capped polyimide oligomers. By the Diels–Alder addition between the bi-epoxide **69** and the anthracene end-capped oligomer **70** in the minimum amount of solvent under 155 °C and a 300–400 psi nitrogen environment, the target copolymers with the molecular weights approximately achievable from 21,000 (**68a**) to 32,000 dalton (**68b**) were obtained (Scheme 6.24). The molecular weights of these polymers could be controlled by the reaction time. Moreover, the copolymers exhibited thermal stability and solubility in common organic solvents, like $CHCl_3$, CH_2Cl_2, and DMF, and the solubility of copolymers **68c** was lower than the other two copolymers. However, the polymers would be at the cross-linked state, which was insoluble above 500 °C. In addition, by the dehydration, the backbone of polymer **71** containing the extended pentiptycene units could be obtained via the retro-Diels–Alder reaction. The target polymers were very stable and without degradation.

6.3 Other Iptycene-Containing Polymers

Compared with other conjugated polymers without the rigid bulk moiety, the steric bulk of pentiptycene skeleton prevented the strong π–π associations of conjugated polymers, which could result in excellent solubility. It was considered that the larger size of iptycene could be more effective to isolate the polymer backbone. Thus,

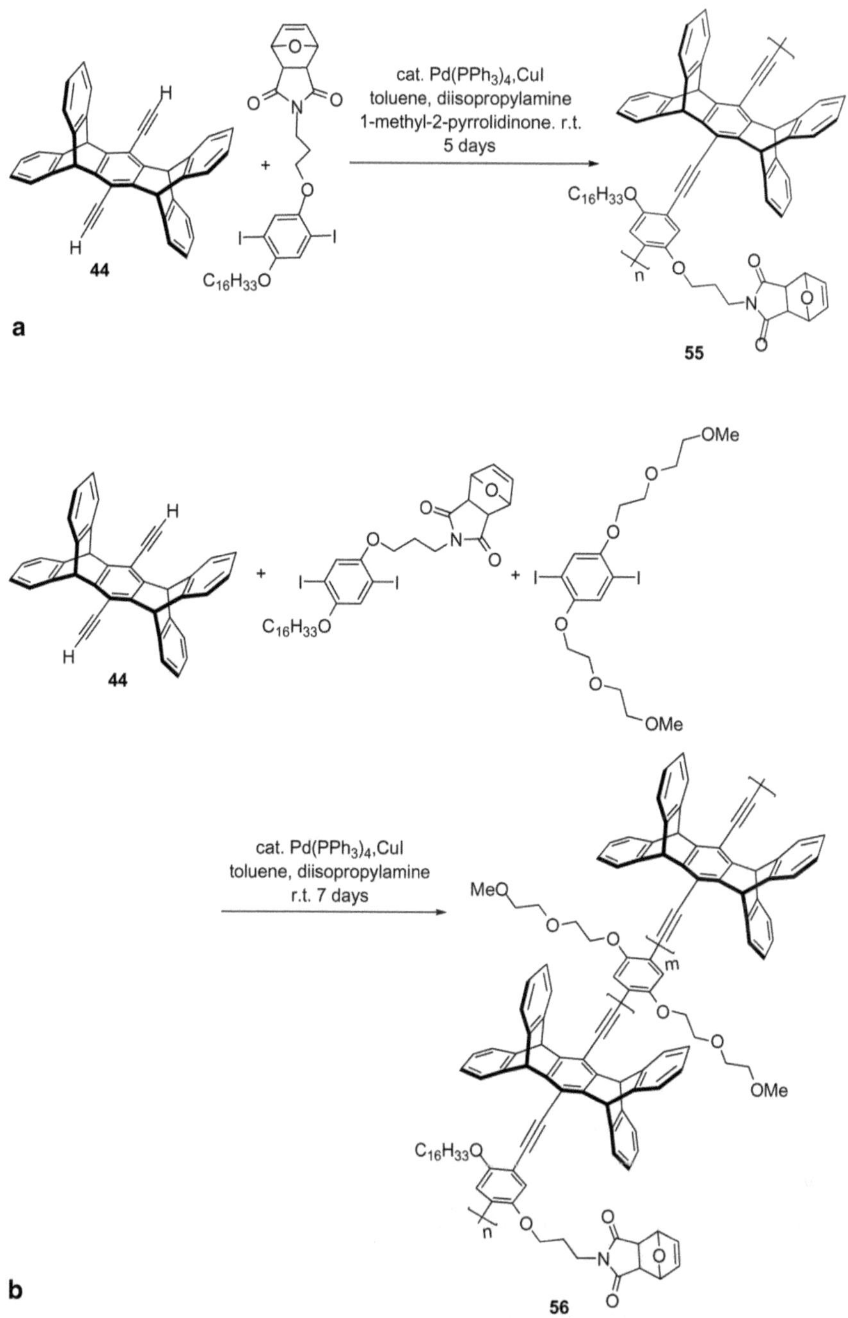

Scheme 6.20 Synthesis of protected maleimide-containing polymers **55** and **56**

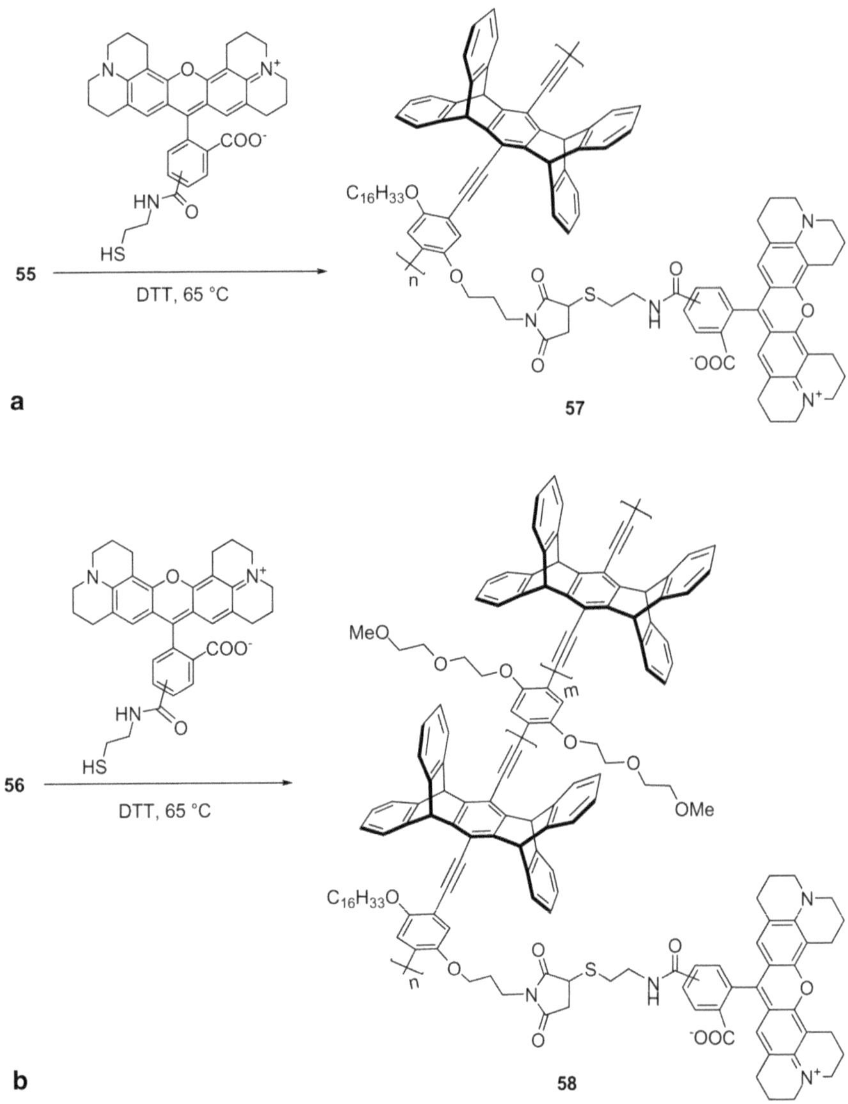

Scheme 6.21 Synthesis of rhodamine-tethered conjugated polymers **57** and **58**

Zhao and Swager [34] further attempted to introduce the more complex iptycene moieties into the PPEs. Consequently, a series of iptycene-based copolymers **72a–c** (Fig. 6.7) could be obtained by the Sonogashira coupling reaction between the iptycene-based diacetylenes and 2,5-di-iodopyridine, 5,5′-diiodo-2,2′-bipyridine, or 1,4-diiodo tetrafluorobenzene, respectively. Moreover, they found that the polymers **72a–c** all expectedly exhibited no evident excimer interactions or self-quenching in the thin films or in solution.

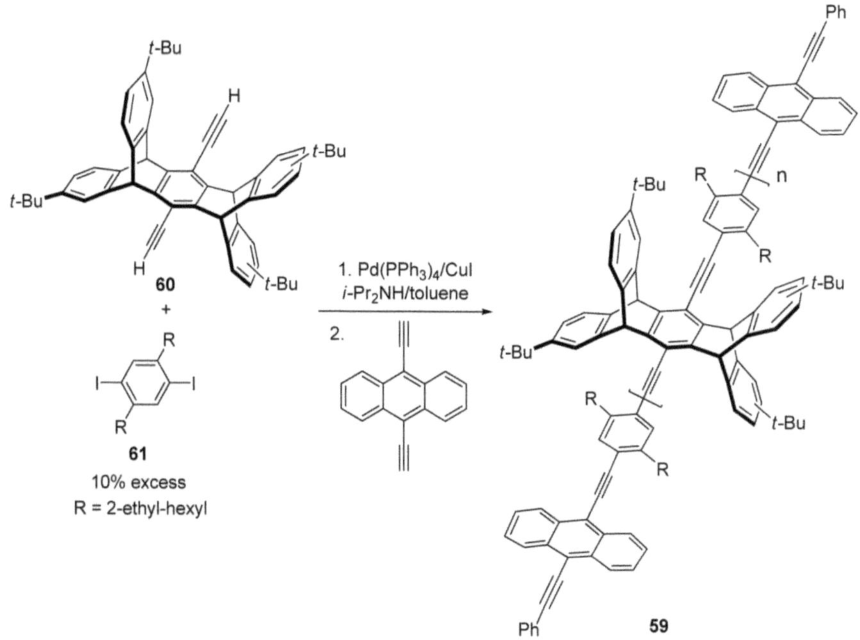

Scheme 6.22 Synthesis of iptycene-containing PPEs **59**

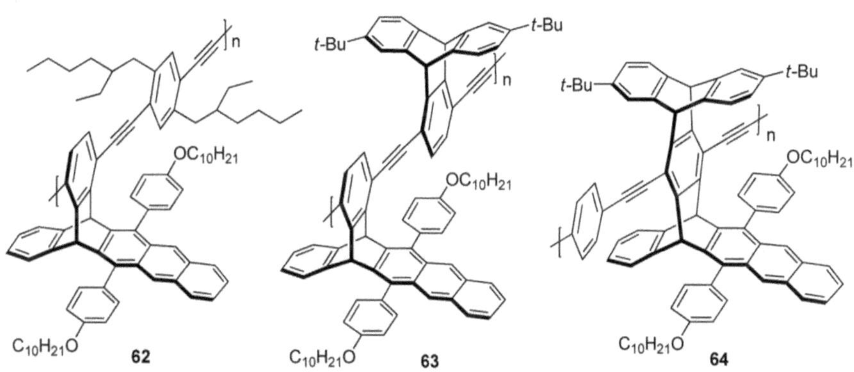

Fig. 6.6 Structures of novel PPEs **62–64**

In 2005, Zao and Swager [34] further synthesized a series of PPDs **73a–c** with the iptycene structure as the core unit of backbones. As shown in Scheme 6.25, the target polymers **73a–c** could be obtained easily by the Pd-catalyzed homocoupling reaction from the iptycene monomers **74a–c**. Likewise, the iptycene moieties incorporated in the PPDs prevented the interchain interactions of solid state, and thus enhanced their solubility in common organic solvents; although, the gel-like substance (the gel could be almost dissolved in $CHCl_3$) would form at the end of polymerization

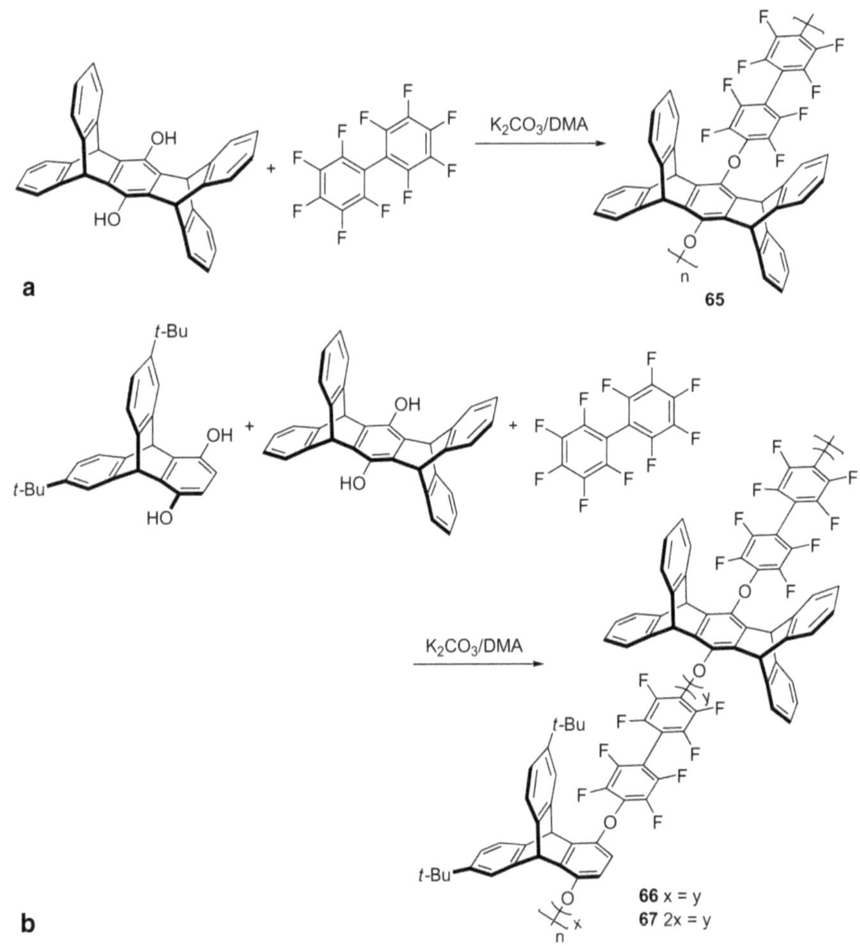

Scheme 6.23 Synthesis of poly(aryl ether)s **65–67** with iptycene moieties

of **73a**. It was noted that the PPDs in solution showed good sensitivity, even better than PPEs towards explosives detection.

6.4 Poly(iptycenes)

In 2003, Wudl and co-workers [35] reported a novel aromatic ladder polymer, poly(iptycene) **75**, which was completely composed of iptycene units. As shown in Scheme 6.26, 2,3,9,10-tetrachloropentacene (Cl_4Pn) was heated up to 350 °C in nitrogen to give the insoluble, hyperbranched ladder polymer **75** through a clean Diels–Alder self-coupling reaction. This reaction process could be monitored by the

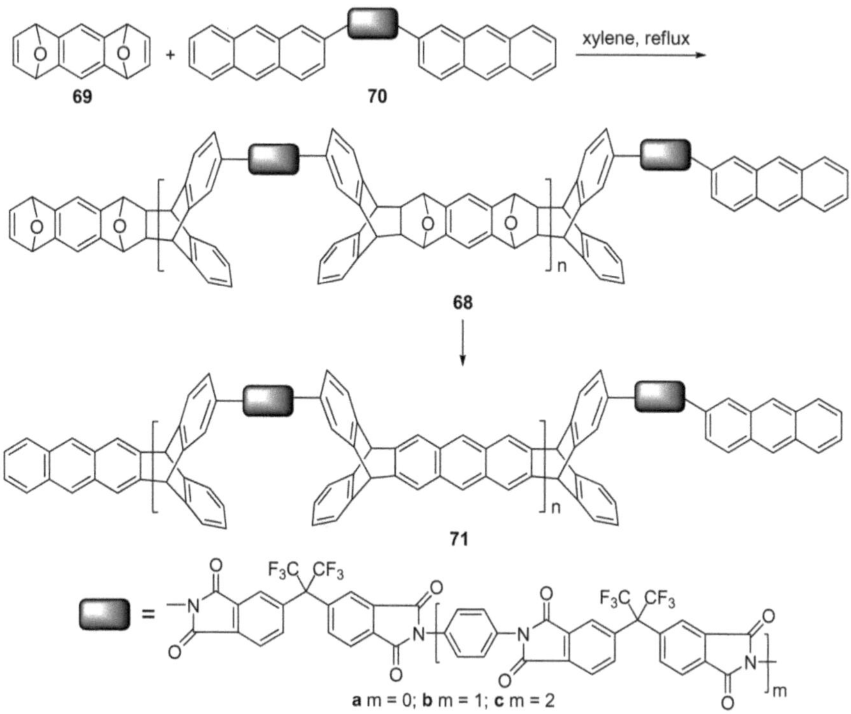

Scheme 6.24 Synthesis of polymer **71** containing the extended pentiptycene units

thermal gravimetric analysis of the thermal decomposition of Cl_4P. Thus, the emergence of a plateau in the thermogravimetric curve could be seen as the formation of polymer **75**. Although both the reactant and target polymers were in a high symmetry, the reaction process probably was in a biradical asymmetric mechanism, which was revealed by the theoretical calculations of this reaction.

In 2005, Swager and co-workers [36] reported the synthesis of a soluble, shape-persistent, 2D ladder-type poly(iptycene). At first, they attempted to synthesize the desired polymer via the traditional Diels–Alder reaction, and found that when the mixture of reactants was in a refluxing solution for 1 week without pressurizing, the result was not fully up to expectations. Thus, they modified the conditions and found when the polymerization of **78** occurred under the hyperbaric conditions in the presence of small amounts of hyperbranching agents **76** and **77** (Fig. 6.8), the higher overall degree of polymerization product **79** could be obtained, along with fewer degradation products (Scheme 6.27). It was interesting that the polymer **79** was probably the first example of two different polymers aligning perpendicular to each other.

Soon after, Swager et al. [37] further synthesized a novel aromatic ladder polymer, poly(iptycene) **80**, by the Diels–Alder reaction of the bifunctional AB-type monomer **81** (1,4-epoxy-5,12-di-hexyloxy-1,4-dihydrotetracene) in the melt phase under the

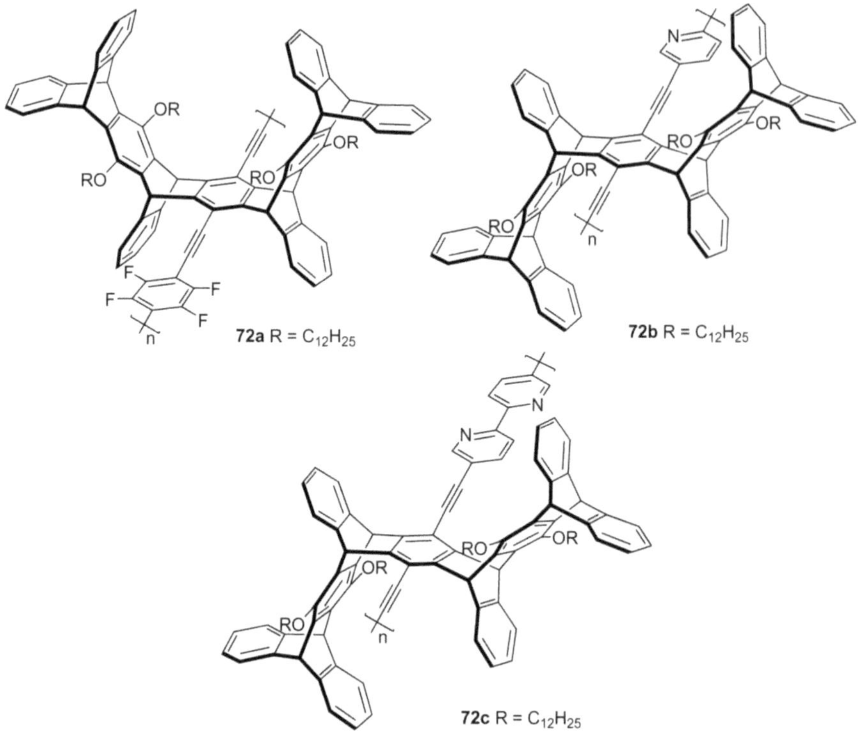

Fig. 6.7 Structures of iptycene-based copolymers **72a–c**

high pressure in solution to reach the high degree of polymerization. And then, the dehydration of **80** by the treatment with pyridinium *p*-toluenesulfonate and acetic anhydride could give the all-iptycene polymer **82** in almost constant molecular weight (Scheme 6.28). As a result, both the side-chain and 3D iptycene units resulted in good solubility of polymer **82** in organic solvents including CH_2Cl_2, $CHCl_3$, and THF.

Recently, Chen and Swager [38] reported the synthesis of a new aromatic polymer, poly(2,6-triptycene) **83**. At first, they found that the target polymer **83** could not be obtained by the palladium(0)-catalyzed Stille coupling polymerization. Thus, they then modified the route to synthesize the polymer by a nickel(0)-mediated Yamamoto-type polycondensation. As shown in Scheme 6.29, the desired polymers could be obtained in a reasonable yield and a good polymerization degree by the homopolymerization of the two monomers of 2,6-dibromotriptycene and 2,6-diiodotriptycene. Likewise, the high content of triptycene units in poly(2,6-triptycene) resulted in the good solubility in common organic solvents. The properties of highly transparent and good thermal stability for the thin film of **83** made it a promising candidate for further applications.

In 2000, Williams and Swager [24] synthesized another novel soluble "all-iptycene," poly(aryleneethynylene) **85** (Fig. 6.9), by the Sonogashira–Hagihara

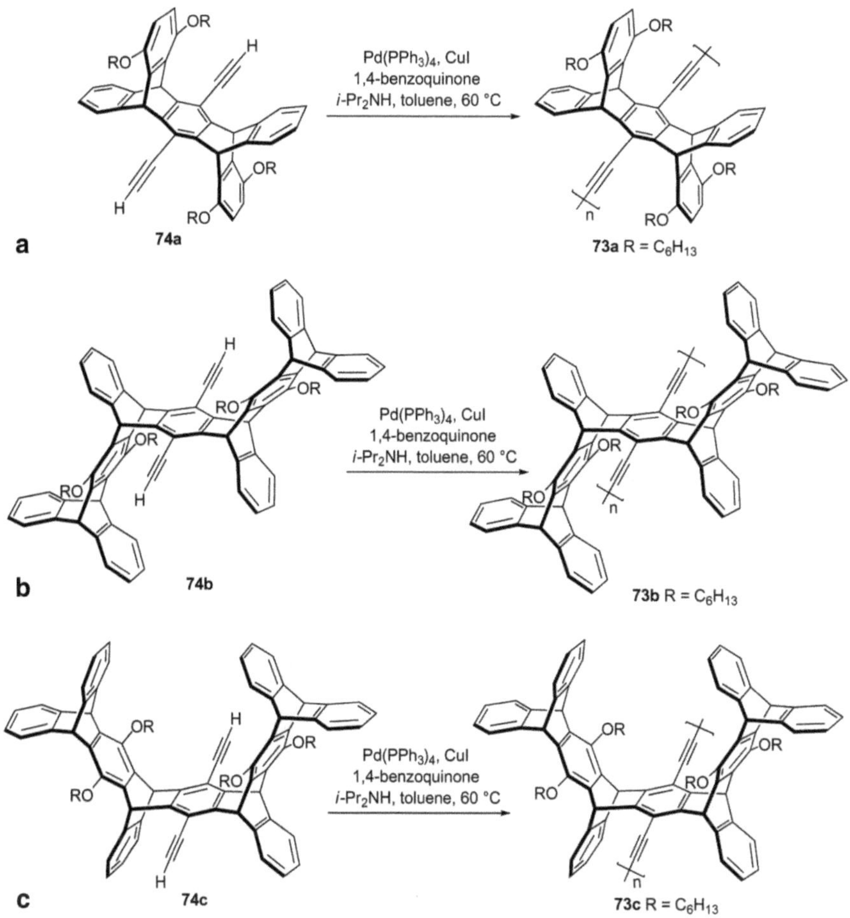

Scheme 6.25 Synthesis of PPD **73**

coupling of the diiodothiophene monomer **84** and pentitypcene-based monomer **46** in a CH_2Cl_2 solution. This all-iptycene polymer **85** displayed the excellent solubility in CH_2Cl_2, whereas it also showed to be slightly soluble in other common solvents including chloroform, toluene, and THF.

By the self-initiated free radical polymerization of butadienes bearing iptycene structures at the heating condition, Amara and Swager [39] also reported the synthesis of a series of all-iptycene polymers **86–88** with high molecular weights in 2004, which is shown in Scheme 6.30. The poly(butadiene)s, especially polymer **86** with large internal free volume, could lead to the low dielectric constants, the enhancement of miscibility, and the reduction of phase separation in polymer blends. However, the π–π interactions between the naphthalene rings on adjacent chains brought an adverse effect on the solubility and porosity of polymer **87**.

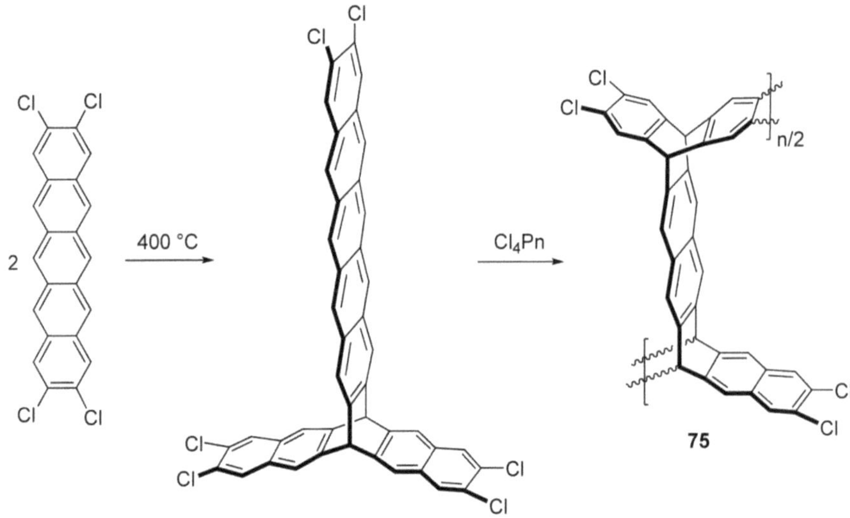

75

Scheme 6.26 Synthesis of aromatic ladder polymer, poly(iptycene) **75**

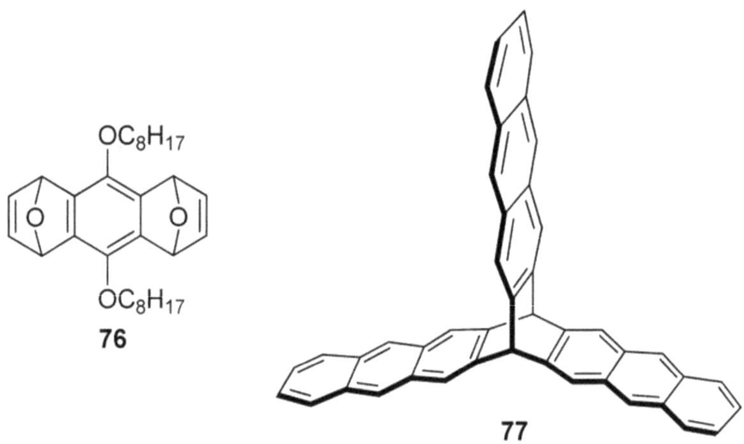

76

77

Fig. 6.8 Structures of hyperbranching agents

6.5 Iptycene-Based Oligomers

0In 2006, Yang and co-workers [40] reported the first example of iptycene-based oligomers, pentiptycene-derived oligo(p-phenyleneethynylene)s **89–91**, which could be synthesized by an iterative addition-aromatization method. In the case of the dimer **89**, the dimerization of the compound **93** gave the intermediate **92**, which was then followed by a sequence of reactions, including nucleophilic carbonyl addition, reductive aromatization, and o-alkylation to give the target dimer **89** (Scheme 6.31).

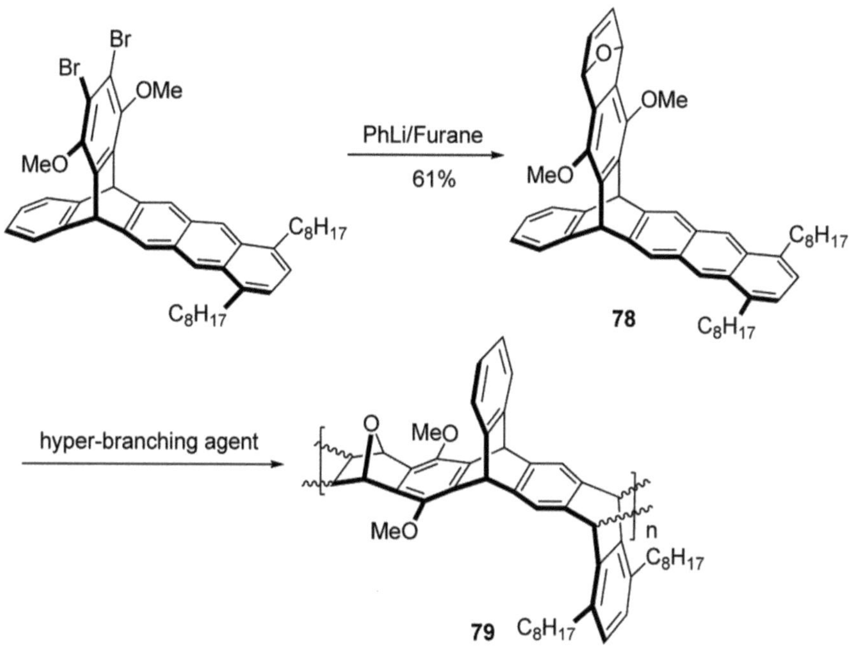

Scheme 6.27 Synthesis of ladder-type poly(iptycene) **79**

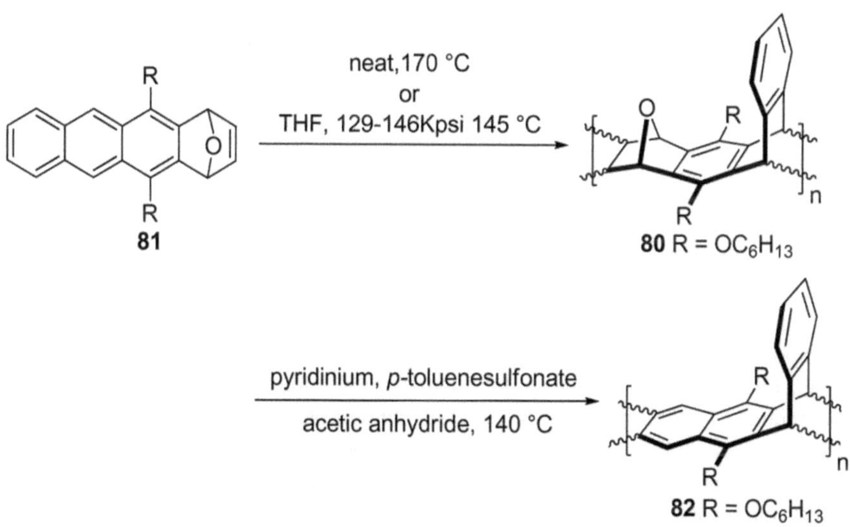

Scheme 6.28 Synthesis of all-iptycene polymer **82**

For the trimer **90**, it was obtained by the repeated reaction between the pentiptycene acetylide **93** and the pentiptycene quinone **94**. Similarly, the tetramer **91** could be synthesized by the coupling reaction of dimer **95** (Fig. 6.10). These oligomers could be served as good models to study the intrachain conformation of PPEs and interchain exciton coupling effects with their unique fluorescence behaviors. It was

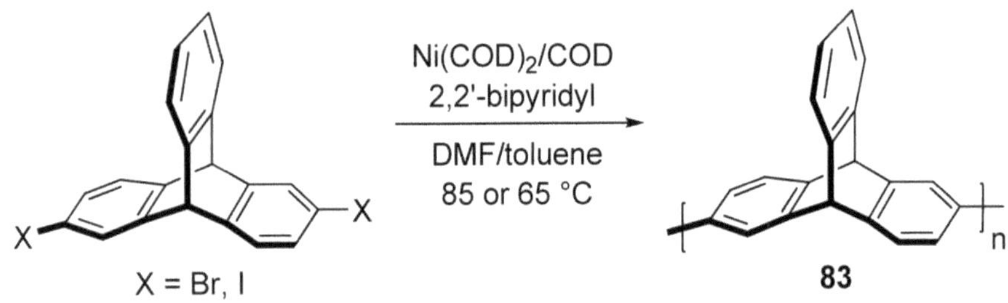

Scheme 6.29 Synthesis of poly(2,6-triptycene) **83**

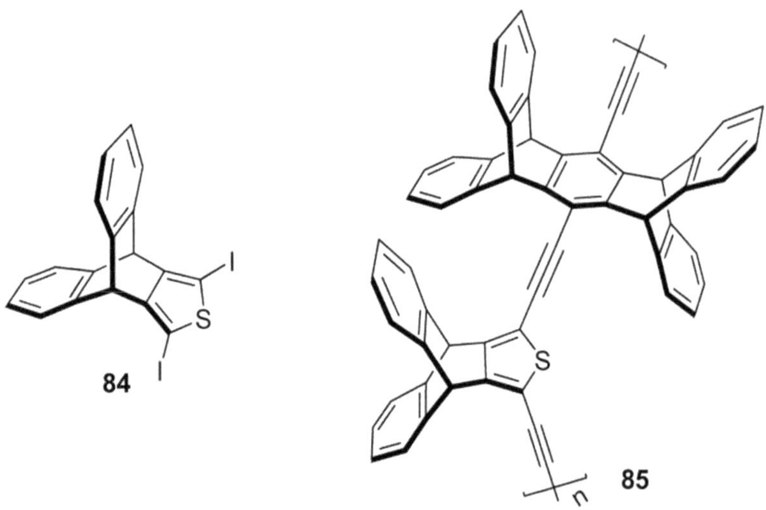

Fig. 6.9 Structures of the compounds **84** and **85**

found that the twisting of the π-conjugated backbones brought blue-shifted absorption, and higher fluorescence quantum yields and longer lifetimes, compared with the phenylene–pentiptycene–phenylene three-ring system.

Soon after, Godt and co-workers [41] reported a kind of longer, monodisperse oligo-PPEs **96a–d** ($n = 1$–4) containing triptycene moieties. These oligomers could be obtained by the repetitive selective alkyne deprotection, followed by the Pd/Cu-catalyzed alkynyl–aryl coupling from the alkyne **97**. The synthesis of the first two members **96a** ($n = 1$) and **96b** ($n = 2$) was depicted in Scheme 6.32. Although the triptycene-based oligomers showed the similar optical properties with the oligo(2,5-di-hexyl-1,4-phenylene-ethynylene)s in dilute solution, markedly difference between the oligomers and the reference in the solid or undiluted state was also observed.

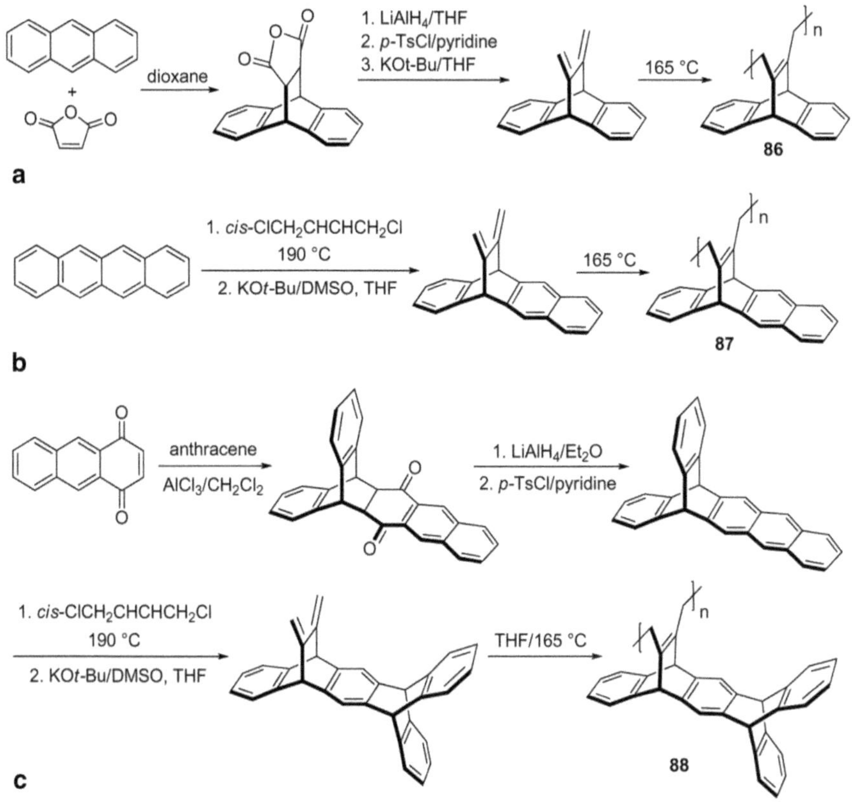

Scheme 6.30 Synthesis of all-iptycene polymers **86–88**

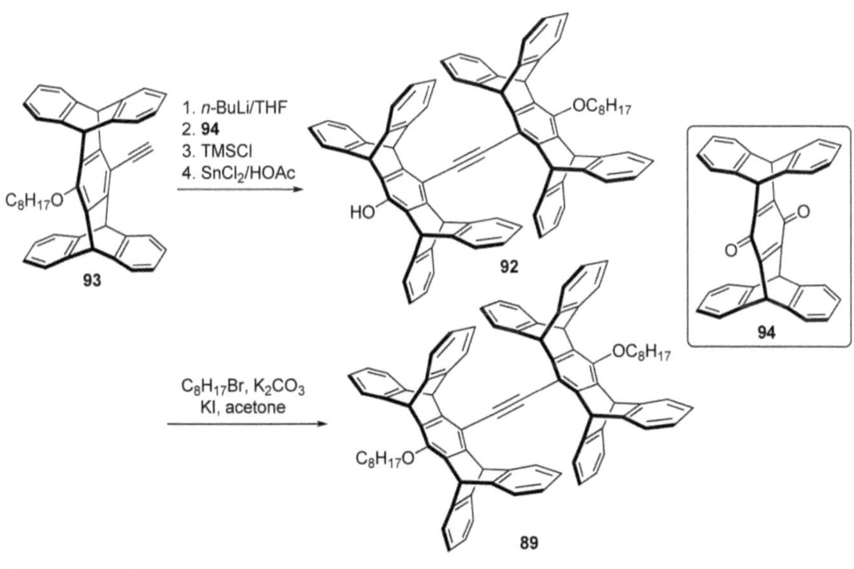

Scheme 6.31 Synthesis of pentiptycene-derived oligo PPE **89**

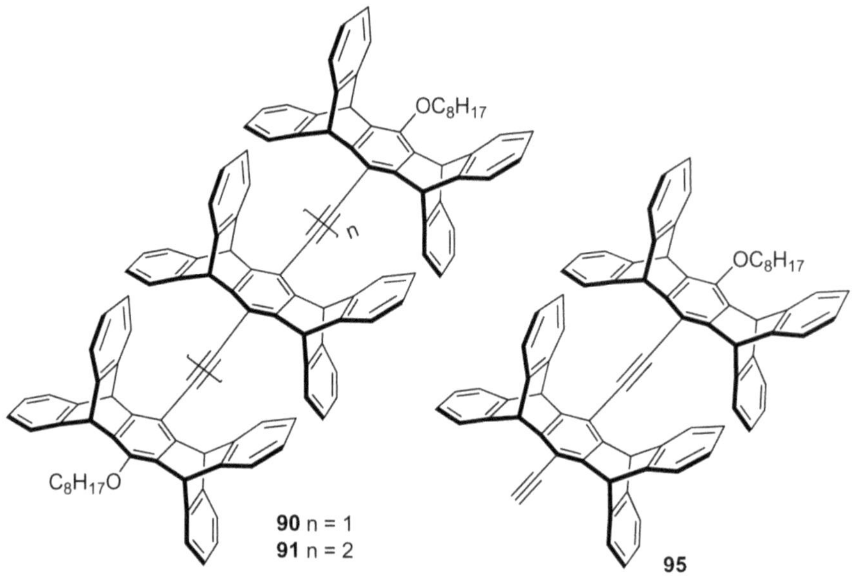

Fig. 6.10 Structures of iptycene-based oligomers **90**, **91**, and **95**

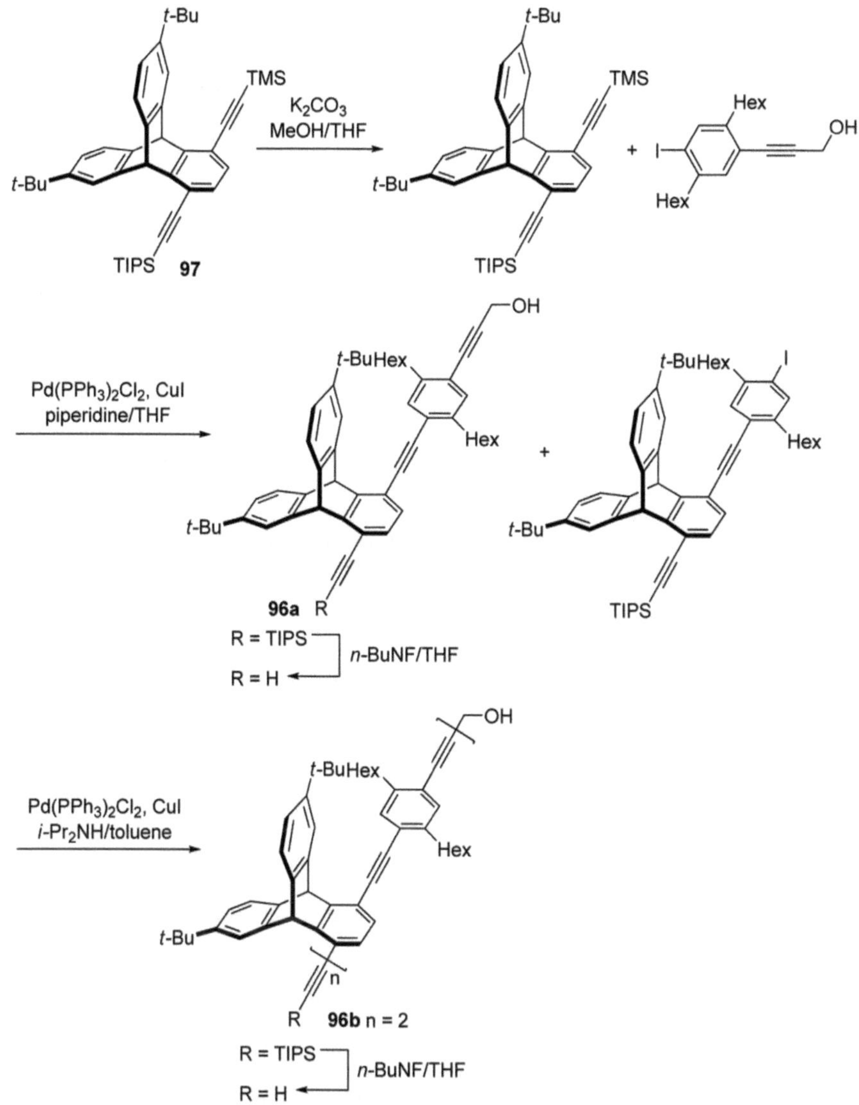

Scheme 6.32 Synthesis of monodisperse oligo-PPE **96**

References

1. Klanderm BH, Faber JWH (1968) Novel bridged anthracene derivatives and polyesters and copolyesters therefrom. J Polym Sci Part A1 Polym Chem 6:2955–2965
2. Hoffmeister E, Kropp JE, McDowell TL, Michel RH, Rippie WL (1969) Triptycene polymers. J Polym Sci Part A1 Polym Chem 7(1):55–72

3. Kasashima Y, Kaneda T, Saito G, Akutsu F, Naruchi K, Miura M (1994) Synthesis and properties of aromatic polyamides from 2,7-triptycenediamine. Macromol Chem Phys 195(8):2693–2697
4. Akutsu F, Inoki M, Kondo M, Inagawa T, Kayaki K, Kasashima Y (1997) Synthesis and properties of novel polyarylates and poly(ether ether ketone)s derived from 2,7-triptycenediol. Polym J 29(12):1023–1028
5. Tsui NT, Paraskos AJ, Torun L, Swager TM, Thomas EL (2006) Minimization of internal molecular free volume: a mechanism for the simultaneous enhancement of polymer stiffness, strength, and ductility. Macromolecules 39(9):3350–3358
6. Tsui NT, Torun L, Pate BD, Paraskos AJ, Swager TM, Thomas EL (2007) Molecular barbed wire: threading and interlocking for the mechanical reinforcement of polymers. Adv Funct Mater 17(10):1595–1602
7. Tsui NT, Yang Y, Mulliken AD, Torun L, Boyce MC, Swager TM, Thomas EL (2008) Enhancement to the rate-dependent mechanical behavior of polycarbonate by incorporation of triptycenes. Polymer 49(21):4703–4712
8. Liu Y, Turner SR, Wilkes G (2011) Melt-phase synthesis and properties of triptycene-containing copolyesters. Macromolecules 44(11):4049–4056
9. Zhang Q, Li S, Li W, Zhang S (2007) Synthesis and properties of novel organosoluble polyimides derived from 1,4-bis 4-(3,4-dicarboxylphenoxy) triptycene dianhydride and various aromatic diamines. Polymer 48(21):6246–6253
10. Hsiao SH, Wang HM, Chen WJ, Lee TM, Leu CM (2011) Synthesis and properties of novel triptycene-based polyimides. J Polym Sci Pol Chem 49(14):3109–3120
11. Sydlik SA, Chen Z, Swager TM (2011) Triptycene polyimides: soluble polymers with high thermal stability and low refractive indices. Macromolecules 44(4):976–980
12. Cho YJ, Park HB (2011) High performance polyimide with high internal free volume elements. Macromol Rapid Commun 32(7):579–586
13. Gong F, Mao H, Zhang Y, Zhang S, Xing W (2011) Synthesis of highly sulfonated poly(arylene ether sulfone)s with sulfonated triptycene pendants for proton exchange membranes. Polymer 52(8):1738–1747
14. Long TM, Swager TM (2003) Molecular design of free volume as a route to low-κ dielectric materials. J Am Chem Soc 125(46):14113–14119
15. Swager TM (2008) Iptycenes in the design of high performance polymers. Acc Chem Res 41(9):1181–1189
16. Ghanem BS, Msayib KJ, McKeown NB, Harris KDM, Pan Z, Budd PM, Butler A, Selbie J, Book D, Walton A (2007) A triptycene-based polymer of intrinsic microposity that displays enhanced surface area and hydrogen adsorption. Chem Commun (Camb) (1):67–69
17. Ghanem BS, Hashem M, Harris KDM, Msayib KJ, Xu M, Budd PM, Chaukura N, Book D, Tedds S, Walton A, McKeown NB (2010) Triptycene-based polymers of intrinsic microporosity: organic materials that can be tailored for gas adsorption. Macromolecules 43(12):5287–5294
18. Zhu ZG, Swager TM (2001) Conjugated polymers containing 2,3-dialkoxybenzene and iptycene building blocks. Org Lett 3(22):3471–3474
19. Zhu ZG, Swager TM (2002) Conjugated polymer liquid crystal solutions: control of conformation and alignment. J Am Chem Soc 124(33):9670–9671
20. Hoogboom J, Swager TM (2006) Increased alignment of electronic polymers in liquid crystals via hydrogen bonding extension. J Am Chem Soc 128(47):15058–15059
21. Chen Z, Bouffard J, Kooi SE, Swager TM (2008) Highly emissive iptycene-fluorene conjugated copolymers: synthesis and photophysical properties. Macromolecules 41(18):6672–6676
22. Yang JS, Swager TM (1998) Porous shape persistent fluorescent polymer films: an approach to TNT sensory materials. J Am Chem Soc 120(21):5321–5322
23. Yang JS, Swager TM (1998) Fluorescent porous polymer films as TNT chemosensors: electronic and structural effects. J Am Chem Soc 120(46):11864–11873
24. Williams VE, Swager TM (2000) Iptycene-containing poly(arylene-ethynylene)s. Macromolecules 33(11):4069–4073
25. Amara JP, Swager TM (2005) Synthesis and properties of poly(phenylene ethynylene)s with pendant hexafluoro-2-propanol groups. Macromolecules 38(22):9091–9094

26. Yamaguchi S, Swager TM (2001) Oxidative cyclization of bis(biaryl)acetyl-enes: synthesis and photophysics of dibenzo[g,p]chrysene-based fluorescent polymers. J Am Chem Soc 123(48):12087–12088
27. Zahn S, Swager TM (2002) Three-dimensional electronic delocalization in chiral conjugated polymers. Angew Chem Int Ed 41(22):4225–4230
28. Zhao XM, Cardolaccia T, Farley RT, Abboud KA, Schanze KS (2005) A platinum acetylide polymer with sterically demanding substituents: effect of aggregation on the triplet excited state. Inorg Chem 44(8):2619–2627
29. Bailey GC, Swager TM (2006) Masked michael acceptors in poly(phenyleneethynylene)s for facile conjugation. Macromolecules 39(8):2815–2818
30. Nesterov EE, Zhu ZG, Swager TM (2005) Conjugation enhancement of intramolecular exciton migration in poly(*p*-phenylene ethynylene)s. J Am Chem Soc 127(28):10083–10088
31. Ohira A, Swager TM (2007) Ordering of poly(*p*-phenylene ethynylene)s in liquid crystals. Macromolecules 40(1):19–25
32. Meador MAB (1988) Processable, high-temperature polymers from 1,4,5,8-tetrahydro-1,4-5,8-diepoxyanthracene and bis-dienes. J Polym Sci Polym Chem 26(11):2907–2916
33. Meador MAB, Meador MA, Ahn MK, Olshavsky MA (1989) Evidence for thermal dehydration occurring in Diels–Alder addition polymers. Macromolecules 22(11):4385–4387
34. Zhao DH, Swager TM (2005) Conjugated polymers containing large soluble diethynyl iptycenes. Org Lett 7(20):4357–4360
35. Perepichka DF, Bendikov M, Meng H, Wudl F (2003) A one-step synthesis of a poly(iptycene) through an unusual Diels–Alder cyclization/dechlorination of tetrachloropentacene. J Am Chem Soc 125(34):10190–10191
36. Thomas SW, Long TM, Pate BD, Kline SR, Thomas EL, Swager TM (2005) Perpendicular organization of macromolecules: synthesis and alignment studies of a soluble poly(iptycene). J Am Chem Soc 127(51):17976–17977
37. Chen ZH, Amara JP, Thomas SW, Swager TM (2006) Synthesis of a novel poly(iptycene) ladder polymer. Macromolecules 39(9):3202–3209
38. Chen Z, Swager TM (2008) Synthesis and characterization of poly(2,6-triptycene). Macromolecules 41(19):6880–6885
39. Amara JP, Swager TM (2004) Incorporation of internal free volume: synthesis and characterization of iptycene-elaborated poly(butadiene)s. Macromolecules 37(8):3068–3070
40. Yang JS, Yan JL, Hwang CY, Chiou SY, Liau KL, Tsai HHG, Lee GH, Peng SM (2006) Probing the intrachain and interchain effects on the fluorescence behavior of pentiptycene-derived oligo(*p*-phenylene-ethynylene)s. J Am Chem Soc 128(43):14109–14119
41. Maag D, Kottke T, Schulte M, Godt A (2009) Synthesis of monodisperse oligo(1,4-phenyleneethynylene-alt-1,4-triptycyleneethynytene)s. J Org Chem 74(20):7733–7742

Part III
Applications of Iptycenes and Their Derivatives

Chapter 7
Iptycenes and Their Derivatives in Molecular Machines

Triptycene with three arene units fused to the [2.2.2]bicyclooctatriene bridgehead system has a D_{3h} symmetric structure, and thus resembles a macroscale gear-wheel. This structural feature makes the triptycene a potential component for the construction of molecular machines.

7.1 Molecular Gearings

Between the 1960s and 1970s, the highly hindered compounds with appreciable rotation barriers between sp^3-hybridized carbon atoms won a considerable attention. Especially, triptycene and its derivatives, in which the rotation around the sp^3–sp^3 single bond between the bridgehead carbon atom and the attached substituent was restricted to give long-lived conformational isomers, became the subject of several reports. In the early work, Iwamura [1] synthesized a triptycene derivative, 9,10-bis(1-cyano-1-methylethyl)triptycene (**1**, Fig. 7.1) and found that the triptycene **1** contained a rotational barrier of 37.7 kcal/mol which was a considerably high barrier for rotation around the sp^3–sp^3 single bond.

At almost the same time, Koukotas et al. [2] and Schwartz et al. [3] synthesized 9,9′-bitriptycyl **2** and 2,2′-di-methyl-9,9′-bitriptycyl **3** (Fig. 7.2). They found that these systems with bulky triptycene groups also showed a high rotational barrier between the central 9 and 9′ sp^3-hybridized carbon atoms, and the barrier value of system **3** could be more than 54 kcal/mol.

In 1980s, Mislow's group [4] and Iwamura's group [5–8] independently designed and synthesized several bis(9-triptycyl)methanes **4a–d** and bis(9-triptycyl)ethers **5a–d** by the addition of excess benzynes to bis(9-anthryl)methane and the thermolysis of 9-triptycyl 9-triptycene-peroxy-carboxylates, respectively (Scheme 7.1). It was found that these bevel-gear-shaped molecules showed a rapid fully coupled rotation around the two C–X bonds ($X = CH_2$, O) in solution on the nuclear magnetic resonance (NMR) time scale. Moreover, the molecules rotated in solution as the frictionless bevel gears, which only needed a few kilocalorie per mole barriers to gearing. It was noteworthy that the energy barrier for gear slipping was measured by the interconversion of the isomers, and for the two 9-triptycyl (Tp) rotors in **4**, **5**

C.-F. Chen, Y.-X. Ma, *Iptycenes Chemistry*,
DOI 10.1007/978-3-642-32888-6_7, © Springer-Verlag Berlin Heidelberg 2013

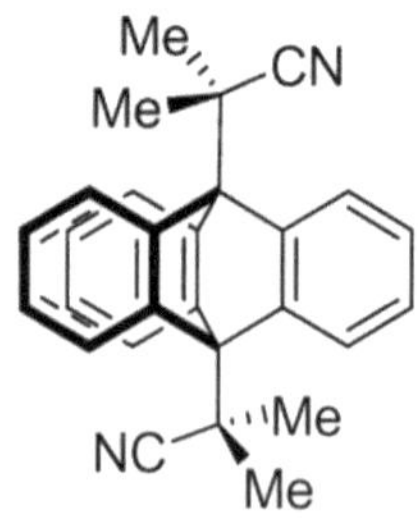

1

Fig. 7.1 Structure of 9,10-bis(1-cyano-1-methylethyl)triptycene **1**

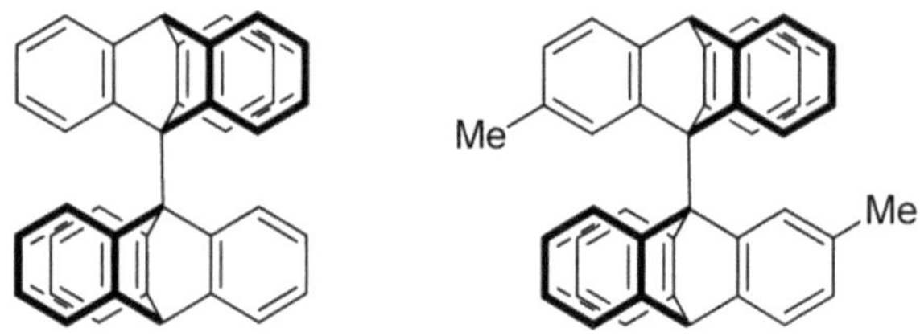

2 3

Fig. 7.2 Structures of 9,9′-bitriptycyl **2** and 2,2′-dimethyl-9,9′-bitriptycyl **3**

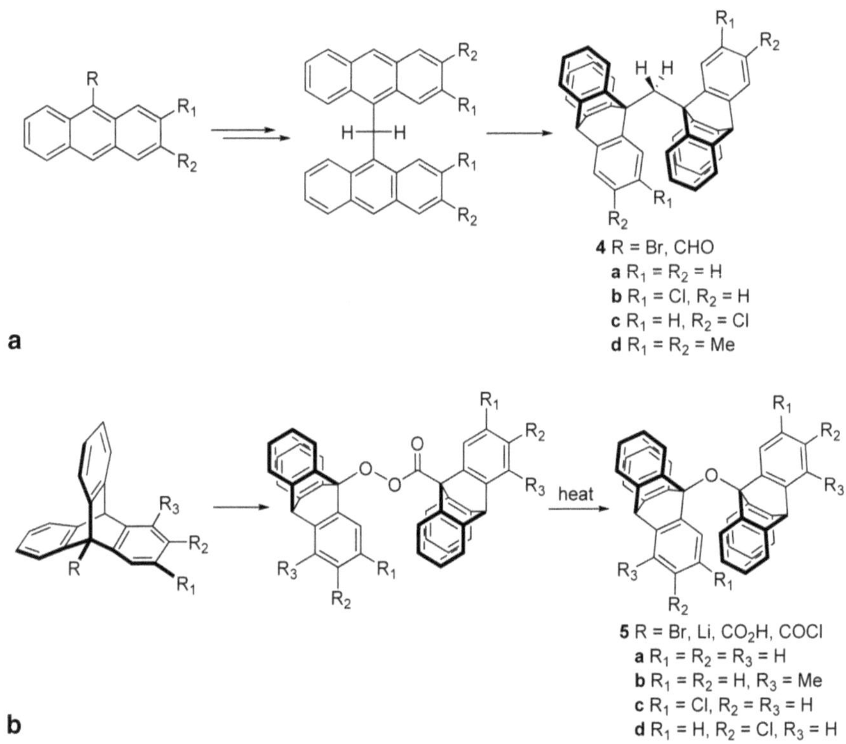

Scheme 7.1 Synthesis of bis(9-triptycyl)methanes **4a–d** and bis(9-triptycyl)ethers **5a–d**

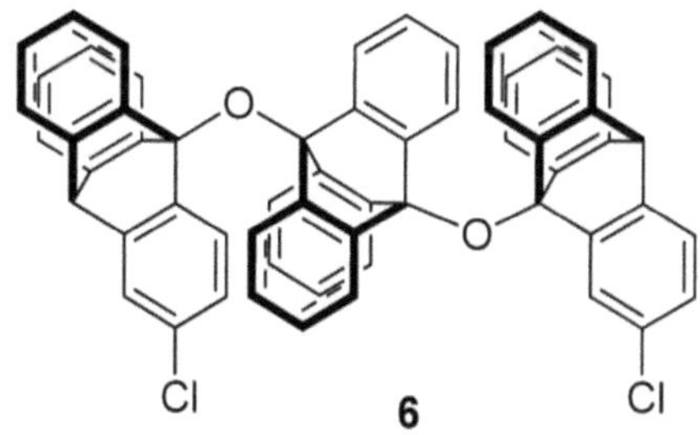

6

Fig. 7.3 Structure of triptycene-based diether **6**

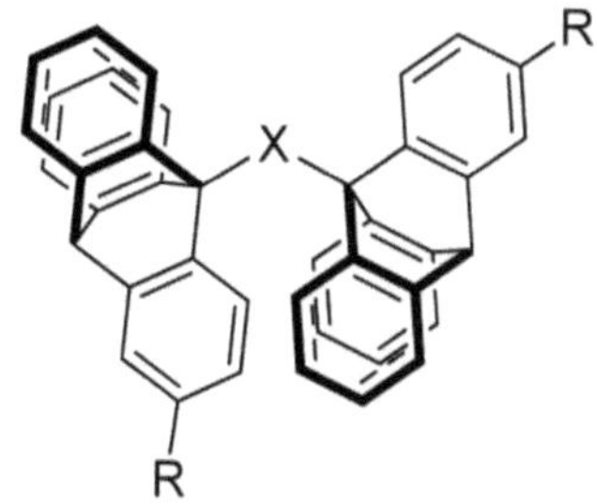

7 X = CH_2, NH, O, SiH_2, PH, S
R = Cl

Fig. 7.4 Structures of bis(9-triptycy1)X compounds **7**

and related molecules, activation energies of 30–40 kcal/mol were generally required [9]. Along with the further studies [6, 8, 10], it was found that there were a pair of stereoisomers of the molecular bevel gear of bis(9-triptycyl)methanes and ethers due to the different phase relationships between the labeled cogs. Also for some systems, the stereoisomers could be separated and characterized by the chiral high-performance liquid chromatography (HPLC).

Furthermore, Koga et al. [7] utilized the similar route to prepare the triptycene-based diether **6** (Fig. 7.3), which had three triptycene fragments. They found that in this system, when the two-sided triptycene substituents rotated rapidly in the geared motion, the rotating information could transfer from one end of the molecule to the other end with the torsional motions of the mid-triptycyl. This rotation mode could achieve the transfer of rotation information in a gear system.

Afterward, Kawada et al. [11–14] further prepared a series of bis(9-triptycy1)X compounds **7** (X = NH [11], S [12], SiH_2 [14], PH [13], Fig. 7.4). Moreover, it was found that the barrier for gear slipping was dependent on the linked group X (X = CH_2, NH, O, SiH_2, PH, S). Among them, the lowest one was 20.4 kcal/mol of the silane, while the highest one was 42.0 kcal/mol of the ether. These results could be attributed to the different angle and/or distance between the two 9-triptycyl groups. Likewise, there was phase isomerism for them, which could be also separated and characterized by the chiral HPLC.

During the same period, Yamamoto [15–21], Mislow [22] and their coworkers investigated the rotational behaviors of the bridgehead-substituted triptycene systems, including benzyl [15, 18, 20]-, phenyl [15, 22]-, and phenoxyl [16–18]-substituted

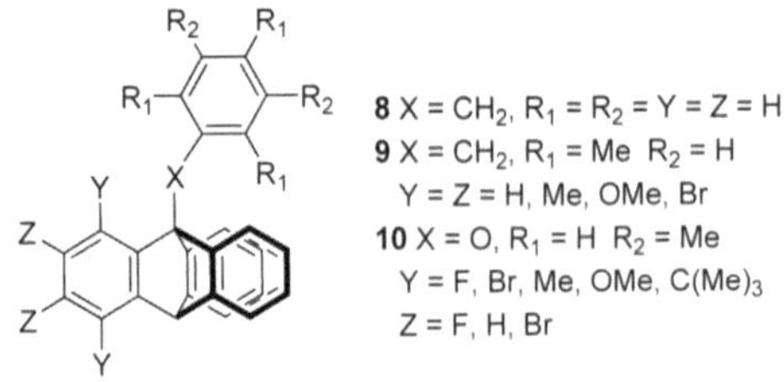

Fig. 7.5 Structures of the molecules **8–10**

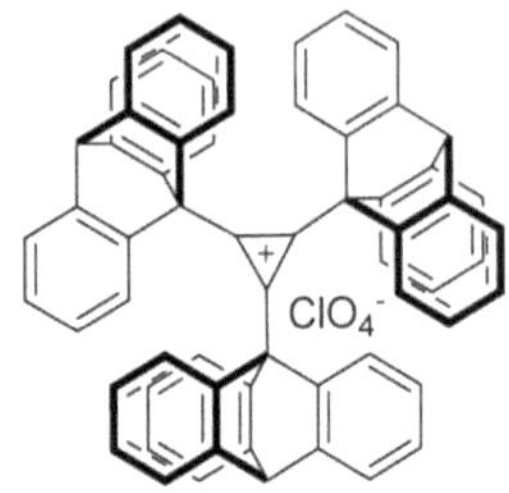

11

Fig. 7.6 Structure of tris(9-triptycyl)cyclopropenylium perchlorate **11**

triptycene at the bridgehead. As a result, it was found that these peri-substituted benzylic triptycenes (**8**, Fig. 7.5) preferred to the isolated rotation with gear slippage barriers (~ 10 kcal/mol) rather than the correlated rotation. While for the systems of **9** (Fig. 7.5), the peri-substituents seemed to contain the enhanced barrier to gear slippage. This slippage mode would lead to frictionless correlated disrotation (dynamic gearing) with 12–17 kcal/mol barriers. In the case of systems **10** (Fig. 7.5), the size of the peri-substituent (Y and Z group) played the key role in the barrier process. When the peri-substituent was fluoro or methoxy group, the barrier to gear rotation was 10–12 kcal/mol; while methyl, bromine, or *t*-butyl group acted as peri-substituent, the barrier was up to 15–17 kcal/mol in an isolated aryl rotation.

In 1990, Chance et al. [23] reported the first triply correlated system, tris(9-triptycyl)cyclopropenylium perchlorate (**11**, Fig. 7.6), in which three triptycene units were connected by a cyclopropenium ring. The C–C bond length in this cyclopropenium ring was 1.379(7) Å. It also revealed that the torsional motion of the triptycyl rotors was frozen on the NMR time scale. This system matched their mechanical counterparts in both static and dynamic properties, which at the molecular level followed the corollary of the mechanical selection rules of macro-world, according to which all motions were disallowed in a closed cyclic array consisting of an odd number of securely meshed macroscopic gears [23].

Soon later, Kawada et al. [13, 14] designed and synthesized *cis*-1,2-bis(9-triptycyl)ethylene (**12**, Scheme 7.2), which contained a deeper meshing of the two triptycyl moiety. It was noteworthy that system **12** could not be directly prepared by the addition reaction of *cis*-1,2-bis(9-anthryl)ethylene (**13**) with benzyne. The resulting compound of the addition reaction of reactant **13** was 1,2-bis(9-triptycyl)acetylene (**14**) instead of the designed compound **12**. However, the

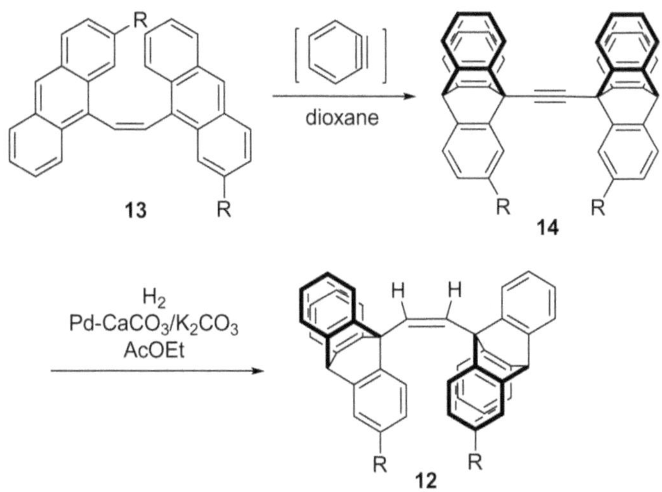

Scheme 7.2 Synthesis of *cis*-1,2-bis(9-triptycyl)ethylene **12**

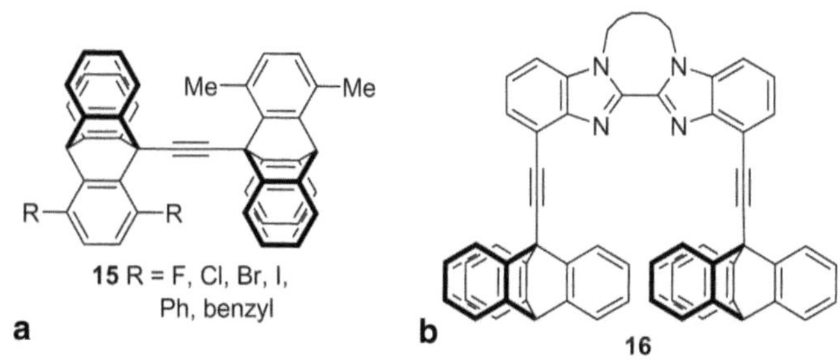

Fig. 7.7 Structures of bis(9-triptycyl) ethynes **15** (**a**) and triptycene-based molecular spur gears **16** (**b**)

palladium-catalyzed hydrogenation of **14** could give the ethylene system **12**. As a gearing model, the system **12** displayed less freedom around the bond, compared with the corresponding methane system (**7**, $X = CH_2$), which did not conform to the desired results that ethylene was structurally more flexible and relaxed the steric congestion more effectively. This unsatisfactory result was probably caused by the more severe overcrowding structure of the ethylene system **12**.

Ten years ago, Toyota et al. [24–26] reported the new series of system, bis(9-triptycyl) ethynes (**15**, Fig. 7.7a) with rotation around C(*sp*)–C(sp^3) bonds. For these systems, the barriers to rotation were varied with different substituents from 11.6 (R = F) to 17.3 kcal/mol (R = I); when the halogen groups were replaced by phenyl or benzyl groups, the barrier to rotation would increase up to 15.8–18.8 kcal/mol. These results revealed that the size, shape, and flexibility of the substituents could determine the rotational barriers. In 2009, Frantz et al. [27] investigated the mechanistic aspects of triptycene-based molecular gears with the parallel axes via the computational

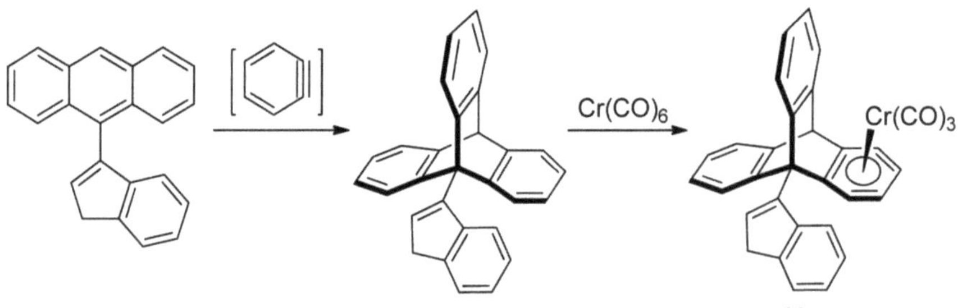

Scheme 7.3 Synthesis of complex **18**

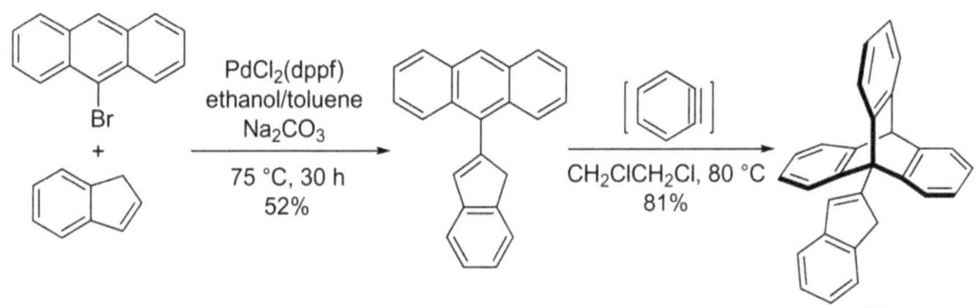

Scheme 7.4 Synthesis of indenyl-triptycene **19**

methods. Soon after, Frantz et al. [28] designed the derivatives of 4,4-bis(triptycen-9-ylethynyl)bibenzimidazole, which were suitable to molecular spur gears with a gearing behavior of collateral triptycene groups. The results of density functional theory (DFT) calculations suggested that these molecules preferred geared rotation to slippage. Moreover, they successfully synthesized the triptycene-based molecular spur gears (**16**), which are shown in Fig. 7.7b.

In 2004, Harrington et al. [29] attempted to take advantage of the metal-based components' migration as the foot pedal to block the rotation of the triptycene moieties. Thereby, they designed and synthesized compound 9-(1*H*-inden-3-yl)triptycene **17** by the treatment of 9-(1*H*-inden-3-yl)anthracene with benzyne. However, it was found that a chromium tricarbonyl unit preferred to coordinate to one of the aromatic blades of triptycene rather than the six-membered ring of the indenyl group (Scheme 7.3). It was probably because the position for the indenyl group was too crowded to directly incorporate a metal carbonyl substituent. According to the X-ray crystallography and variable-temperature NMR spectroscopy, they found that there existed a 2:1 mixture of two rotamers for complex **18** at low temperature, and the rotation barriers for **17** and **18** were 12 and 13 kcal mol^{-1}, respectively.

Subsequently, Nikitin et al. [30, 31] further synthesized the indenyl-triptycenes **19** (Scheme 7.4), in which the 2-indenyl group connected to the 9-position of the triptycene. In this route to the target molecule, the palladium-catalyzed coupling reaction was the key step.

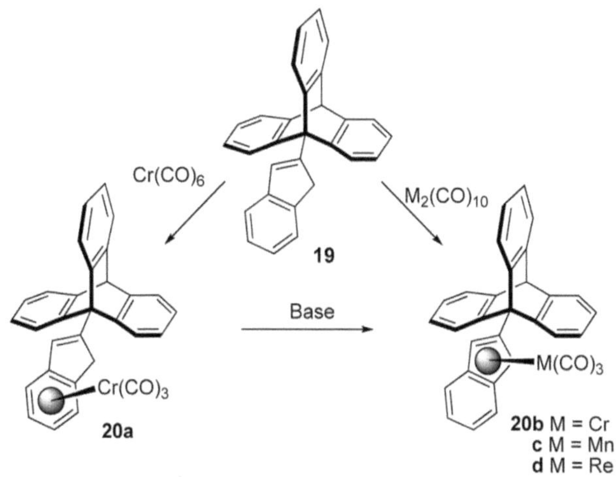

Scheme 7.5 Representation of the molecular brake

With two indenyl-triptycenes (**17** and **19**) in hand, Nikitin et al. also studied their structures and dynamic behaviors. On the NMR time scale, system **17** displayed that the barrier to rotation about the indenyl-triptycene linkage was 12 kcal/mol; while system **19** was almost in an essentially free rotation at ambient temperature. Obviously, the system **19** was fitter as the attractive molecular machinery prototype system. Treatment of compound **19** with the complex [$Cr(CO)_6$] would give the chromium complex **20a** in a η^6-mode. Further X-ray crystallography study revealed that its continued free rotation could be maintained without adverse effects of the tripodal $Cr(CO)_3$ fragment, which was concordant with the result from ^{1}H and ^{13}C NMR spectral analyses. However, when **19** was treated with $M(CO)_{10}$ (M = Cr, Mn, Re), the resulting complexes **20b**, **c,** and **d** would be in the η^5-mode. As the result of a shorter distance between the triptycyl paddlewheel and tripodal $M(CO)_3$ fragment, molecular rotation was blocked with steric hindrance. On the basis of those results, they further prepared a short-stroke molecular mechanical shuttle or latch with complexes of **19**, which was controlled by a strong base. In this system, complex **20a** played as the "ON" state, while the chromium complex **20b**, manganese complex **20c**, and rhenium complexes **20d**, which were all in a η^5-fashion, played as "OFF" states. In these "ON–OFF" processes, the ML_3 moiety in complex **20a** would slide toward the five-membered ring driven by the charge effect, along with the participation of strong base. After an approximately 2 Å movement, the ML_3 moiety finally attached to the five-membered aromatic ring of indene as a η^5-fashion complex, and this molecular mechanical shuttle was at "OFF" state with blocked rotation (Scheme 7.5).

In 1995, Gakh et al. [32] synthesized the first triptycene-based crown ethers (**21**, Scheme 7.6) via the facile two-step procedure, in which the condensation of 9,10-bis(chloromelhyl)anthracene with polyethylene glycols was the key reaction step. Moreover, it was found that the triptycene moiety in this molecule tended to

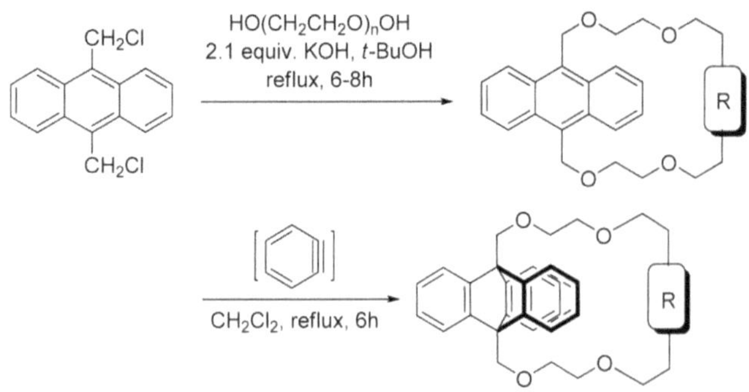

21

R = -O(CH_2CH_2O)$_2$-, -OCH_2CH_2O-

Scheme 7.6 Synthesis of triptycene-based crown ethers **21**

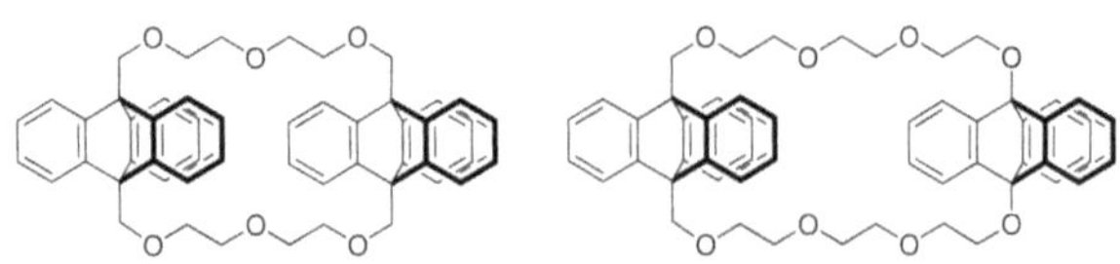

22 **23**

Fig. 7.8 Structures of the molecules **22** and **23**

rotate relating to the crown ether ring. Soon afterward, Bryan et al. [33] further obtained bis(9,10-triptyceno)-26-crown-6 (**22**) and bis(9,10-triptyceno)-32-crown-8 (**23**, Fig. 7.8) by the reaction of the substituted anthracene with diethylene glycol or triethylene glycol, followed by the similar synthetic strategy. With the X-ray crystallography analysis, both of the crown rings in the two molecules were forced open by the bulky triptycene groups. Also, molecule **22** seemed to have a crowed space structure, and the whole crown cavity was occupied by the π-electron clouds of two arene rings; while for molecule **23** with the longer linkages, the structure appeared much free. In addition, the triptycene "gears" in these two systems could be in a fast rotation above 60 °C, which could be demonstrated by the variable temperature ^{1}H NMR. The two molecules were the first systems containing two triptycene groups simultaneously linked through their two aliphatic carbon atoms, as a simple molecular paddlewheel mechanism.

In 2008, Tsutsumi et al. [34] prepared a kind of triptycene-based molecular rotors (**24a**, **b**, Fig. 7.9) bearing a pair of ionophores at both bridgeheads, in which **24a** contained two linear oxyethylene chains, while there were two cyclic oxyethylene units (aza-15-crown-5 ether) with flexible alkyl spacers in molecule **24b**. The result of DFT calculations revealed that the introduction of the flexible ionophores never increased, but generally decreased the rotational barrier. Since these two flexible ionophores could serve as the receptors of the external stimulus. Thus, they attempted to take use of *s*-block metal cations as an external stimulus to control the rotational behaviors of these **24a**, **b** systems. Using the dynamic NMR spectroscopy, it was found that systems **24a**, **b** could rotate freely without the pressure of *s*-block metal

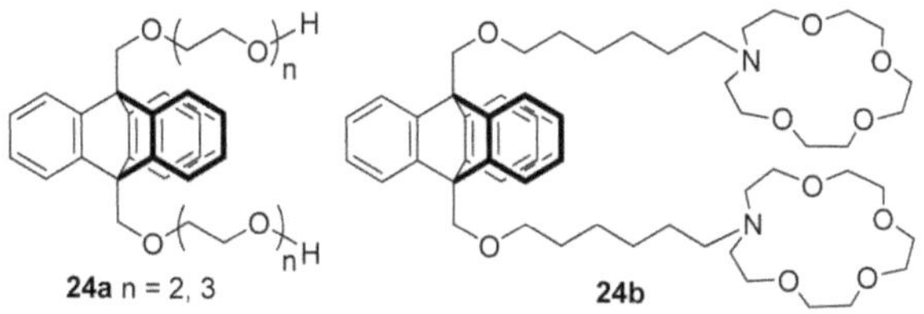

Fig. 7.9 Structures of the molecules **24a, b**

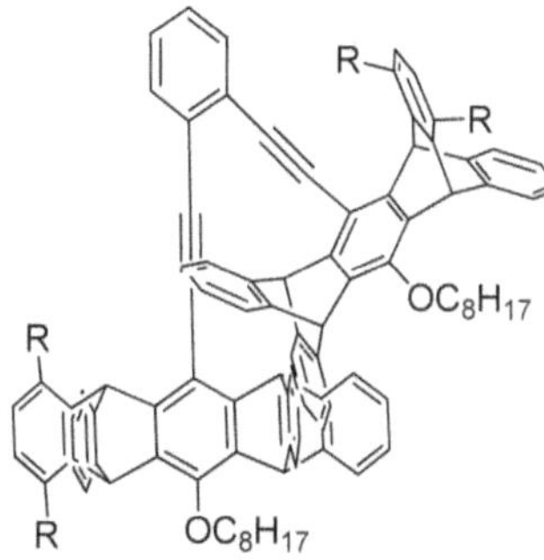

Fig. 7.10 Structure of four-toothed molecular bevel gear *S, S*-**25**

cations; while the aid of both ionophores with cations resulted in the restricted rotation. The restricted degrees of the systems were related to the stability of the complexes of ionophores with cations. Thus, the addition of the Ba^{2+} cation could effectively control the rate (activation energy) of the internal rotation, due to the higher stability of the complexes.

More recently, Yang, Chao and their co-workers [35] firstly reported the design and synthesis of the pentiptycene-based molecular gears (*S, S*-**25**, Pp_2X, Fig. 7.10). As shown in Scheme 7.7, the nucleophilic addition of lithium trimethylsilylacetylide to the pentiptycene quinone gave the quinone **26**, then followed by trapping of the phenoxide intermediate with trimethylsilyl (TMS) chloride and the reductive aromatization of the central ring afforded a pair of racemic isomers **27′a, b** (*R/S*) with TMS-protected groups. The mixture of protected isomers without separation was directly deprotected at a basic condition to give a mixture of **27a, b** in 33 % yield. Subsequently, the mixture of **27a, b** underwent the *O*-alkylation, and the resulting product **28** finally gave the targets **25** as the racemic isomers in a yield of 74 % by Sonogashira reaction with 1,2-diiodobenzene. With the interrotor rotation modes in the teeth-unlabeled (R = H) and teeth-labeled (R = OMe) states, they evaluated the potential utility of the pentiptycene in a four-toothed molecular bevel gear (*S, S*-**25**) to be acted as a four-toothed rotor in details. Although the mismatch in teeth number and molecular symmetry seemed to bring about some bafflement, their NMR spectra revealed a fast-gear slippage in these Pp_2X systems, which was also in agreement with the results of DFT and Austin model 1 (AM1) calculations.

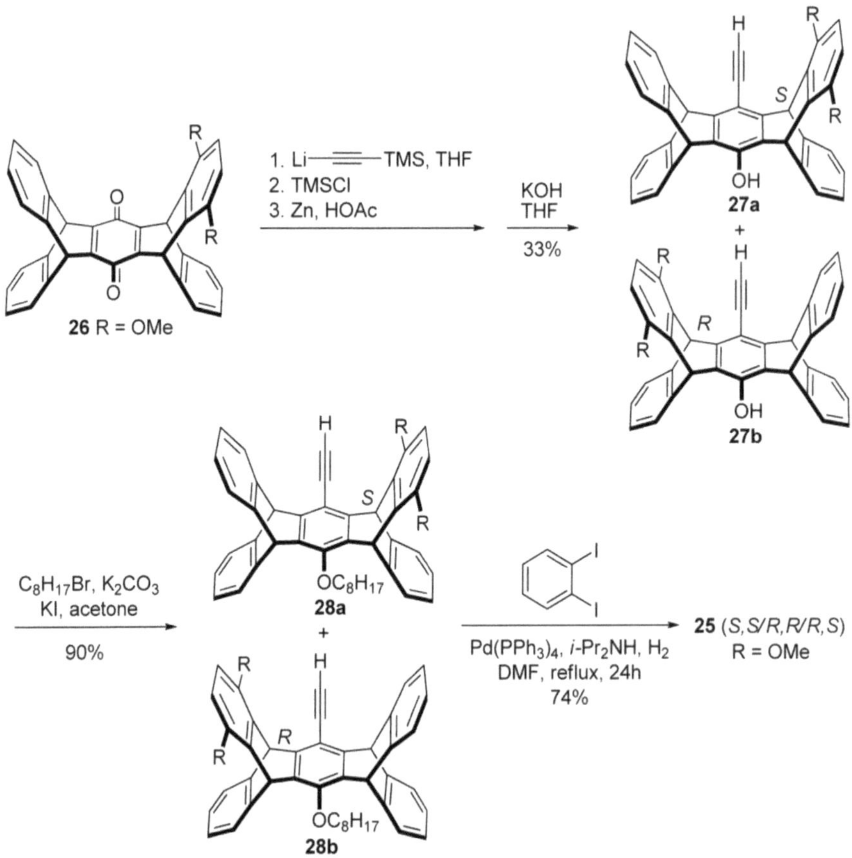

Scheme 7.7 Synthesis of pentiptycene-based molecular gears **25**

7.2 Molecular Brakes and Ratchets

In 1994, Kelly et al. [36] designed and constructed the first example of molecular brake containing a triptycene moiety, which was connected to a 2,2′-bipyridyl group. This system was operated by adding the metal ions, and the conformational change led to the reversibly halt rotation of the gear via the coordination of metal ion at the remote site (Fig. 7.11a). As the continuation of this work, Kelly et al. [37] further attempted to design a novel ratchet (**29**) by employing a triptycene as the toothed ratchet wheel, while a [4]helicene instead of bipyridyl was applied as the pawl and the spring (Fig. 7.11b). Although model **29** had all essential components of simple ratchet in a single molecule, the spin polarization transfer NMR results revealed that its rotation was bidirectional at 160 °C. Moreover, the further theoretical and thermodynamic analyses [38, 39] also revealed that this ratchet model showed a bidirectional rotation to some extent, since the helicene moiety just acted as a friction brake, which partly inhibited the *anti*clockwise rotation of the triptycene rotor. As a result

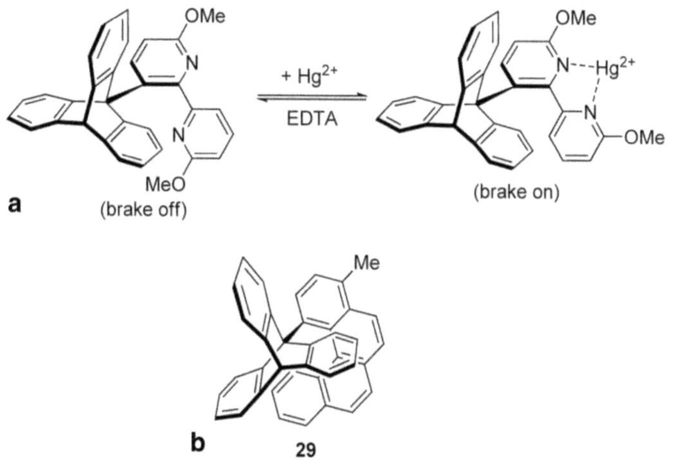

Fig. 7.11 **a** Actual depiction of the operation of molecule brake. **b** Structure of model **29**

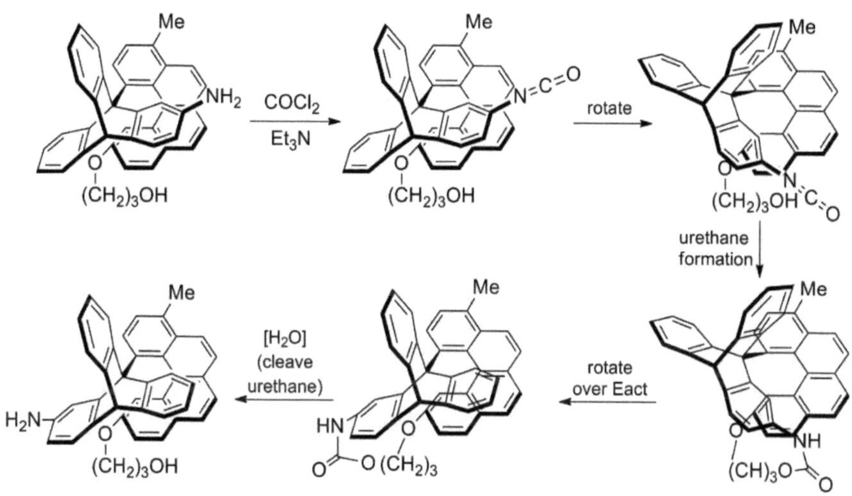

Scheme 7.8 Representation of molecular ratchet system

of the second law of thermodynamics and the principle of microscopic reversibility, model **29** could not be a unidirectional ratchet under an isothermal environment.

The essential thing in a real unidirectional rotation ratchet is blocking one certain-direction rotation (such as *anti*clockwise rotation) to the utmost extent, simultaneously lowering the energy barrier against rotation of the other direction (such as clockwise rotation), which resulted in the unidirectional rotation of the rotor. Thus, Kelly's group [40, 41] improved their system by introducing a tether into the helicene unit and using carbonyl dichloride (phosgene) as a chemical energy provider. Consequently, they achieved a 120° unidirectional rotation, which was driven by chemical energy in a molecular ratchet system (Scheme 7.8). However, this molecule could not achieve continuous and fast 360° rotation. So, Kelly et al. [41–45] attempted to design and

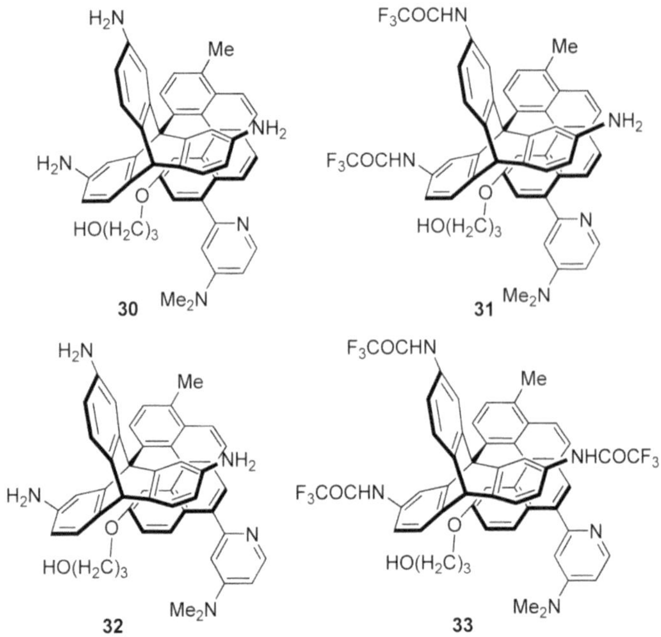

Fig. 7.12 Structures of the molecules **30**–**33**

prepare the repeated rotations with a continuous unidirectional rotation based on the triptycenyl–helicene system. To achieve synthetic molecular motors with a unidirectional rotary motion, several aspects should be considered: (1) three blades of the triptycene unit were required to be selectively bonded, including proper spacial location and timing, (2) there should be a transporter to capture the phosgene and then deliver to a certain blade, (3) the urethane should be readily formed and cleaved, and last but not least (4) all the modifications should not inhibit the procedure, even if they might not facilitating the rotation. On the basis of these requirements, Kelly et al. [41–45] further attempted to build a unimolecular motor with repeated rotations by modifying the triptycenyl–helicene system. It was found that *N,N*-4-dimethylaminopyridine (DMAP) was a good delivery agent, which directed the phosgene molecule exclusively to the nearest amino group. Systems **30** and **31** (Fig. 7.12) were rationally designed, which could be promising candidates as chemically powered, molecular-scale motor. However, due to the unexpected Bürgi–Dunitz or similar interactions, the designed motors **32** and **33** still failed to achieve 360° unidirectional rotations.

Recently, Sun et al. [46], Chen et al. [47], and Yang et al. [48] designed and synthesized a series of pentiptycene-derived molecular brake systems. Pentiptycene-derived stilbene brake systems [46] with different-sized brake components could achieve the switching of "ON–OFF" states via the *cis/trans* (*Z–E*) photoisomerization. Another pentiptycene-derived molecular brake [47] with 2-methyl-eneindanone showed a highly efficient control of "ON–OFF" states, as the result of the combined action of light-driven *E–Z* and electrically driven *Z–E* isomerization. In addition, Yang et al.

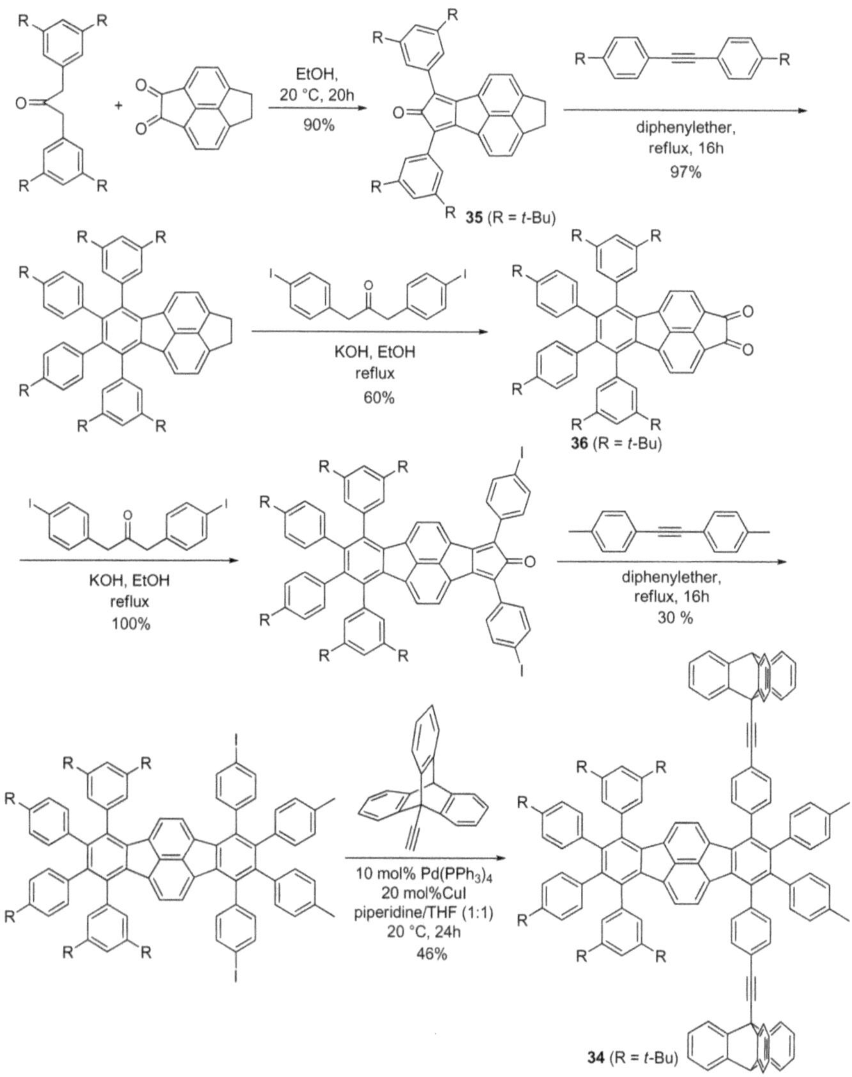

Scheme 7.9 Synthesis of wheelbarrow **34**

[48] further utilized redox processes to achieve acceleration stop of pentiptycene rotor in pentiptycene-derived *p*-phenylenediamine.

7.3 Molecular Wheelbarrows

In 2003, Jimenez-Bueno and Rapenne [49] designed and synthesized molecular analog of a wheelbarrow **34**, which contained 3,5-di-*t*-butyl phenyl groups as legs and ethynyl–tripycene groups as wheels connected to a polycyclic aromatic hydrocarbon platform. As shown in Scheme 7.9, the double Knoevenagel reaction of

Fig. 7.14 Structures of models **39e**, **f**

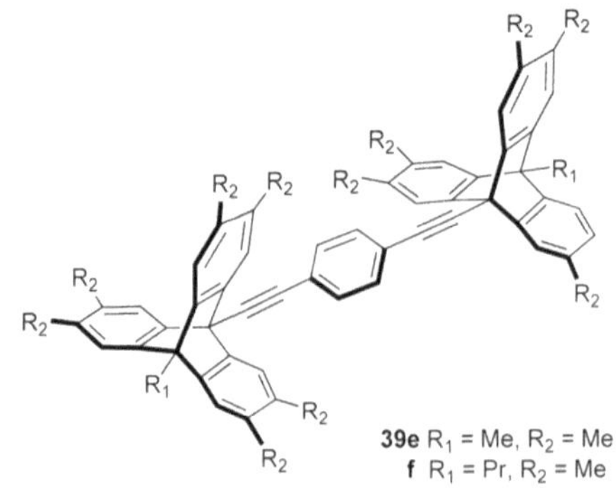

analysis of quadrupolar echo ^{2}H NMR, the gyroscopic rotation in crystals of **39e** occurred through a 180° site exchange (two-fold flip) with a very low rotational barrier of 4.4 kcal/mol, which was slightly higher than the internal barrier for ethane (3.0 kcal/mol) in the gas phase.

7.5 Miscellaneous

In 2010, Nikitin et al. [55] developed a novel triptycene-based molecular machine (**40**) with two ferrocenyl groups attached to the 9- and 10-positions of triptycene (Fig. 7.15), which was named "molecular dial". The ^{1}H and ^{13}C NMR spectra of **40** showed that there was a mixture (1:2) of *meso-* and *rac*-emic rotamers over the – 20–50 °C range. Moreover, these interconversions proceed in a stepwise manner such as *rac–meso–rac*, as a set of molecular dials, which was confirmed by the two-dimensional-EXSY NMR spectroscopic data recorded with different mixing time.

At the same year, Jiang et al. [56] designed and synthesized a bifunctionalized [3]rotaxane based on the triptycene-derived macrotricyclic host (**41**) and found that the shuttle process between two different stations of this [3]rotaxane molecular machine could be reversibly achieved by acid (trifluoroacetic acid (TFA)) and base (1,8-diazabicyclo[5.4.0]undec-7-ene (DBU)) control (Fig. 7.16). To the

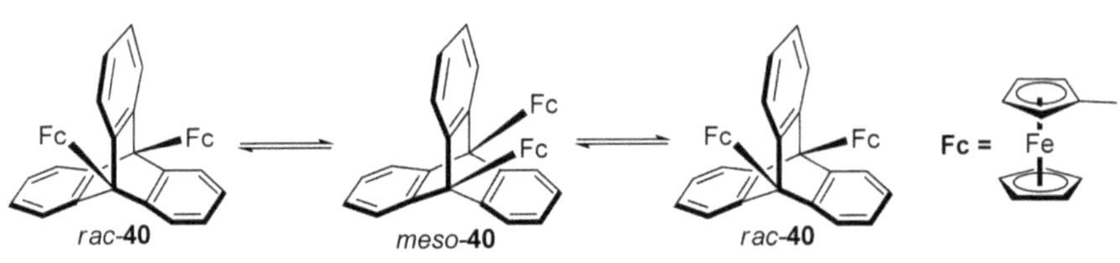

Fig. 7.15 Dynamic behavior of 9, 10-diferrocenyltriptycne **40**

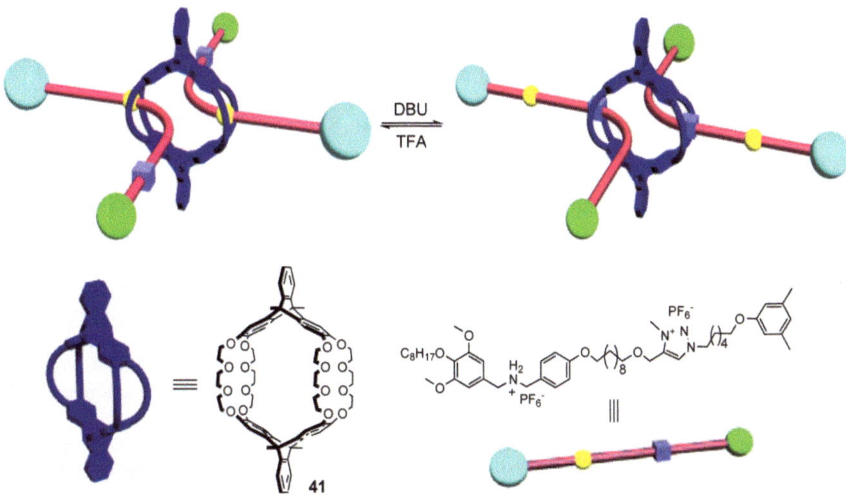

Fig. 7.16 [3]Rotaxane-based molecular machine controlled by acid and base

best of our knowledge, this was the first example of [3]rotaxane-based molecular machine with two threads passing through one host. This work also provided a novel route to construct molecular machine based on the various host–guest systems.

References

1. Iwamura H (1973) Stable isomers of 1,6-bis-(1-cyano-1-methyl-ethyl)triptycenes. J Chem Soc Chem Commun 1973(7):232
2. Koukotas C, Mehlman SP, Schwartz LH (1966) Bitriptycyl. J Org Chem 31(6):1970–1971
3. Schwartz LH, Koukotas C, Yu CS (1977) 2,2′-Dimethyl-9,9′-bitriptycyl: case of an enormous rotational barrier between sp^3-hybridized carbon-atoms. J Am Chem Soc 99(23):7710–7711
4. Cozzi F, Guenzi A, Johnson CA, Mislow K, Hounshell WD, Blount JF (1981) Stereoisomerism and correlated rotation in molecular gear systems: residual diastereomers of bis(2,3-dimethyl-9-triptycyl)methane. J Am Chem Soc 103(4):957–958
5. Kawada Y, Iwamura H (1980) Unconventional synthesis and conformational flexibility of bis(1-triptycyl) ether. J Org Chem 45(12):2547–2548
6. Kawada Y, Iwamura H (1981) Bis(4-chloro-1-triptycyl) ether: separation of a pair of phase isomers of labeled bevel gears. J Am Chem Soc 103(4):958–960
7. Koga N, Kawada Y, Iwamura H (1983) Recognition of the phase relationship between remote substituents in 9,10-bis(3-chloro-9-triptycyloxy)triptycene molecules undergoing rapid internal-rotation cooperatively. J Am Chem Soc 105(16):5498–5499
8. Kawada Y, Iwamura H (1983) Correlated rotation in bis(9-triptycyl) methanes and bis(9-triptycyl) ethers. Separation and inter-conversion of the phase isomers of labeled bevel gears. J Am Chem Soc 105(6):1449–1459
9. Iwamura H, Mislow K (1988) Stereochemical consequences of dynamic gearing. Acc Chem Res 21(4):175–182
10. Kawada Y, Iwamura H (1981) Phase isomerism in gear-shaped molecules. Tetrahedron Lett 22(16):1533–1536

11. Kawada Y, Yamazaki H, Koga G, Murata S, Iwamura H (1986) Bis(9-triptycyl)amines, a missing link between the corresponding methanes and ethers: an unconventional synthesis and influence of nitrogen configurational inversion on the coupled disrotatory trajectory. J Org Chem 51(9):1472–1477
12. Kawada Y, Ishikawa J, Yamazaki H, Koga G, Murata S, Iwamura H (1987) Correlation of the two torsional degrees of freedom about the bonds connecting the bridgehead carbons to the sulfur atom in bis (9-triptycyl) sulfide. Tetrahedron Lett 28(4):445–448
13. Kawada Y, Kimura Y, Yamazaki H, Ishikawa J, Sakai H, Oguri M, Koga G (1994) Further studies on the dependence of the barriers for gear slippage on the joint group in bis(9-triptycyl)X type molecules. bis(9-triptycyl) phosphine, the missing link in the series along the third row of the periodic table. Chem Lett 23(7):1311–1314
14. Kawada Y, Sakai H, Oguri M, Koga G (1994) Preparation of and dynamic gearing in *cis*-1,2-bis(9-triptycyl)ethylene. Tetrahedron Lett 35(1):139–142
15. Yamamoto G, Ōki M (1983) Restricted rotation involving the tetrahedral carbon. 46. correlated rotation in 9-(2,4,6-trimethylbenzyl)triptycenes: direct and roundabout enantiomerization-diastereomerization processes. J Org Chem 48(8):1233–1236
16. Yamamoto G, Ōki M (1986) Restricted rotation involving the tetrahedral carbon. 59. stereodynamics of singly *peri*-substituted 9-(3,5-dimethyl-phenoxy)triptycene derivatives. Bull Chem Soc Jpn 59(11):3597–3603
17. Yamamoto GK (1989) Detailed dynamic NMR-study of a molecular gear, 1-methoxy-9-(3,5-dimethylbenzyl)triptycene. Bull Chem Soc Jpn 62(12):4058–4060
18. Yamamoto G (1990) Use of benzene rings as parts of rigid rotors: dynamic stereochemistry of 9-(aryl-x) triptycene derivatives. Pure Appl Chem 62(3):569–574
19. Nemoto T, Ono T, Uchida A, Ohashi Y, Yamamoto G (1994) Structures of *ap*-8-bromo-1,4-dimethyl-9-(2-methylbenzyl)triptycene and *sc**(9*S**)-8-bromo-1,4-dimethyl-9-(2-methylbenzyl)triptycene. Acta Crystallogr C 50:297–300
20. Nemoto T, Ohashi Y, Yamamoto G (1996) *ap*-8-Chloro-1,4-dimethyl-9-(2-methylbenzyl) triptycene and *sc**(9*S**)-8-chloro-1,4-dimethyl-9-(2-methyl benzyl)triptycene. Acta Crystallogr C 52:716–720
21. Yamamoto G, Higuchi H, Yonebayashi M, Nabeta Y, Ojima J (1996) Stereodynamics of *N, N*-dialkyl-9-triptycylamines. Tetrahedron 52(38):12409–12420
22. Nachbar RB, Hounshell WD, Naman VA, Wennerstrom O, Guenzi A, Mislow K (1983) Application of empirical force-field calculations to internal dynamics in 9-benzyltriptycenes. J Org Chem 48(8):1227–1232
23. Chance JM, Geiger JH, Okamoto Y, Aburatani R, Mislow K (1990) Stereochemical consequences of a parity restriction on dynamic gearing in tris(9-triptycyl)germanium chloride and tris(9-triptycyl)cyclopropenium perchlorate. J Am Chem Soc 112(9):3540–3547
24. Toyota S, Yamamori T, Asakura M, Ōki M (2000) Rotational isomerism involving an acetylenic carbon: synthesis of and barrier to rotation around C(*sp*)-C(*sp*^{3}) bonds in bis(1,4-disubstituted 9-triptycyl)ethynes. Bull Chem Soc Jpn 73(1):205–213
25. Toyota S, Yamamori T, Makino T, Ōki M (2000) Rotational isomerism involving an acetylenic carbon II: effect of 1-halogen substituents on rotamer population and rotational barrier around C(*sp*)-C(*sp*^{3}) bonds in bis(9-triptycyl)ethynes. Bull Chem Soc Jpn 73(11):2591–2597
26. Toyota S, Yamamori T, Makino T (2001) Effects of aryl and arylethynyl substituents at the 1-position on rotational barrier around C(*sp*)-C(*sp*^{3}) bonds and bending deformation of acetylenic carbons in bis(9-triptycyl)ethynes. Tetrahedron 57(17):3521–3528
27. Frantz DK, Baldridge KK, Siegel JS (2009) Application of structural principles to the design of triptycene-based molecular gears with parallel axes. Chimia 63(4):201–204
28. Frantz DK, Linden A, Baldridge KK, Siegel JS (2012) Molecular spur gears comprising triptycene rotators and bibenzimidazole-based stators. J Am Chem Soc 134(3):1528–1535
29. Harrington LE, Cahill LS, McGlinchey MJ (2004) Toward an organometallic molecular brake with a metal foot pedal: synthesis, dynamic behavior, and X-ray crystal structure of (9-indenyl)triptycene chromium tricarbonyl. Organometallics 23(12):2884–2891

30. Nikitin K, Mueller-Bunz H, Ortin Y, McGlinchey MJ (2007) Joining the rings: the preparation of 2- and 3-indenyl-triptycenes, and curious related processes. Org Biomol Chem 5(12):1952–1960
31. Nikitin K, Mueller-Bunz H, Ortin Y, McGlinchey MJ (2009) A molecular paddlewheel with a sliding organometallic latch: syntheses, X-ray crystal structures and dynamic behaviour of $Cr(CO)_3\{\eta^6$-2-(9-triptycyl)indene$\}$, and of $M(CO)_3\{\eta^5$-2-(9-triptycyl)indenyl$\}$ (M = Mn, Re). Chem Eur J 15(8):1836–1843
32. Gakh AA, Sachleben RA, Bryan JC, Moyer BA (1995) A facile synthesis and X-ray structure determination of the first triptycenocrown ethers. Tetrahedron Lett 36(45):8163–8166
33. Bryan JC, Sachleben RA, Gakh AA, Bunick GJ (1999) Molecular gears: structures of (9,10-triptyceno) crown ethers. J Chem Crystallogr 29(5):513–521
34. Tsutsumi O, Suzuki F, Oowaki M, Okazaki Y, Takeda K, Mazaki Y, Yamamoto G (2008) Stimulus-responsive molecular rotors: control of rotary motion in triptycene-ionophore systems with *s*-block metal cations. Thin Solid Films 517(4):1428–1433
35. Kao CY, Hsu YT, Lu HF, Chao I, Huang SL, Lin YC, Sun WT, Yang JS (2011) Toward a four-toothed molecular bevel gear with C_2-symmetrical rotors. J Org Chem 76(14):5782–5792
36. Kelly TR, Bowyer MC, Bhaskar KV, Bebbington D, Garcia A, Lang FR, Kim MH, Jette MP (1994) A molecular brake. J Am Chem Soc 116(8):3657–3658
37. Kelly TR, Tellitu I, Sestelo JP (1997) In search of molecular ratchets. Angew Chem Int Ed 36(17):1866–1868
38. Davis AP (1998) Tilting at windmills? The second law survives. Angew Chem Int Edit 37(7):909–910
39. Kelly TR, Sestelo JP, Tellitu I (1998) New molecular devices: in search of a molecular ratchet. J Org Chem 63(11):3655–3665
40. Kelly TR, De Silva H, Silva RA (1999) Unidirectional rotary motion in a molecular system. Nature 401(6749):150–152
41. Kelly TR, Silva RA, Silva HD, Jasmin S, Zhao Y (2000) A rationally designed prototype of a molecular motor. J Am Chem Soc 122(29):6935–6949
42. Kelly TR (2001) Progress toward a rationally designed molecular motor. Acc Chem Res 34(6):514–522
43. Kelly TR, Cavero M (2002) Selective monoacylation of a diamine using intramolecular delivery by a DMAP unit. Org Lett 4(16):2653–2656
44. Kelly TR, Cai X, Damkaci F, Panicker SB, Tu B, Bushell SM, Cornella I, Piggott MJ, Salives R, Cavero M, Zhao Y, Jasmin S (2007) Progress toward a rationally designed, chemically powered rotary molecular motor. J Am Chem Soc 129(2):376–386
45. Markey MD, Kelly TR (2008) Troubleshooting a molecular motor: a remarkably stable *N*-acyl pyridinium salt. Tetrahedron 64(36):8381–8388
46. Sun WT, Huang YT, Huang GJ, Lu HF, Chao I, Huang SL, Huang SJ, Lin YC, Ho JH, Yang JS (2010) Pentiptycene-derived light-driven molecular brakes: substituent effects of the brake component. Chem Eur J 16(38):11594–11604
47. Chen YC, Sun WT, Lu HF, Chao I, Huang GJ, Lin YC, Huang SL, Huang HH, Lin YD, Yang JS (2011) A pentiptycene-derived molecular brake: photochemical $E \rightarrow Z$ and electrochemical $Z \rightarrow E$ switching of an enone module. Chem Eur J 17(4):1193–1200
48. Yang CH, Prabhakar C, Huang SL, Lin YC, Tan WS, Misra NC, Sun WT, Yang JS (2011) A redox-gated slow-fast-stop molecular rotor. Org Lett 13(20):5632–5635
49. Jimenez-Bueno G, Rapenne G (2003) Technomimetic molecules: synthesis of a molecular wheelbarrow. Tetrahedron Lett 44(33):6261–6263
50. Grill L, Rieder KH, Moresco F, Rapenne G, Stojkovic S, Bouju X, Joachim C (2007) Rolling a single molecular wheel at the atomic scale. Nat Nano 2(2):95–98
51. Rapenne G, Jimenez-Bueno G (2007) Molecular machines: synthesis and characterization of two prototypes of molecular wheelbarrows. Tetrahedron 63(30):7018–7026
52. Godinez CE, Zepeda G, Garcia-Garibay MA (2002) Molecular compasses and gyroscopes. II. Synthesis and characterization of molecular rotors with axially substituted bis[2-(9-triptycyl)ethynyl]arenes. J Am Chem Soc 124(17):4701–4707

53. Godinez CE, Zepeda G, Mortko CJ, Dang H, Garcia-Garibay MA (2004) Molecular crystals with moving parts: synthesis, characterization, and crystal packing of molecular gyroscopes with methyl-substituted triptycyl frames. J Org Chem 69(5):1652–1662
54. Garcia-Garibay MA, Godinez CE (2009) Engineering crystal packing and internal dynamics in molecular gyroscopes by refining their components. Fast exchange of a phenylene rotator by 2H NMR. Cryst Growth Des 9(7):3124–3128
55. Nikitin K, Müller-Bunz H, Ortin Y, Muldoon J, McGlinchey MJ (2010) Molecular dials: hindered rotations in mono- and diferrocenyl anthracenes and triptycenes. J Am Chem Soc 132(49):17617–17622
56. Jiang Y, Guo JB, Chen CF (2010). A new [3]rotaxane molecular machine based on a dibenzylammonium ion and a triazolium station. Org Lett 12(19):4248–4251

Chapter 8
Iptycenes and Their Derivatives in Material Science

8.1 Liquid Crystals

In general, all compounds with liquid crystalline properties would be consisted of a rigid core and long flexible chains, which blocked the formation of completely ordered system. Thus, the systems containing the rigid iptycene core and the flexible long alkyl or alkoxy chains seemed to be potential candidates as liquid crystalline materials [1, 2]. In the early 1990s, Simon and Norvez [3, 4] synthesized triptycene derivative **1** containing five long paraffinic chains (Fig. 8.1) and found that it showed the mesomorphic behavior at room temperature. The rigid triptycene core of this pentasubstituted derivative (**1**) regularly arrayed in a hexagonal lamellar structure, along with the long chains extending above and below the layer, which was indicated by its X-ray diffraction patterns. The triptycene derivative **1** exhibited the mesomorphic property at room temperature, probably because the cell areas were big enough to hold the chains at disordered state. On the other hand, it was found that the triptycene derivative **2** (Fig. 8.1) with six alkoxy chains and an aromatic core would form the symmetric compatible lamellar lattices. However, the cell areas in these lamellar lattices were too small to be available for all chains in a disordered state, which led to the crystalline state instead of the mesomorphic state.

Afterwards, Long and Swager [5] reported two series of triptycene-based compounds **3** and **4** with liquid crystal (LC) properties. The synthetic pathways for the two triptycene derivatives with two alkynyl groups were outlined in Scheme 8.1a and b, respectively. Compound **3** was a symmetric mesogen, in which the triptycene moiety positioned at the center ring of a *p*-bis(phenylethynyl)benzene core; while for **4**, it had the relatively asymmetric structure with the triptycene core linking at the outer aromatic ring. The structure of these molecules (**3a–g**) with the triptycene rigid core and the flexible side chains of alkynyl moieties made them prefer to display monotropic nematic phases with a head-to-tail manner package at room temperature. Results from the further experiments showed that phase-transition temperatures decreased as the length of the alkoxy chains increased. For the LCs of molecule **4**, the monotropic nematic phases could be observed as well. Moreover, there were no apparent crystallization peaks on cooling for **4**, which was favorable toward capturing

C.-F. Chen, Y.-X. Ma, *Iptycenes Chemistry*,
DOI 10.1007/978-3-642-32888-6_8, © Springer-Verlag Berlin Heidelberg 2013

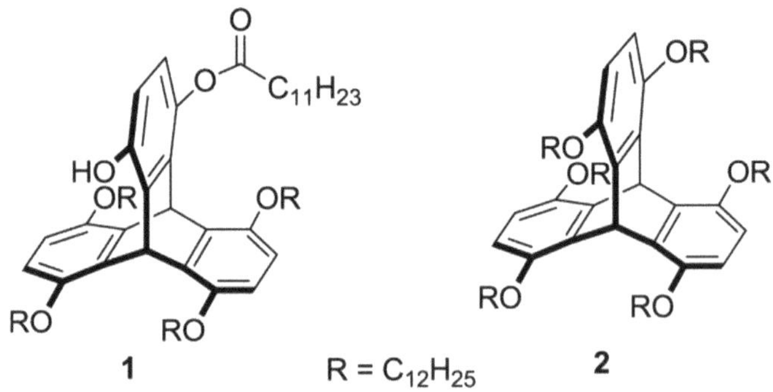

Fig. 8.1 Structures of the compounds **1** and **2**

the intermediate state via rapid cooling. However, the molecule with chiral side chains was crystalline, only doping into another LC could exhibit a chiral nematic phase.

It was usually regarded that LCs with high optical anisotropic properties were based on the structures with high aspect ratios (AR = length/width) [6]. Likewise, the dye solutes in LCs only containing this sort of higher or larger aspect ratio alignment could display a higher optical anisotropy. Thus, Long and Swager [7] considered to control the order and alignment of dye solutes with lower aspect ratio by introducing the unique iptycene (**5**) with rigid three-dimensional structures. As expected, the rigid skeleton of iptycene was doped into the LCs, which created a unique solute orientation. Consequently, the threading of LCs through the internal free volumes (IFVs) between the three aromatic triptycene faces avoided the formation of a vacuum. The alignment with iptycene stretched the direction of the polymer host to create the solute orientations, and thus most effectively filled the voids (Fig. 8.2).

Soon afterward, Long and Swager [8] further investigated the behaviors of several fluorescent and dichroic dye models (**6**–**9**, Fig. 8.3) doped in LC hosts and extended the model dye to the aminoanthraquinone to afford the dichroic dye **10**. Typically, the ordered alignment of LCs would increase the number of triptycene moieties to create the larger IFVs. Nevertheless, the pentiptycene-derived dye **8** with greater IFVs surprisingly seemed to fail to create a new order, but decrease the alignment of LCs. It was probably caused by the huge IFVs those pentiptycene moieties created, which would block the clefts inserting and filling the void spaces of LC hosts. Both of the models (**5**–**7** and **9**) would show that the increase of switch times of order in LCs. Also, the position of iptycene group seemed to influence the respond time to certain extent. When the triptycene served as the end fragment of the guest molecule similar to the dye models **7** and **8**, they could disrupt the local order to create the new cylindrical shape in a relatively short time.

In 2002, Zhu and Swager [9] reported a class of highly soluble triptycene-based conjugated polymers (**11a** and **11b**, Fig. 8.4) with the extended conjugation length, which could form the well-aligned structures in nematic LCs. Interestingly, they further discovered that the LCs may be redirected with electric fields; this feature made it a potential candidate for molecular electronic devices.

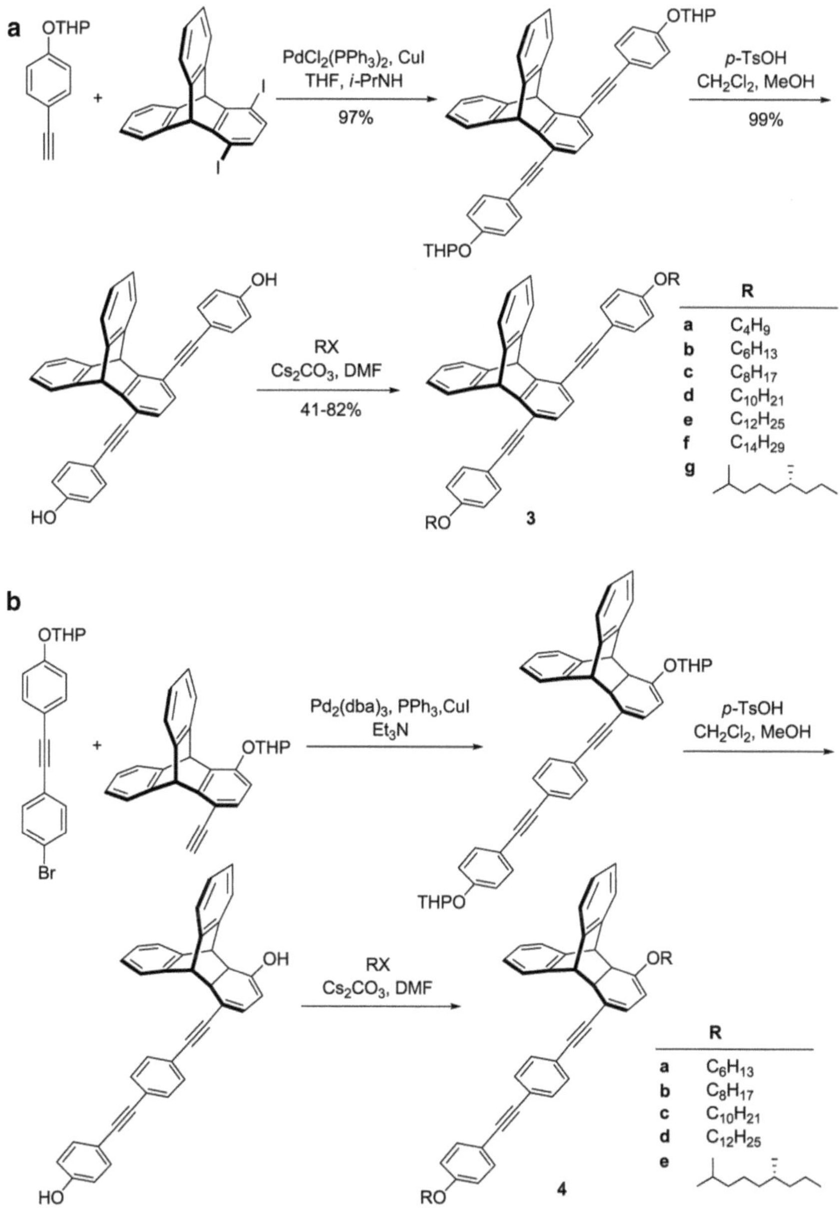

Scheme 8.1 Synthesis of triptycene derivatives **3** and **4**

In addition, Araoka et al. [10] also reported that the cholesteric LCs doped with the triptycene-derived polymeric dye **12** (Fig. 8.5a) could produce laser at the lower energy edge of the photonic gap. The low threshold lasing with well-defined lasing conditions was attributed to the unique property of dye. Moreover, the doping

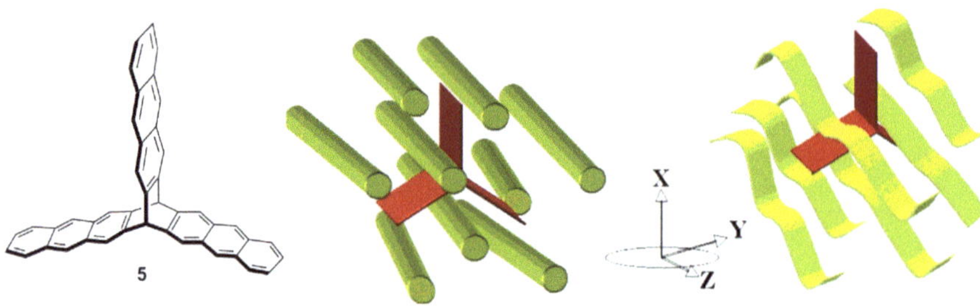

Fig. 8.2 Iptycene **5** aligns in LCs and stretched polymers in an orientation that most effectively fills space. (Reprinted with the permission from [6]. Copyright 2008 American Chemical Society)

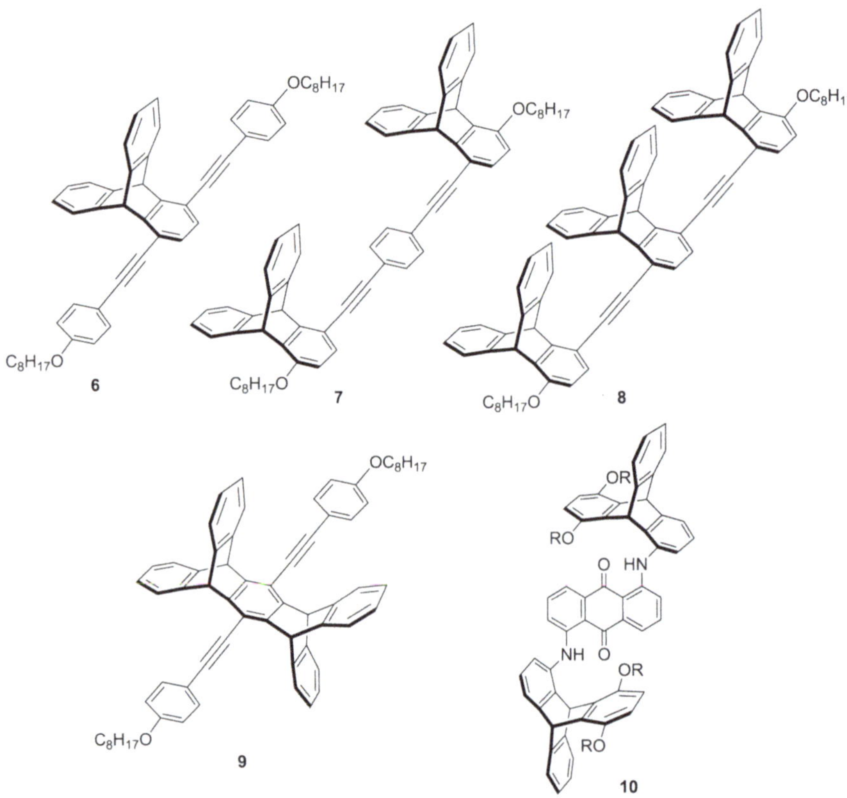

Fig. 8.3 Structures of the compounds **6–10**

triptycene-derived polymeric dye could enhance the degree of the host LC molecules' order, which aligned along the local director with the transition moment of the emission parallel to the conjugation direction (Fig. 8.5b).

In order to verify that increasing the electronic conjugation length of polymers could enhance the intrachain energy transfer that resulted in a higher intrachain

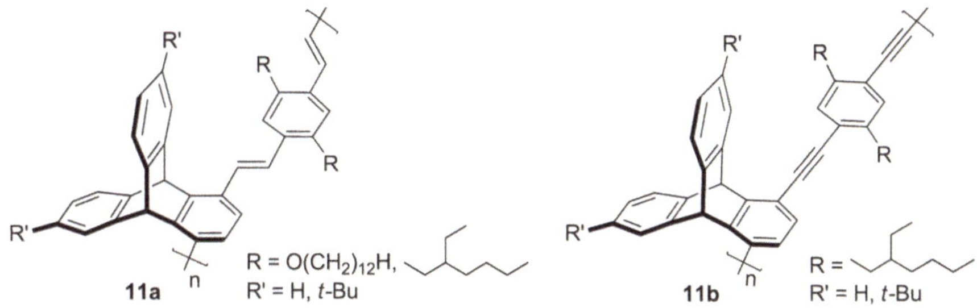

Fig. 8.4 Structures of tripytcene-based conjugated polymers **11a, b**

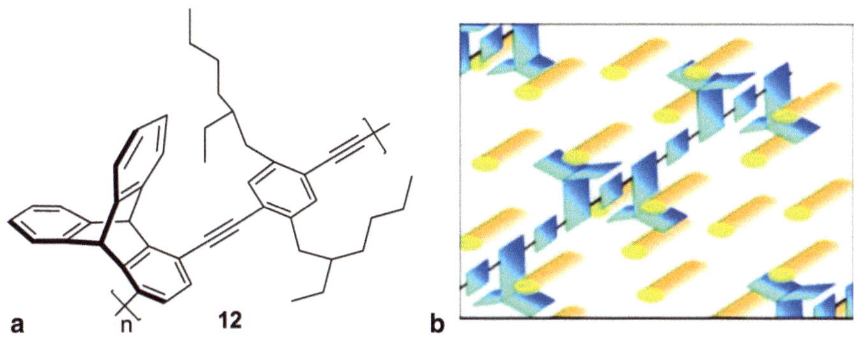

Fig. 8.5 **a** Structure of polymer **12**. **b** Schematic diagram of the doped LCs. (Reprinted with the permission from [6] Copyright 2008 American Chemical Society)

exciton migration rate in the nematic LC solutions, Nesterov et al. [11] designed and synthesized the pentiptycene-containing poly(phenyleneethynylene) (PPE) **13** (Fig. 8.6), which has the low-energy emissive traps due to the anthracenyl end groups. Consequently, the polymer (**13**) dissolved in a nematic LC phase showed a chain-extended, highly conjugated planarized conformation, as a result of the pentiptycene moieties. Moreover, the higher fluorescence quantum yield and reduced Stokes shift could be interpreted as the more rigid and planar conformation of **13** in LCs. The use of this sort of incorporating polymers, which improved the chemosensory performance without sacrificing the fluorescence efficiency, provided a new possibility for the novel materials of electronics, photovoltaics, and even the sensor applications.

As stated above, the incorporation of rigid iptycene group not only effectively prevented the aggregation of chains and improved the solubility, but also extended the alignment and enhanced the order in LCs. Thus, Hoogboom and Swager [12] attempted to enhance the dichroic ratio (DR) and the ordering of PPEs in a nematic LC through the introduction of hydrogen-bonding end groups. As expected, they found that by capping with the ureidopyrimidinone (Upy) end groups, polymer chains **14** (Fig. 8.7) actually exhibited higher DR in LCs system, due to the hydrogen-bonding interactions.

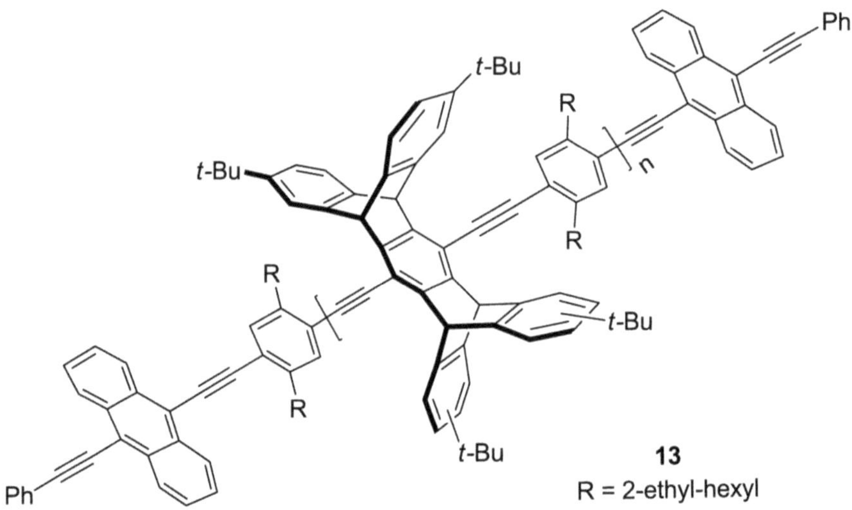

Fig. 8.6 Structure of pentiptycene-containing PPE **13**

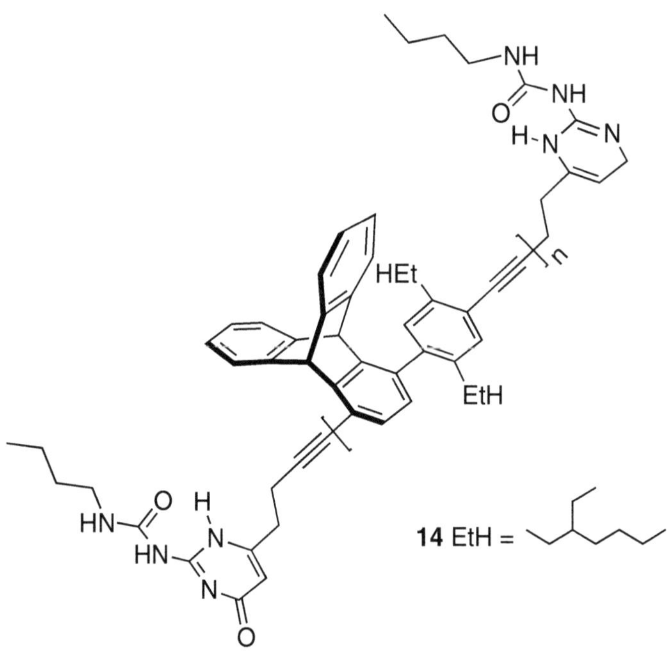

Fig. 8.7 Structure of polymer **14**

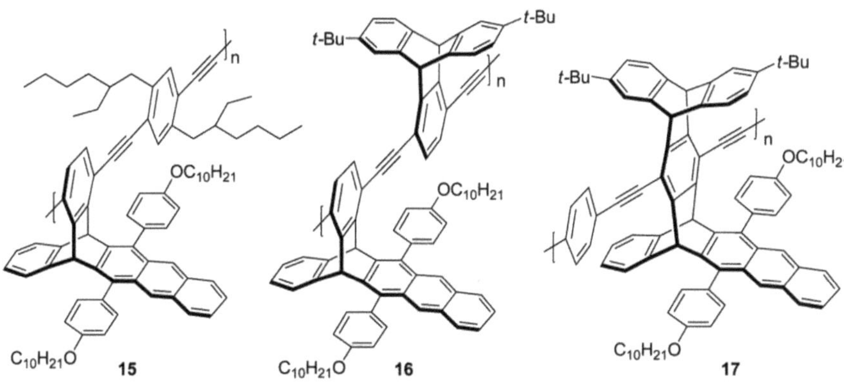

Fig. 8.8 Structures of extended iptycene-containing polymers **15–17**

In 2007, Ohira and Swager [13] further reported a class of more elaborate iptycene-containing polymers with anthracyl group (**15–17**, Fig. 8.8), which had molecular weights (gel permeation chromatography (GPC), relative to polystyrene (PS)) of more than 100,000. As expected, it was found that the novel PPEs with elaborate iptycene scaffold, which was dissolved in LCs, obviously exhibited the greater alignment with a higher unidirectional order. The data of optical spectroscopy showed that these polymers (**15–17**) were in a chain-extended rod conformation, and the orientation of the polymer chains maintained a high degree of unity with the LC director, along with an enhanced conjugation length. Moreover, the increasing molecular weight made the global and local chain conformation extend and align uniaxially well. The different results of these three PPEs (**13–15**) revealed that the steric congestion about the polymer main chain seemed to be favorable to the high-order parameter to reach the optimal properties of LCs. Thus, it reflected that the realization of facile and precise alignment control of high molecular weight conjugated polymers was crucial to achieve optimal properties [13].

8.2 Optical and Electronic Materials

The use of materials with low dielectric constant as the interlayer dielectrics could enhance the speed of integrated circuits via reducing the capacitive cross talk. In order to afford the best insulating materials, two key aspects should be considered: (1) the materials can create empty space and effectively vacuum in their structures, and (2) the dielectric can maintain the structural integrity necessary to hold multiple layers of interconnecting wires in place over a broad temperature range [6]. Consequently, Long and Swager [14] investigated that the incorporation of rigid iptycene moieties into polymers **18** and **19** had lower dielectric constants than the nontriptycene reference polymers **20** and **21**, which might be due to the formation of nanometer-sized

Table 8.1 The polymers with low dielectric constants

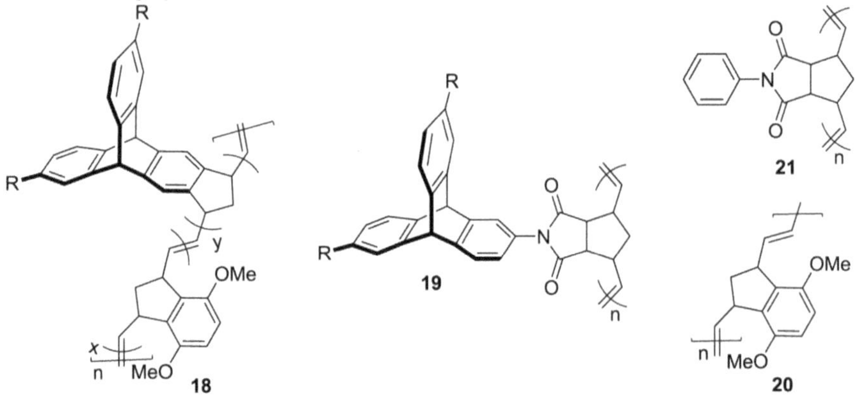

Entry	Polymer	Dielectric constant (ε, at 10 kHz)[a]
1	**18a** R = H; $x = 0$, $y = 1$	2.78 (0.09)
2	**18b** R = *t*-Bu; $x = 0$, $y = 1$	2.59 (0.08)
3	**18c** R = H; $x = 1$, $y = 1$	2.82 (0.06)
4	**18d** R = *t*-Bu; $x = 1$, $y = 1$	2.69 (0.06)
5	**20** (refer polymer) $x = 1$, $y = 0$	3.35 (0.15)
6	**19**	2.98 (0.13)
7	**21** (refer polymer)	3.00 (0.11)

[a]Error limits in parentheses

pores. As shown in Table 8.1, it revealed that the restraint rotation of triptycene moieties in **18** seemed to be favored to afford the lower dielectric constants compared with the polymer **19,** which attached to triptycene group with a single bond. Thus, the rigid triptycene groups in **18** were restricted by a two-point connection, which subsequently made the polymer maintain not only the high thermal stability, low water absorption, and high modulus, but also show lower dielectric constants.

Based on the method of introducing the free volumes to achieve the lower dielectric constants, Amara and Swager [15] further synthesized a series of high molecular weight iptycene-containing polymers (**22**–**24**, Fig. 8.9) via the self-initiated free radical polymerization by heating butadienes-bearing iptycenes. These poly-(butadiene)s with large IFVs, especially polymer **24**, showed low dielectric constants with excellent mechanical characters, high thermal stability, and a high T_g (≥ 500 °C).

In addition, Sydlik et al. [16] also synthesized the aromatic polyimides (**25** and **26**, Table 8.2) by the polymerization of the 2,6-diaminotriptycene derivatives with various five- and six-membered ring anhydrides. According to Table 8.2, it was found that the incorporation of triptycene units into the polyimide chains resulted in the lower refractive index. Especially, the refractive indexes for **25e**, **26a**, and **25b** were found to be 1.31, 1.27, and 1.19, respectively, which showed even better results than the commercial materials (e.g. Kapton(R)HN with a refractive index of 1.70). Apart from that, the polyimides based on iptycenes showed good solubility and thermal stability along with no glass transition temperatures below 450 °C, which thus made them potential as good spin-on dielectric materials.

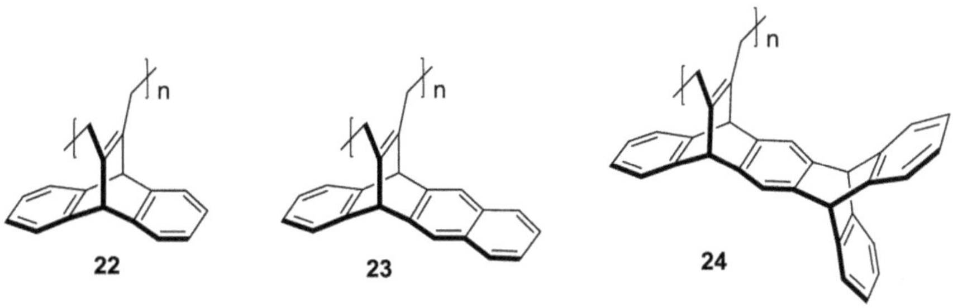

Fig. 8.9 Structures of iptycene-containing polymers **22–24**

Likewise, Hsiao et al. [17] synthesized a class of fluorinated iptycene-based polyimides (**27** and **28**, Fig. 8.10). These polyimides could also serve as the promising candidates for the applications in microelectronics and optoelectronics. Moreover, the polyimides with low dielectric constants (2.05–2.89 at 1 MHz) could be easily utilized for preparing the pale yellow to nearly colorless films via the solution casting, and the films also kept the local excellent thermal stability, flexibility, or mechanical properties of the polyimides.

In 2008, Chen et al. [18] reported a series of the donor–acceptor copolymers (**29** and **30**, Fig. 8.11) from two triptycene-type quinoxaline and thienopyrazine acceptor monomers and found that the copolymers showed the interesting electronic properties. The diverse donor–acceptor charge transfer interactions and conformation change of the chains' backbone led to the different absorption/emission natures of the copolymers with blue, greenish-blue, and red color in solution and in the solid state, respectively. In addition, the rigid iptycene moieties also prevented the intermolecular interactions and the formation of excimers in the solid state, due to the insulation effect. Thus, these copolymers could be promising materials for electroluminescent devices.

Furthermore, Chou et al. [19] synthesized compounds **31–33** (Fig. 8.12) with a triptycene core, which exhibited hole and electron-transporting characteristics. For **31**, it also exhibited high brightness, good current, and power efficiency, which made it a promising host material for blue, green, and red electrophosphorescence devices. Likewise, triptycene derivatives **32** and **33** showed the 10.1 and 16.9 % external quantum efficiency as the host for the blue phosphorescence emitter FIrpic (Ir^{III} bis(4,6-difluorophenyl-pyridinato)-picolinate). The normal operation brightness of the device would be greatly enhanced to 500–1,000 cd m^{-2} with the cohost system of **32** and **33**, due to the lower driving voltage and the balance of hole and electron transport.

In 2011, Polishak et al. [20] reported that through doping the triptycene-modified chromophore **34** (Fig. 8.13) into poly(bisphenol A carbonate) (BPAPC), the material performance of the systems could be improved. By comparison with the chromophore **35**, the absorption spectrum of **34** showed blue shift in both the solution state and the solid state; this was probably because the triptycene groups created a relatively nonpolar nanoenvironment for the donor. Due to the threading and IFVs effect of

Table 8.2 Refractive indexes of the aromatic polyimides **25** and **26**

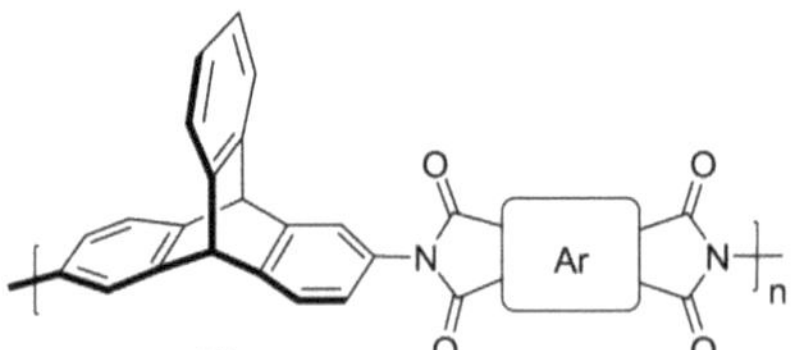

25

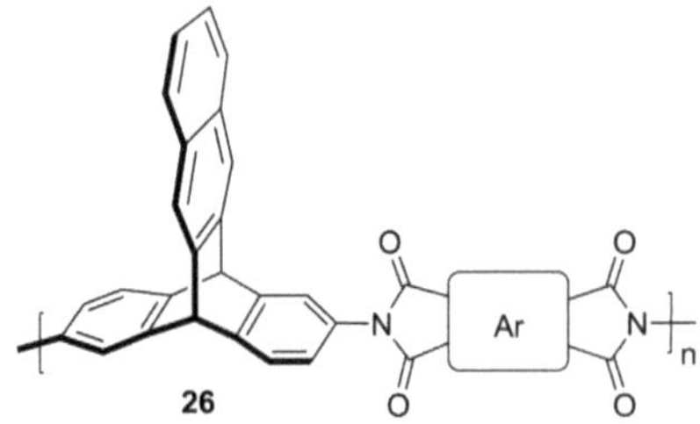

26

Entry	Ar	Polymer	Dielectric constant (ε, at 1 MHz)	Refractive index (n)*
1	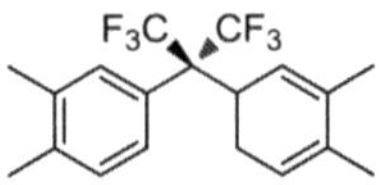	**25a**	2.22	1.49
2	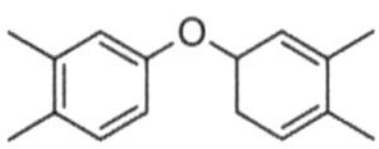	**25b**	2.76	1.66
3	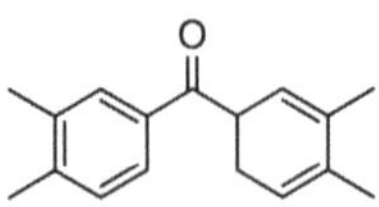	**25c**	2.86	1.69
4	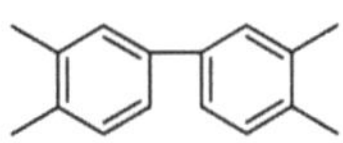	**25d**	2.86	1.69
5	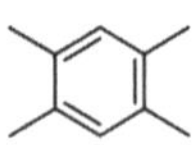	**25e**	1.61	1.27
6	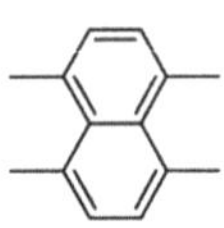	**25f**	3.20	1.79
7	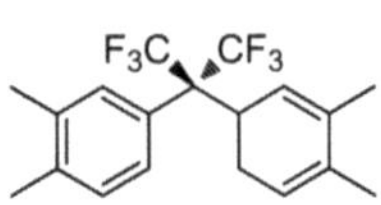	**26a**	1.42	1.19
8	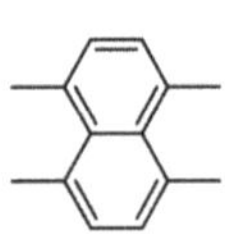	**26b**	1.72	1.31

* Estimated from the refractive index using $n = \varepsilon^2$

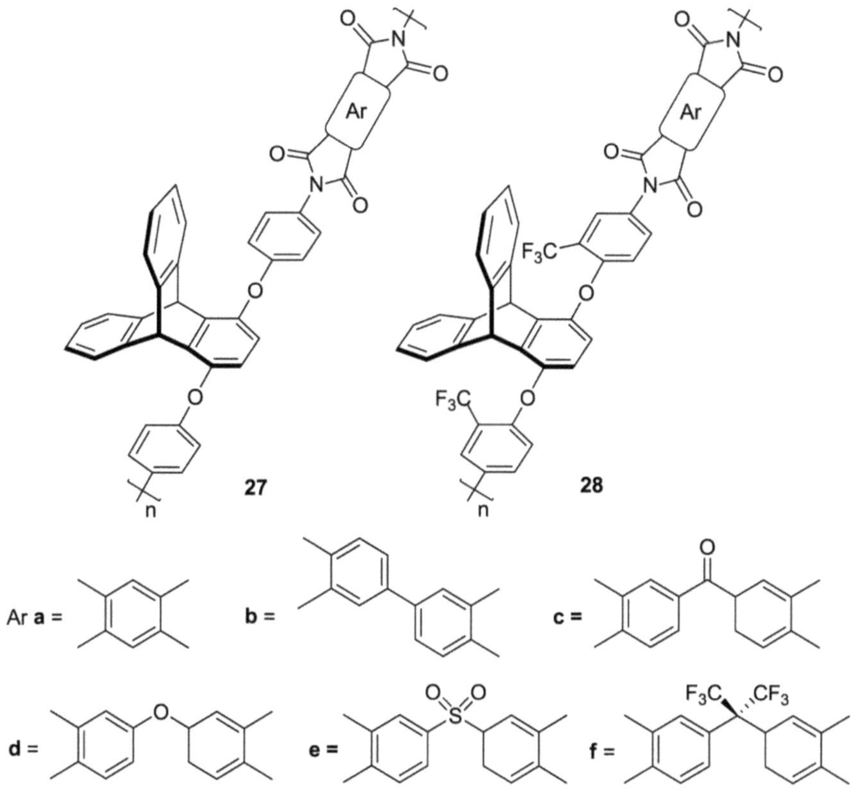

Fig. 8.10 Structures of the iptycene-based polyimides **27** and **28**

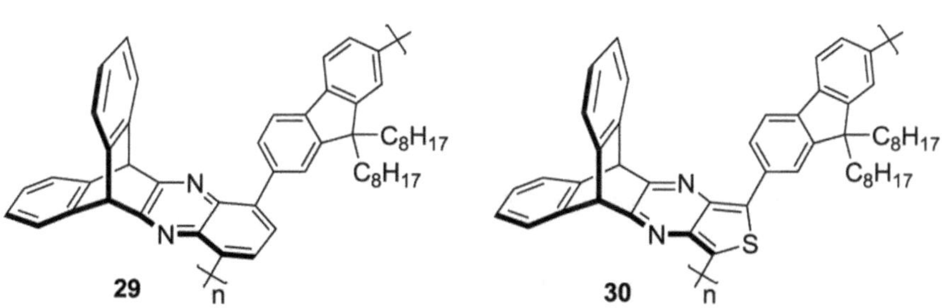

Fig. 8.11 Structures of donor–acceptor copolymers **29** and **30**

triptycene moiety, the system showed the excellent properties such as high thermal stability and good compatibility, which made it potential candidate for device applications. Moreover, they also put forward a new possible poling mechanism for triptycene-containing polymeric materials. There were two processes during poling, including the orientational rotation of the chromophores and the local reorganization of segments.

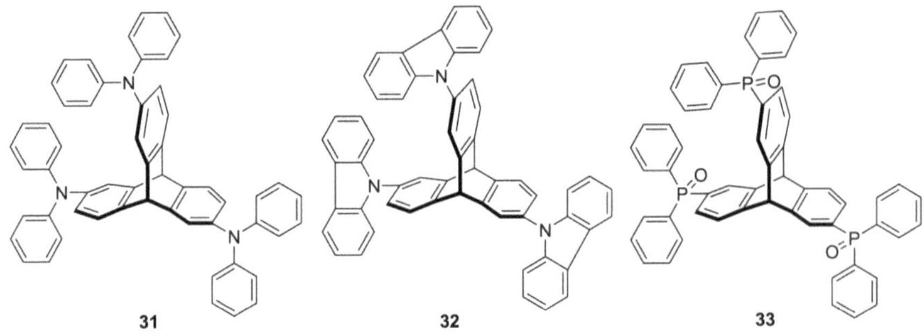

Fig. 8.12 Structures of the compounds **31** and **33**

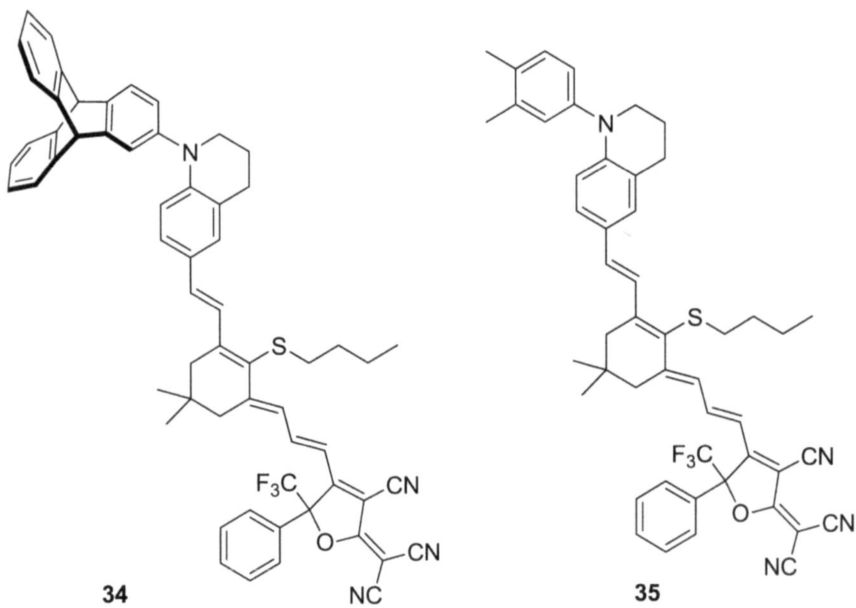

Fig. 8.13 Structures of compounds **34** and **35**

8.3 Porous Materials for Adsorption and Separation

The polymers of intrinsic microporosity (PIMs) with larger surface areas and pore volumes along with ultramicroporous structure seemed to be promising candidates for gas storage. Thus, Ghanem et al. [21] investigated the hydrogen adsorption of the novel triptycene-based polymer (**36**, R = Et, Fig. 8.14), which contained intrinsic microporosity and great surface area. The Brunauer–Emmett–Teller (BET) surface area of **36** would reach to 1,064 $m^2\ g^{-1}$ and 1,930 $m^2\ g^{-1}$ displayed by the nitrogen and hydrogen adsorption measurements at 77 K, respectively. Although a gap still existed when it was compared with the best carbons or metal-organic frameworks

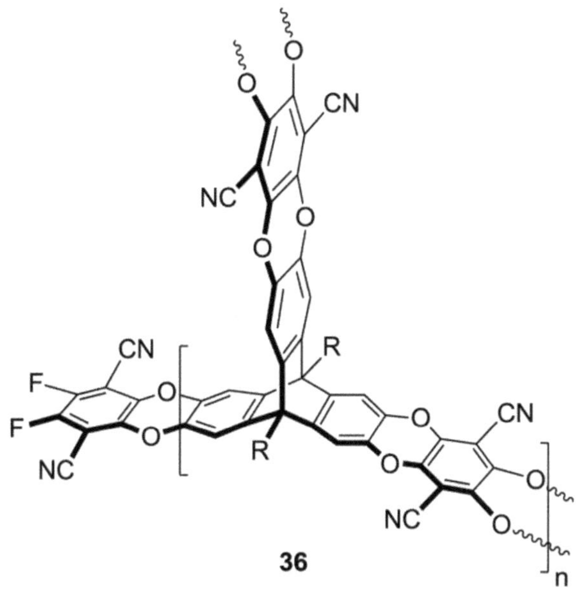

Fig. 8.14 Structure of triptycene-based polymer **36**

(MOFs) adsorption materials, the PIMs **36** showed the better ability of adsorption (1.65 % by mass at 1 bar) at low pressures than a common hypercross-linked polymer, even any microporous material with the similar surface area. It was noteworthy that the size of pores of PIMs **36** was strongly biased toward subnanometer pores, which was revealed by the analysis about the low-pressure nitrogen adsorption data via the Horvath–Kawazoe method [22]. Furthermore, Horvath and Kawazoe [23] also studied the adsorption capacity of other PIMs, which had different alkyl groups at the bridgehead positions. As a result, the PIMs (**36**) with the shorter groups such as methyl, or the branched groups such as isopropyl substituents, showed highest hydrogen-adsorption capacities at low or moderate pressures (1.83 % by mass at 1 bar/77 K; 3.4 % by mass at 18 bar/77 K) compared with another purely organic materials. Further analysis of data revealed that the high capacities of gas adsorption related to a high concentration of subnanometer micropores, which could be controlled by the varied length and the degree of grafting of the two side-moieties at the bridgehead positions of the triptycene subunits.

Besides the triptycene-based polymers, Chong et al. [24] also designed a novel kind of porous solid-state materials, triptycene-based nickel salphen complexes **37–45** with different IMFVs (Fig. 8.15 and Fig. 8.16). With the comparison of the varied salphen complexes for nitrogen adsorption studies, there were several aspects which should be considered: (1) the requirement of triptycene moiety to form the porosity in salphens, (2) the increasing number of salphen moieties, (3) the extended triptycene to further create porosity and enhance the IMFV, and (4) the end-capped bulky salicylaldehyde to enhance the porosity. Thus, only the complexes **39** and **41–45** with accessible IMVFs exhibited the discernible adsorption ability, and the majority of their pores were microporous, which were obtained by the pore size

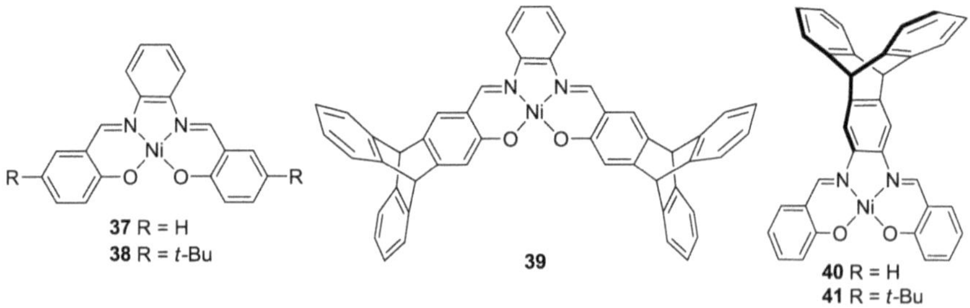

Fig. 8.15 Structures of triptycene-based nickel salphen complexes **37–41**

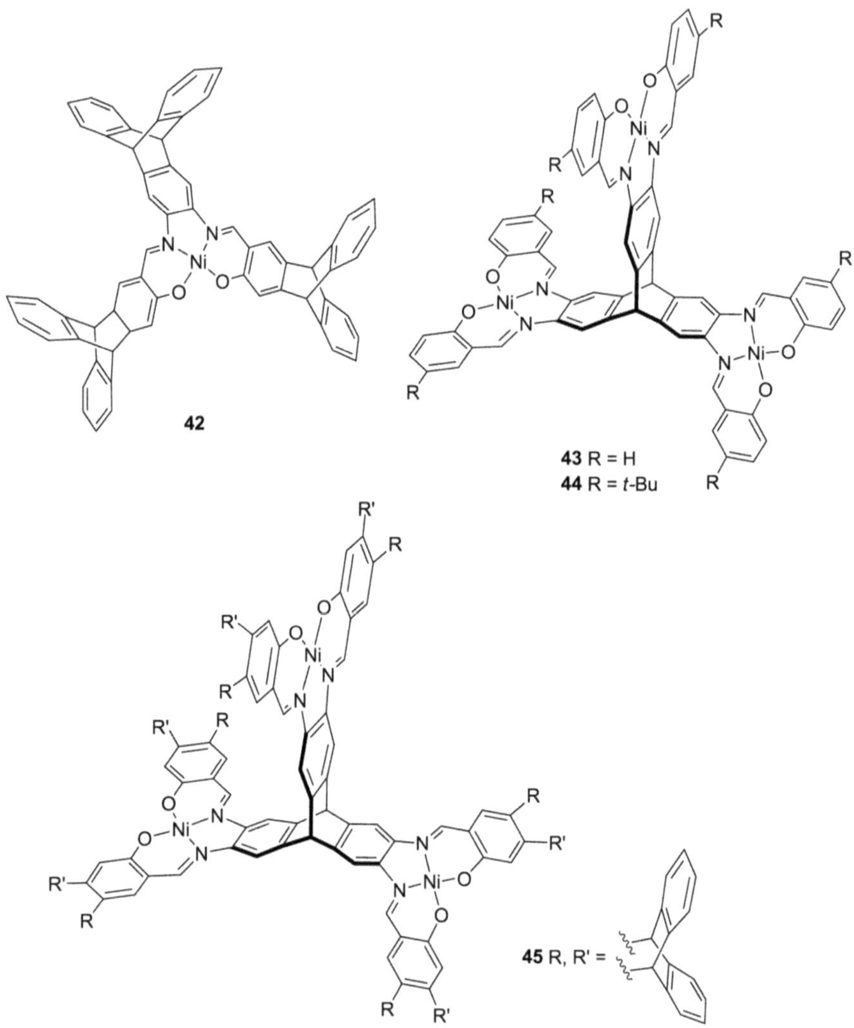

Fig. 8.16 Structures of triptycene-based nickel salphen complexes **42–45**

Table 8.3 H_2 adsorption properties of complexes **39–45**

Entry	Complexes	H_2 adsorption [$cm^3\ g^{-1}$][a]	Weight H_2 adsorbed [%][a]
1	**39**	6.7	0.1
2	**41**	21.2	0.2
3	**42**	52.4	0.5
4	**43**	49.8	0.4
5	**44**	84.3	0.8
6	**45**	121.0	1.1

[a]Measured at 1 atm. and 77 K

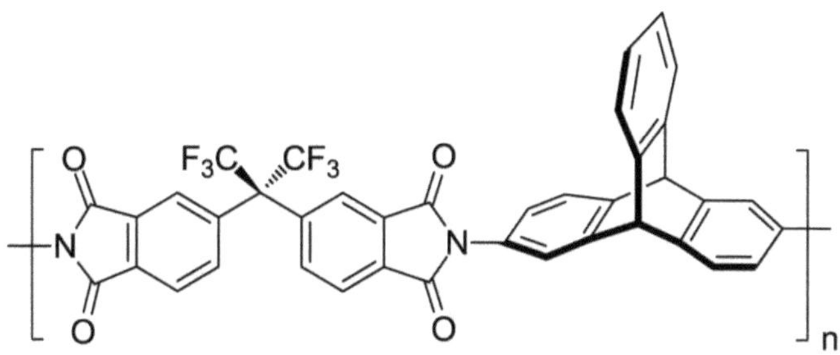

46

Fig. 8.17 Structure of fluorinated triptycene-based polyimide **46**

distributions via the Horvath–Kawazoe analyses. However, the case of H_2 differed from that of the N_2 which related to surface area, the effect of aromatic rings also determined the ability of H_2 adsorption. Thus, **45** with higher density of aromatic rings showed greater capacity of H_2 (1.1 % by mass at 1 bar/77 K) than that of **44**, even though **44** with the larger surface area (Table 8.3).

In 2011, Cho and Park [25] reported that the fluorinated triptycene-based polyimide **46** (Fig. 8.17) showed the higher gas permeability, which was attributed to the bulky 6 F^- groups to lead to the less packing efficiency and irregular free spaces. With the studies of gas separation, it was found that **46** showed a high permeation and excellent air-separation performance in the case of CO_2/N_2 and CO_2/CH_4, even beyond the upper bound of permeability–selectivity relations in polymeric thin films which was established by Robeson [26]. In addition, **46** also showed strong tolerance to CO_2 plasticization without significantly reduced selectivity, when the CO_2 fugacity increased. The superior properties of gas separation along with the maintenance of the local good mechanical characters made it a promising material for specific gas separation.

Recently, Gong et al. [27] reported that the triptycene-based poly(aryl ether sulfone)s (**47**, Fig. 8.18) showed good water-swelling resistance with high thermal stability and good mechanical strength. Moreover, the membranes of **47** maintained high proton conductivity in a wide range of humidity (34–94 % relative humidity (RH)) at 80°C. It was notable that the conductivity of membrane with low ion-exchange capacity (IEC < 2.09) displayed more dependence on humidity than the

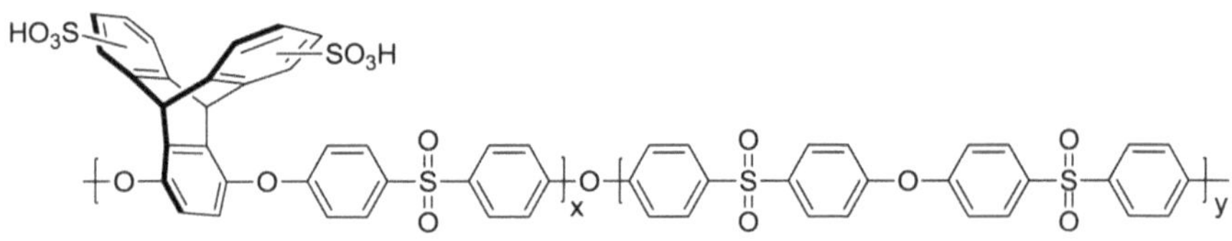

47

Fig. 8.18 Structure of triptycene-based poly(aryl ether sulfone)s **47**

Fig. 8.19 Structures of the star-like triptycene-based microporous polymers **48a, b**

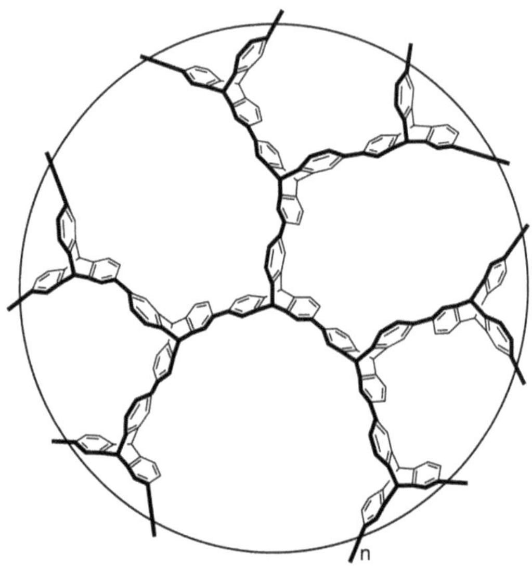

48a from 2,6,14-triiodotriptycene
48b from 2,6,14-tribromotriptycene

high IEC one. The stable conductivity of high IEC membrane in a wide range of humidity probably resulted in their good connectivity of proton paths and excellent water-holding capability. They also tested the performance of the material **47** in H_2/O_2 fuel cell. As expected, the high proton conductivity at low humidity led to good cell performance, the current density of membrane could reach to 222 mA cm^{-2}, under the 0.6 V cell voltages in 40°C, with 30 % RH. The performance made it a promising material for the proton-exchange membrane fuel cell.

More recently, Zhang et al. [28] designed and synthesized a novel kind of star-like triptycene-based microporous polymers (**48a** and **48b**, Fig. 8.19) via the nickel^0-catalyzed Ullmann cross-coupling reactions from 2,6,14-triiodotriptycene or 2,6,14-tribromotriptycene, respectively. These polymers with high BET surface areas displayed the reversibly absorbable ability for H_2 and CO_2, and the detail data are shown in Table 8.4. Moreover, these gas-absorbing porous polymers also showed good thermal stability. Thus, they could serve as the promising gas-adsorbent candidates.

Table 8.4 Gas adsorption properties of polymers **48a, b**

Entry	Polymer	S_{BET} ($m^2\ g^{-1}$)	Weight H_2 adsorbed [%][a]	Weight CO_2 adsorbed [%][b]
1	**48a**	1,305	1.60	16.15
2	**48b**	1,990	1.92	18.20

[a]Measured at 1 atm. and 77 K
[b]Measured at 1 atm. and 273 K

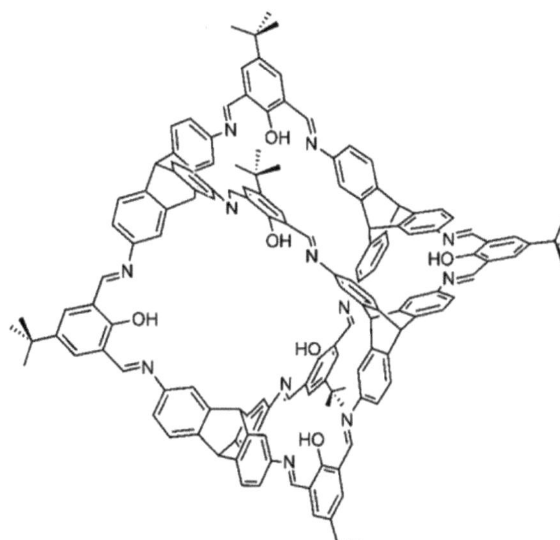

49

Fig. 8.20 Structure of triptycene-based adamantoid cage **49**. (Reproduced from [29], with the permission of John Wiley and Sons)

Apart from PIMs based on iptycene moieties, the functionalized triptycene-based adamantoid cage compound (**49**, Fig. 8.20), which was synthesized by Mastalerz et al. [29] showed the CO_2/CH_4 adsorption ability as well. This purely organic compound could selectively take up 9.4 wt% CO_2 and 0.94 wt% methane at 1 atm. and 273 K. According to the further study, it was found that the selectivity probably caused by the polar functional groups inside the cavity. Moreover, it was found that the molecular cage also absorbed 5.6 mmol/g H_2 at 1 atm. and 77 K. In addition, molecular cage **49** had a 1,566 m^2/g BET surface area, which was the highest one for a discrete purely organic compound material reported till date.

More recently, Zhao et al. [30] reported the first triptycene-based microporous poly(benzimidazole) networks (**50a**, **b**), which was prepared by a one-pot reaction of triptycene–hexaone derivative and dialdehyde in glacial acetic acid in the presence of ammonium acetate (Scheme 8.2). It was noteworthy that the process of condensation occurred smoothly without catalyst or template. Both of the polymers **50a** and **50b** showed good hydrogen storage (1.57 wt% at 77 K and 1.0 bar) and carbon dioxide affinity (14.0 wt% at 273 K and 1.0 bar) properties with the high BET-specific surface area (600 m^2/g).

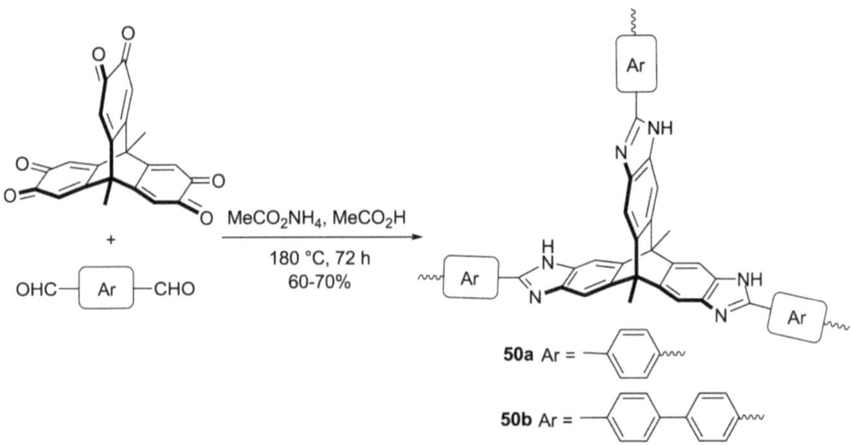

Scheme 8.2 Synthesis of triptycene-based microporous poly(benzimidazole) networks **50**

References

1. Chong JH, MacLachlan MJ (2009) Iptycenes in supramolecular and materials chemistry. Chem Soc Rev 38(12):3301–3315
2. Jiang Y, Chen CF (2011) Recent developments in synthesis and applications of triptycene and pentiptycene derivatives. Eur J Org Chem 2011(32):6377–6403
3. Norvez S, Simon J (1990) Epitaxygens: mesophases based on the triptycene molecular subunit. J Chem Soc Chem Commun 1990(20):1398–1399
4. Norvez S (1993) Liquid-crystalline triptycene derivatives. J Org Chem 58(9):2414–2418
5. Long TM, Swager TM (2002) Triptycene-containing bis(phenylethynyl) benzene nematic liquid crystals. J Mater Chem 12(12):3407–3412
6. Swager TM (2008) Iptycenes in the design of high performance polymers. Acc Chem Res 41(9):1181–1189
7. Long TM, Swager TM (2001) Minimization of free volume: alignment of triptycenes in liquid crystals and stretched polymers. Adv Mater 13(8):601–604
8. Long TM, Swager TM (2002) Using "internal free volume" to increase chromophore alignment. J Am Chem Soc 124(15):3826–3827
9. Zhu ZG, Swager TM (2002) Conjugated polymer liquid crystal solutions: control of conformation and alignment. J Am Chem Soc 124(33):9670–9671
10. Araoka F, Shin KC, Takanishi Y, Ishikawa K, Takezoe H, Zhu ZG, Swager TM (2003) How doping a cholesteric liquid crystal with polymeric dye improves an order parameter and makes possible low threshold lasing. J Appl Phys 94(1):279–283
11. Nesterov EE, Zhu ZG, Swager TM (2005) Conjugation enhancement of intramolecular exciton migration in poly(*p*-phenylene ethynylene)s. J Am Chem Soc 127(28):10083–10088
12. Hoogboom J, Swager TM (2006) Increased alignment of electronic polymers in liquid crystals via hydrogen bonding extension. J Am Chem Soc 128(47):15058–15059
13. Ohira A, Swager TM (2007) Ordering of poly(*p*-phenylene ethynylene)s in liquid crystals. Macromolecules 40(1):19–25
14. Long TM, Swager TM (2003) Molecular design of free volume as a route to low-κ dielectric materials. J Am Chem Soc 125(46):14113–14119
15. Amara JP, Swager TM (2004) Incorporation of internal free volume: synthesis and characterization of iptycene-elaborated poly(butadiene)s. Macromolecules 37(8):3068–3070

16. Sydlik SA, Chen Z, Swager TM (2011) Triptycene polyimides: soluble polymers with high thermal stability and low refractive indices. Macromolecules 44(4):976–980
17. Hsiao SH, Wang HM, Chen WJ, Lee TM, Leu CM (2011) Synthesis and properties of novel triptycene-based polyimides. J Polym Sci Polym Chem 49(14):3109–3120
18. Chen Z, Bouffard J, Kooi SE, Swager TM (2008) Highly emissive iptycene-fluorene conjugated copolymers: synthesis and photophysical properties. Macromolecules 41(18):6672–6676
19. Chou HH, Shih HH, Cheng CH (2010) Triptycene derivatives as high-T_g host materials for various electrophosphorescent devices. J Mater Chem 20(4):798–805
20. Polishak BM, Huang S, Luo J, Shi Z, Zhou XH, Hsu A, Jen AKY (2011) A triptycene-containing chromophore for improved temporal stability of highly efficient guest–host electrooptic polymers. Macromolecules 44(6):1261–1265
21. Ghanem BS, Msayib KJ, McKeown NB, Harris KDM, Pan Z, Budd PM, Butler A, Selbie J, Book D, Walton A (2007) A triptycene-based polymer of intrinsic microposity that displays enhanced surface area and hydrogen adsorption. Chem Commun 2007(1):67–69
22. Horvath G, Kawazoe K (1983) Method for the calculation of effective pore-size distribution in molecular-sieve carbon. J Chem Eng Jpn 16(6):470–475
23. Ghanem BS, Hashem M, Harris KDM, Msayib KJ, Xu M, Budd PM, Chaukura N, Book D, Tedds S, Walton A, McKeown NB (2010) Triptycene-based polymers of intrinsic microporosity: organic materials that can be tailored for gas adsorption. Macromolecules 43(12):5287–5294
24. Chong JH, Ardakani SJ, Smith KJ, MacLachlan MJ (2009) Triptycene-based metal salphens-exploiting intrinsic molecular porosity for gas storage. Chem Eur J 15(44):11824–11828
25. Cho YJ, Park HB (2011) High performance polyimide with high internal free volume elements. Macromol Rapid Commun 32(7):579–586
26. Robeson LM (2008) The upper bound revisited. J Membrane Sci 320(1–2):390–400
27. Gong F, Mao H, Zhang Y, Zhang S, Xing W (2011) Synthesis of highly sulfonated poly(arylene ether sulfone)s with sulfonated triptycene pendants for proton exchange membranes. Polymer 52(8):1738–1747
28. Zhang C, Liu Y, Li B, Tan B, Chen CF, Xu HB, Yang XL (2011) Triptycene-based microporous polymers: synthesis and their gas storage properties. ACS Macro Lett 1(1):190–193
29. Mastalerz M, Schneider MW, Oppel IM, Presly O (2011) A salicylbisimine cage compound with high surface area and selective CO_2/CH_4 adsorption. Angew Chem Int Ed 50(5):1046–1051
30. Zhao YC, Cheng QY, Zhou D, Wang T, Han BH (2012) Preparation and characterization of triptycene-based microporous poly(benzimidazole) networks. J Mater Chem 22(23):11509–11514

Chapter 9
Iptycenes and Their Derivatives in Host–Guest Chemistry

At the outset, the studies on iptycenes in host–guest chemistry were only confined to the interactions between the phenyl rings of iptycene and guest molecule, and such examples were also very limited. In 1998, Rathore and Kochi [1] reported the reversibly binding of triptycene-based radicals toward nitric oxide (NO), and found that the process could be controlled by a simple temperature modulation. Moreover, the process for the uptake of NO molecule went along with a distinct color change from yellow to green or purple. However, it was also found that the NO was tightly bound only at −30 °C due to the weak interactions between cofacial phenylene units of the triptycene derivatives and NO. Soon after, Konarev et al. [2] investigated the complexation of triptycene toward fullerene C_{60}, and found that the C_{60} molecule in the complex showed the freezing of free rotation even at 87 °C, which was revealed by the infrared (IR) spectrum. The optical absorption spectrum also showed the low intensity of the C_{60} transitions in the 420–500 nm range, probably as a result of the separation of C_{60} molecules by the triptycenes along with the increasing C_{60}–C_{60} distances.

Actually, iptycene molecules have the three-dimensional rigid structures and rich reactive positions, and it is believed that these features could make them promising useful building blocks for the design and synthesis of novel hosts. Thus, Chen and coworkers paid much attention to the synthesis and applications of novel triptycene-derived hosts several years ago. Consequently, Chen's group [3, 4] have succeeded to develop several kinds of novel iptycene-derived hosts including triptycene-derived crown ethers, calixarenes, heterocalixarenes, pentiptycene-derived hosts, and other iptycene-derived macrocyclic hosts, which not only greatly enrich the research of the iptycene chemistry, but also open the door for iptycene-based host–guest chemistry.

9.1 Triptycene-Derived Crown Ethers

9.1.1 Triptycene-Derived Cylindrical Macrotricyclic Polyethers

Cylindrical macrotricyclic polyethers [5] with one central cavity and two lateral circular cavities have attracted much attention in host–guest chemistry for their specific

C.-F. Chen, Y.-X. Ma, *Iptycenes Chemistry*,
DOI 10.1007/978-3-642-32888-6_9, © Springer-Verlag Berlin Heidelberg 2013

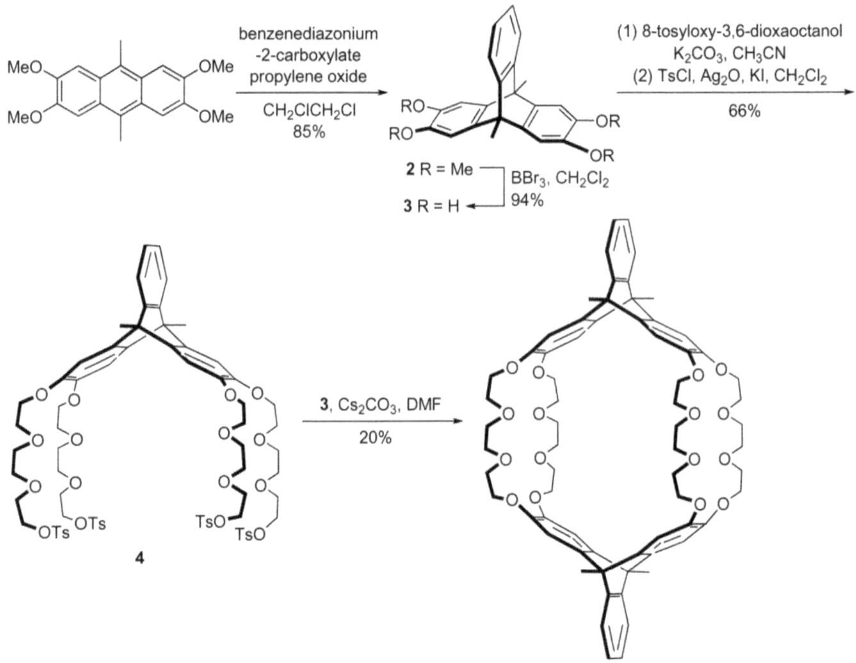

Scheme 9.1 Synthesis of triptycene-derived macrotricyclic polyether **1**

topological features. However, the previously reported macrotricyclic polyethers are all composed of two macrocycles linked by two bridges, which make their structures be so flexible that their complexation with guests may be influenced. It is considered that by the combination of four crown chains with the triptycene with three-dimensional rigid structure, a new kind of macrotricyclic polyethers with specific structures and properties could be developed. Thus, Zong and Chen [6] first designed and synthesized a novel triptycene-derived macrotricyclic polyether **1** containing two dibenzo-[24]crown-8 (DB24C8) moieties. As shown in Scheme 9.1, reaction of 2,3,6,7-tetramethoxyl-9,10-dimethyl-anthracene and benzenediazonium-2-carboxylate in the presence of propylene oxide gave the triptycene derivative **2** in 85 % yield. The demethylation of **2** with boron bromide produced triptycene bis(catechol) **3** in 94 % yield, which was then reacted with 8-tosyloxy-3,6-dioxaoctanol in the presence of K_2CO_3, and followed by a reaction with *p*-toluenesufonyl chloride to give compound **4**. The target molecule **1** was obtained in a 20 % yield by the further reaction of **4** and **3** under a high dilution condition in the presence of cesium carbonate. The X-ray crystal structure of compound **1** exhibited not only two lateral crown ether cavities but also one rich-electron central cavity with a size of ca. 10.2×13.9 Å^2, which could thus show specific complexation with different guests.

Zong and Chen [6] first tested the complexation between macrocycle **1** and paraquat derivatives **5–7** (Fig. 9.1), and found that they could all form 1:1 complexes

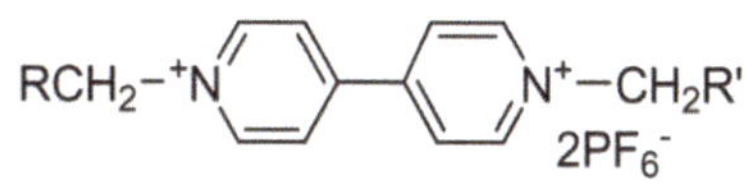

5 R = R' = H **6** R = R' = CH_2CH_3 **7** R = R' = $(CH_2)_6CH_3$

8 R = R' = CH_3 **9** R = R' = CH_2OH **10** R = R' = CH_2CH_2OH

11 R = R' = $(CH_2)_5OH$ **12** R = CH_2OH, R' = CH_2OMe **13** R = R' = CH_2OCH_3

14 R = R' = $CH_2NHCOCH_3$ **15** R = R' = $CONHCH_2Ph$ **16** R = R' = 9-anthracyl

Fig. 9.1 Structures of paraquat derivatives **5–16**

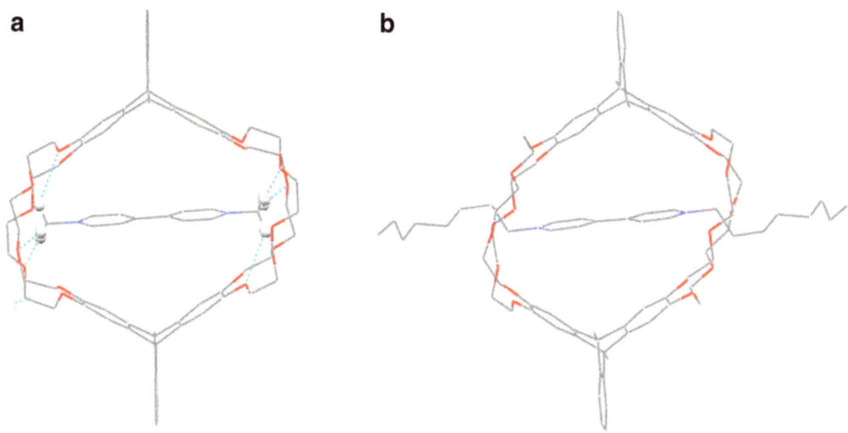

Fig. 9.2 Crystal structures of the complexes **a** **1•5** and **b** **1•7**

in solution, which were determined by the molecular ratio plot based on nuclear magnetic resonance (NMR) data. For the complex **1•5**, the rates of complexation and decomplexation were both slow at room temperature, and the association constant (K_a) was determined to be $4 \times 10^5\ M^{-1}$. Its single crystal structure (Fig. 9.2) further showed that paraquat **5** was included in the center of host **1**, while the two *N*-methyl groups were positioned in the two DB24C8 cavities, which led to a symmetrical pseudorotaxane-like structure. Moreover, the complex was stabilized by multiple hydrogen (H) bonds and the π–π stacking interactions, which was consistent with the result in solution. For the guests **6** and **7**, it was found that their complexation with host **1** all belonged to the fast-exchange system, which was different from that of complex **1•5**, and showed relatively low association constants. However, the X-ray crystal structure of complex **1•7** showed that the two *N*-octyl groups of guest **7** threaded the two lateral DB24C8 cavities of host **1** to form a new [2]pseudorotaxane-like complex, which was similar to that of complex **1•5** (Fig. 9.2).

Furthermore, Zhao et al. [7] found that the macrotricyclic host **1** could form stable 1:1 or 1:2 complexes with different functional paraquat derivatives **8–16** (Fig. 9.1) via the different complexation modes in both solution and solid state. Firstly, it was

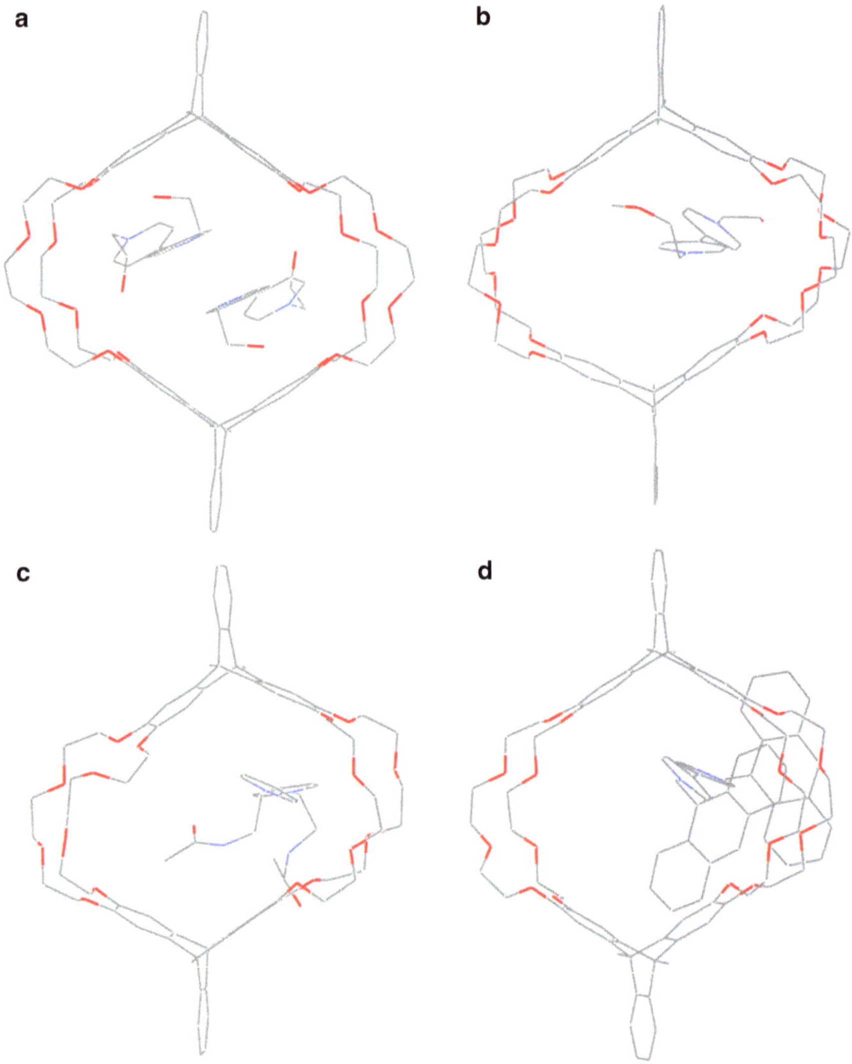

Fig. 9.3 Crystal structures of the complexes **a 1•9**, **b 1•12**, **c 1•14**, and **d 1•16**

found that host **1** and the paraquat derivative **9** with two β-hydroxyethyl groups formed a 1:2 complexes $\mathbf{1{\bullet}9_2}$ in 1:1 chloroform/acetonitrile solution, along with the charge transfer between the electron-rich aromatic rings of the host and the electron-poor pyridinium rings of the guest, which resulted in a deep orange solution immediately. The X-ray crystal structure of complex $\mathbf{1{\bullet}9_2}$ showed that the two guest molecules were included in the central cavity of the host to form a 1:2 complex, and the two N-β-hydroxyethyl groups were located outside the cavity of the host (Fig. 9.3), which was consistent with the result in solution. It was noteworthy that

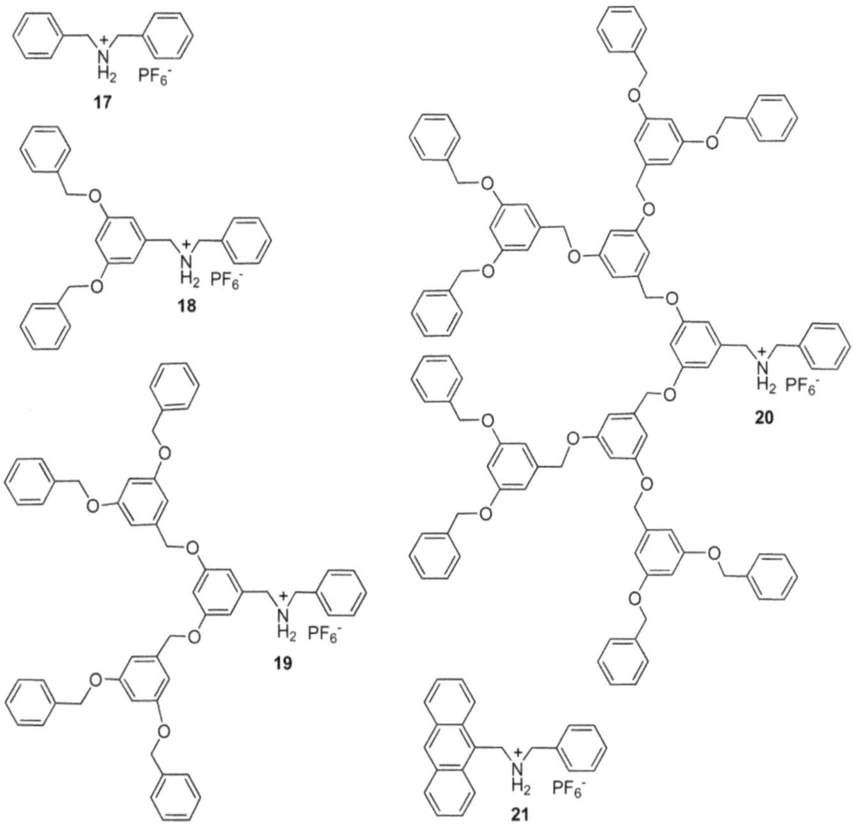

Fig. 9.4 Structures of secondary ammonium salts **17–21**

there existed not only C–H· · ·O hydrogen-bonding interaction and π–π stacking interaction between the host and the guest, but also C–H· · ·O hydrogen-bonding interactions between the two guests, which played an important role in the formation of the 1:2 stable complex. Likewise, host **1** and the guest **10** containing two N-γ-hydroxypropyl groups also formed a 1:2 complex under the same condition. However, the guest **11** containing two N-ω-hydroxyhexyl groups did not form 1:2 but 1:1 complex with host **1**; meanwhile the guest **12** with one methoxy group and guest **13** with two terminal methoxy groups could all form 1:1 complexes with host **1**. Moreover, other paraquat derivatives **14–16** containing terminal amide and 9-anthracyl-methyl group also formed the 1:1 complexes with host **1**, in which the guest molecule was included in the cavity of the host. Interestingly, the crystal structures of the complexes **1•12** and **1•16** also showed that the guest molecules were only positioned at one side of the cavity of the host (Fig. 9.3).

Due to the macrotricyclic host **1** containing two DB24C8 moieties, Chen et al. interestingly found that the host could also form 1:2 complexes with secondary ammonium salts **17–21** (Fig. 9.4) in both solution and solid state. At first, Zong et al. [8]

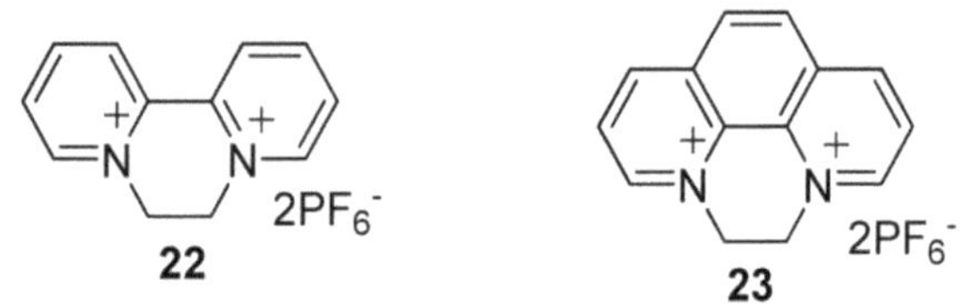

22 23

Fig. 9.5 Structures of diquat **22** and diquaternary salt of phenanthroline **23**

investigated the complexation between host **1** and the dibenzylammonium salt **17** and found that they could form a novel [3]pseudorotaxane-type stable complex **1•17**$_2$ in solution. The X-ray crystal structure showed that two guest molecules of **17** were threaded symmetrically through the center of the DB24C8 cavities of host **1**, which led to a "gull-wing" structure. There existed not only multiple noncovalent interactions between the host and the guests, but also the π–π stacking interactions between the two guests, which played important roles in the stability of the complex. Similarly, they also obtained a series of new dendritic [3]pseudorotaxane-type complexes from host **1** and the dendritic ammonium salts **18–20**, which were evidenced by their NMR spectra and mass spectra results. Moreover, it was interestingly found that host **1** could also form a 1:2 complex with (9-anthracylmethyl)benzylammonium salt **21**, in which the 9-anthracyl groups were selectively positioned outside the lateral crown ether cavities [7].

It was known that the association and disassociation of the complex between DB24C8 and secondary ammonium salt could be chemically controlled by pH [9], which inspired Chen and coworkers to further study the competitive binding abilities of host **1** toward propyl-substituted paraquat derivative **8** and dibenzylammonium salt **17**. Consequently, it was found that there existed a guest-exchange complexation process between host **1** and the two different guests, which could be chemically controlled by the simple addition of trifluoroacetic acid (TFA) and tributylamine (TBA) [7].

To take advantage of electron-rich cavity of host **1**, the complexation between **1** and the diquat **22** or the diquaternary salt of phenanthroline **23** (Fig. 9.5) were also investigated. Consequently, it was found that both **22** and **23** could efficiently form the 1:1 complexes with host **1** in solution and in the solid state. Since host **1** containing two DB24C8 moieties could form a complex with K^+ ions, and the consequent complexation of the cations would introduce extraelectrostatic repellent force to the cationic diquaternary guest molecules and dissociate the previously formed host–guest complex. Moreover, it was also known that 18-crown-6 was a very strong sequestering agent for K^+ ion. Thus, Han et al. [10] further investigated the potassium (K) ion-controlled binding and release of the guest molecules in the above complexes by the ^{1}H NMR experiments. Consequently, the authors found that the binding and release of the guests in the complexes could be easily controlled by adding and removing the K ions (Fig. 9.6).

The X-ray crystal structure of complex **1•22** showed that the diquat molecule threaded from the central cavity of host **1** and occupied only half of the cavity, which implied that there was still free volume available in the complex for the further inclu-

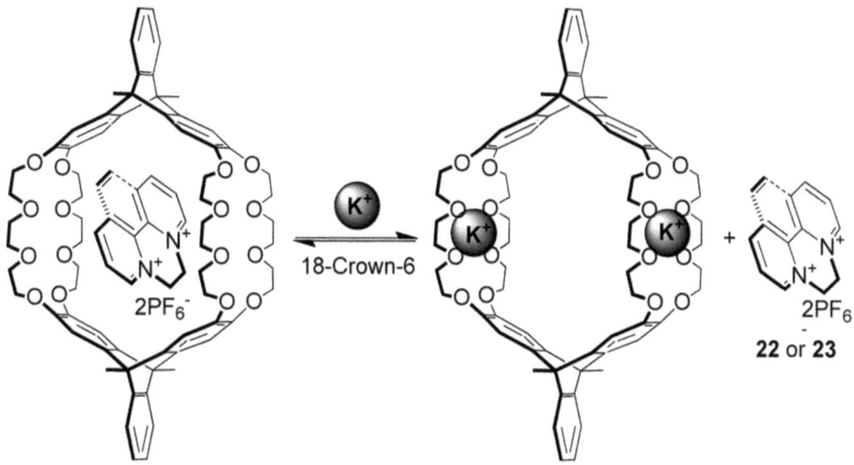

Fig. 9.6 Process of ion-controlled binding and release of the guest in complex

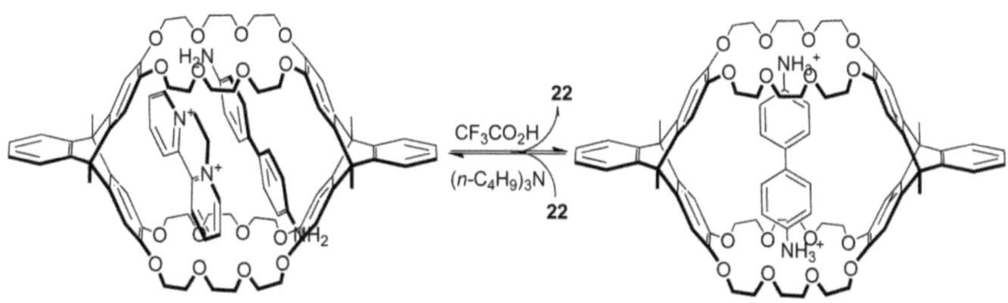

Fig. 9.7 Acid-base controlled selective complexation process between the ternary complex containing benzidine and a binary complex

sion of another guest within the cavity of host **1**. Consequently, Han and Chen [11] investigated the complexation between host **1** and two different kinds of guests. As expected, the authors found that host **1** could simultaneously bind diquat and electron-rich aromatic molecules (benzidine, biphenyl-4,4′-diol, or 4,4′-dibutoxybiphenyl) to form stable ternary complexes in both solution and solid state. The crystal structure of the ternary complex revealed that both the charge-transfer interaction between the electron-rich and electron-deficient guests, and the face-to-face π-stacking interactions between the host and the two guests played an important role in stabilizing the complex. Moreover, it was also found that the selective complexation process between the ternary complex containing benzidine and a binary complex between host **1** and biphenyl-4,4′-diaminium salt could be effectively controlled by the addition of TFA and TBA (Fig. 9.7).

In 2007, Han and Chen [12] designed and synthesized a new triptycene-based cylindrical macrotricyclic host **24** (Scheme 9.2) containing an anthracene unit and two DB24C8 moieties. Similar to the synthetic method for macrocycle **1**, host **24**

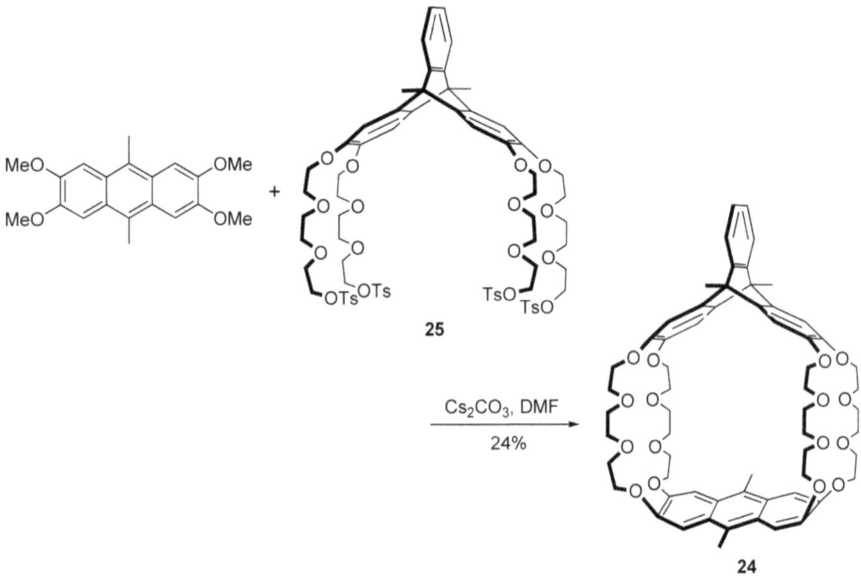

Scheme 9.2 Synthesis of triptycene-derived cylindrical macrotricyclic host **24**

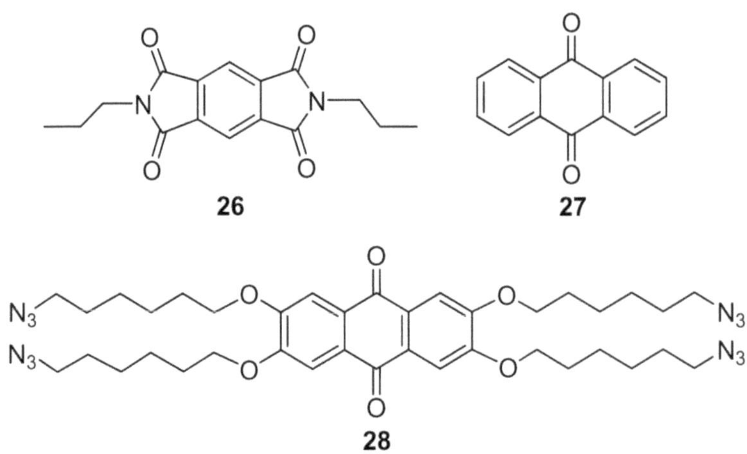

Fig. 9.8 Structures of guests **26–28**

was easily synthesized in 24 % yield by the reaction between 9,10-dimethyl-2,3,6,7-tetrahydroxy-anthracene and the compound **25** under a high dilution condition in the presence of cesium carbonate. First, they tested the complexation between host **24** and pyromellitic diimide **26** (Fig. 9.8) in solution by the ^{1}H NMR spectral method and found that there was no obvious complexation between the host and the guest. However, when lithium (Li) salt was added to the above system, the result showed that

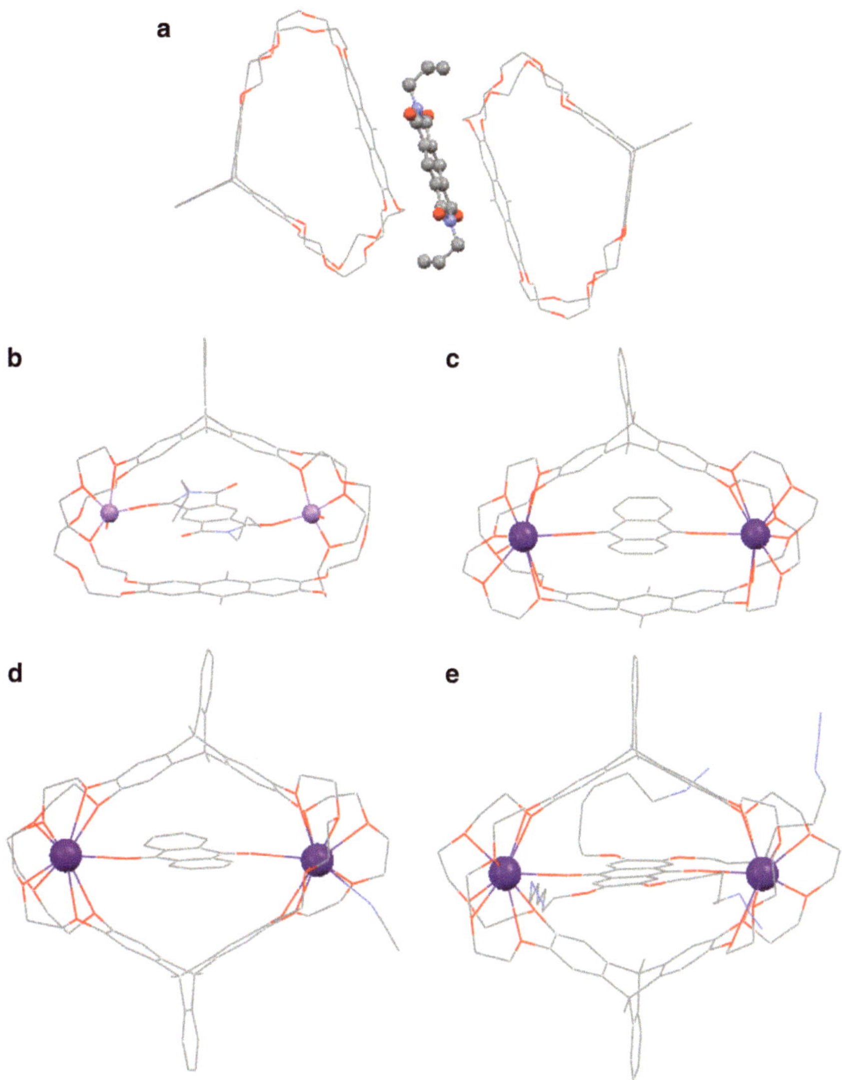

Fig. 9.9 Crystal structures of complexes **a 24•26•24**, **b 24•26**•2Li$^+$•2H$_2$O, **c 24•27•2K**$^+$, **d 1•27**•2**K**$^+$•CH$_3$CN, and **e 1•28•2K**$^+$

a novel complex formed, along with the color change of the solution from yellow to blue. The crystal structures showed that in the absence of the Li ions, guest **26** lied out of the cavity of host **24** to form an interesting sandwich structure with two molecules of host **24**; but in the presence of the Li ions, a novel cascade complex **24•26•2Li**$^+$ was formed in the solid state (Fig. 9.9). Similarly, the authors proved that host **24** and anthraquinone **27** could also form a stable cascade complex in the presence of K ions in both solution and solid state. Similar to the complexation of host **1**, macrocycle

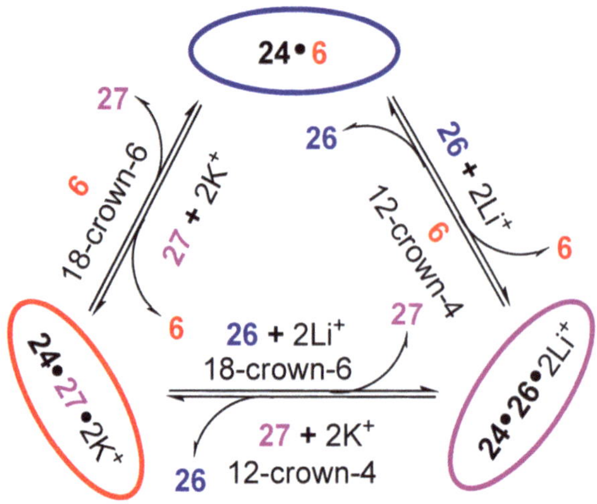

Fig. 9.10 Scheme representation of switch processes between host **24** and the guests **6**, **26**, and **27** controlled by adding and removing cations

24 could also form a stable 1:1 complex with paraquat derivative **6**, and due to the strong π–π stacking interactions between the anthracene ring of host **24** and the aromatic rings of guest **6**, the association constant ($K_a = (4.4 \pm 0.7) \times 10^3\ M^{-1}$) of the 1:1 complex between host **24** and guest **6** was almost twice as much as the K_a of the complex between host **1** and guest **6**. Based on the fact that host **24** could selectively complex with three different kinds of guests **26**, **27**, and **6** in different modes, Chen et al. further investigated the switchable processes between different complexation systems by ^{1}H NMR-titration experiments. Consequently, they found that the three switch processes could be efficiently performed by the addition and removal of the cations, which is shown in Fig. 9.10.

Similarly, Han and Chen [12] found that macrotricyclic host **1** could also form the pseudorotaxane-type cascade complexes with both anthraquinone **27** and its tetra-azide terminal functionalized derivative **28** in the presence of K ions in solution and in the solid state (Fig. 9.9).

According to the same method for the synthesis of host **1**, Zhao et al. [14] conveniently synthesized a new triptycene-derived macrotricyclic host **29** (Fig. 9.11a) containing two dibenzo-30-crown-10 (DB30C10) moieties. As expected, host **29** and guest **17** could form a 1:2 stable complex in both solution and solid state. Interestingly, it was also noteworthy that host **29** could selectively form a 1:2 stable complex **29•21$_2$** with two guests of **21** in solution and in the solid state, and the 9-anthracyl groups of both of the molecules **21** in complex **29•21$_2$** were all selectively positioned inside the lateral crown ether cavities (Fig. 9.11). It was further found that the complexation and disassociation of complex **29•21$_2$** could be chemically controlled by the addition of base and acid. Moreover, the complex **29•21$_2$** could serve as a selective supramolecular fluorescence probe for Ba^{2+} ion, because two Ba^{2+} ions could form a 1:2 stable complex with the macrocyclic host **29**, and the presence of Ba^{2+} ion could subsequently induce the considerable fluorescence enhancement of the complex **29•21$_2$** [13].

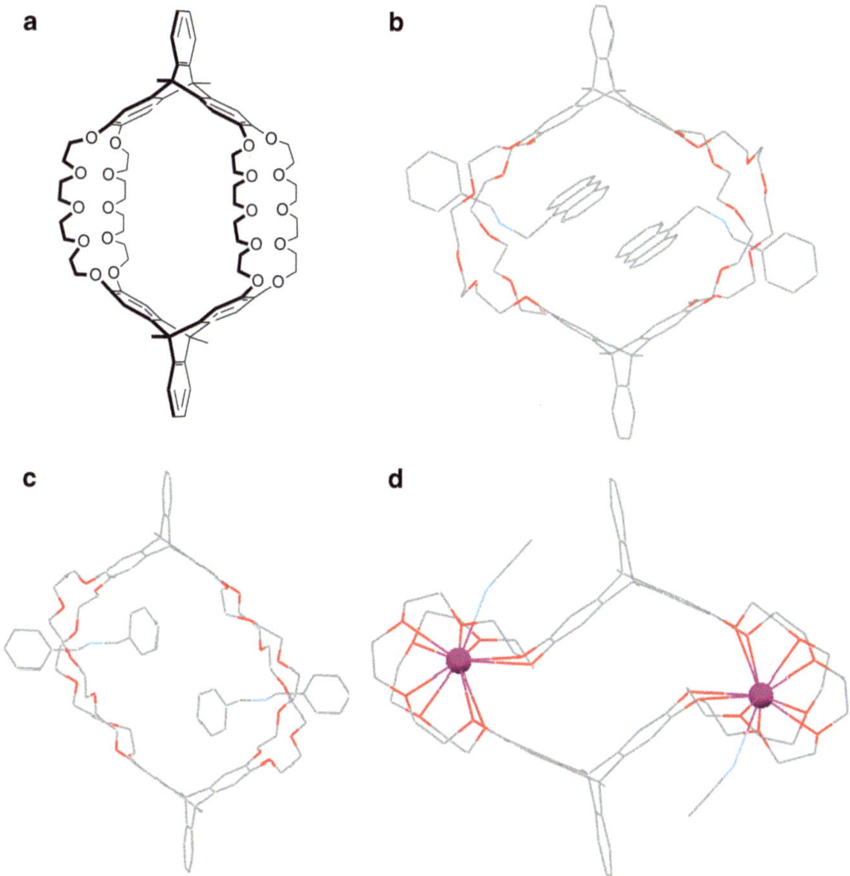

Fig. 9.11 Chemical structure of **a** macrotricyclic host **29**, the crystal structures of complexes **b** $\mathbf{29{\bullet}21_2}$, **c** $\mathbf{29{\bullet}17_2}$, and **d** $\mathbf{29}{\bullet}\mathbf{2Ba^{2+}}{\bullet}2CH_3CN$

Afterward, Guo et al. [13] further found that the host **29** could form the stable 1:2 complexes with the paraquat derivatives **5**, **6**, **9**, or **13** in solution and in the solid state, which were obviously different from the 1:1 complexation mode of host **1** containing two DB24C8 moieties with the same paraquat derivatives. According to the analyses of ^{1}H NMR spectra, electrospray ionization mass spectra (ESI-MS) and the X-ray crystal structures (Fig. 9.12), it was interestingly found that not only the multiple intermolecular noncovalent interactions but also the anion–π interactions between PF_6^- and the bipyridinium rings of the guests played important roles in the formation of the complexes. This might represent the first examples of the host–guest complexes which were stabilized by the anion–π interactions. Moreover, it was further found that the binding and release of the guest molecules in the complexes could be easily controlled by the addition and removal of K ions.

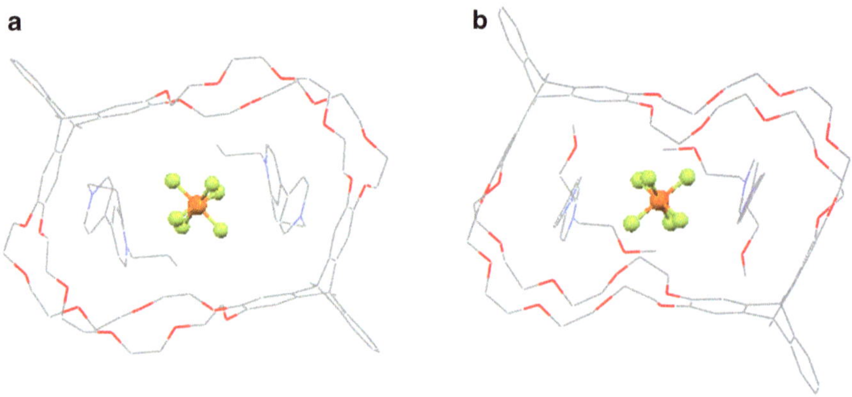

Fig. 9.12 Crystal structures of complexes **a 29•6**$_2$ and **b 29•13**$_2$

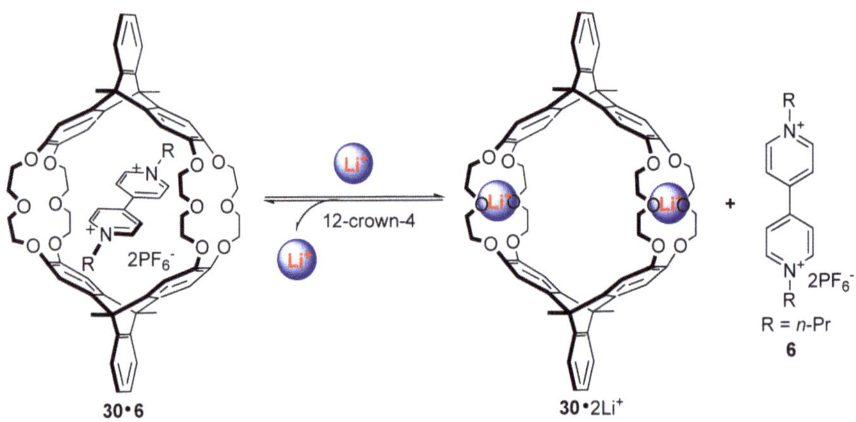

Fig. 9.13 Graphic representation of Li$^+$ ion-controlled binding and releasing of guest **6** in complex **30•6**

Recently, Han et al. [16] also synthesized a new triptycene-derived macrotricyclic host **30** containing two dibenzo-[18]-crown-6 moieties and found that host **30** could form 1:1 complexes with different functional paraquat derivatives in solution, and the association constants of the complexes were determined to be from $(2.0 \pm 0.1) \times 10^2$ to $(1.4 \pm 0.1) \times 10^3$. However, it was interestingly found that depending on the paraquat derivatives with different functional groups, the host could form stable 1:1 or 1:2 complexes with different complexation modes in the solid state, which was significantly different from those macrotricyclic hosts containing two dibenzo-[24]-crown-8 moieties [7] and two dibenzo-[30]-crown-10 moieties [15]. The formation of the complexes was also proved by the ESI-MS and electrochemical experiments. Moreover, it was found that the binding and release of the guests in the complexes could be easily controlled by the addition and removal of Li ions (Fig. 9.13).

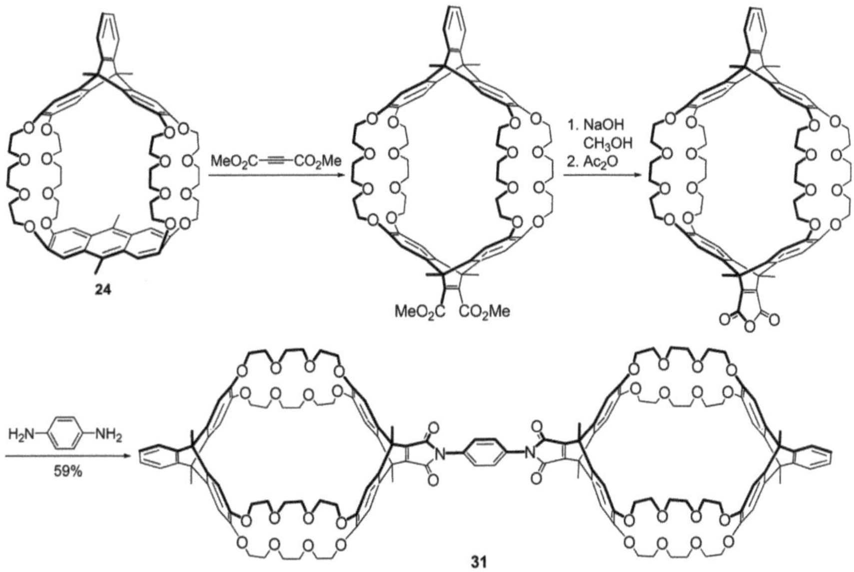

Scheme 9.3 Synthesis of triptycene-derived bis-macrotricyclic host **31**

In 2010, Guo et al. [17] also synthesized a novel triptycene-derived bis-macrotricyclic host **31** containing two symmetrical macrotricyclic moieties. As shown in Scheme 9.3, the macrocyclic compound **24** reacted with dimethyl acetylenedicarboxylate, followed by dehydrolysis in acetic anhydride to give the corresponding anhydride in 83 % yield for the two steps, which further reacted with *p*-phenylenediamine to afford **31** in 59 % yield. As expected, host **31** could form a stable 1:4 complex with 4 equivalents of the dibenzylammonium salt **17** in both solution and solid state. The single crystal structure of complex **31•17**$_4$ showed that two dibenzylammonium ions were threaded through the centers of the DB24C8 cavities of each macrotricyclic moiety in **31**.

9.1.2 Tweezer-Like Triptycene-Derived Crown Ethers

In view of the fact that molecular tweezers [18–20] with a specific structure containing a tether and two flat could exhibit a wide range of potential applications in biological and supramolecular chemistry, Peng et al. [21] designed and synthesized a tweezer-like host **32**, which was composed of two triptycene units linked by two crown ether chains. The synthetic route to host **32** was outlined in Scheme 9.4. Starting from the triptycene derivative **33**, the ester product **34** was obtained by the reaction of **33** with 8-tosyloxy-3,6-dioxaoctanol in CH_3CN in the presence of K_2CO_3, then followed by the treatment with TsCl in CH_2Cl_2 in the presence of Ag_2O and KI.

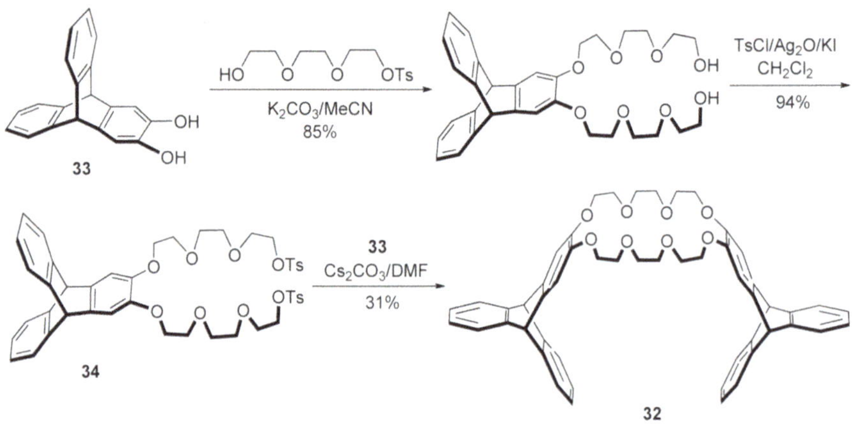

Scheme 9.4 Synthesis of tweezer-like host **32**

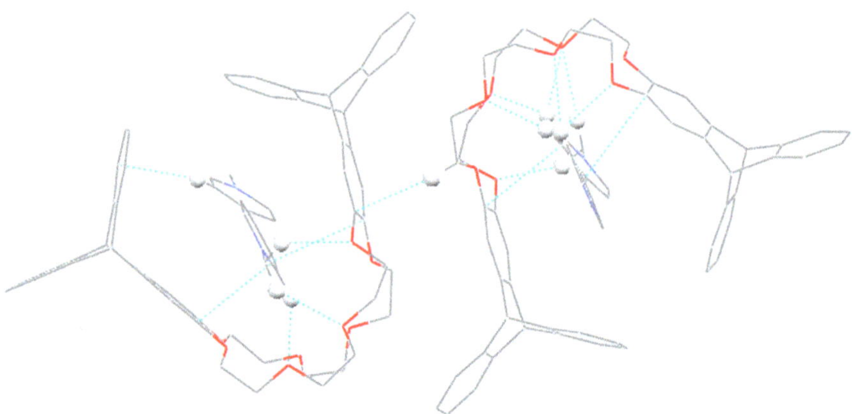

Fig. 9.14 The crystal structure of complex **32•5**

Finally, the target host **32** was obtained in 31 % yield by the reaction of **34** and **33** in *N,N*-dimethylformamide (DMF) in the presence of Cs_2CO_3.

They first investigated the complexation ability of host **32** toward paraquat derivatives with different functional groups in solution. Consequently, it was found that host **32** could form a 1:1 stable complex with the paraquat **5**, along with a swift color change from pale yellow to a light red-brown probably as a result of the charge transfer between the electron-rich aromatic rings of **32** and the bipyridinium rings of guest **5**. Similar to the complex **32•5**, the alkyl substituted paraquat derivatives **6** and **8** could also form 1:1 complexes with host **32**, while for the paraquat derivatives containing terminal vinyl group, hydroxyl, and anthracyl group, a similar but slightly weak complexation with host **32** was observed. The single crystal of complex **32•5** was also obtained, and the X-ray crystal structure (Fig. 9.14) showed that in one

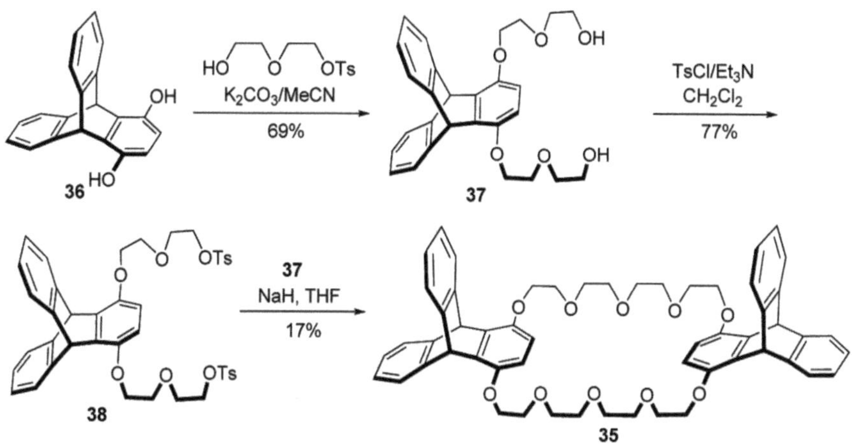

Scheme 9.5 Synthesis of triptycene-derived host **35**

crystal cell, there were two tweezer-like hosts with different orientation, and each host tweezered a bipyridinium guest by different complexation modes. Moreover, the interactions between the bipyridinium rings and phenyl rings of triptycene units led to a waveform structure, which could further form the two-dimensional layer viewed along the *c*-axis and three-dimensional microporous network. Due to the guests containing the well-known 4,4′-bipyridinium electro-active unit, their electro-chemical behaviors in the absence and presence of the host were also studied. As a result, it was found that the formation of the complexes could be caused by a charge-transfer interaction, and the complexes could dissociate upon two one-electron reduction of the bipyridinium salt.

In 2010, Jiang et al. [22] also synthesized a new triptycene-derived host **35** containing a bisparaphenylene-34-crown-10 moiety by a sequence of reactions from triptycene derivative **36** (Scheme 9.5). Studies on the binding properties of host **35** showed that it could form the corresponding 1:1 stable complexes with both the paraquat **5** and the cyclo-bis(paraquat-*p*-phenylene) **39** via the different complexation modes in solution and in the solid state. For guest **5**, it was found that the paraquat ring was located in about a crystallographic center of symmetry to form a pseudosandwiched structure with the two triptycene moieties. However, in the case of guest **39**, the two triptycene moieties in host **35** were positioned in the same side of the crown ether to form a tweezer-like cavity, and the two molecules (**35** and **39**) mutually acted as not only the host but also the guest to form a pseudoternary complex (Fig. 9.15). Moreover, it was further revealed that both complexes **35•5** and **35•39** were formed by charge-transfer interactions, without the dissociation upon the first one-electron reduction process of the bipyridinium ring by their electrochemical behaviors.

Based on the triptycene building block, Han and Chen [23] also reported a new tweezer-like host **40** containing two DB24C8 moieties, which could be conveniently

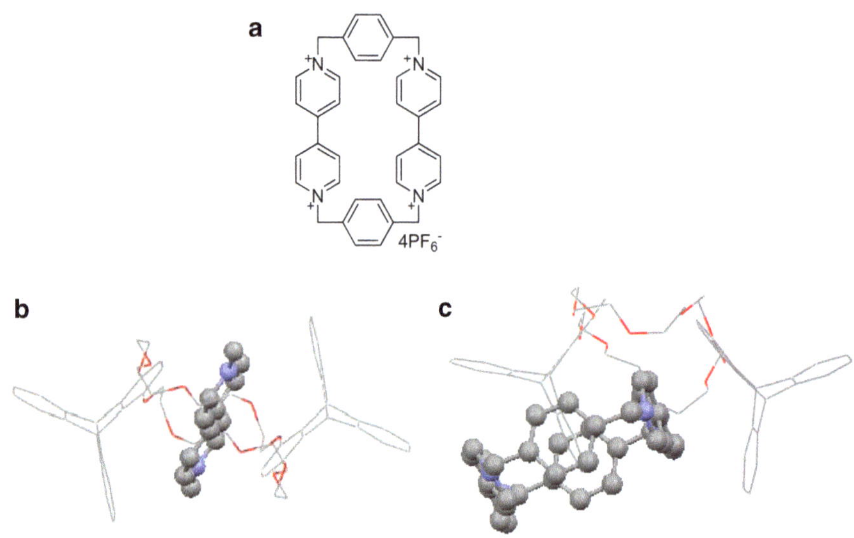

Fig. 9.15 **a** Chemical structure of cyclo-bis(paraquat-*p*-phenylene) **39**, the crystal structures of complexes **b** **35•5**, and **c** **35•39**

synthesized by the reaction of the triptycene derivative **41** and 2 equivalents of 1,2-bis[2-[2-(2-tosyloxyethoxy)ethoxy]ethoxy]benzene under a high dilution condition (Scheme 9.6). The host **40** exhibited the complexation toward both dibenzylammonium salts and paraquat derivatives to form a bis[2]pseudorotaxane complex and a stable clip-shaped complex in solution and in the solid state, respectively. Moreover, the competitive binding ability of host **40** toward the two different kinds of guests was also examined, and the result showed that the complexation process between host **40** and guests **5** and **17** could be chemically controlled by simply changing the solution pH with TBA and TFA (Fig. 9.16).

9.2 Triptycene-Derived Calixarenes

Due to the unique structural properties, convenient preparation, easy chemical modification, and versatile platform in developing complexing agents for ions and molecules, calixarenes [24] are also called as "the third generation of host molecules" after crown ethers and cyclodextrins by Gutsche [25]. However, the calix[4]arene has too small cavity to complex with ordinary organic molecules except for some solvents, while calix[6]arene and larger calixarenes have so many conformations that their cavities are actually difficult to be utilized as well. Thus, Chen and coworkers envisioned that taking the place of one or more phenol groups in the calix[4]arene by the triptycene moiety with the three-dimensional rigid structure could thus result in a class of novel calixarenes with specific structures and properties.

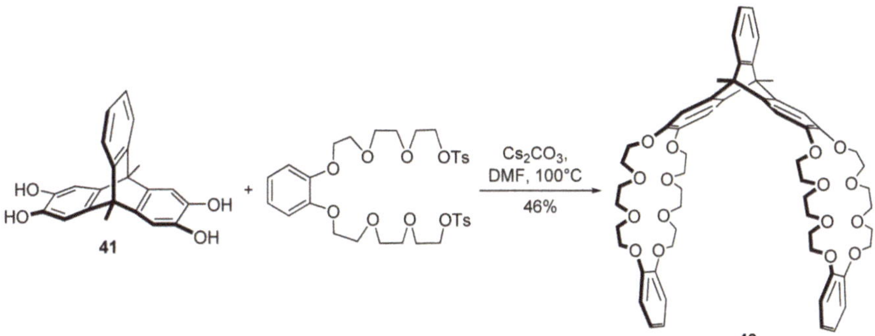

Scheme 9.6 Synthesis of tweezer-like host **40**

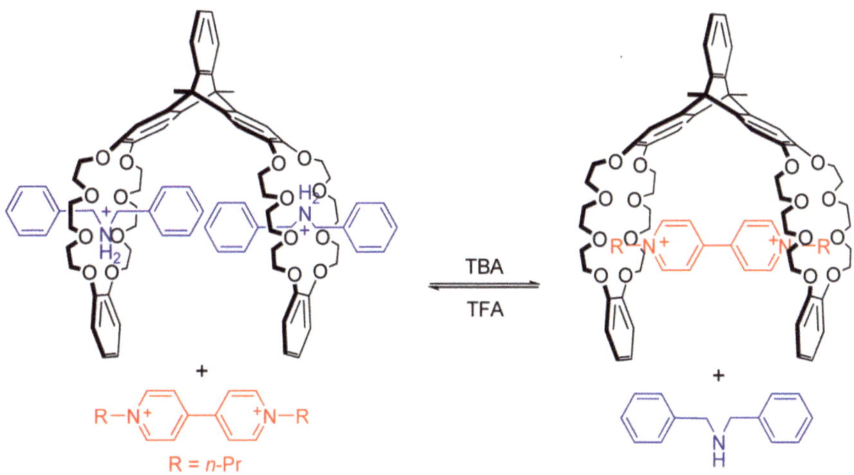

Fig. 9.16 The competitive complexation process between host **40** and guests **5** and **17** controlled by the addition of TBA and TFA

Firstly, Chen and coworkers [26, 27] reported the synthesis of a series of triptycene-derived calix[6]arene derivatives **42** and **43**. As shown in Scheme 9.7, by the one-pot reaction of the triptycene derivative **44** and *p*-*t*-butylphenol in *o*-dichloro benzene in the presence of 4-methylbenzenesulfonic acid, a pair of triptycene-derived calix[6]arenes **42a** and **43a** could be obtained. ^{1}H NMR spectra of macrocycles **42a** and **43a** showed a big difference from each other, suggesting that they are a pair of diastereomers. The crystal structures (Fig. 9.17) further proved that **42a** was a *cis* isomer with cone or boat conformation, and **43a** was a *trans* isomer with partial cone or chair conformation. Similarly, a pair of the triptycene-derived calix[6]arenes **42b** and **43b** were also obtained in 17 and 11 % yield, respectively, by the one-pot reaction of **44** and *p*-phenyl-phenol. On the other hand, the macrocycles **42** and **43** could also be synthesized by a two-step fragment-coupling approach via the compounds **45a**

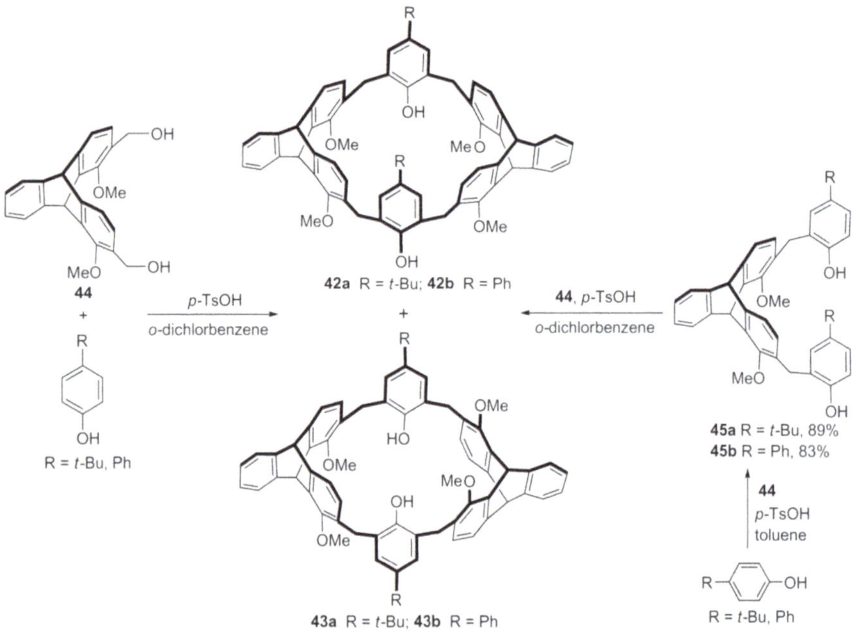

Scheme 9.7 Synthesis of triptycene-derived calix[6]arene derivatives **42** and **43**

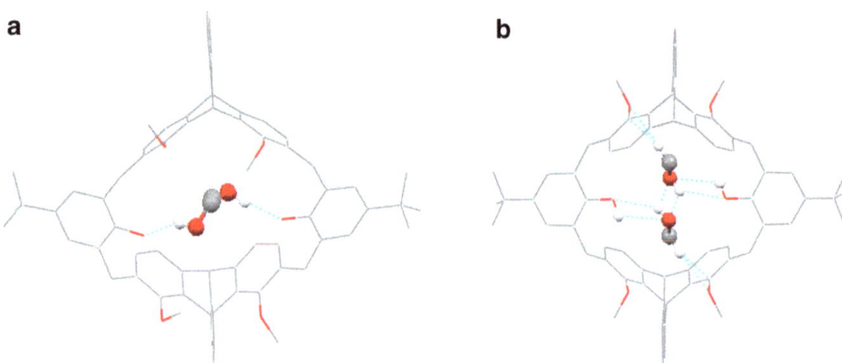

Fig. 9.17 Crystal structures of **a** **42a**@2CH_3OH and **b** **43a**@2H_2O

and **45b**, respectively. Compared with the one-pot reaction, the ready accessibility of the [1 + 2] product **45**, and the subsequent higher yields of cyclization reactions displayed more practicality for the synthesis of the macrocycles.

The triptycene-derived calix[6]arenes not only have large enough cavities to serve as hosts, but also have specific fixed conformations, thus they might easily encapsulate small neutral guest molecules within their cavities in the solid state. Consequently, it was found that **42a** could encapsulate two CH_3OH molecules in its cavity, while two H_2O molecules were found in the cavity of **43a** (Fig. 9.17). These results had not been found in their corresponding classical calix[6]arene analogues. Similarly, the other triptycene-derived calix[6]arenes could all accommodate different small neutral molecules in their cavities in the solid state by the multiple noncovalent interactions between the solvents and the macrocycle. Moreover, due to their well-defined fixed conformations and sufficiently large electron-rich cavities, the macrocycles **42b** and **43b** displayed the complexation properties toward fullerenes C_{60} and C_{70}, which were revealed by the fluorescence method. The results showed that **42b** could form a 1:1 complex with C_{60}, and the association constant K_a was determined to be $(6.90 \pm 0.17) \times 10^4\ M^{-1}$. Under the same conditions, macrocycle **42b** and C_{70} formed a 1:1 complex with the association constant of $(5.22 \pm 0.20) \times 10^4\ M^{-1}$. Similar to the case of **42b**, macrocycle **43b** could also form 1:1 complexes with both C_{60} and C_{70}, with the corresponding association constants of $(8.68 \pm 0.30) \times 10^4\ M^{-1}$ and $(5.90 \pm 0.38) \times 10^4\ M^{-1}$, respectively.

In addition, when the macrocycles **42a, b** and **43a, b** were treated with BBr_3 in dry CH_2Cl_2, the demethylated compounds **46** and **47** could be obtained with the high yields, respectively. Moreover, the treatment of **46a** and **47a** with $AlCl_3$ in toluene at room temperature resulted in the debutylated products **48** (Scheme 9.8a) and **49** (Scheme 9.8b) in 61 and 54 % yield, respectively. Under the same conditions, it was further found that the macrocycles **42a** and **43a** could be simultaneously debutylated and demethylated to directly afford **48** (Scheme 9.8a) and **49** (Scheme 9.8b) in moderate yields. Furthermore, the X-ray crystal structures (Fig. 9.18) showed that a $CHCl_3$ molecule was accommodated in the cavity of **43b** by virtue of a pair of C–H$\cdots\pi$ interactions between the proton of $CHCl_3$ and the phenyl rings of the triptycene moiety. One water molecule could also be encapsulated within the cavity of macrocycle **48** by the multiple H-bonding interactions.

Soon afterward, Tian and Chen [28] further synthesized a series of novel calix[5]arenes **50a–c** containing one 1,8-dimethoxytriptycene moiety by the heat-induced fragment-coupling reactions (Scheme 9.9). Moreover, it was found that when macrocycles **50a–c** were treated with BBr_3 in CH_2Cl_2, the demethylated compounds **51a–c** were obtained in high yields. Then, the debutylated products **52a, b** could be further obtained in 82 and 75 % yields from **51a, b**, respectively. Treatment of macrocycles **50a, b** with $AlCl_3$ in toluene at room temperature straight produced the debutylated and demethylated products **52a, b** in high yields, respectively. Similarly, it was also found that the triptycene-derived calix[5]arenes with fixed cone conformations and large enough cavities could encapsulate the solvent molecules of dichloromethane and methanol inside their cavities in solid state (Fig. 9.19).

Recently, Li and Chen [29] designed and synthesized a class of novel triptycene-derived calix[6]resorcinarene-like macrocycles containing two triptycene moieties and two *p*-substituted phenol moieties. As shown in Scheme 9.10, 2,7-dimethoxyltriptycene reacted with hexamethylenamine in refluxed trifluoroacetic acid, followed by the reduction of the aldehyde with sodium borohydride in the mixture solution of MeOH and tetrahydrofuran (THF) to afford compound **56**, which was then reacted with the excess of *p*-substituted-phenols in toluene in the presence of *p*-toluenesulfonic acid to give the compounds **55a–c** in the yields of 54–78 %.

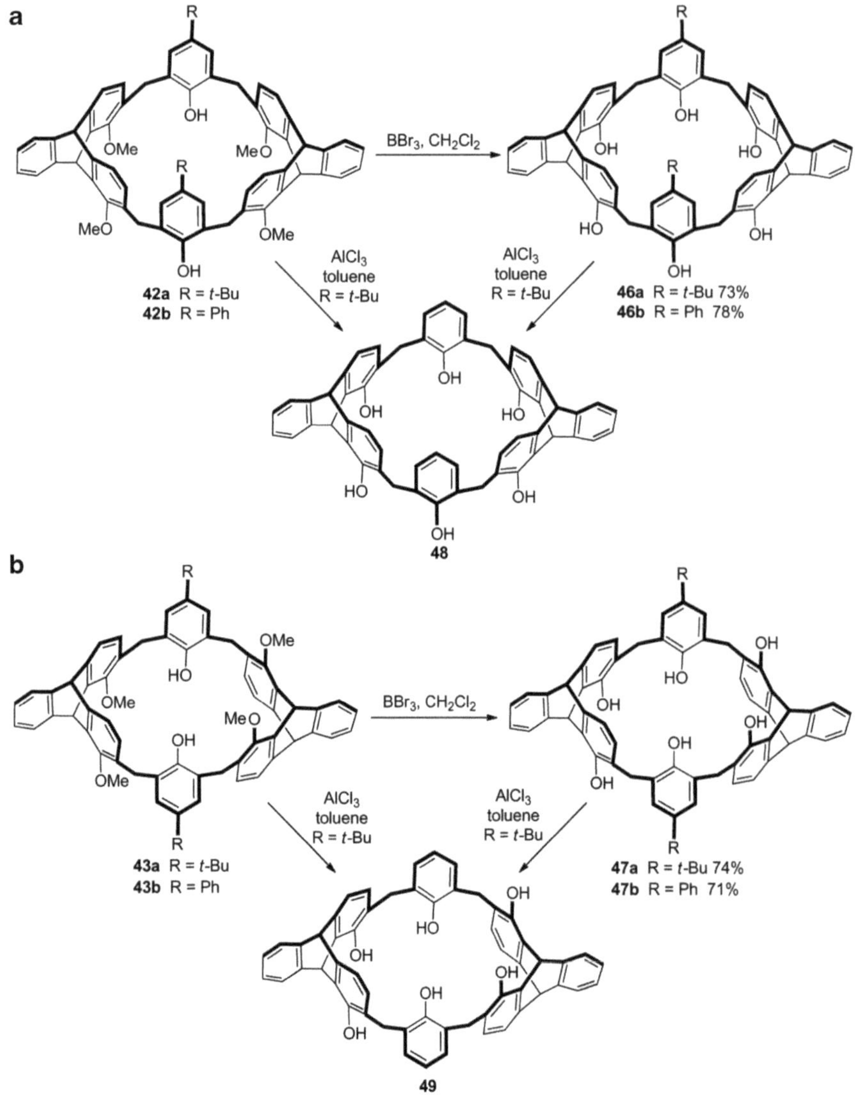

Scheme 9.8 Demethylation and debutylation of macrocycles **42** and **43**

Finally, the target calix[6]resorcinarene-like macrocycles **53a–c** could be obtained in 20–22 % yields by the reaction of compound **55a–c** and 2,7-dimethoxyltriptycene derivative **56**, respectively. Moreover, the demethylated compounds **56a–c** could be afforded by the treatment of **55a–c** with BBr_3 in CH_2Cl_2. Based on the ^1H NMR spectra and crystal structures, it was also found that all of the macrocycles **55a–c** and the corresponding demethylated macrocycles **56a–c** were the *cis* isomers with the fixed cone conformation in solution and in the solid states. Furthermore, the

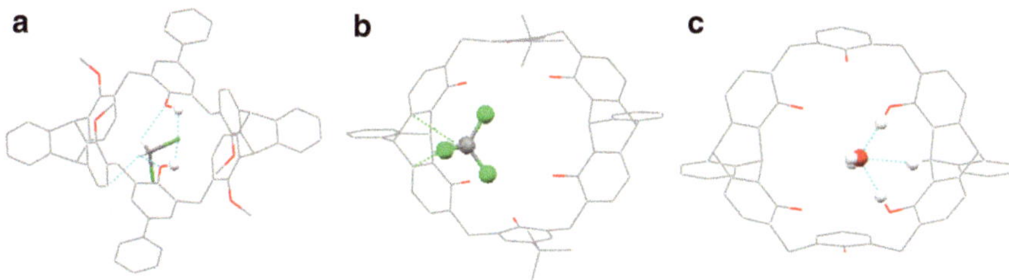

Fig. 9.18 Crystal structures of **a** **43b**@CH_2Cl_2, **b** **46a**@$CHCl_3$, and **c** **48**@H_2O

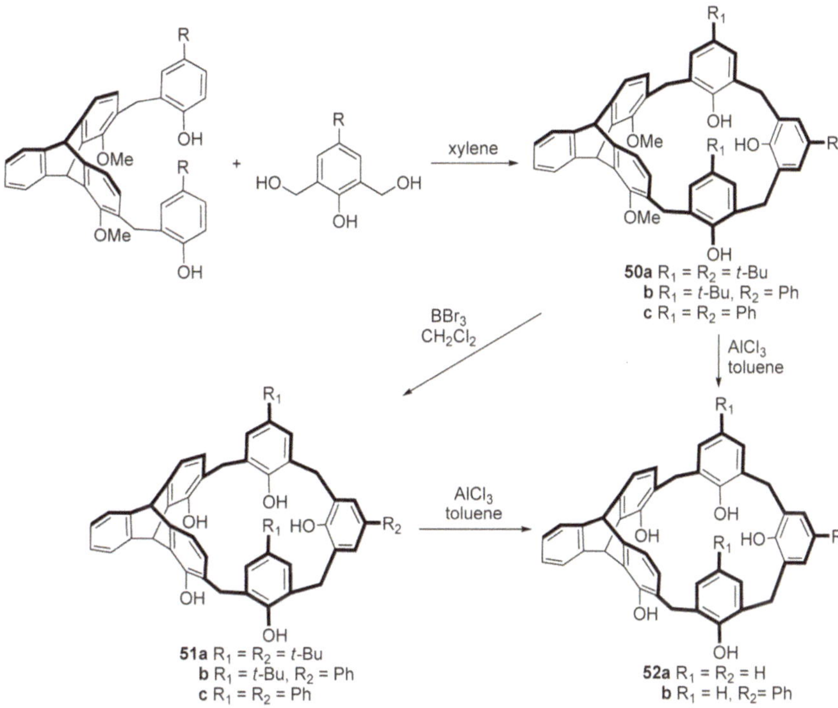

Scheme 9.9 Synthesis of triptycene-derived calix[5]arene derivatives **50–52**

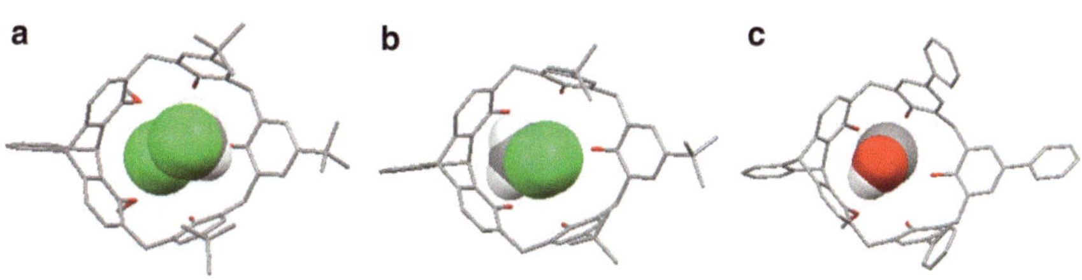

Fig. 9.19 Views of the crystal structures of **a** **50a**@CH_2Cl_2, **b** **51a**@CH_2Cl_2, and **c** **50c**@CH_3OH

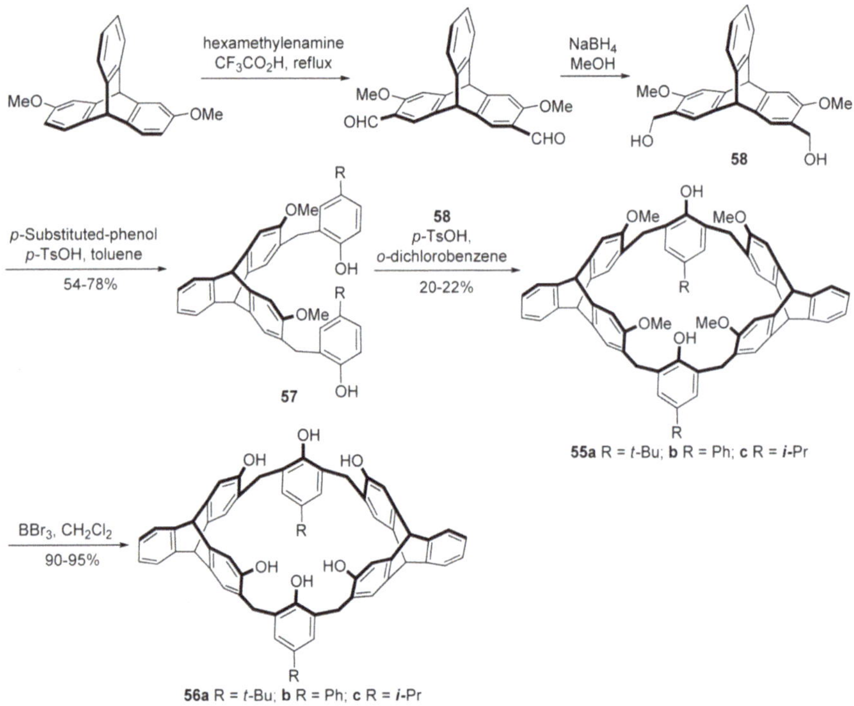

Scheme 9.10 Synthesis of triptycene-derived calix[6]resorcinarene-like macrocycles

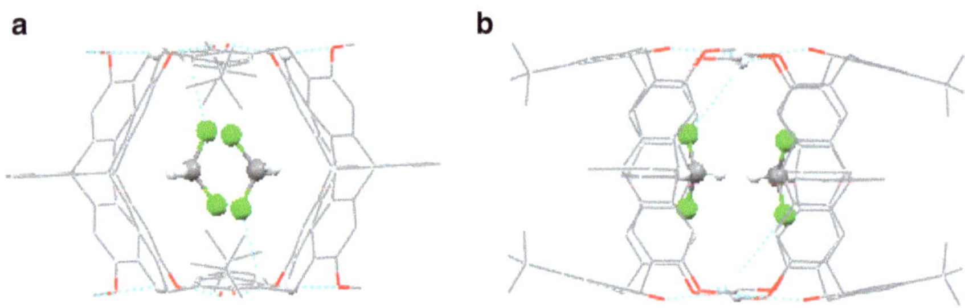

Fig. 9.20 **a** *Top* view and **b** *side* view of the head-to-head dimer of **55a** with CH_2Cl_2 molecules situated in the cavity

X-ray crystal structure showed that compound **55a** could form a head-to-head dimeric capsule with two CH_2Cl_2 molecules inside the dimeric cavity, suggesting that the C–H···O hydrogen-bonding interactions between the macrocycles and the C–H···Cl interactions between the macrocycles and solvent molecules played important roles in the formation of the dimeric structure (Fig. 9.20).

9.3 Triptycene-Derived Oxacalixarenes and Azacalixarenes

Heterocalixarenes [30–33], in which the carbon linkages between the aromatic units are replaced by the heteroatoms, have attracted much attention in recent years for their readily availability, structural diversities, tunable cavities, and potential applications in supramolecular chemistry. However, most of the known heterocalixarenes have the small cavities or flexible conformations, which subsequently limit their applications in supramolecular chemistry. In recent years, Chen and coworkers revealed that when a proper functionalized triptycene was used as the nucleophilic reagent to react with proper electrophilic reagents, a new kind of novel heterocalixarenes with large cavities and fixed conformations could be achieved, and they would also show specific properties in supramolecular chemistry.

In 2007, Zhang and Chen [34] reported the synthesis of a series of extended oxacalixarenes **59–61** with large cavities by the nucleophilic aromatic substitution reactions of 2,7-dihydroxytriptycene **62** with proper electrophilic reagents in DMSO or acetone in the presence of Cs_2CO_3 or K_2CO_3 (Scheme 9.11). The expanded oxacalixarenes could be obtained by a two-step method as well. As a result, when compound **62** reacted with 2 equivalents of cyanuric chloride in THF in the presence of diisopropylethylamine (DIPEA), and then followed by the further reaction with **62** in acetone at room temperature with DIPEA as a base, the target expanded oxacalixarenes **61a** and **61b** could be obtained in 12 and 15 % yield, respectively. The further structural studies showed that the macrocyclic compounds **61a** and **61b** were a pair of diastereomers, in which **61a** was a *cis* isomer with boat conformation, and **61b** was a *trans* isomer in chair conformation.

Recently, Hu and Chen [35] reported a new triptycene-derived oxacalixarene **63a** (Fig. 9.21) containing two triptycene subunits and two naphthyridine subunits, which was conveniently synthesized by a one-pot coupling reaction of 2,7-dihydroxytriptycene and 2,7-dichloro-1,8-naphthyridine in a refluxed 1,4-dioxane solution in the presence of Cs_2CO_3. The structural studies indicated that the macrocycle adopted a fixed boat conformation even up to 380 K in solution, although its cavity was expanded compared with those previously reported oxacalixarenes. Moreover, the crystal structure (Fig. 9.21) showed that macrocycle **63a** was a *cis* isomer with a boat conformation and a cavity of 13.29×10.99 Å^2 (wider rim) and 8.56×8.84 Å^2 (narrower rim), and the nitrogen atoms of the 1,8-naphthyridine were all positioned inside the cavity. These structural features indicated that the macrocycle could be a suitable host for encapsulating organic guests such as C_{60} and C_{70}. Consequently, it was found that **63a** could form 1:1 complexes with both C_{60} and C_{70} with the association constant K_a of $(7.5 \pm 0.3) \times 10^4$ M^{-1} and $(9.0 \pm 0.3) \times 10^4$ M^{-1}, respectively. The stronger binding of **63a** with C_{70} was most probably due to the giant boat-like cavity of the host that complements the oval-shaped C_{70}. This high-efficient complexation ability toward C_{60} and C_{70} also represented the first example of the complexation of the oxacalixarene with the fullerenes.

The reaction conditions were important to the macrocyclization; therefore, Hu and Chen [36] further found that when the reaction was carried out in a dimethyl-

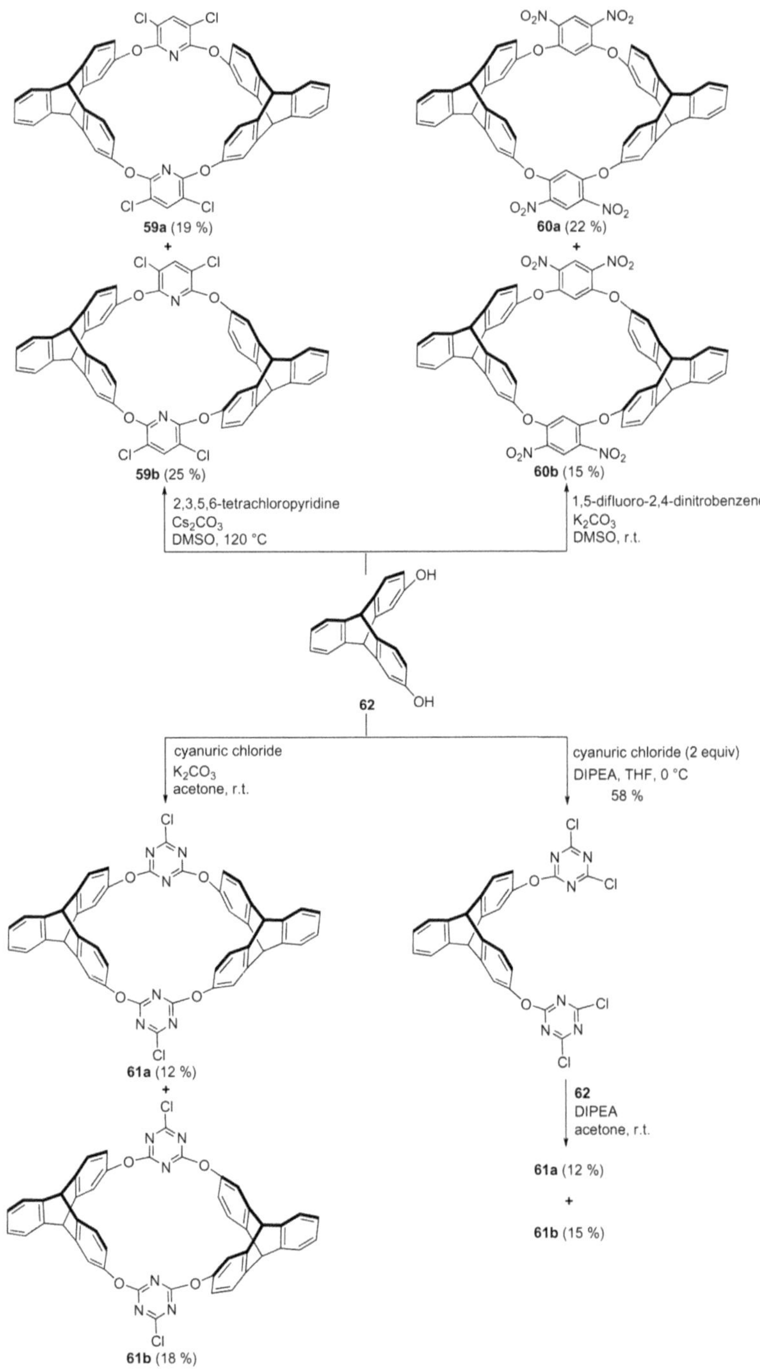

Scheme 9.11 Synthesis of triptycene-derived extended oxacalixarenes **59–61**

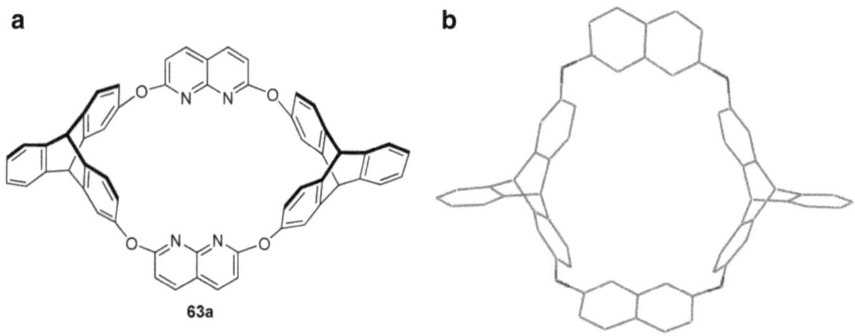

Fig. 9.21 **a** Chemical structure and **b** crystal structure of triptycene-derived oxacalixarene **63a**

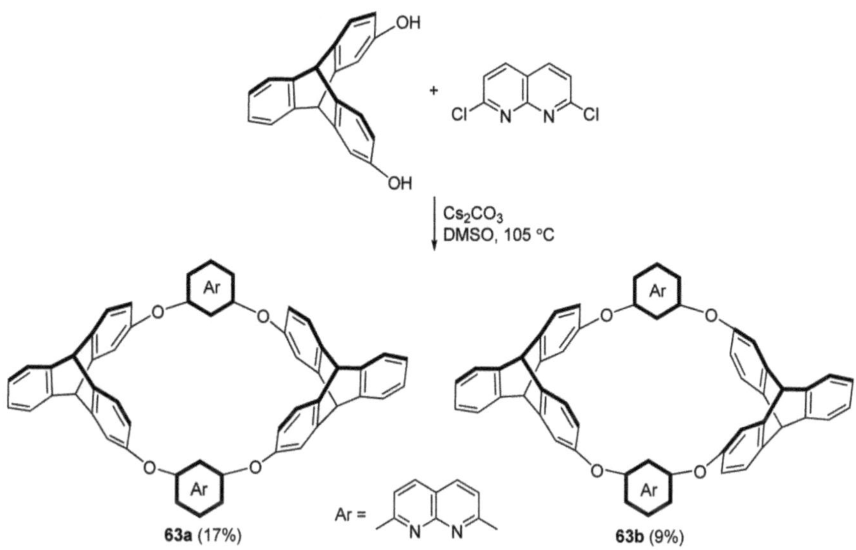

Scheme 9.12 Synthesis of triptycene-derived oxacalixarenes **63a, b**

sulfoxide (DMSO) solution at 105 °C, a pair of triptycene-derived oxacalixarenes **63a** and **63b** (Scheme 9.12) could be isolated in 17 and 9 % yield, respectively. The structural studies of the macrocycles showed that they were a pair of diastereomers, in which **63a** was a *cis* isomer with a boat-like 1,3-alternate conformation, and **63b** was a *trans* isomer with a curved boat-like conformation in the solid state. Moreover, both **63a** and **63b** had the fixed conformation in solution at room temperature, which was revealed by the variable-temperature ^{1}H NMR experiments. Due to the unique structural features of the macrocycles, both of the oxacalixarenes exhibited the abilities to form 1:1 complexes with paraquat derivatives containing different terminal functional groups in solution and in the solid states, which resulted in formation of a series of [2]pseudorotaxane-type complexes. Interestingly, it was further found that

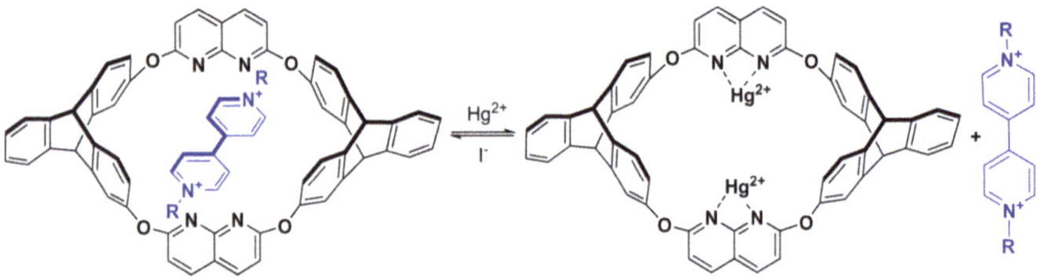

Fig. 9.22 The controlled dethreading/rethreading process of the [2]pseudo-rotaxane under Hg^{2+} association/dissociation

the controlled dethreading/rethreading processes of the resulting [2]pseudorotaxanes could be easily achieved by the acid-base stimuli or Hg^{2+} association/dissociation (Fig. 9.22).

Azacalixarenes, another important class of heterocalixarenes with the unique feature of bridging nitrogen atoms are potential candidates for molecular recognition and assembly. However, the properties of small cavities or nonfixed conformations limit their further applications as well. Therefore, based on the building blocks of 2,7-diaminotriptycene **64a** and its analogue **64b** containing an *o*-dimethoxybenzene subunit, Xue and Chen [37] designed and synthesized several pairs of novel triptycene-derived *N(H)*-bridged azacalixarenes **65–68** by both the one-pot (Scheme 9.13) and two-step fragment-coupling (Scheme 9.14) approaches. Due to the three-dimensional rigid structure of triptycene, the macrocyclic compounds all adopted fixed conformations in solution and in the solid state. Moreover, the X-ray crystallographic analyses further revealed that the *cis* isomers with boat conformations and large enough cavities all exhibited the capability of encapsulating methanol or acetone molecules inside their cavities. Interestingly, it was also found that two molecules of **66a** could form a dimer through the $O^{\delta-}\cdots N^{\delta+}$ and $O^{\delta-}\cdots C^{\delta+}$ interactions, and two acetone molecules and two dichloromethane molecules were situated in its cavity (Fig. 9.23).

9.4 Other Triptycene-Derived Macrocyclic Hosts

In 2008, Xue and Chen [37] reported the synthesis of two novel triptycene-derived tetralactam macrocycles **69a, b** (Fig. 9.24) by a one-pot [2 + 2] cyclization reaction of pyridine-2,6-dicarbonyl dichloride and 2,7-diamino-triptycene in dry THF in the presence of Et_3N. Macrocycles **69a, b** were a pair of diastereomers with highly symmetric structures, in which **69a** was a *cis* isomer with cone conformation, and **69b** was a *trans* isomer with chair conformation. With these macrocycles **69a, b** in hand, Chen et al. first tested their complexation with squaraine **70** (Fig. 9.24), which was a kind of fluorescent near-IR dyes with specific photophysical properties for wide potential applications. Consequently, it was found that both of macrocycles **69a, b** were highly efficient hosts for complexation toward squaraine to form a new

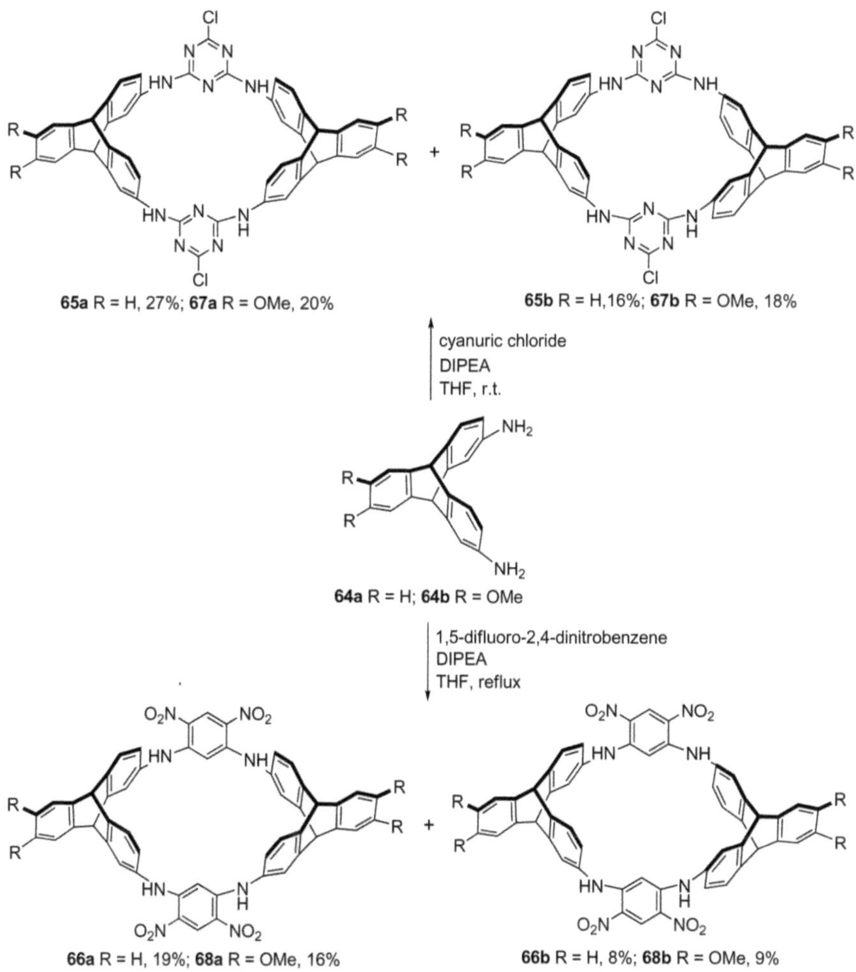

Scheme 9.13 Synthesis of triptycene-derived *N(H)*-bridged azacalixarenes **65–68** by a one-pot approach

kind of stable pseudorotaxane-type complex in both solution and solid state, which could subsequently protect the dye from polar solvents as well. The association constants between the hosts and the squaraine in chloroform were determined to be $(6.8 \pm 0.3) \times 10^5\ M^{-1}$ for host **69a** and $(1.3 \pm 0.3) \times 10^5\ M^{-1}$ for host **69b**. Interestingly, it was further found that in the complex **69a•70a** the guest protons showed two sets of signals, which implied that half of the guest molecule was positioned at the wide rim of the macrocycle, whereas the other half was at the narrow rim. Thus, the guest could display environment-induced asymmetry upon complexation with the host **69a**, due to the cone conformation of **69a**.

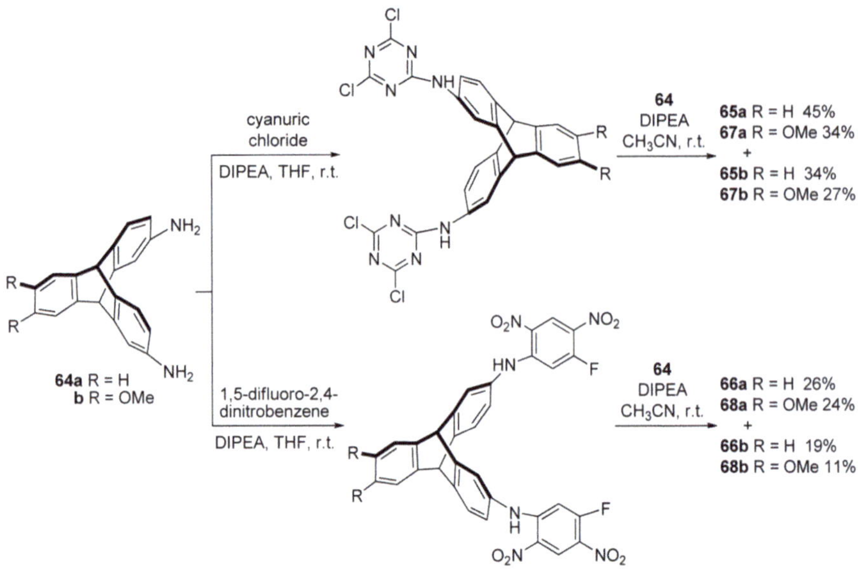

Scheme 9.14 Synthesis of triptycene-derived *N(H)*-bridged azacalixarenes **65–68** by a two-step fragment-coupling approach

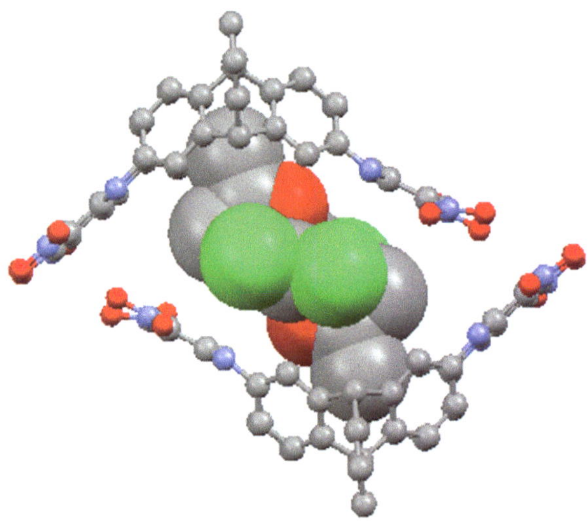

Fig. 9.23 The crystal structure of $(\mathbf{66a})_2@2CH_2Cl_2@2$acetone

Furthermore, Xue et al. [39] studied the complexation between the two macrocycles and more squaraine dyes containing different terminal groups in details, and then designed and constructed a series of squaraine-based [2]pseudo-rotaxanes and [2]rotaxanes. Consequently, it was found that squaraine **70b** containing smaller terminal groups could thread the wheels **69a**, **b** to form [2]pseudorotaxane complexes.

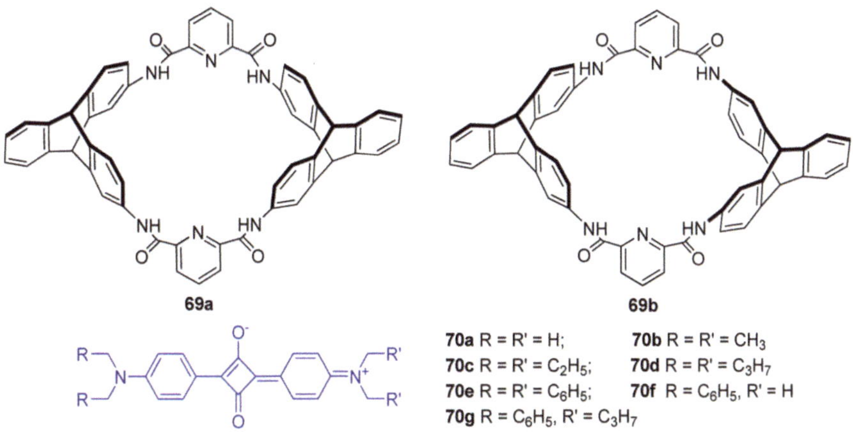

Fig. 9.24 Structures of triptycene-derived tetralactam macrocycles **69a**, **b** and squaraines **70a–g**

For **70c**, it could penetrate through macrocycle **69b** to afford a [2]pseudorotaxane-type complex **69b•70c**, while no insertion process between **69a** and **70c** was found unless the mixture of **69a** and **70c** in $CDCl_3$ was heated at 60 °C for more than 6 days. However, there were no complextion behaviors between hosts **69a**, **b** and squaraines **70d** and **e** even after the mixtures were heat to 60 °C for several days, which suggested that the *N, N*-bis(*n*-butyl) and *N, N*-bis(benzyl) groups were large enough to prevent the squaraines to form [2]rotaxanes with the macrocycles **69a**, **b** via the slippage method. Similarly, the squaraine **70g** with two different bulky stoppers could not form [2]pseudorotaxane complexes with hosts **69a**, **b** as well. Moreover, for the mixed system of host **69a** and guest **70f**, it was found that there were two new sets of resonances in the ^{1}H NMR spectrum, and the intensity of one set was higher than the other one, suggesting that two isomeric [2]pseudorotaxane complexes based on **69a** and **70f** were obtained, and the complexation showed a slight selectivity.

According to the above results, squaraines **70d**, **70e**, and **70g** containing bulky stopper groups were further chosen as the templates to synthesize [2]rotaxanes via the clipping reactions. Consequently, it was found that when **70d** was used as the template, two [2]rotaxanes **71a**, **b** as a pair of diastereomers could be formed by the condensation reaction of pyridine-2,6-dicarbonyl dichloride and 2,7-diamino-triptycene (Scheme 9.15). Similarly, a pair of isomeric [2]rotaxanes **72a** and **b** were obtained with squaraine **70e** as the template. In the case of the nonsymmetrical squaraine **70g**, three isomers **73a–c** (Scheme 9.16) were simultaneously obtained by the templated synthetic approach. Moreover, it was found that the chemical stabilities of [2]rotaxanes **71–73** were all increased relative to the free squaraines, and they could further self-assemble into a secondary arrangement, which will be depicted in the following section about self-assembly with details. It was also found that the macrocyclic compounds could efficiently protect the squaraine dyes from polar

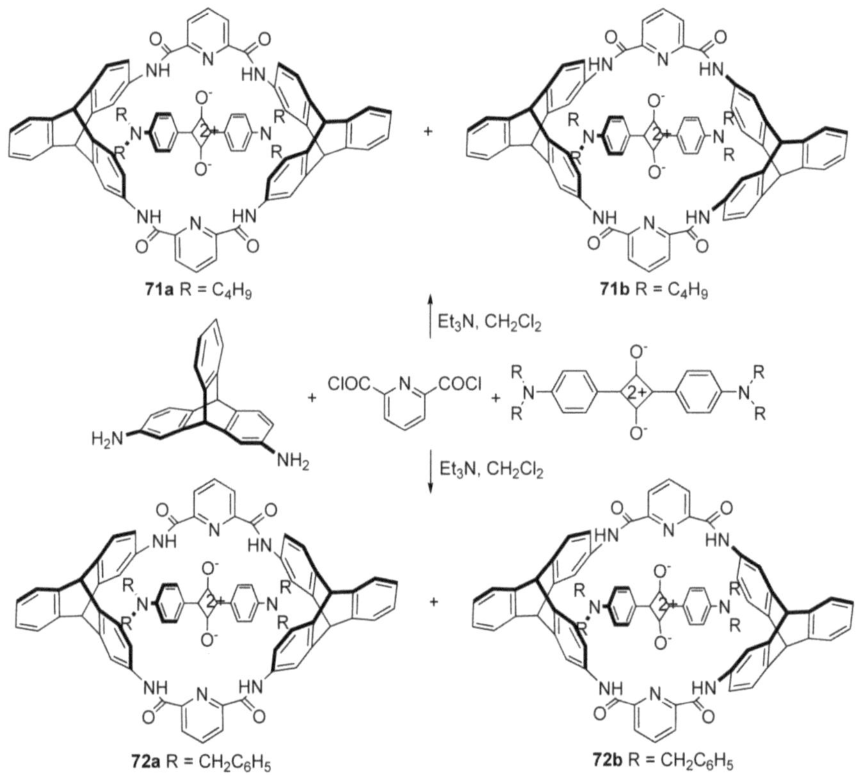

Scheme 9.15 Synthesis of [2]rotaxanes **71** and **72**

solvents, and the attack by the nucleophilic species. Furthermore, the *trans* isomers seemed to be protected more efficiently than those of the *cis* isomers, partially due to the location of macrocycle **69b** on both sides of the cyclobutene core of squaraine dyes [39].

In 2007, Zhang and Chen [40] designed and synthesized a novel nanosized cage **74** by the copper-mediated modified Eglington–Glaser oxidative-coupling method. As shown in Scheme 9.17, by the palladium-catalyzed coupling reaction of 2,7,14-triiodotriptycene with the singly protected 1,3-diethynlbenzene, and then deprotection with tetrabutylammonium fluoride (TBAF) in THF, the molecular cage **74** was finally obtained by the oxidative coupling of **75** in pyridine in the presence of $CuCl/Cu(OAc)_2$. The structure of the molecular cage was determined by NMR, mass spectrometry (MS), and X-ray analysis. Moreover, the crystal structure of **74** showed that the cage molecules with a helical chiral feature could complex with 1,3,5-trimethylbenzene in the solid state.

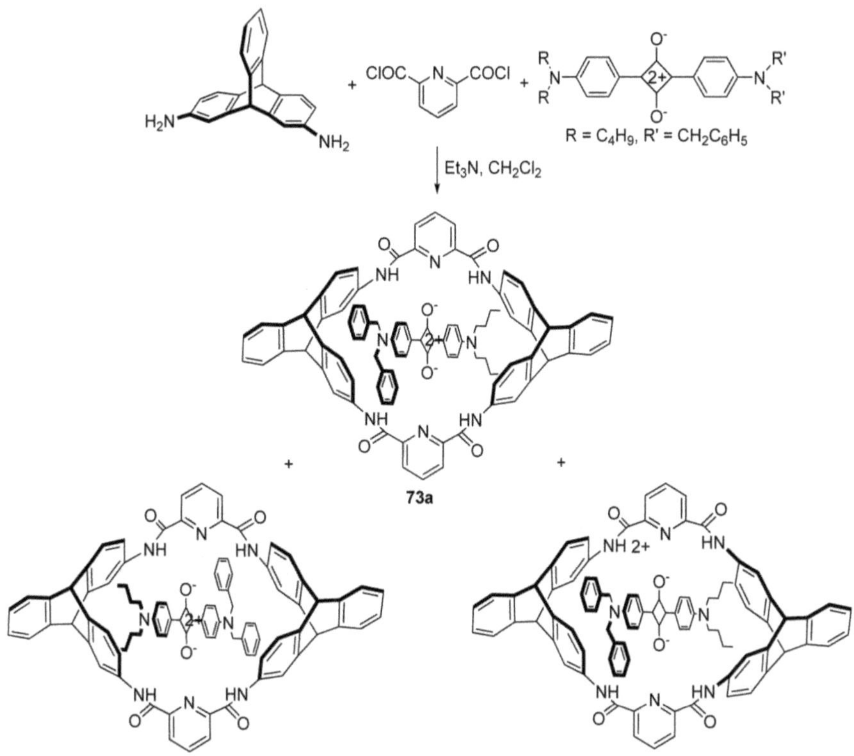

Scheme 9.16 Synthesis of [2]rotaxanes **73a–c**

9.5 Pentiptycene-Derived Hosts

Compared with triptycene, the synthesis of pentiptycene-derived hosts and their applications in host–guest chemistry are still much less explored. However, the pentiptycene has the unique H-shaped structure, which is inherently different from the Y-shaped scaffold of triptycene. Thus, pentiptycene derivatives can be used as a new class of useful building blocks for construction of novel hosts as well. As a result, a series of novel pentiptycene-derived hosts have been designed and synthesized by the use of peripherally substituted pentiptycene derivatives as building blocks.

In 2009, Cao et al. [41] first reported the design and synthesis of a novel pentiptycene-derived bis(crown ether) **76** with two DB24C8 moieties in *cis* position by the reaction between the pentiptycene bis(catechol) **77** and bistosylate **78** under high dilution conditions in the presence of cesium carbonate (Scheme 9.18). It was interestingly found that host **76** exhibited high ability to bind tetracationic cyclobis(paraquat-*p*-phenylene) (CBPQT) **39** to form a 1:1 stable complex **76•39** in both solution and solid state, which represented the first example of a synthetic host that had the cavity to accommodate the $CBPQT^{4+}$ ring. The single crystal structure

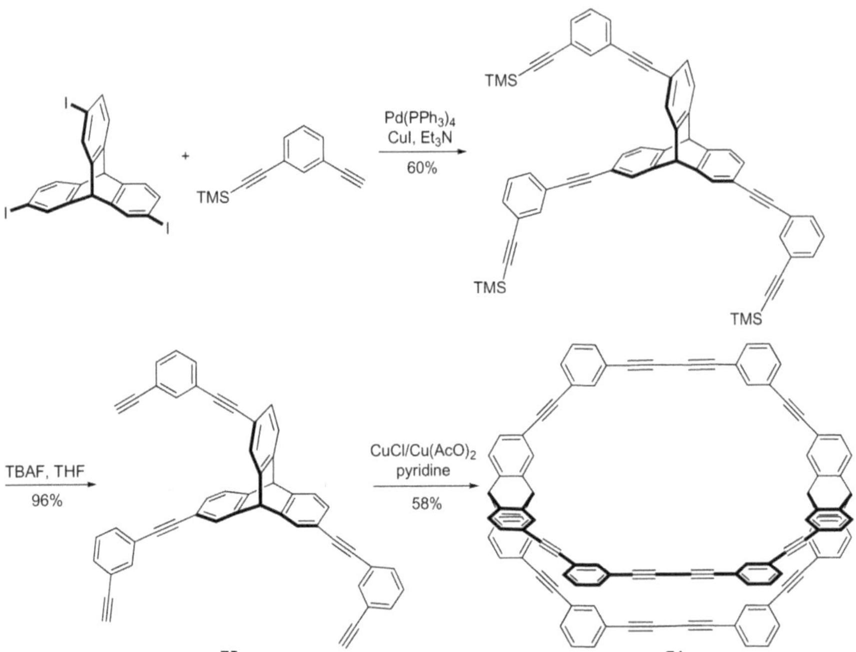

Scheme 9.17 Synthesis of nanosized cage **74**

of the complex revealed that the $CBPQT^{4+}$ ring could act as both a guest toward the pentiptycene-based bis(crown ether) and a host toward one aromatic ring of the DB24C8 moieties, which resulted in the formation of a pseudoternary complex in the solid state. Moreover, it was also found that formation of complex **76•39** was caused by charge-transfer interactions, which revealed by the electrochemical behavior of $CBPQT^{4+}$ in the absence and in the presence of the pentiptycene-derived crown ether.

According to the similar synthetic approach to that of **76**, Cao et al. [42] also synthesized the pentiptycene-derived mono(crown ether)s **79** and **80** (Fig. 9.25), and studied their complexation with $CBPQT^{4+}$ **39**. Similar to the pentiptycene-derived bis(crown ether) **76**, it was found that both the mono(crown ether)s **79** and **80** could bind $CBPQT^{4+}$ to form the 1:1 complexes in solution with the association constants of $(2.52 \pm 0.04) \times 10^2\ M^{-1}$ and $(1.07 \pm 0.06) \times 10^3\ M^{-1}$, respectively. Especially, the crystal structures showed that in the complexes **79•39** and **80•39** (Fig. 9.26), half of the $CBPQT^{4+}$ ring was included in the cavity of **79** or **80**, while one benzene ring of the pentiptycene framework was positioned inside the cavity of the $CBPQT^{4+}$ ring, which resulted in not only the pentiptycene-derived mono(crown ether), but also the $CBPQT^{4+}$ moiety in the complexes to be acted as both hosts as well as the guests. Likewise, the electrochemical experiments suggested that both complexes were stabilized by the charge-transfer interactions.

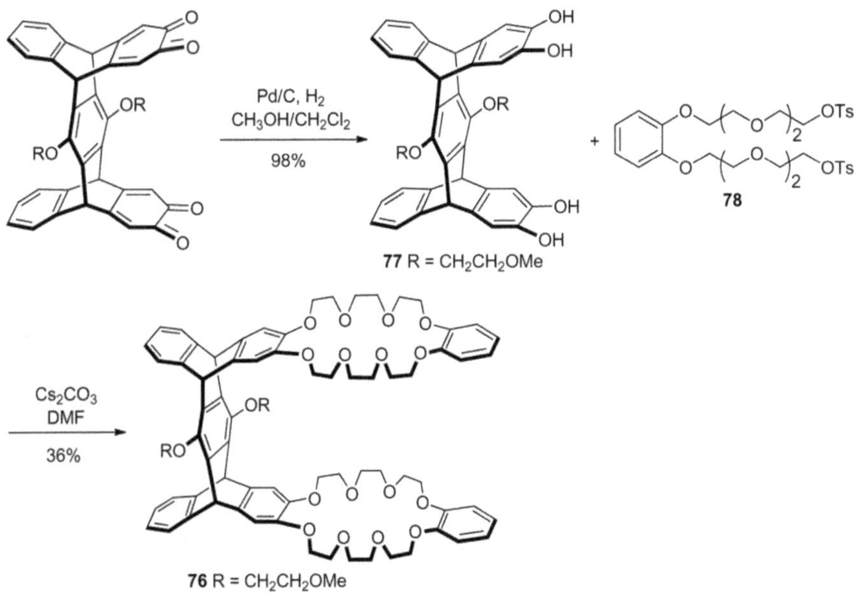

Scheme 9.18 Synthesis of pentiptycene-derived bis(crown ether) **76**

Besides the CBPQT, Cao et al. [43] also found that the tweezer-like pentiptycene-derived mono(crown ether) **79** could bind paraquat derivatives with different functional groups to form the 1:1 stable complexes in solution and in the solid state. The X-ray crystal structure of complex **79•16** (Fig. 9.27) showed that the guest positioned in the cavity of the tweezer-like pentiptycene-derived mono (crown ether), and was adjacent to the DB24C8 moiety. The bipyridinium unit in the complex was distorted by an 18.76° dihedral angle between the bipyridinium rings. The electrochemical experiments further suggested that the complexes between **79** and the paraquat derivatives were formed by the charge-transfer interactions, and the complexes were dissociated upon two one-electron reduction of the bipyridinium ring. Moreover, it was also found that the binding and release of the paraquat could be easily controlled by addition and removal of K ion (Fig. 9.28).

Similar to the synthetic method to **76**, Cao et al. [44] recently also synthesized the pentiptycene-derived bis(crown ether)s **81** and **82** (Fig. 9.29) containing two 24-crown-8 moieties in the *cis* or *trans* position, respectively. At first, the authors found that the bis(crown ethers) **81** could bind $CBPQT^{4+}$ salt to form a 1:1 complex **81•39** in solution with the association constant of $(1.4 \pm 0.1) \times 10^3$ M^{-1}. However, a 1:2 complex **82•39$_2$** between the bis(crown ethers) **82** and $CBPQT^{4+}$ salts was formed, with an average association constant of $(7.4 \pm 0.2) \times 10^3$ M^{-2} in a different complexation mode. The single crystal structure of complex **82•39$_2$** showed that the triptycene-like crown ether moiety of one side of **82** and the pentriptycene scaffold of its adjacent molecule formed an open cavity, in which one $CBPQT^{4+}$ ring was included. Meanwhile, the terminal benzene ring of each DB24C8 moiety of **82**

Fig. 9.25 Structures of pentiptycene-derived mono(crown ether)s **79** and **80**

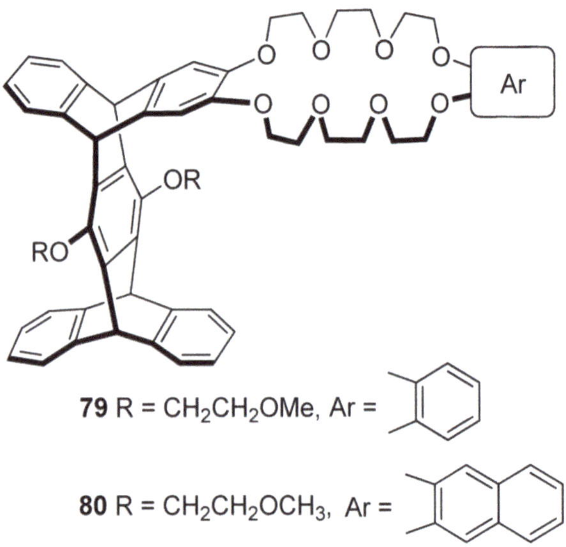

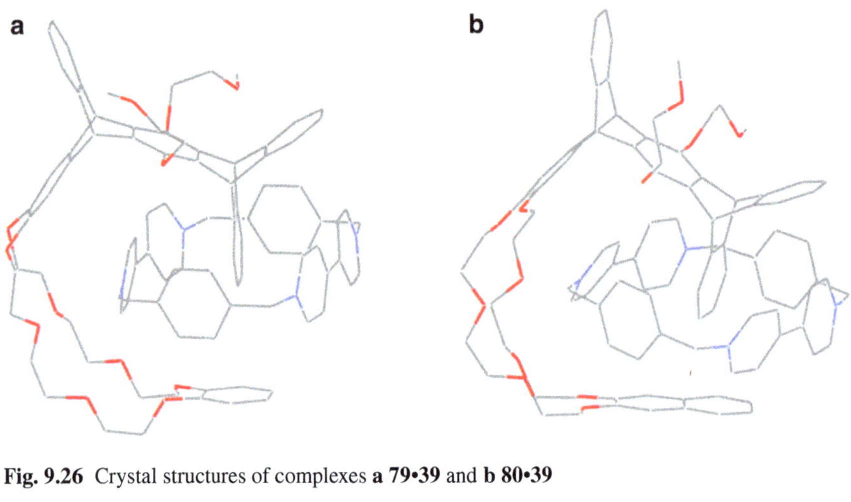

Fig. 9.26 Crystal structures of complexes **a 79•39** and **b 80•39**

was positioned inside the cavity of the $CBPQT^{4+}$ ring. Moreover, it was also found that the binding and release of the $CBPQT^{4+}$ ring in the complexes based on the pentiptycene-derived bis(crown ether)s could be efficiently controlled by adding and removing K ions, and this complexation behavior with metal ions played important roles in carrying out this process. Interestingly, it was further found that this switchable process between two different complexes **81•39** and **39•83** could be efficiently performed by adding and releasing the K ions, and the role of the $CBPQT^{4+}$ ring acting as a host or a guest could thus be changeable in the supramolecular system (Fig. 9.30).

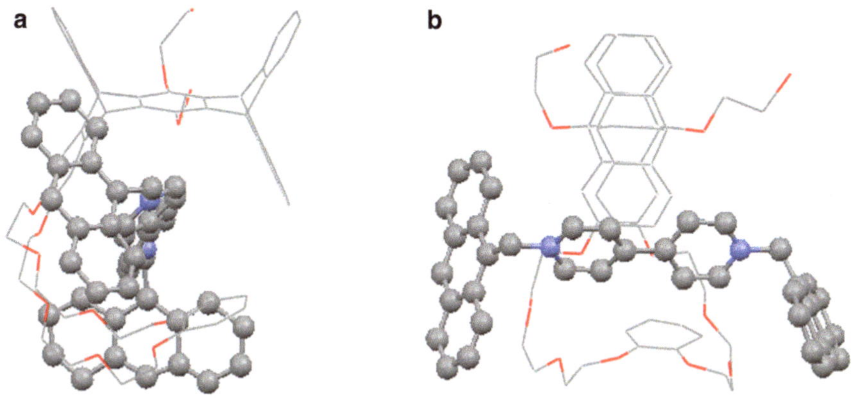

Fig. 9.27 Views of the crystal structure of complex **79•16**

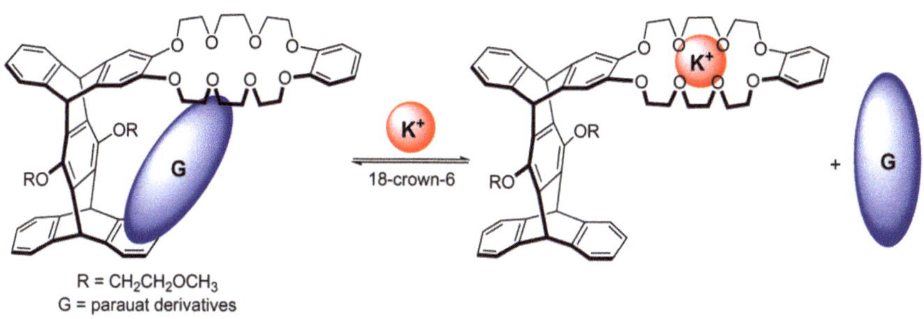

Fig. 9.28 K^+ ion-controlled binding and release of the guest in the complex

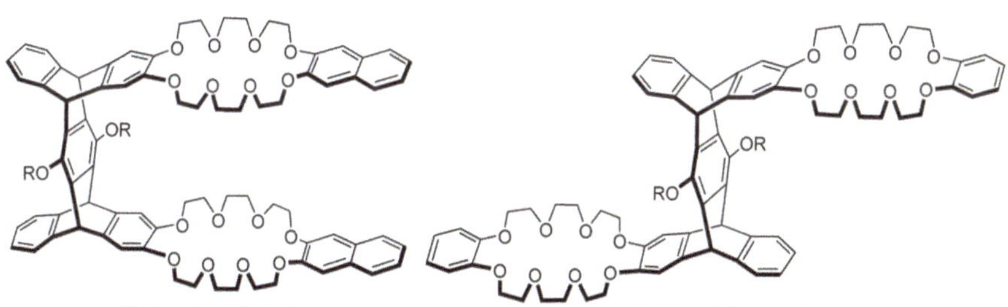

81 R = CH_2CH_2OMe **82** R = CH_2CH_2OMe

Fig. 9.29 Structures of pentiptycene-derived bis(crown ether)s **81** and **82**

As mentioned in the previous chapter, Cao et al. [45] synthesized a series of pentiptycene-derived rigid tweezer-like molecules incorporating the nitrogen-containing heterocyclic rings. Especially, molecular tweezer **84** (Fig. 9.31) was found to show the complexation ability toward C_{60} in solution with an association constant of 3.5×10^3 M^{-1} for the 1:1 complex **84**@C_{60}. The π–π interactions and the van der Waals forces between the sterically fitted concave and convex π surfaces of the pentiptycene and triptycene moieties and C_{60} probably played important roles in the formation of complex **84**@C_{60}.

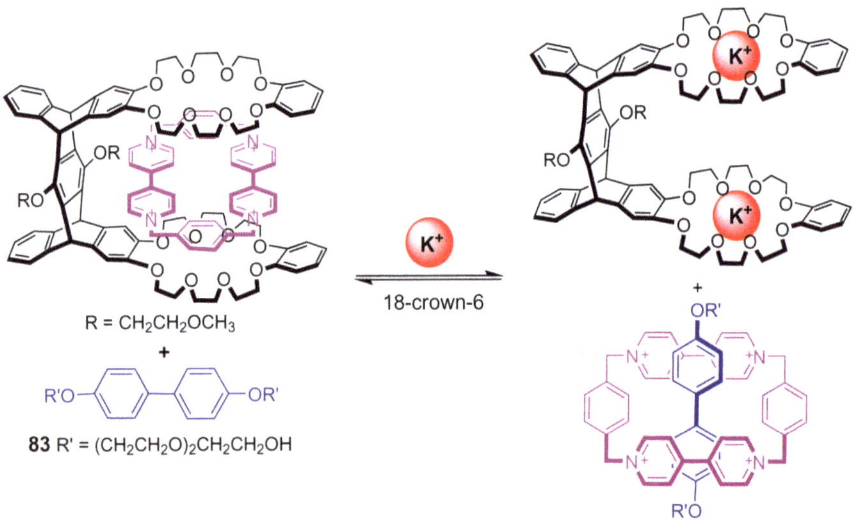

Fig. 9.30 Schematic representation of the changeable role of $CBPQT^{4+}$ ring controlled by adding and removing the K ions

Fig. 9.31 Structure of molecular tweezer **84**

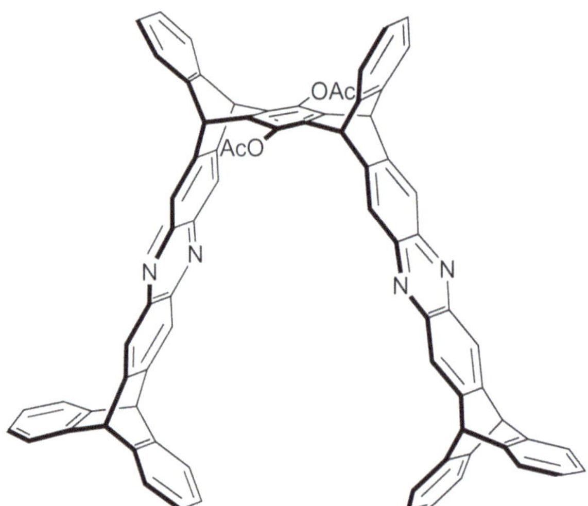

84

References

1. Rathore R, Kochi JK (1998) Cofacial phenylene donors as novel organic sensors for the reversible binding of nitric oxide. J Org Chem 63(24):8630–8631
2. Konarev DV, Drichko NV, Lyubovskaya RN, Shul'ga YM, Litvinov AL, Semkin VN, Dubitsky YA, Zaopo A (2000) Donor-acceptor interaction of fullerene C_{60} with triptycene in molecular complex TPC•C_{60}. J Mol Struct 526:25–29

3. Chen CF (2011) Novel triptycene-derived hosts: synthesis and their applications in supramolecular chemistry. Chem Commun 47(6):1674–1688
4. Jiang Y, Chen CF (2011) Recent developments in synthesis and applications of triptycene and pentiptycene derivatives. Eur J Org Chem 2011(32):6377–6403
5. An HY, Bradshaw JS, Izatt RM (1992) Macropolycyclic polyethers (cages) and related-compounds. Chem Rev 92(4):543–572
6. Zong QS, Chen CF (2006) Novel triptycene-based cylindrical macrotricyclic host: synthesis and complexation with paraquat derivatives. Org Lett 8(2):211–214
7. Zhao JM, Zong QS, Han T, Xiang JF, Chen CF (2008) Guest-dependent complexation of triptycene-based macrotricyclic host with paraquat derivatives and secondary ammonium salts: a chemically controlled complexation process. J Org Chem 73(17):6800–6806
8. Zong QS, Zhang C, Chen CF (2006) Self-assembly of triptycene-based cylindrical macrotricyclic host with dibenzylammonium ions: construction of dendritic [3]pseudorotaxanes. Org Lett 8(9):1859–1862
9. Ashton PR, Ballardini R, Balzani V, Baxter I, Credi A, Fyfe MCT, Gandolfi MT, Gomez-Lopez M, Martinez-Diaz MV, Piersanti A, Spencer N, Stoddart JF, Venturi M, White AJP, Williams DJ (1998) Acid-base controllable molecular shuttles. J Am Chem Soc 120(46):11932–11942
10. Han T, Zong QS, Chen CF (2007) Complexation of triptycene-based cylindrical macrotricyclic polyether toward diquaternary salts: ion-controlled binding and release of the guests. J Org Chem 72(8):3108–3111
11. Han T, Chen CF (2007) Formation of ternary complexes between a macrotricyclic host and hetero-guest pairs: an acid-base controlled selective complexation process. Org Lett 9(21):4207–4210
12. Han T, Chen CF (2007) Selective templated complexation of a cylindrical macrotricyclic host with neutral guests: three cation-controlled switchable processes. J Org Chem 72(19):7287–7293
13. Han T, Chen CF (2008) Efficient potassium-ion-templated synthesis and controlled destruction of [2]rotaxanes based on cascade complexes. J Org Chem 73(19):7735–7742
14. Zhao JM, Zong QS, Chen CF (2010) Complexation of triptycene-based macrotricyclic host toward (9-anthracylmethyl)benzylammonium salt: a Ba^{2+} selective fluorescence probe. J Org Chem 75(15):5092–5098
15. Guo JB, Han Y, Cao J, Chen CF (2011) Formation of 1:2 host–guest complexes based on triptycene-derived macrotricycle and paraquat derivatives: anion-π interactions between PF_6^- and bipyridinium rings in the solid state. Org Lett 13(20):5688–5691
16. Han Y, Lu HY, Zong QS, Guo JB, Chen CF (2012) Synthesis of triptycene-derived macrotricyclic host containing two dibenzo-[18]-crown-6 moieties and its complexation with paraquat derivatives: Li^+-ion-controlled binding and release of the guests in the complexes. J Org Chem 77(5):2422–2430
17. Guo JB, Xiang JF, Chen CF (2010) Synthesis of a bis-macrotricyclic host and its complexation with secondary ammonium salts: an acid-base switchable molecular handcuff. Eur J Org Chem (26):5056–5062
18. Chen CW, Whitlock HW (1978) Molecular tweezers: simple-model of bifunctional intercalation. J Am Chem Soc 100(15):4921–4922
19. Klarner FG, Kahlert B (2003) Molecular tweezers and clips as synthetic receptors. Molecular recognition and dynamics in receptor-substrate complexes. Acc Chem Res 36(12):919–932
20. Harmata M (2004) Chiral molecular tweezers. Acc Chem Res 37(11):862–873
21. Peng XX, Lu HY, Han T, Chen CF (2007) Synthesis of a novel triptycene-based molecular tweezer and its complexation with paraquat derivatives. Org Lett 9(5):895–898
22. Jiang Y, Cao J, Zhao JM, Xiang JF, Chen CF (2010) Synthesis of a triptycene-derived bisparaphenylene-34-crown-10 and its complexation with both paraquat and cyclobis(paraquat-*p*-phenylene). J Org Chem 75(5):1767–1770
23. Han T, Chen CF (2006) A triptycene-based bis(crown ether) host: complexation with both paraquat derivatives and dibenzylammonium salts. Org Lett 8(6):1069–1072
24. Asfari Z, Böhmer V, Harrowfield J, Vicens J (eds) (2001). Calixarenes 2001. Kluwer, Dordrecht

25. Gutsche CD (1998) Calixarenes revisited. In: Stoddart JF (ed) Monographs in supramolecular chemistry. Royal Society of Chemistry, Cambridge
26. Tian XH, Hao X, Liang TL, Chen CF (2009) Triptycene-derived calix[6]arenes: synthesis, structure and tubular assemblies in the solid state. Chem Commun (44):6771–6773
27. Tian XH, Chen CF (2010) Triptycene-derived calix[6]arenes: synthesis, structures, and their complexation with fullerenes C60 and C70. Chem Eur J 16(27):8072–8079
28. Tian XH, Chen CF (2010) Synthesis, structures, and conformational characteristics of triptycene-derived calix[5]arenes. Org Lett 12(3):524–527
29. Li PF, Chen CF (2011) Triptycene-derived calix[6]resorcinarene-like hosts: synthesis, structure and self-assemblies in the solid state. Chem Commun 47(44):12170–12172
30. König B, Fonseca MH (2000) Heteroatom-bridged calixarenes. Eur J Inorg Chem 2000(11):2303–2310
31. Maes W, Dehaen W (2008) Oxacalix[n](het)arenes. Chem Soc Rev 37(11):2393–2402
32. Wang MX (2008) Heterocalixaromatics, new generation macrocyclic host molecules in supramolecular chemistry. Chem Commun 2008(38):4541–4551
33. Tsue H, Ishibashi K, Tamura R (2009) Azacalixarenes Chemistry. J Synth Org Chem Jpn 67(9):898–908
34. Zhang C, Chen CF (2007) Triptycene-based expanded oxacalixarenes: synthesis, structure, and tubular assemblies in the solid state. J Org Chem 72(10):3880–3888
35. Hu SZ, Chen CF (2010) Triptycene-derived oxacalixarene with expanded cavity: synthesis, structure and its complexation with fullerenes C_{60} and C_{70}. Chem Commun 46(23):4199–4201
36. Hu SZ, Chen CF (2011) Triptycene-derived oxacalixarenes as new wheels for the synthesis of [2]rotaxanes: acid-base- and metal-ion-switchable complexation processes. Chem Eur J 17(19):5423–5430
37. Xue M, Chen CF (2009) Triptycene-derived *N(H)*-bridged azacalixarenes: synthesis, structure, and encapsulation of small neutral molecules in the solid state. Org Lett 11(22):5294–5297
38. Xue M, Chen CF (2008) Triptycene-based tetralactam macrocycles: synthesis, structure and complexation with squaraine. Chem Commun 2008(46):6128–6130
39. Xue M, Su YS, Chen CF (2010) Isomeric squaraine-based [2]pseudorotaxanes and [2]rotaxanes: synthesis, optical properties, and their tubular structures in the solid state. Chem Eur J 16(28):8537–8544
40. Zhang C, Chen CF (2007) Synthesis and structure of a triptycene-based nanosized molecular cage. J Org Chem 72(24):9339–9341
41. Cao J, Jiang Y, Zhao JM, Chen CF (2009) A pentiptycene-based bis(crown ether) host: synthesis and its complexation with cyclobis(paraquat-*p*-phenylene). Chem Commun 2009(15):1987–1989
42. Cao J, Lu HY, Xiang JF, Chen CF (2010) Complexation between pentiptycene-based mono(crown ether)s and tetracationic cyclobis(paraquat-*p*-phenylene): who is the host or the guest? Chem Commun 46(20):3586–3588
43. Cao J, Lu HY, You XJ, Zheng QY, Chen CF (2009) Complexation of a pentiptycene-based tweezer-like receptor with paraquat derivatives: ion-controlled binding and release of the guests. Org Lett 11(19):4446–4449
44. Cao J, Guo JB, Li PF, Chen CF (2011) Complexation between pentiptycene derived bis(crown ether)s and $CBPQT^{4+}$ salt: ion-controlled switchable processes and changeable role of the $CBPQT^{4+}$ in host–guest systems. J Org Chem 76(6):1644–1652
45. Cao J, Zhu XZ, Chen CF (2010) Synthesis, structure, and binding property of pentiptycene-based rigid tweezer-like molecules. J Org Chem 75(21):7420–7423

Chapter 10
Iptycenes and Their Derivatives in Molecular Self-Assembly

10.1 Self-Assembly in Crystal

The unique three-dimensional rigid structural features of iptycenes and their derivatives make them promising candidates as the building blocks for the formation of a variety of supramolecular structures in the solid states via self-assembly.

In 1986, Bashir-Hashemi et al. [1] found that the tritriptycene **1** could pack into a channel-shaped three-dimensional structure filled by the disordered acetone molecules, in which the π–π interactions between the phenyl rings of molecule **1** played the key roles. Similarly, Venugopalan et al. [2] reported the crystallographic investigation of heptiptycene **2**, which was crystallized from its chlorobenzene solution. The authors found that molecule **2** could pack into the zigzag ribbons in the solid state, which further self-assembled into the channels separated by the chlorobenzene molecules via the T-shaped fashion interactions between the hydrogen atoms of chlorobenzene and the aromatic rings of **2**. In addition, Lu et al. [3] found that the dodecaphenyltriptycene **3** with the enhanced size of the clefts could also self-assemble into the well-ordered extended structure in the solid state. The channel-shaped void spaces were almost filled with the benzene molecules, and the π–π interactions between the solvent molecules and the phenyl rings of **3** stabilized this crystal packing (Fig. 10.1).

In 1999, Hashimoto et al. [4] investigated the crystal structures of triptycene quinone **4a** (TPQ) and its 6,7-dimethyl derivative **4b**. They found that there existed weak intermolecular π–π interactions between the quinone and benzene moieties in **4b**, which resulted in its ribbon-like self-assembled structure. Meanwhile in the crystal of **4a**, there were also the similar π–π interactions and the similar ribbon-like arrangement as those of **4b**. However, the donor part of **4b** containing higher electron density resulted in the shorter interplanar distance, which might be caused by the enhanced donor–acceptor interactions. Both the ribbon-like self-assembled layer structures of **4a** and **4b** could further form the three-dimensional structures by the connection of C–H$\cdots$O hydrogen bonds between the adjacent triptycenes (Fig. 10.2).

Soon later, Hashimoto et al. [5] further studied the self-assembling of two TPQ's derivatives **5a** and **5b** (Fig. 10.3), and found that both compounds exhibited

C.-F. Chen, Y.-X. Ma, *Iptycenes Chemistry*,
DOI 10.1007/978-3-642-32888-6_10, © Springer-Verlag Berlin Heidelberg 2013

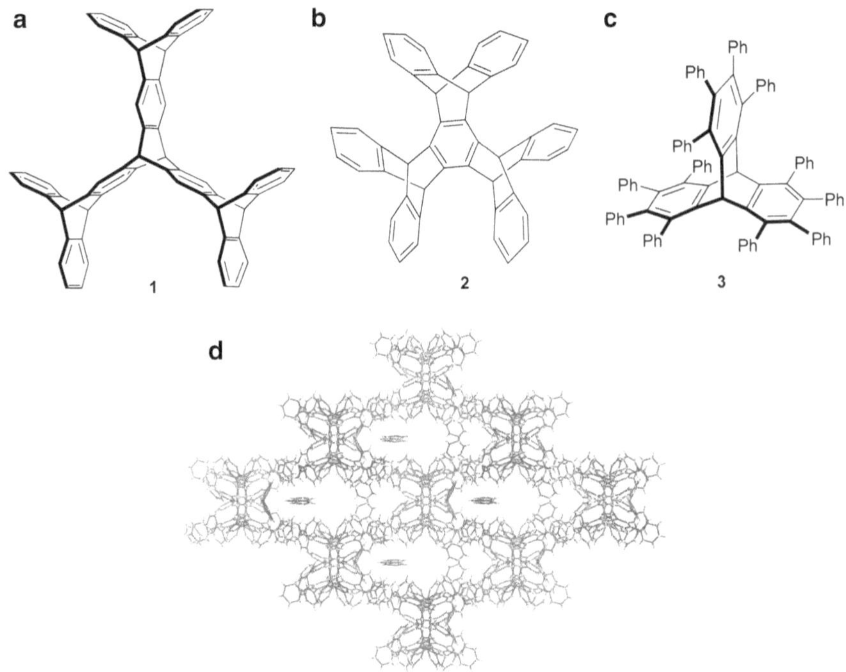

Fig. 10.1 Structures of **a** tritriptycene **1**, **b** heptiptycene **2**, **c** dodecaphenyltriptycene **3**, and **d** packing diagrams of **3**

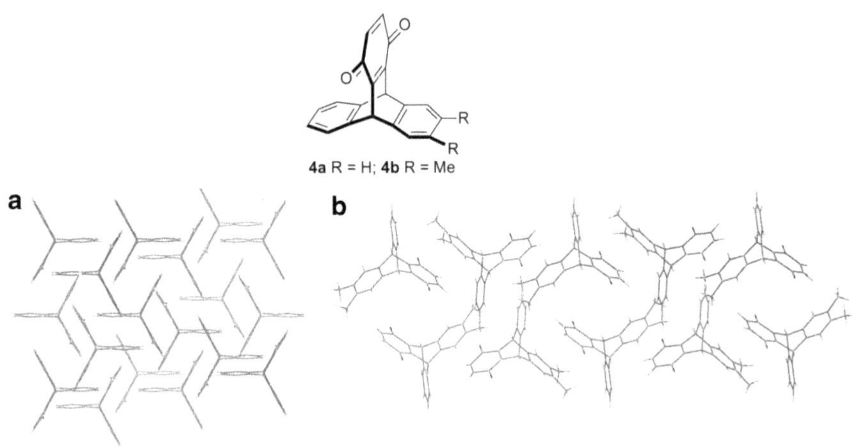

Fig. 10.2 Solid-state extended structures of **a 4a** and **b 4b**

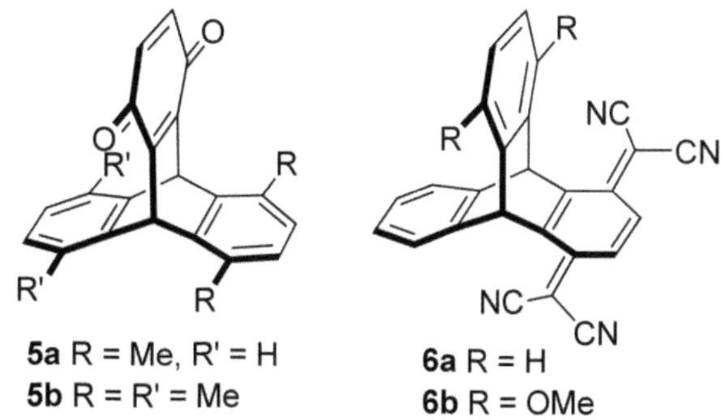

Fig. 10.3 Structures of compounds **5** and **6**

the ribbon-like structures formed by the weak face-to-face π–π interactions between the electron donor (D, substituted benzene) and electron acceptor (A, quinone). However, it was found that 9,10-dihydro-9,10-*o*-benzeno-1,4-bis (dicyanomethylene)anthracene **6** (TP-TCNQ, Fig. 10.3) showed the weak intermolecular charge-transfer (CT) interaction in the crystals along with the significant color change (wine red in solution, whereas dark blue in the solid state). In the case of the clathrates of 5,8-dimethyl-TPQ **5a** and 5,8-dimethoxy-TP-TCNQ **6b**, it was also found that there were D–A interactions which led to the ribbon-like stacks in the solid state. Moreover, all of the ribbon-like structures could further be connected by the hydrogen-bonding interactions between the adjacent molecules to afford the three-dimensional extended supramolecular structures.

In addition, Yamamura et al. [6] also studied the binary crystal of TPQ **4a** (R = H) and TPHQ **7**, which were found to be a substitutional solid solution of **4a** doped by compound **7**. It was further found that there was an intermolecular CT interaction between *p*-benzoquinone and hydroquinone moieties in the binary crystal of $\mathbf{4a}_{0.8}\mathbf{7}_{0.2}$, which was revealed by the bathochromic changes of color (from yellow to brown, Fig. 10.4) of the crystal.

Besides the weak D–A stacking, Veen et al. [7] took advantage of the π–π interactions between the three concave faces of triptycene and the convex surface of C_{60} to produce a specific self-assembled structure as well. There were different self-assembled modes in the mixture systems of triptycene•C_{60} and azatriptycene•C_{60}. In the crystal of the triptycene•C_{60}, it was found that the two triptycene molecules encapsulated one C_{60} molecule. The triptycene molecules could arrange in a sheet-like layer structure with the antiparallelism, and the *o*-xylene molecules filled the voids of this double layer between the triptycene molecules (Fig. 10.5a). For the latter, there were no solvents included in the crystal structure, but it was found that three azatriptycenes surrounded one C_{60} molecule; meanwhile three C_{60} surrounded one azatriptycene, which resulted in the hexagonal close-packing pattern of a particular layer. Moreover, the layers further arrayed in an *ab*-repeating arrangement to form a supramolecular structure as shown in Fig. 10.5b.

In 2007, Chong and MacLachlan [8] reported the well-ordered solid-state extended structures of a series of iptycene derivatives **8a–e** (Fig. 10.6) containing pyrazine groups with different sizes. They found that these compounds **8a–d** could pack into the layered structures by the intermolecular π–π stacking in the solid state.

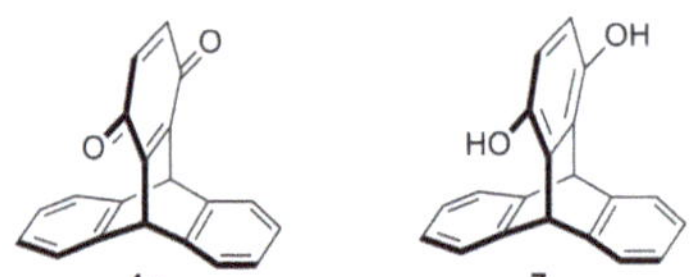

4a **7**

TPQ $(TPQ)_{0.8}(TPHQ)_{0.2}$ TPHQ

Fig. 10.4 Photographs of crystals of pure TPQ, solid solution, and pure TPHQ. (Reprinted from [6], with permission from Elsevier)

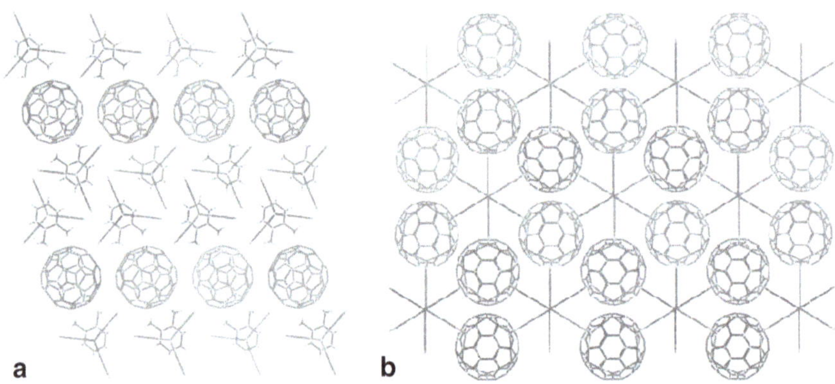

Fig. 10.5 Packing arrangements for (*left*) triptycene•C_{60}•*o*-xylene viewed along the *b*-axis and (*right*) azatriptycene•C_{60} viewed along the *c*-axis

Also, the increased pyrazine groups or triptycenyl moieties could lead to the comparative orderless package, while the enhanced length of the wings was beneficial to the effective packing. Furthermore, the voids between the iptycenes in the solid state were filled with solvent molecules, especially for compound **8d**. However, for the compound **8b**, the filled package of solvent molecules displayed the dependency on the varied solvents. When it was crystallized from acetonitrile solution, the crystal would contain the guest solvent molecules; while the compound **8b** crystallized from tetrahydrofuran (THF) was solvent free.

In 2000, Yang et al. [9] designed and synthesized the triptycene-derived and pentiptycene-derived secondary diamides **9a** and **9b** (Fig. 10.7), and also studied their self-assemblies in the solid state. For the triptycene-derived secondary diamide **9a**, the intermolecular edge-to-face arene–arene interactions led to the folded conformation, meanwhile the amide–amide hydrogen-bonding interactions between the

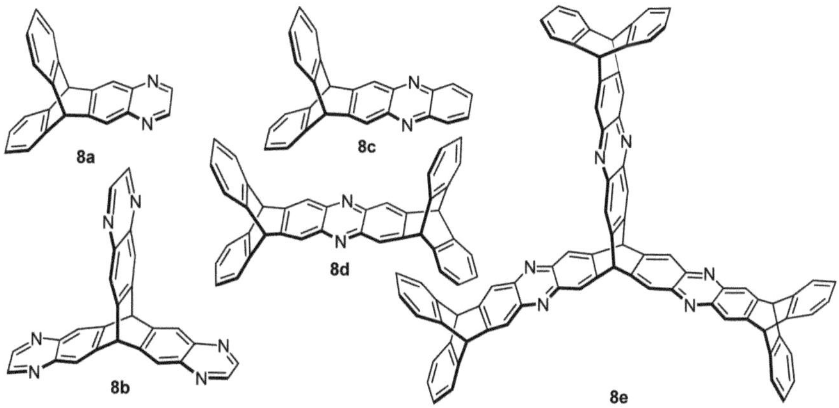

Fig. 10.6 Structures of iptycene derivatives **8a–e**

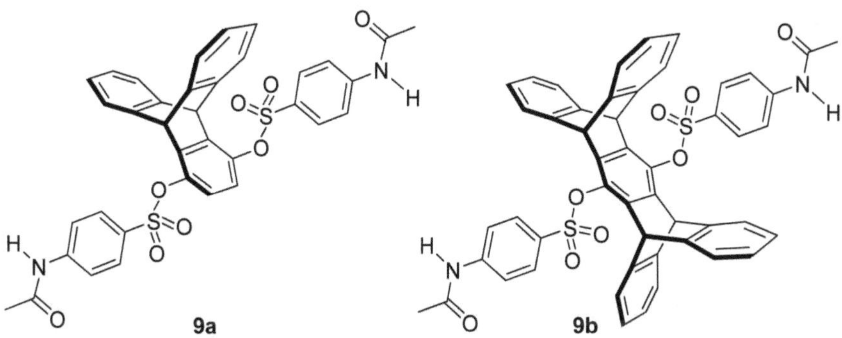

Fig. 10.7 Structures of iptycene-based secondary diamides **9a** and **9b**

adjacent molecules resulted in a chain-like extended structure. Furthermore, it was found that the packing arrangement was different in varied solvents, such as in methanol-toluene, the toluene molecule could fill in the channels of the crystal; whereas in pure methanol system, the crystals were found to be solvent-free. On the other hand, for pentiptycene-derived secondary diamide **9b**, it had not shown the similar hydrogen-bonding networks to that of **9a**. The self-assembled structure of **9b** in the solid state seemed to be almost dominated by the molecular stacking interactions: firstly, the upright stacking of the pentiptycene groups led to one-dimensional pillars, then the pillars further assembled to form the zigzag pentiptycene walls and channels by the stacking of *N*-acetylsulfanilyl substituents. The void spaces in this novel grid-like channel network of **9b** were filled by the guest solvent molecules (methanol) as well.

Soon afterward, Yang's group [10] further investigated the self-assemblies of a series of triptycene-derived or extended triptycene-derived secondary dicarboxamides **10–14** (Fig. 10.8). The triptycene derivatives **10, 11,** and **17–18** were

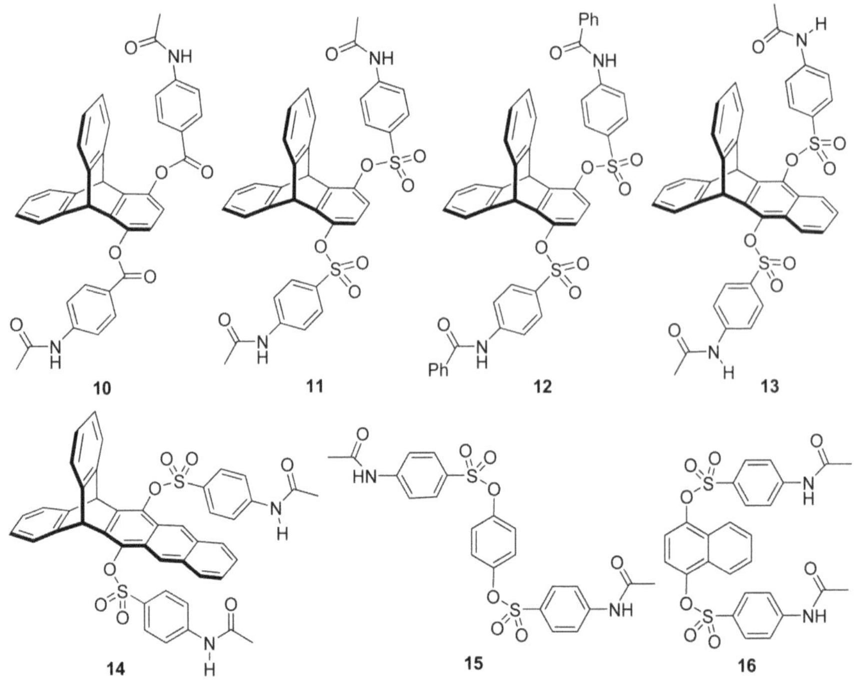

Fig. 10.8 Structures of triptycene-derived or extended triptycene-derived secondary dicarboxamides **10–14**

synthesized by the reaction of precursor **7** with the corresponding 4-acetamidobenzoic acid in the presence of dicyclohexylcarbodiimide (DCC) and the catalytic amount of *N,N*-4-dimethylaminopyridine (DMAP) or the reaction of **7** with substituted benzenesulfonyl chlorides in the presence of K_2CO_3, respectively (Scheme 10.1). Compound **19** was obtained by the reduction of nitro group in **18** with tin^{II} chloride, and then the resulting compound **19** underwent the reaction with benzoyl chloride to afford the target compound **12** (Scheme 10.2). The extended triptycene derivatives **13** and **14** were prepared starting from the corresponding extended triptycene quinones. The single crystals of compounds **12–14** were obtained by the slow evaporation of the solvent at room temperature, and the crystal structures displayed that both of the compounds **12–14** could self-assemble into the folded and compact structures. Compared with the reference compounds **15–16** (Fig. 10.8), it was found that both the amide–amide hydrogen bonding, edge-to-face arene–arene interactions between the triptycene and the *N*-acetylsulfanilyl groups, and the sulfonyl ester turn units played the essential roles in the formation of folded packings in the solid state. Moreover, the solvent molecules could serve as guests to fill in the voids of the lattices of **10–12**.

Recently, Zhang and Chen [11] reported the selective synthesis of a series of adducts **23–24** between 2,7-dihydroxytriptycene **20** and 2,7,14-trihydroxytriptycene

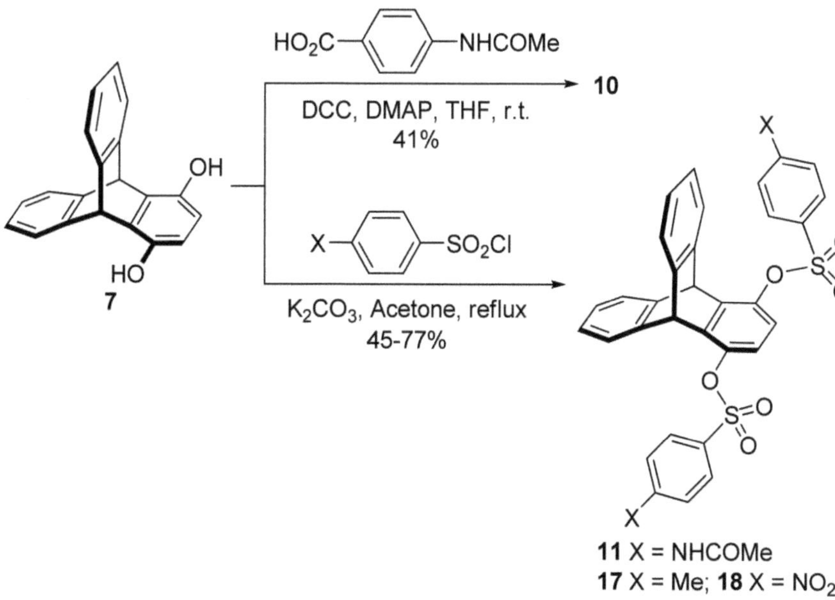

Scheme 10.1 Synthesis of triptycene derivatives **10, 11,** and **17–18**

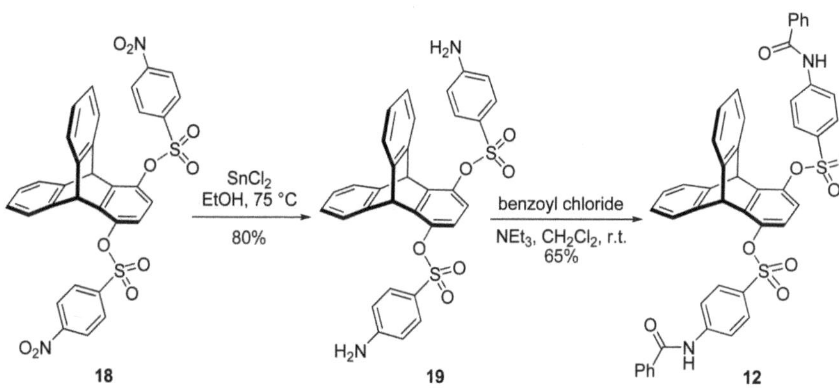

Scheme 10.2 Synthesis of triptycene-derived secondary dicarboxamide **12**

21 with acetone and/or 4,4′-bipyridine **22** (Fig. 10.9), and their self-assembling into high-order supramolecular structures in the solid state. It was found that the components of the adduct **23b** including two molecules of **20**, two molecules of **22,** and one molecule of acetone were independent on the stoichiometries between **20** and **22** to a certain extent, meanwhile different stoichiometries between 2,7,14-trihydroxytriptycene **21** and 4,4′-bipyridine **22** in acetone resulted in different adducts **24b** and **24c**, respectively. Moreover, the crystal structures showed that all of the adducts could self-assemble into ladder-like and/or three-dimensional

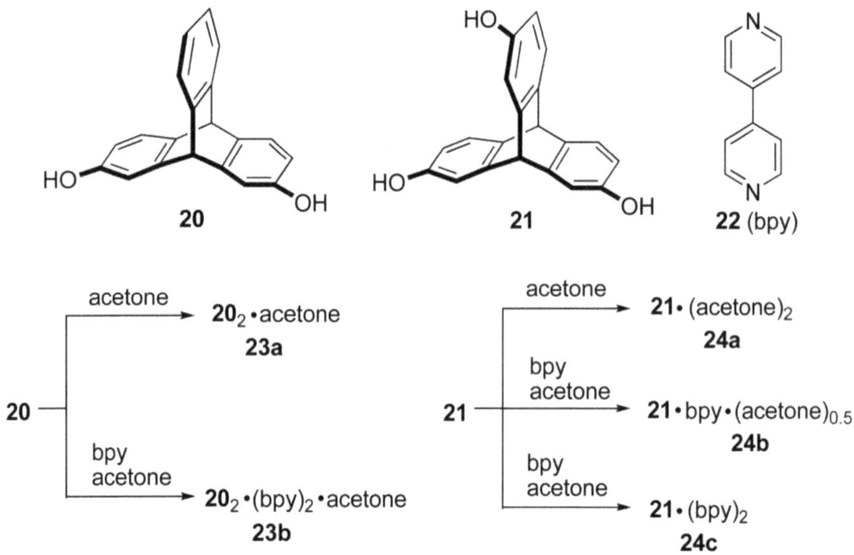

Fig. 10.9 *Top*: chemical structures of **20–22**. *Bottom*: representation of selective synthesis of adducts **23a**, **b** and **24a–c**

network supramolecular structures (Fig. 10.10), and the multiple noncovalent interactions including the competitive O–H···O and O–H···N hydrogen bonds between the hydroxyl-substituted triptycenes with 4, 4′-bipyridine and the solvent molecules play important roles in controlling the formation of superstructures.

In 2007, Zhang and Chen [12] synthesized a series of expanded oxacalixarenes **25–27** (Fig. 10.11) by utilizing 2,7-dihydroxytriptycene as the nucleophilic reagent instead of *m*-diphenol for the S_NAr reaction in dimethyl sulfoxide (DMSO) in the presence of Cs_2CO_3 or K_2CO_3. The S_NAr reaction of 2,7-dihydroxy-triptycene with a proper electrophilic component gave a pair of diastereomers **a** and **b** (Fig. 10.11), in which **a** was a *cis* isomer with boat conformation, and **b** was a *trans* isomer with chair conformation. In the case of **26b**, a dynamic interconversion between boat and chair conformations could be proposed based on the studies of the variable-temperature NMR spectra and X-ray crystal structure. Moreover, it was also found that the expanded oxacalixarenes **25a, 25b**, and **27a** could all form tubular assemblies and further porous architectures (Fig. 10.12). In the case of **25a**, it could assemble into organic tubular structure with the aromatic rings as the wall, and the nitrogen atoms of the pyridine rings all point inward the tube, and the multiple chlorine bonding including C–Cl···Cl, C–Cl···O, and C–Cl···π interactions played important roles in formation of the assembly. Likewise, the macrocyclic molecules **25b** and **27a** could also assemble into organic tubular structures and further porous architectures via multiple intermolecular chlorine-bonding interactions.

Actually, Zong et al. [13] also found the similar self-assembled tubular structure in the crystal of complex **28** (Fig. 10.13a) formed by the triptycene-derived macrotricyclic host and the dibenzylammonium salts in 2006. As shown in Fig. 10.13b,

a

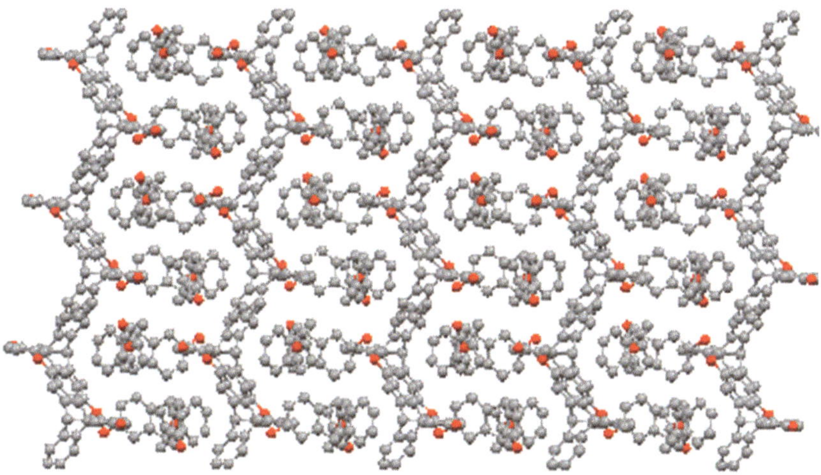

b

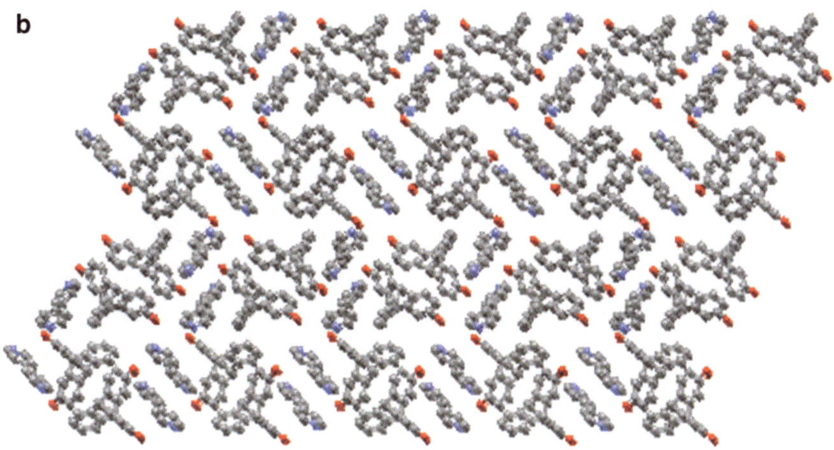

Fig. 10.10 Crystal packings of **a 23a** and **b 23b**

the macrocyclic molecules could pack into tubular channels with the four aromatic rings as the wall, which further extended in the crystallographic *a* direction with the ammonium ions threaded inside. Similarly, Guo et al. [14] also found that the macrotricyclic host **29** could pack into a tubular channel, which further extended in the crystallographic *b* direction with two guests of the paraquat derivative **30** threaded inside the central cavity of the host (Fig. 10.13c).

In 2009, Tian et al. [15] reported a class of novel triptycene-derived calix[6]arenes **31** and **32** (Fig. 10.14) containing two 1,8-dimethoxytriptycene moieties and two *p*-*t*-butylphenol moieties, and also investigated their self-assembled behaviors in the solid state. As shown in Fig. 10.15, both the *cis* isomer **31** with cone conformation and the *trans* isomer **32** with a chair conformation could self-assemble into tubular

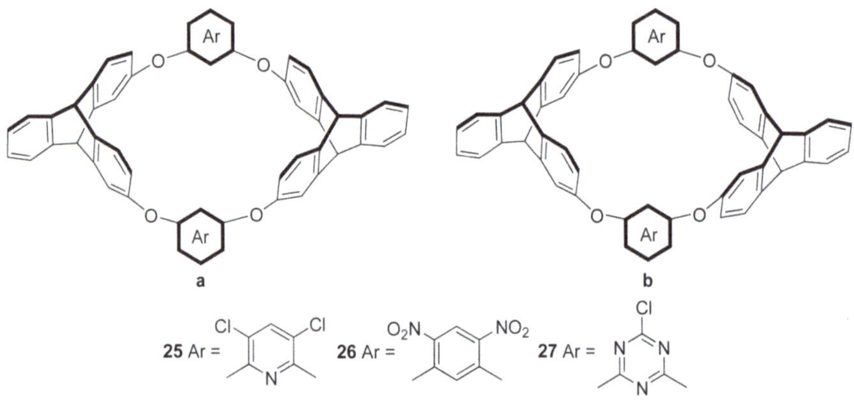

Fig. 10.11 Structures of expanded oxacalixarenes **25**–**27**

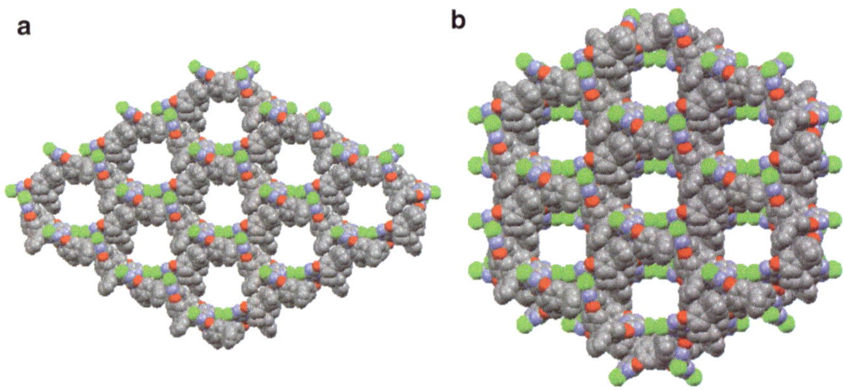

Fig. 10.12 Porous architectures self-assembled by **a 25a** and **b 27a**

structures with the aromatic rings as the walls and phenolic oxygen atoms situated in their cavities. Moreover, the tubular assemblies could stack into two-dimensional superstructure, and further three-dimensional microporous architecture. In addition, it was also found that there were solvent molecules situated in the channels, and the multiple noncovalent interactions between the solvent molecules and the macrocyclic molecules played important roles in formation of the tubular assemblies, and the further high-order self-assembled architecture.

Furthermore, Chen and Tian [16] also investigated the self-assembled behaviors of triptycene-derived calix[6]arene **33a**, and the demethylated calix[6]arenes **33b** and **33c** (Fig. 10.16) in the solid state. It was found that *trans*-isomeric macrocycle **33a** could not form a self-assembled tubular structure similar to that of **32**. However, by virtue of a pair of C–H$\cdots$O hydrogen bondings between the aromatic proton of the phenol ring in one macrocycle and the phenolic oxygen of its adjacent macrocycle,

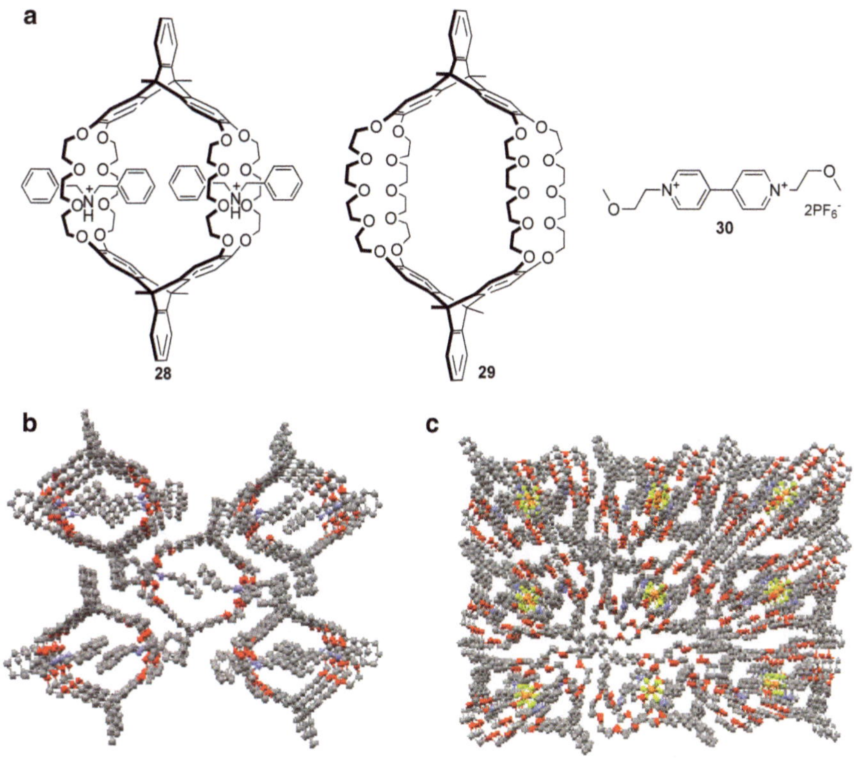

Fig. 10.13 **a** Chemical structures of complexes **28**, host **29** and guest **30**. **b** Tubular channel-like assemblies of complex **28** viewed along the *a*-axis. **c** The self-assembled tubular structure of macrocyclic host **29** with guest **30** situating inside the channels viewed along the *b*-axis

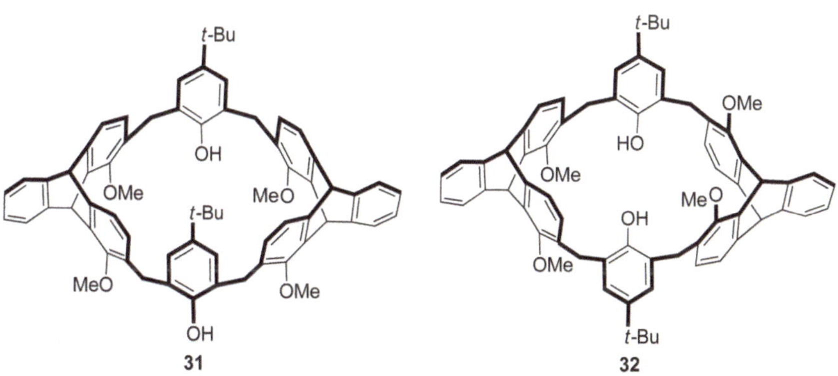

Fig. 10.14 Structures of triptycene-derived calix[6]arenes **31** and **32**

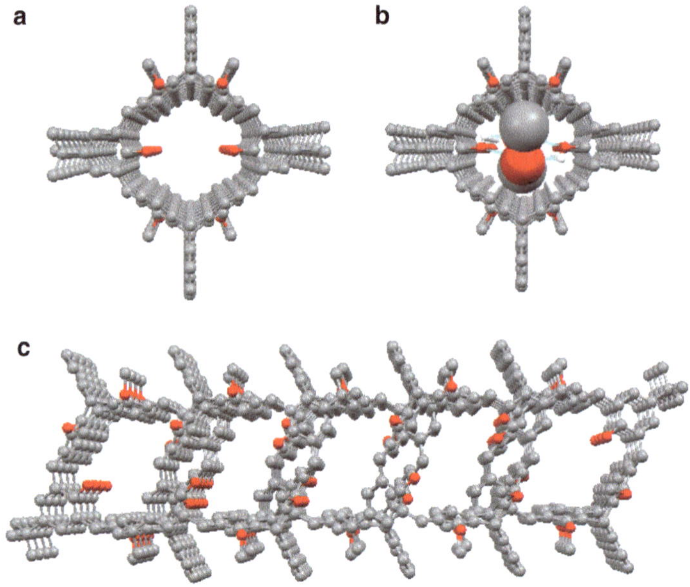

Fig. 10.15 *Top* views of tubular self-assemblies of **31 a** without and **b** with the solvents situated inside the channel and **32 c**

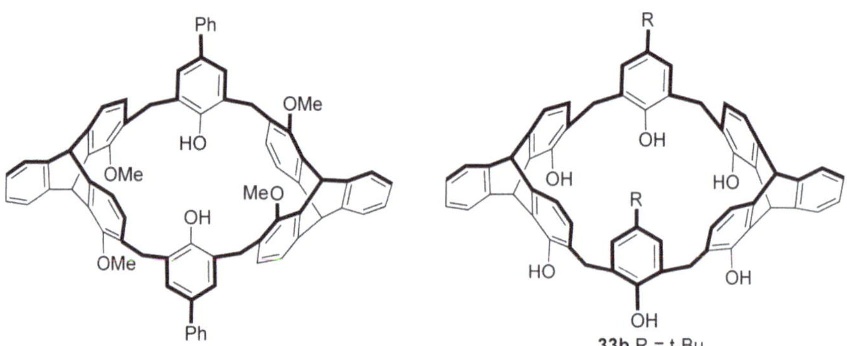

Fig. 10.16 Structures of triptycene-derived calix[6]arenes **33a–c**

and the π–π interaction between the phenyl rings of the adjacent triptycene moieties, the molecule **33a** could self-assemble into a one-dimensional supramolecular structure, and further microporous architecture (Fig. 10.17a). Similarly, *cis*-isomeric calix[6]arenes **33b** and **33c** could also self-assemble into the tubular structure similar to that of **32a**. However, it was found that **33b** with a typical cone conformation and a highly symmetrical C_{2v} space group could form three-dimensional microporous architecture (Fig. 10.17b) via the multiple noncovalent interactions between the

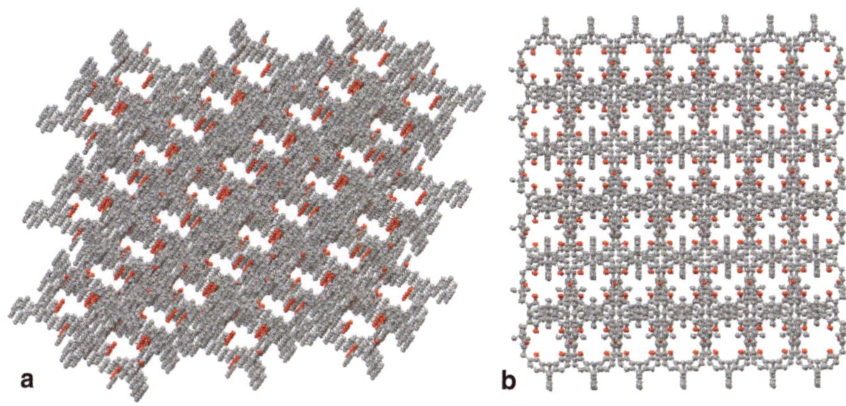

Fig. 10.17 The three-dimensional microporous structures of **a 33a** viewed along the *b* axis and **b 33b** viewed along the *a* axis

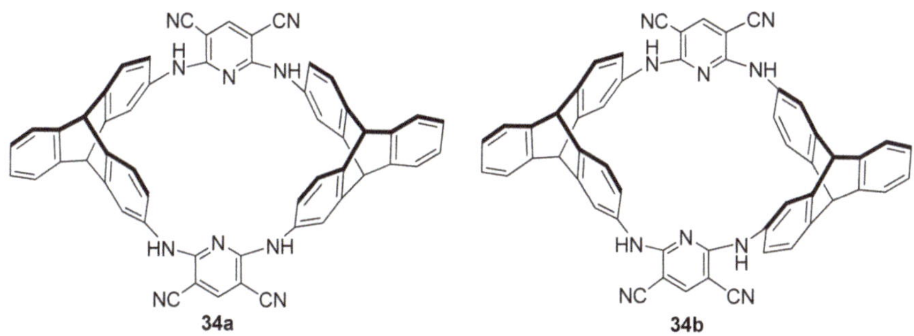

Fig. 10.18 Structures of *NH*-bridged azacalix[2]triptycene[2]pyridines **34a** and **34b**

adjacent macrocyclic molecules, and the self-assembled structure was different from that of the *trans*-isomeric macrocycle **33a** with a chair conformation.

Recently, Xue and Chen [17] designed and synthesized a pair of novel triptycene-derived azacalixarenes, *NH*-bridged azacalix[2]triptycene[2]pyridines **34a** and **34b** (Fig. 10.18) by a one-pot reaction between 3,5-dicyano-2,6-dichloro-pyridine and 2,7-diaminotriptycene in dry acetonitrile in the presence of diisopropylethylamine (DIPEA). They also obtained the crystals of **34a** suitable for X-ray crystallographic analysis by slow evaporation of a mixture solution of **34a** in THF and CH_3OH. The crystal structure of **34a** showed that the azacalixarene molecule could self-assemble into a novel aromatic single-walled organic nanotube. As shown in Fig. 10.19a, there existed two molecules of **34** (denoted as **34a** and **34b**) in the boat-like conformation with a similar cavity in the unit cell. Symmetric expansion around the pyridine rings of the macrocycle revealed that two molecules of **34a** were located in opposite positions and generated a rectangular geometry with the four cyano groups and four NH sites in the same direction. Meanwhile, two molecules of **34b** were also located

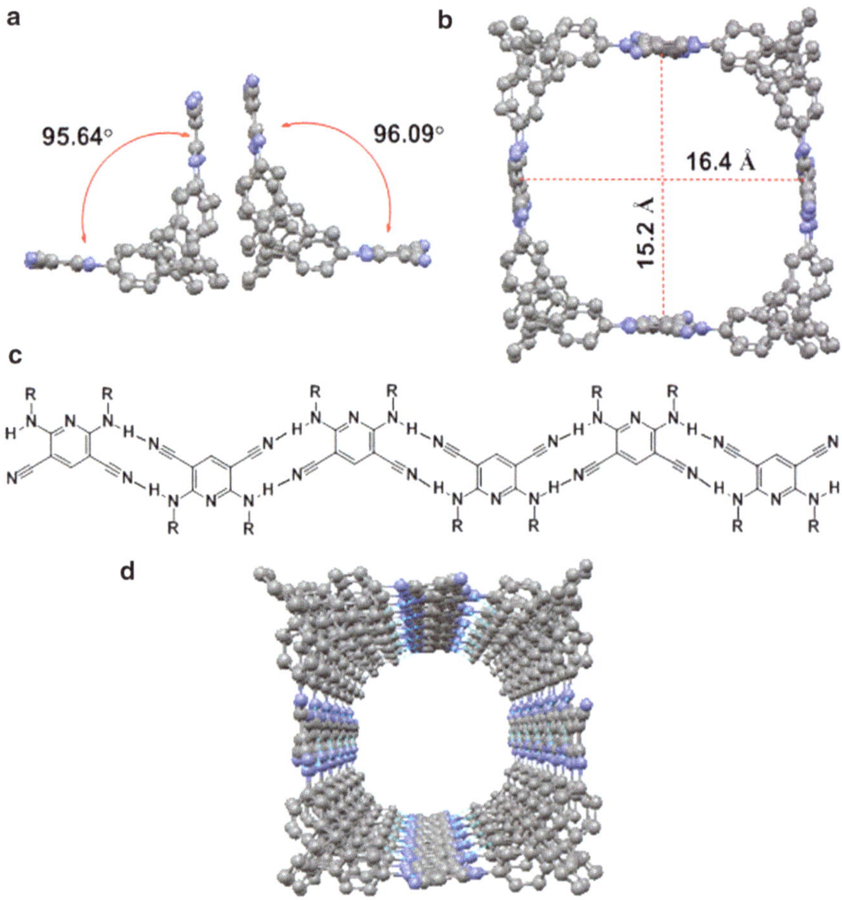

Fig. 10.19 **a** Crystal structure with the asymmetric unit of **34a**. **b** *Top* view of the tetramer. **c** Molecular structure representation of a one-dimensional H-bond chain. **d** Self-assembled square organic nanotube of **34a**

face-to-face and generated a rectangular geometry rotated with respect to the former one by 90°, due to the multiple H-bonds of cyano groups and NH sites between **34a** and **34b** (Fig. 10.19b). Furthermore, the two pyridine rings in one molecule of **34a** could provide four pairs of hydrogen bonds to participate in two zigzag H-bond chains with other molecules (Fig. 10.19c). Then, the four infinite one-dimensional hydrogen bond chains further self-assembled into an aromatic single-walled square organic nanotube with diameters in excess of 15 Å (Fig. 10.19d), which represented the first example that azacalixarene was used as the building block to form an aromatic single-walled organic nanotube. Moreover, the nanotubes could further assemble into a three-dimensional honeycombed architecture, and THF molecules were found to be located in each corner of the square channels (Fig. 10.20). In addition, it was

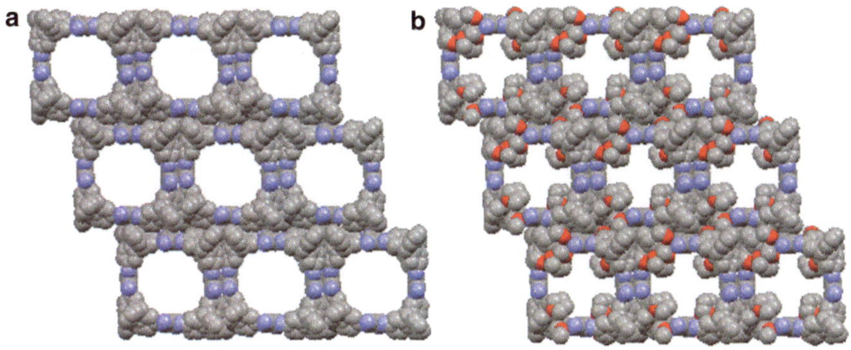

Fig. 10.20 Honeycombed architecture of **34a** without solvents (**a**), and with THF and methanol molecules positioned in the channel (**b**) viewed along the *a* axis

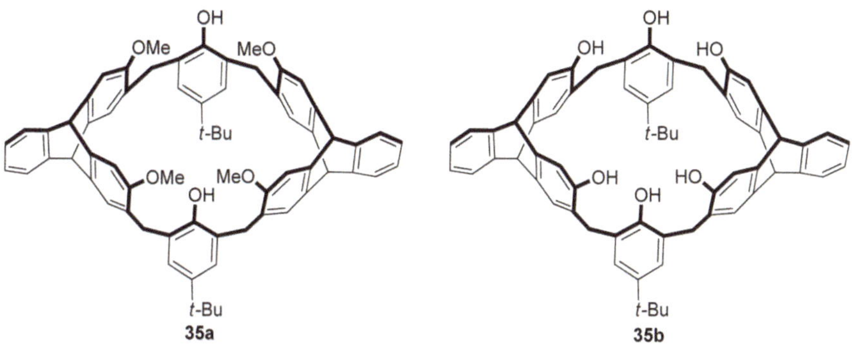

Fig. 10.21 Structures of triptycene-derived calix[6]resorcinarene-like hosts **35a** and **b**

also found that the self-assembly could retain its tubular structure and further three-dimensional honeycombed architecture upon the exchange of THF-methanol guests with acetone-methanol, which suggested that the organic nanotube assembled by **34a** through four one-dimensional H-bond chains is very stable, and the role of solvents in the formation of this nanotubular structure might be less efficient.

More recently, Li and Chen [18] reported a class of novel triptycene-derived calix[6]resorcinarene-like hosts **35** (Fig. 10.21), which contained two triptycene moieties and two *p*-substituted phenol moieties. Based on the ^{1}H NMR spectra and crystal structures, it was found that both of the macrocycle **35a** and the corresponding demethylated macrocycle **35b** were the *cis* isomers in the fixed cone conformation in solution and in the solid state. Moreover, it was found that compound **35a** could form a head-to-head dimeric-capsule with two CH_2Cl_2 molecules inside the dimeric cavity, suggesting that the C–H$\cdots$O hydrogen-bonding interactions between the macrocycles, and the C–H$\cdots$Cl interactions between the macrocycles and solvent molecules played important roles in the formation of this dimeric structure. Furthermore, the dimers could stack in pillars to form a tubular structure with aromatic

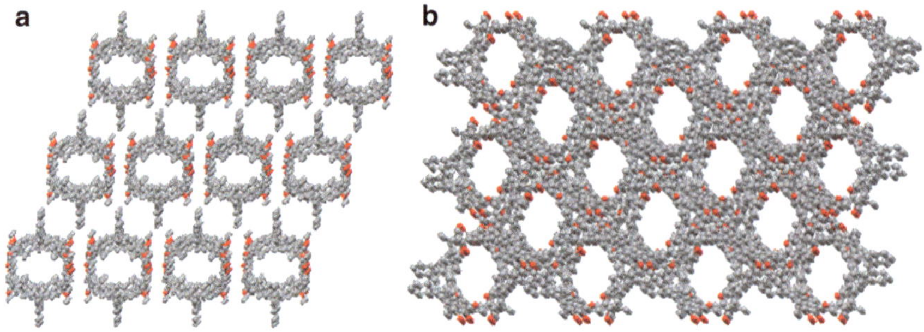

Fig. 10.22 Self-assembled microporous architectures of **a 35a** viewed along the *c*-axis, and **b 35b** viewed along the *b*-axis

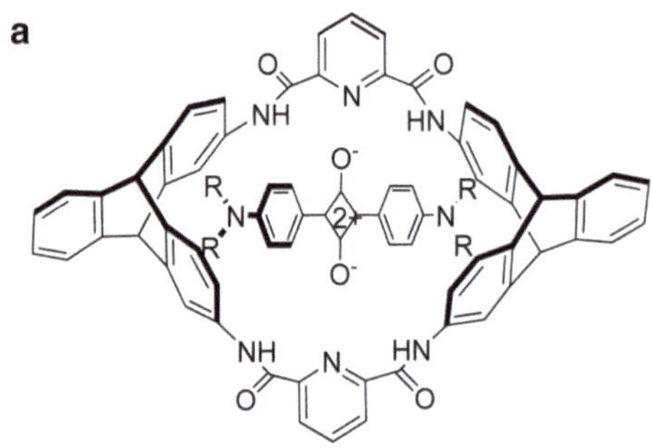

36 R = $CH_2C_6H_5$

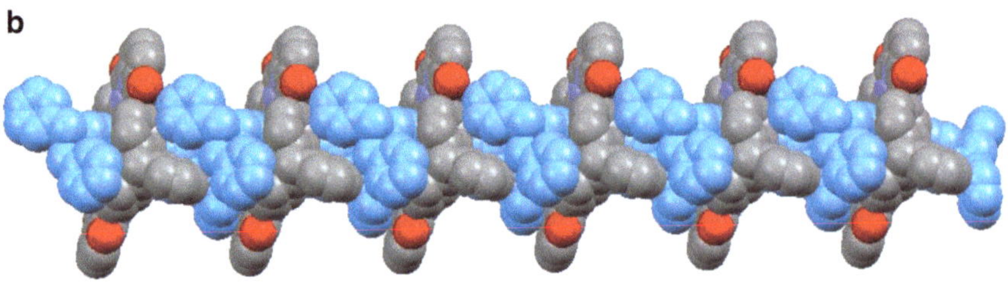

Fig. 10.23 **a** Chemical structure of [2]rotaxane **36**. **b** The oriented nonsymmetrical channel-like structure of **36**

rings as the wall, and then assemble into three-dimensional microporous architecture with the CH_2Cl_2 molecules included in each channel. Similarly, the demethylated molecule **35b** could also self-assemble into the head-to-head dimer, and further three-dimensional microporous architectures by the multiple noncovalent interactions between the macrocycle and its adjacent molecules and solvents, along with solvent molecules were situated in the channels (Fig. 10.22b).

In addition, Chen's group [19] also found that the [2]rotaxane molecule **36** (Fig. 10.23a) based on triptycene-derived tetralactam macrocycle could self-assemble into a tubular structure in the solid state through the multiple noncovalent interactions, including the typical hydrogen bonds between the amide protons of the hosts and the carbonyl oxygen atoms of the guests, and the multiple π–π stacking interaction and C–H$\cdots\pi$ interactions between triptycene subunits and the aromatic rings of the guests. It was noteworthy that the squaraine dyes were positioned inside of the channels. Since the honeycombed suprastructures could not be seen in the free macrocycles, the noncovalent interactions between the host and the guest, as well as the solvent interactions played important roles in the arrangement of extended channels. Moreover, it was interestingly found that the [2]rotaxane **36** with nonsymmetrical macrocycles could self-assemble into an oriented nonsymmetrical channel-like structure (Fig. 10.23b).

In 2007, Zhang and Chen [20] also reported a molecular cage **37**, and found that it could self-assemble into an interlaced one-dimensional structure by the C–H$\cdots\pi$ interactions between the homochiral cage molecules, and the similar C–H$\cdots\pi$ interactions between the heterochiral molecules. Then, the adjacent one-dimensional supramolecular assemblies could be connected with each other by the C–H$\cdots\pi$ interactions between the adjacent molecules to form a two-dimensional-layer structure and a further microporous architecture (Fig. 10.24). Moreover, it was further found that the 1,3,5-trimethyl benzene molecules were located in the channels, and there existed multiple C–H$\cdots\pi$ interactions between the trimethylbenzene molecules and the molecular cage.

10.2 Self-Assembly on Surface

It is known that self-assembled monolayers (SAMs) with functionality molecules have wide potential applications. In 2000, Yang et al. [21] designed and synthesized the pentiptycene-incorporated α, ω-alkanedithiol **39** from the precursor of pentiptycene hydroquinone **38**. As shown in Scheme 10.3, the reaction of compound **38** with 10-undecenyl tosylate, and then followed by the reaction with ethanethioic *S*-acid in THF/MeOH in the presence of AIBN gave the dithioester. Finally, the acid-catalyzed hydrolysis of the dithioester afforded the target **39** in an excellent yield. According to the studies of the α, ω-alkanedithiol **37** system on the metal surface, it was found that compound **37** could array into a looped conformation with two thiol chains bound to the Pt^{III} or Au^{III} surface, which was totally different from the rod-shaped α, ω-dithiol system with a stand-up conformation (Fig. 10.25). This folded conformation could be revealed by the results of MM2 modeling, the scanning tunneling microscopy (STM), and the surface plasmon resonance (SPR). The authors further considered that the bulky pentiptycene groups blocked the close contact of the aliphatic chains, and the folding conformation seemed to be stabilized by the intramolecular chain–chain (dispersion) interactions, which resulted in a looped structure with 5 Å chain-to-chain separation.

In 2003, Wolpaw et al. [22] designed and synthesized 1,5-bis-hexadecyltriptycene and triptycene dithiols containing the pendant groups, which could provide the "surface adsorbing element" for triptycene to orient adsorption on the graphite or gold

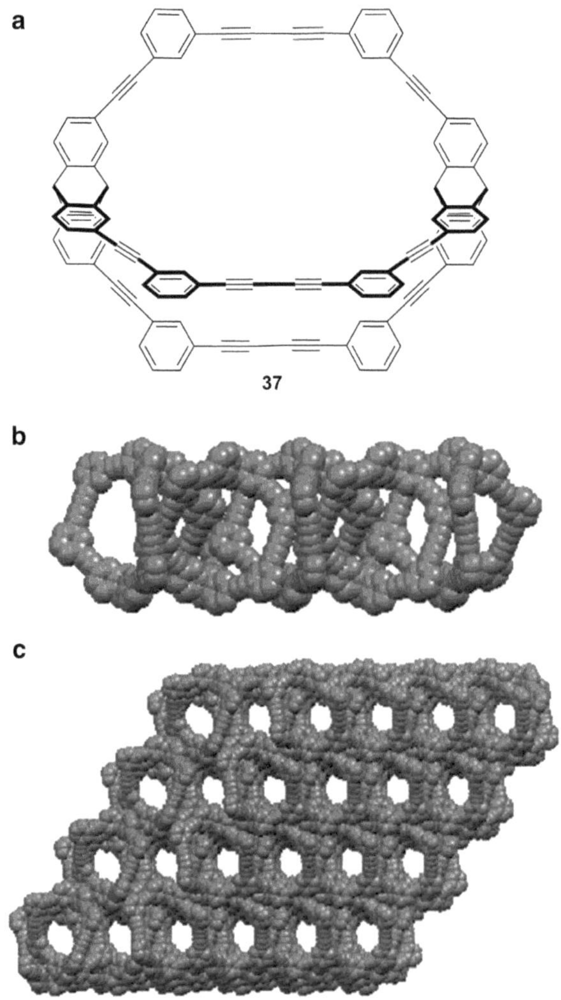

Fig. 10.24 **a** Chemical structure of molecular cage **37**. **b** View of an interlaced one-dimensional supramolecular structure along the *b*-axis. **c** A self-assembled microporous architecture of cage **37**

surface. With the rough STM studies, they found that there were a 2 nm spatial repeat and the triangular ridges. However, there were no intermolecular chain–chain van der Waals interactions between the isolated molecules on highly oriented pyrolytic graphite (HOPG), the weak interactions between the alkyl chains and HOPG seemed to limit the adsorption to form a stable orientation. In 2006, Fernandez-Torrente et al. [23] also reported a particularly stable structure formed from the three-dimensional triptycene clusters. Since the noncovalent interactions among triptycene molecules were much stronger than those ones between triptycene and the Au^{III} surface, the triptycene molecules preferred to spontaneously form the stable three-dimensional supramolecular structures. Then the three-dimensional supramolecule could be used as basic building blocks for the further molecular thin-film growth. According to the

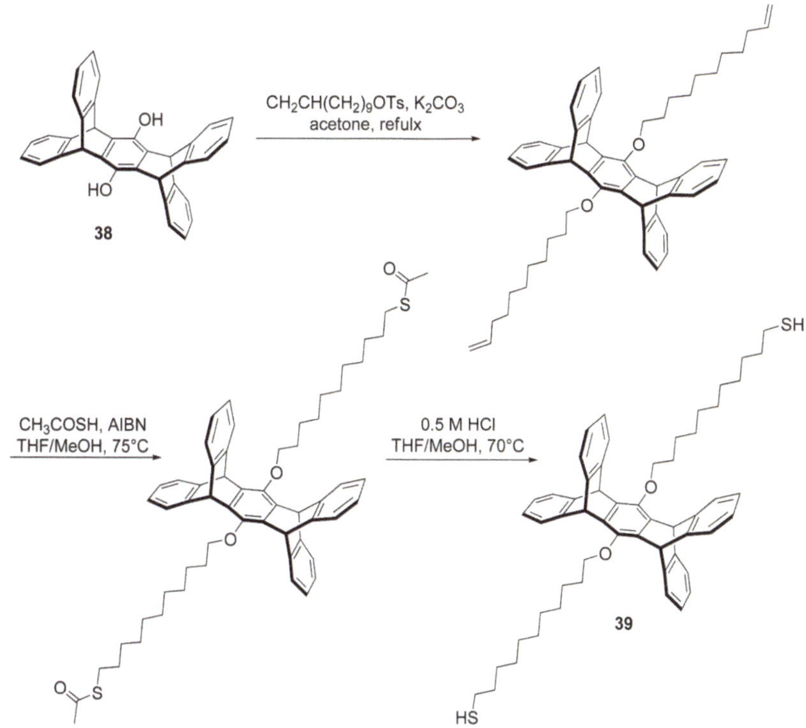

Scheme 10.3 Synthesis of pentiptycene-incorporated α, ω-alkanedithiol **39**

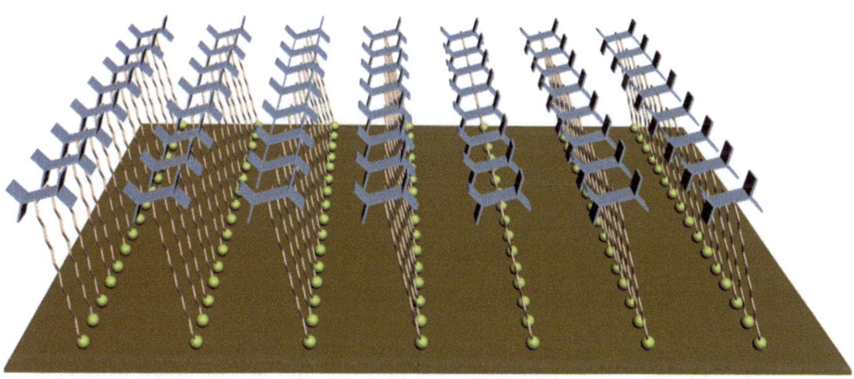

Fig. 10.25 Schematic representations of **39** in folded conformation

high-resolution STM images and computer modeling, it was found that there existed the C–H$\cdots\pi$ intermolecular forces instead of the similar face-to-edge packing as shown in bulk, in addition of the three-dimensional structure of triptycene resulted in the stable three-dimensional structure at the gold surface (Fig. 10.26).

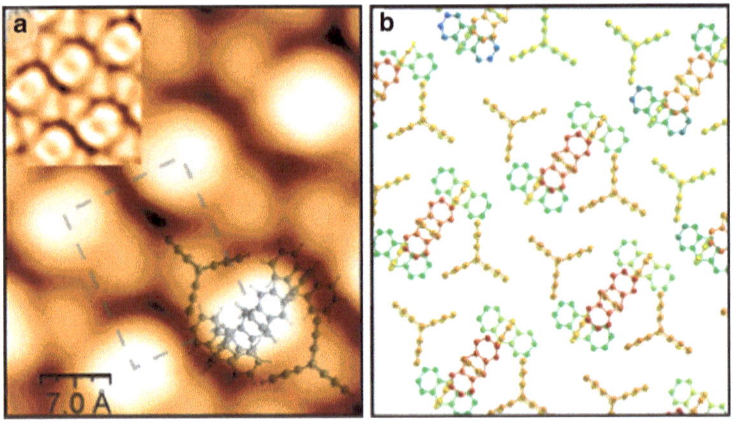

Fig. 10.26 **a** STM image of an ordered island with a rectangular unit cell. **b** View of the proposed cluster structure based on the STM image and on force-field molecular dynamics (MM3) calculations. (Reprinted with permission from [23]. Copyright 2006 American Chemical Society)

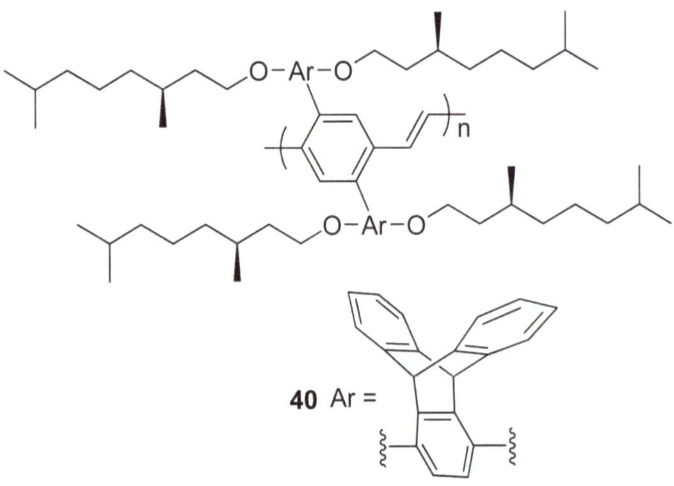

Fig. 10.27 Structure of chiral poly-(*p*-phenylenevinylene) **40**

In 2005, Satrijo and Swager [24] reported that the use of bulky triptycene as interlocking side groups incorporated into the chiral π-conjugated polymer, like chiral poly-(*p*-phenylenevinylene) **40** (Fig. 10.27), could obtain the well-ordered, chiral architectures in the thin-films via spin-coating process. It was reasoned that the iptycene moieties self-assembled into the interlocking structures, which were good for the formation of an ordered, chiral assembly by the spin-cast from the nonpolar chloroform solutions.

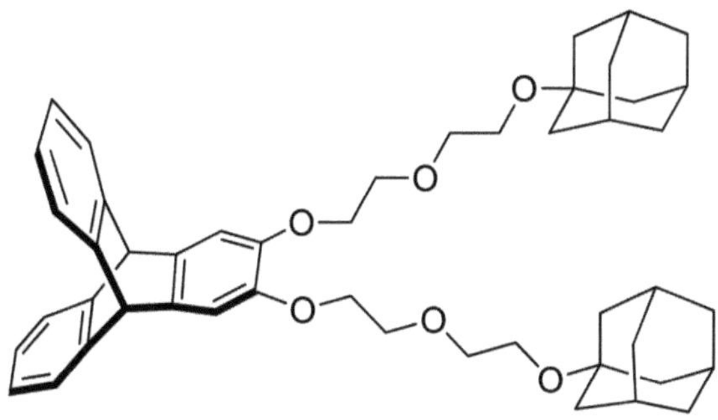

Fig. 10.28 Structure of shape-fitted triptycene moiety

In 2009, Zu et al. [25] reported a new approach to increase the water-solubility of SWCNTs via supramolecular surface modification. This method utilized the shape-fitted triptycene moiety, which linked with two adamantane: the triptycene moiety as the binding group attached to the sidewalls of SWCNTs, and the adamantine moieties in this molecule could be included in water-soluble cyclodextrin. Due to the formation of this supramolecular complex, the SWCNTs with supramolecular surface modification could dissolve in the aqueous phase, which could be characterized by absorbance, static and time-resolved fluorescence spectroscopy, Raman spectroscopy, and TEM. Moreover, it was also found that triptycene moiety seemed to favor the 1.0 nm cylindrical SWCNTs (Fig. 10.28).

10.3 Self-Assembly in Solution

In 2008, Han and Chen [26] found that macrotricyclic host **41** could form the pseudorotaxane type cascade complexes with both anthraquinone and its tetra-azide terminal functionalized derivative in the presence of potassium ions. On the basis of the previous work, the authors further synthesized a series of novel potassium ion templated [2]rotaxanes **42a–c** in the middle or good yields by the "slippage" and/or "threading followed by stoppering" approaches (Scheme 10.4). Moreover, the authors also studied the corresponding deslipping behaviors of the [2]rotaxanes with different triazole stoppers by peeling off the potassium ions with 18-crown-6, since the K^+ acted as both the templates during the stoppering reactions and the non-slipping chocks to shrink the inner diameter of the wheel cavity. As a result, it was found that both rotaxanes **42a** and **42b** could be destructed easily through deslippage of the wheel over one stopper by peeling off the alkali-metal ions with 18-crown-6; however, under the same conditions the dumbbell and ring components of rotaxane **42c** still remain interlocked (Scheme 10.5).

The high-order [n]rotaxanes with aesthetic structures have attracted much attention during the last decade for their potential applications in nanotechnology and molecular machines. In 2010, Jiang et al. [27] designed and synthesized a bifunctionalized [3]rotaxane based on the triptycene-derived macrotricyclic host **41** and the

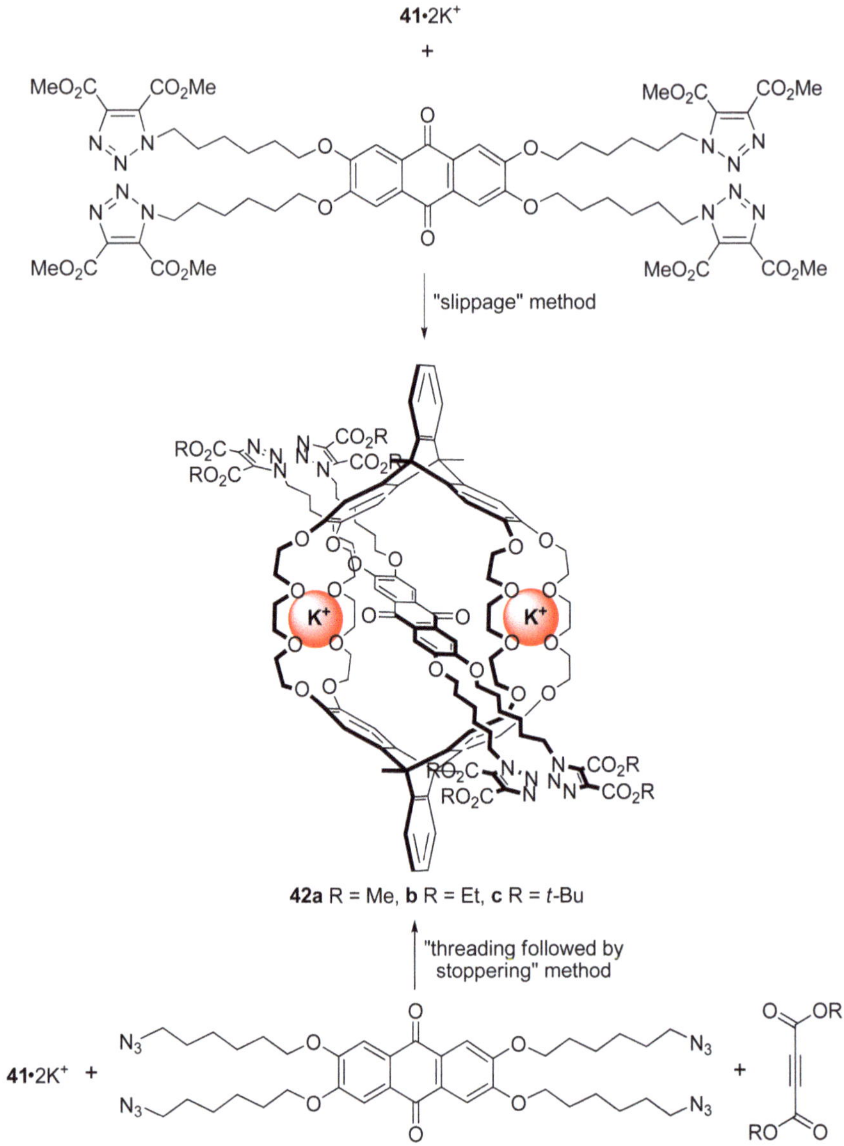

Scheme 10.4 Synthesis of K^+ templated [2]rotaxanes **42a–c** by the "slippage" and/or "threading followed by stoppering" approaches

functionalized dibenzylammonium salt. With the [3]rotaxane **43a** in hand, the authors further obtained the mechanically interlocked linear main-chain poly[3]rotaxane **43b** in high yield by the highly efficient Huisgen 1,3-dipolar cycloaddition (Scheme 10.6). Furthermore, Jiang et al. [28] reported a novel acid-base controllable [3]rotaxane

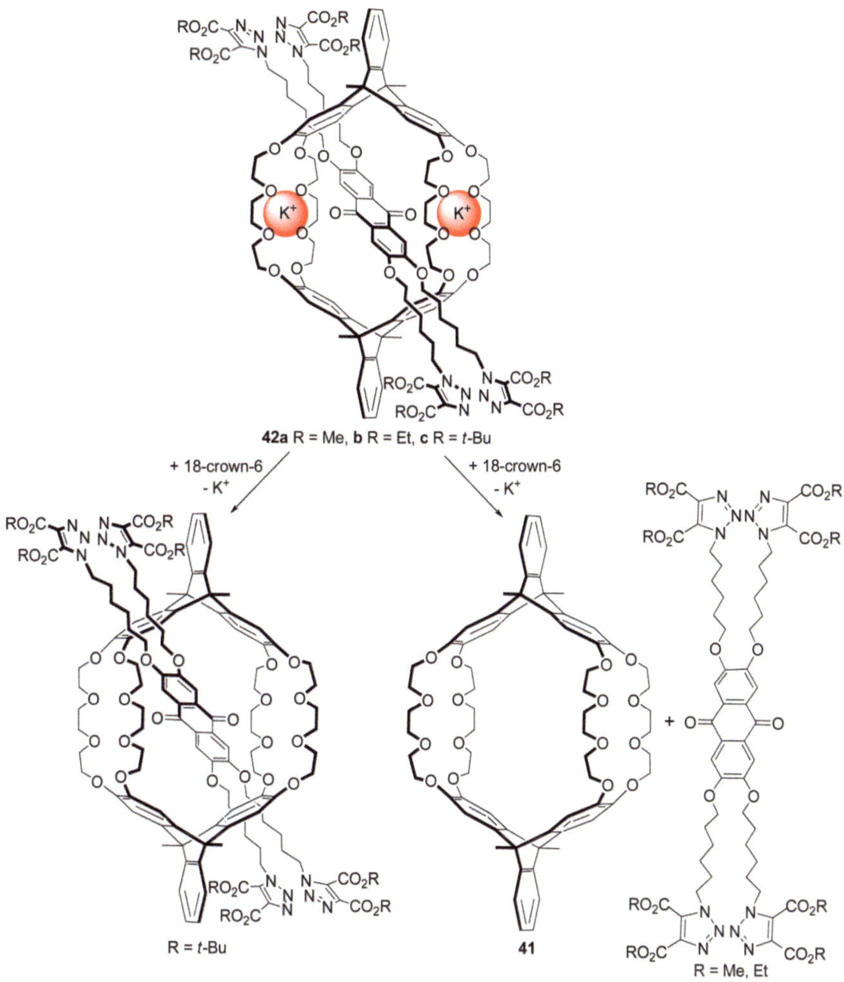

Scheme 10.5 Illustration of the controlled dethreading and threading of the rotaxanes **42a–c**

molecular machine based on the macrotricyclic host **41** by the Huisgen 1,3-dipolar cycloaddition, and the subsequent alkylation of the 1,2,3-triazole group. Consequently, the target [3]rotaxane **44** (Fig. 10.29) had a dibenzylammonium ion and a *N*-methyltriazolium station with different affinity for the DB24C8 subunits of host **41**, and the shuttle process between the two different stations could be efficiently and reversibly controlled by the use of 1,8-diazabicyclo[5.4.0]undec-7-ene (DBU) and trifluoroacetic acid (TFA). This represented the first example of [3]rotaxane molecular machines with two threads passing through one host.

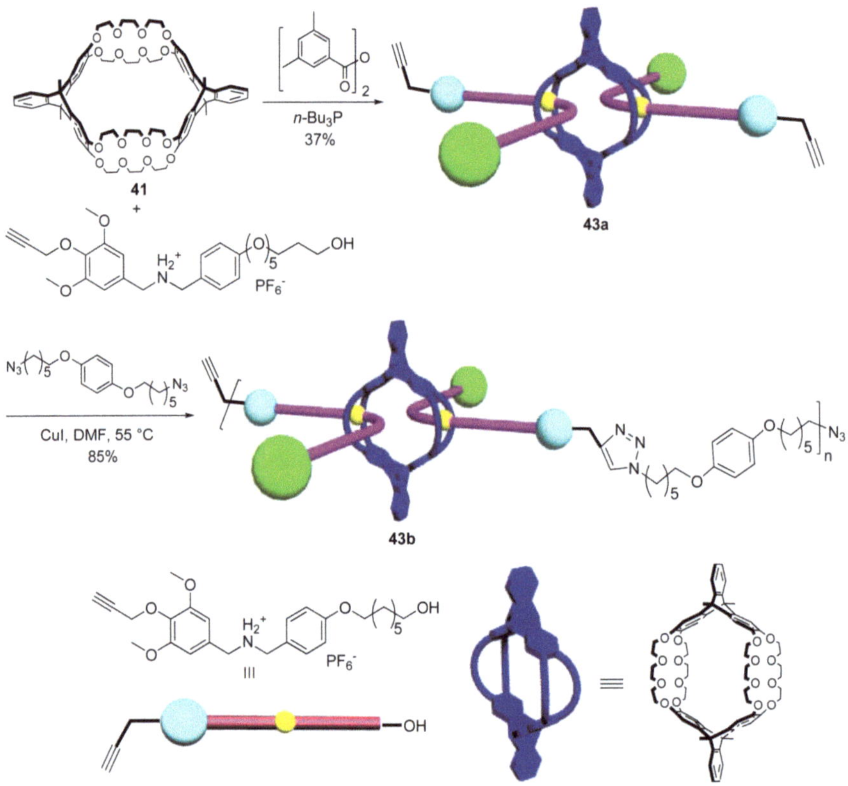

Scheme 10.6 Synthesis of mechanically interlocked linear main-chain poly[3]rotaxane

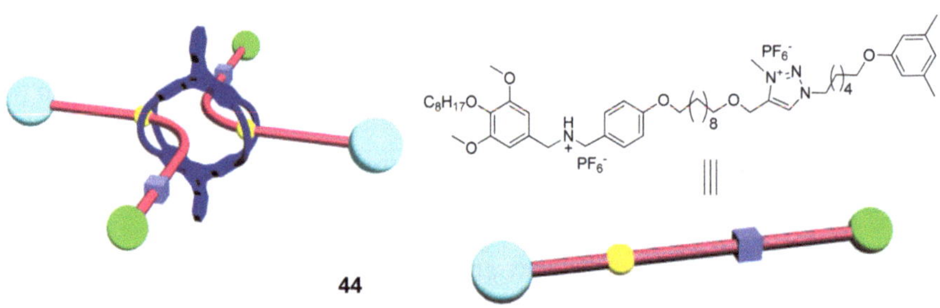

Fig. 10.29 Schematic drawing of [3]rotaxane **44**

At the same year, Chen's group [29] also designed and synthesized a novel triptycene-derived bis-macrotricyclic host **45** (Fig. 10.30) containing two symmetrical macrotricyclic moieties, and found that the host **45** could expectedly form a stable 1:4 complex with 4 equivalents of the dibenzylammonium salts. The single crystal structure of the complex showed that two dibenzylammonium ions were

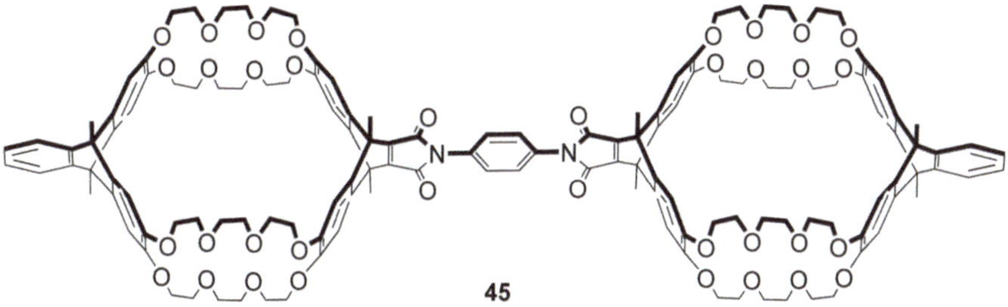

45

Fig. 10.30 Structure of triptycene-derived bis-macrotricyclic host **45**

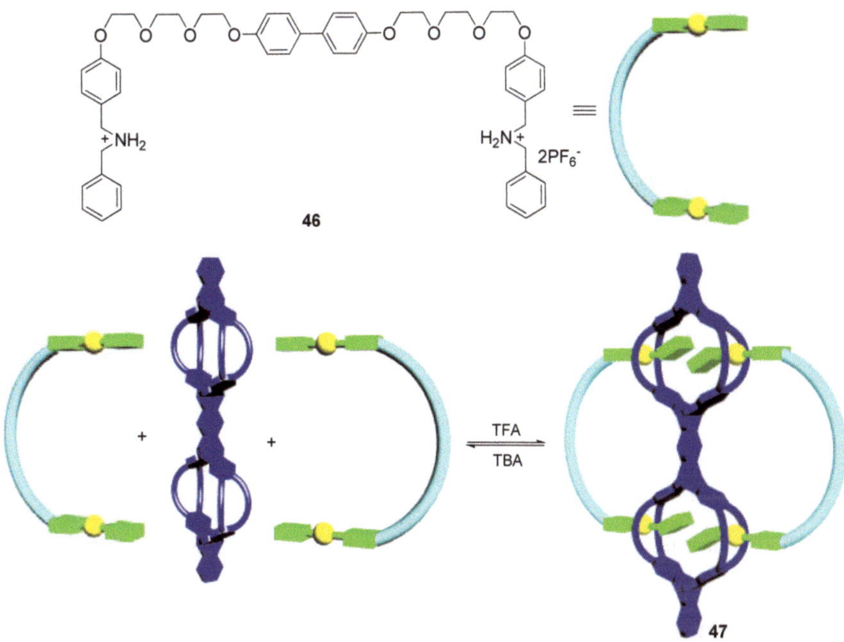

Fig. 10.31 Structure of bis-secondary ammonium salt **46**, and schematic representation of the acid-base switchable molecular handcuff **47**

threaded through the centers of the DB24C8 cavities of each macrotricycle in **45**. Inspired by this result, the authors further synthesized a bis-dibenzylammonium salt **46**, and found that host **45** with multicavities could self-assemble with two equivalents of bis-secondary ammonium salt **46** to form a handcuff-like complex **47**. Interestingly, the complexation and decomplexation of the handcuff-like assembly could be chemically controlled by the stimuli of acid and base (Fig. 10.31).

In 2010, Guo et al. [30] attempted to prepare the interwoven cages based on the triptycene-derived bis-macrotricyclic host **45** and the branched paraquat derivatives via self-assembly in solution. Consequently, they found that the host **45** could assemble with the branched paraquat derivatives **48** and **49** to form the novel supramolecular

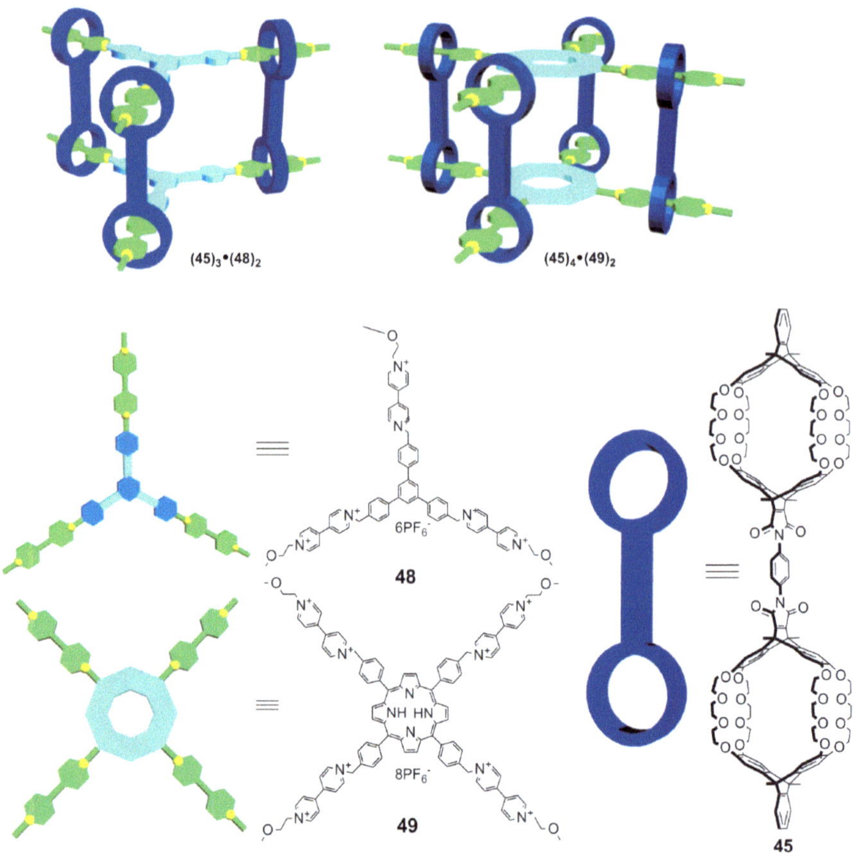

Fig. 10.32 Schematic representations of novel supramolecular cages **45$_3$•48$_2$** and **45$_4$•49$_2$**

cages **45$_3$•48$_2$** and **45$_4$•49$_2$** by the [3 + 2] and [4 + 2] self-assembly, respectively (Fig. 10.32). The formation of these supramolecular cages could be confirmed by the results of ^{1}H NMR and two-dimensional NMR spectroscopy, and electrospray ionization mass spectra (ESI-MS). Moreover, the hydrodynamic radii (or Stokes radii) of the two supramolecular cages and the uniparted building block were also in agreement with the calculated results. It was believed that these current studies could offer opportunities for the further design and construction of novel self-assembled organic containers and molecular machines for potential applications in nano-materials and supramolecular chemistry.

Inspired by the formation of a [2]pseudorotaxane via a dibenzylammonium ion threading the cavity of the dibenzo-24-crown-8 [31], Zhu and Chen [32] designed and synthesized a novel triptycene-derived tris(crown ether) host **50** (Fig. 10.33) by the reaction between triptycene tri(catechol) (**51**) and the bistosylate in *N,N*-dimethylformamide (DMF) in the presence of Cs_2CO_3 under a high-dilution condition. As expectedly, this D_{3h} symmetrical homotritopic host **50** could form

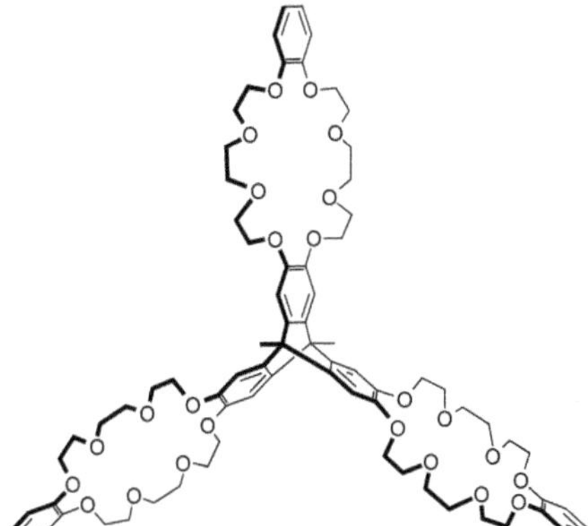

50

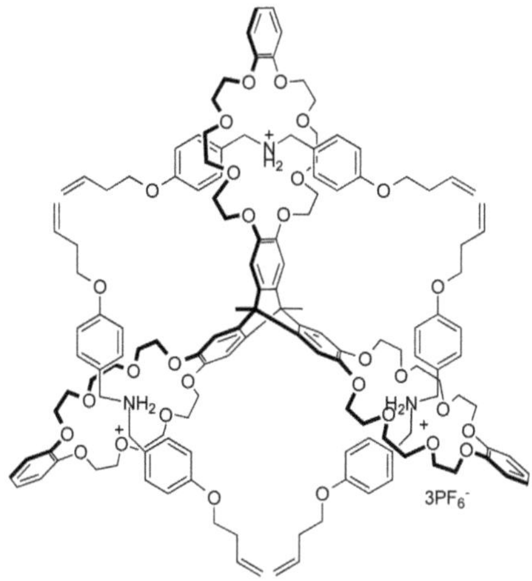

52

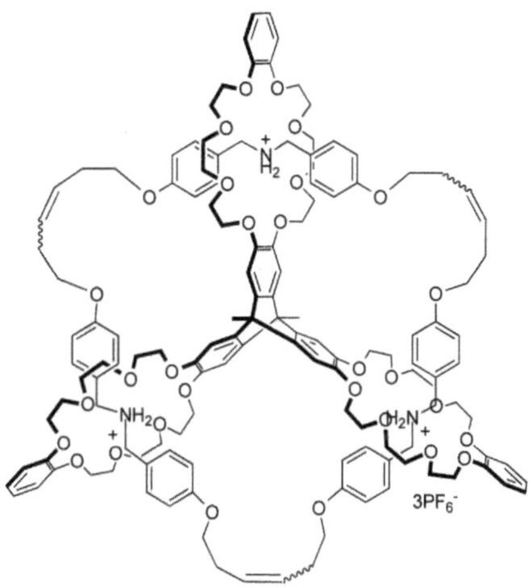

53

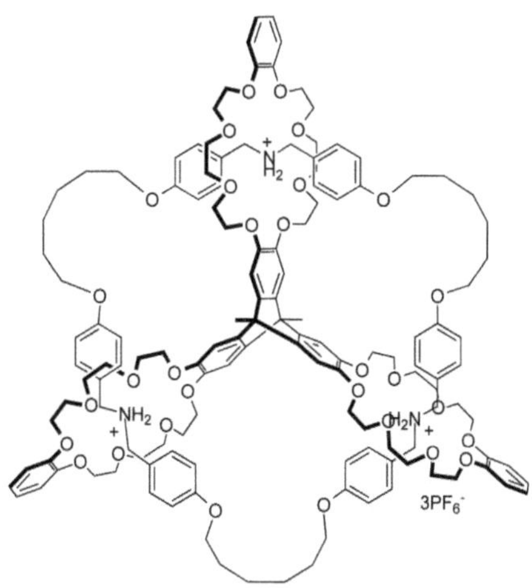

54

Fig. 10.33 Structures of triptycene-derived tris(crown ether) host **50**, and the assemblies **52–54**

a 1:3 stable tris[2]pseudo-rotaxane-type complex **52** with 3 equivalents of the secondary ammonium salts, which was revealed by the ^{1}H NMR spectroscopy and ESI mass spectrometry. Due to the ammonium ions containing two terminal vinyl groups, the authors further performed the olefin metathesis reaction by the second generation Grubbs' catalyst. The reaction was found to proceed smoothly, and exclusively give the triple metathesis product **53** in 82 % yield. Furthermore, hydrogenation of the metathesis product with Adam's catalyst afforded a novel interlocked molecule **54**, which was called as [2](3)catenane or [4]pseudocatenane, in quantitative yield. Structure of assembly **54** was further confirmed by its X-ray crystal structure.

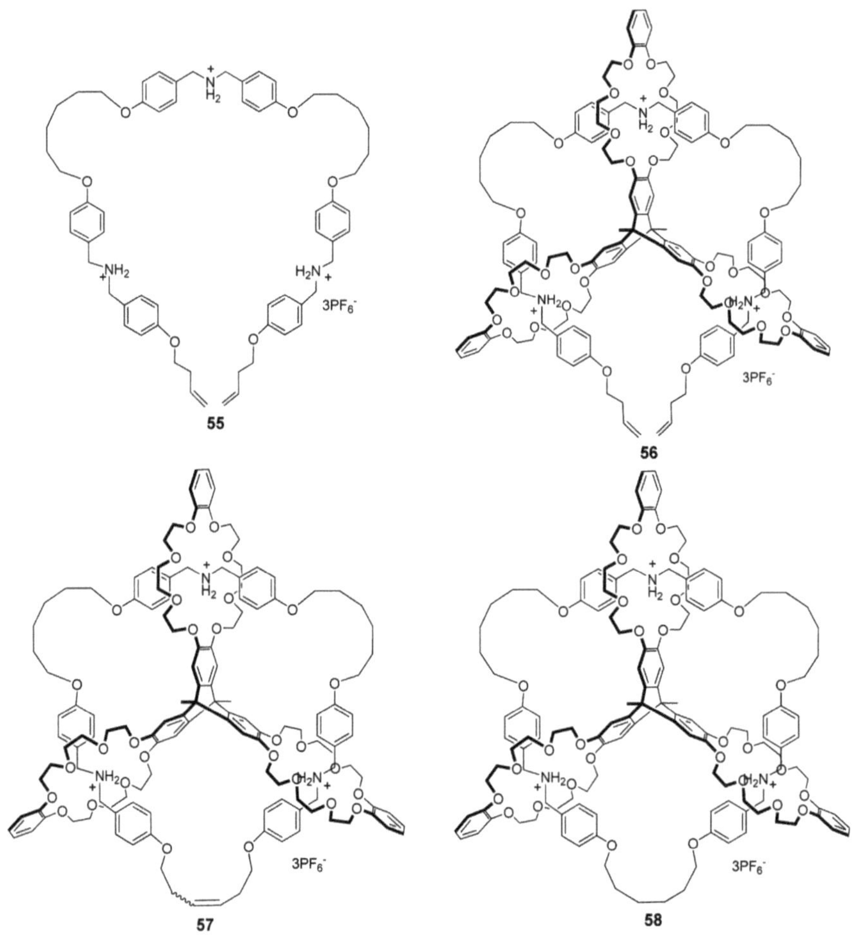

Fig. 10.34 Structures of tri-dibenzylammonium salt **55** and the assemblies **56–58**

Recently, Chen's group [33] also found that the macrocycle **50** and the tri-dibenzyl-ammonium salt **55** could form a 1:1 multivalency-directed complex **56** (or **50•55**), which then afforded the [2](3)catenane **57** through an olefin metathesis reaction in high yield. Likewise, the hydrogenation of **57** with the Adam's catalyst afforded [2](3)catenane **58** in quantitative yield (Fig. 10.34). Furthermore, the authors designed and synthesized a trisdialkyl-ammonium macrocycle ion **59** with one terminal vinyl group and found that an interesting magic-ring [2](3)catenane could be obtained by the reversible olefin metathesis of the macrocycle ion **59** (Scheme 10.7). This result suggested that the combination of multivalence and olefin metathesis could provide a powerful tool for the synthesis of complex-interlocked molecules.

Followed the similar synthetic strategy of host **50**, Zhu and Chen [34] also synthesized a chiral triptycene-derived tri(crown ether) host **60** (Fig. 10.35). And then,

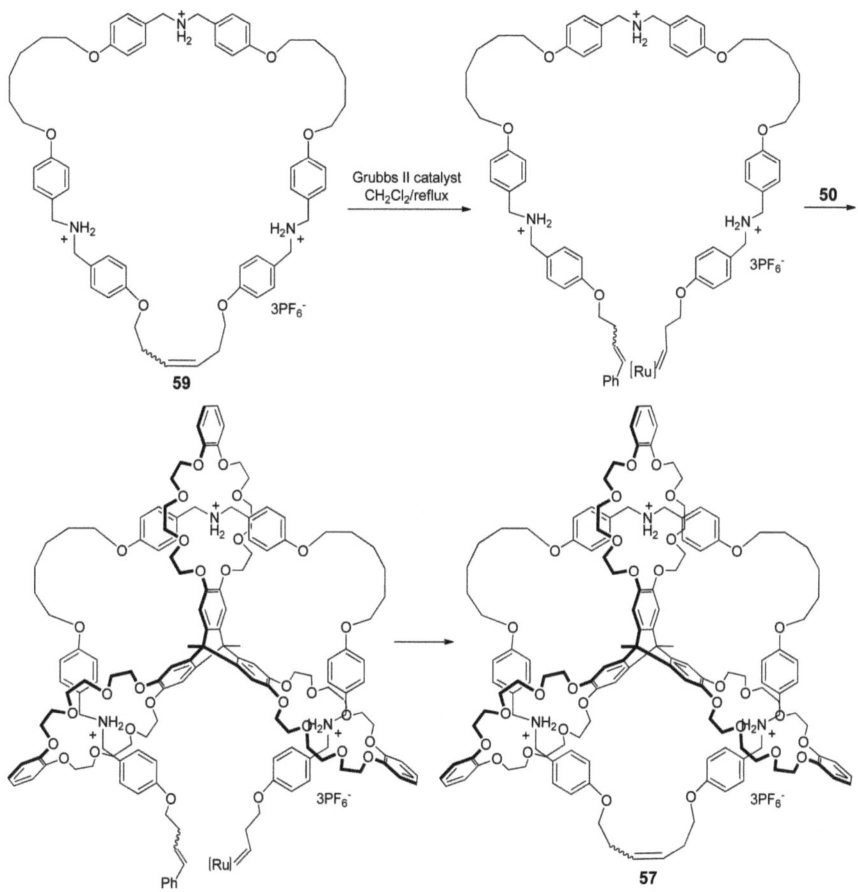

Scheme 10.7 Synthesis of magic-ring [2](3)catenane **57**

the authors obtained a chiral [2](3)catenane **61** based on the chiral host **60**. Although it was difficult to neutralize the ammonium groups in **60** with triethyl-amine, tributylamine, or diisopropylethylamine, the authors found that it could be deprotonated to the neutral interlocked molecule by DBU in DMSO, which was an important precursor for the further functionalization to a new class of neutral chiral-interlocked molecules. As a result, the authors further synthesized a series of new [2](3)catenane derivatives **62–65** by the sequence of reactions as shown in Scheme 10.8. The CD spectral investigations for host **60** and complexes **61–65** showed that the Cotton effect of the (*R*)-1,1′-binaphthyl chromophore at 241 nm was greatly reduced in these complexes through a comparison with host **60**, whereas a new positive Cotton effect at 248 nm was found in these [4]pseudocatenanes, which was probably attributed to the chirality transfer from the binaphthyl units to the macrocycle lying in the cavities of host **60**.

Fig. 10.35 Structure of chiral triptycene-derived tri(crown ether) **60**

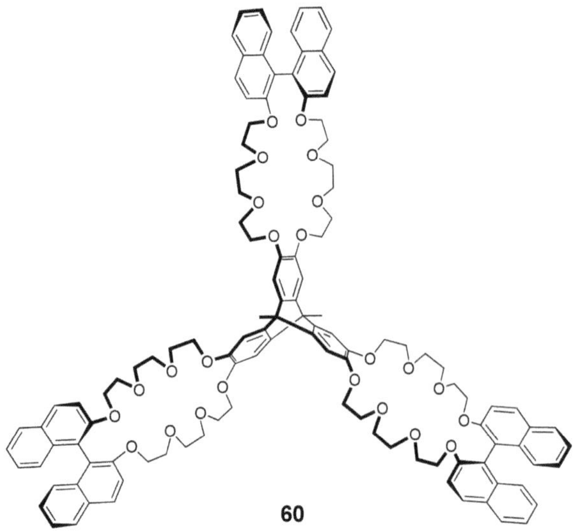

60

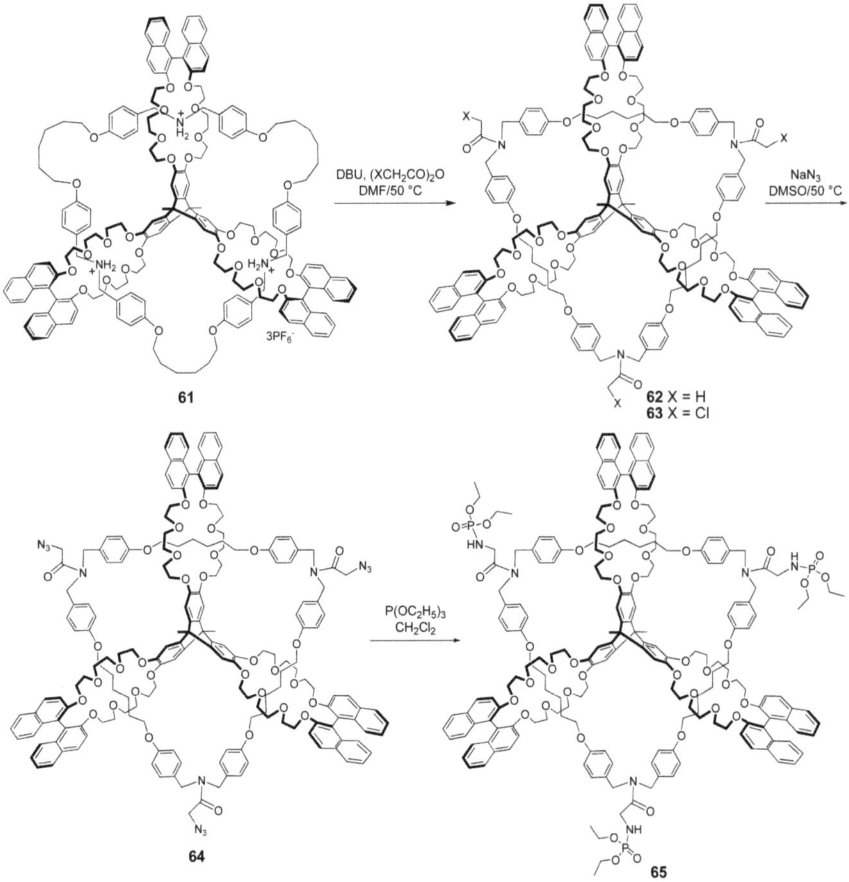

Scheme 10.8 Synthesis of [2](3)catenane derivatives **62–65**

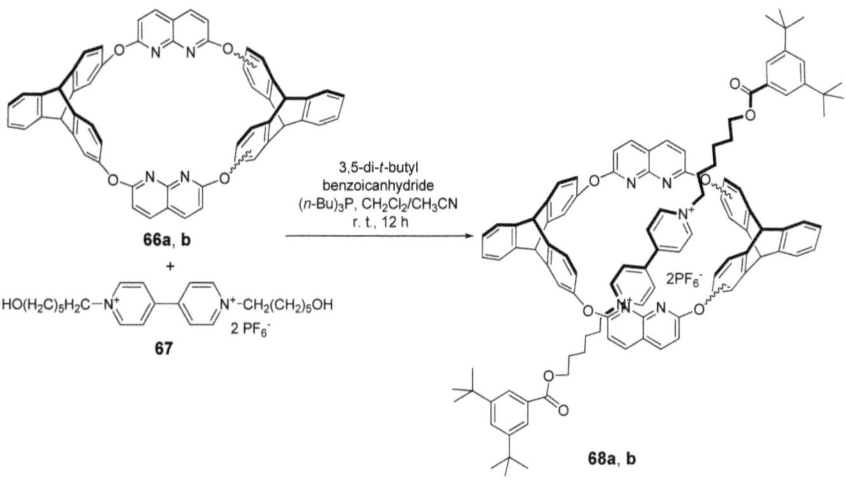

Scheme 10.9 Synthesis of mechanically interlocked molecules **68a**, **b**

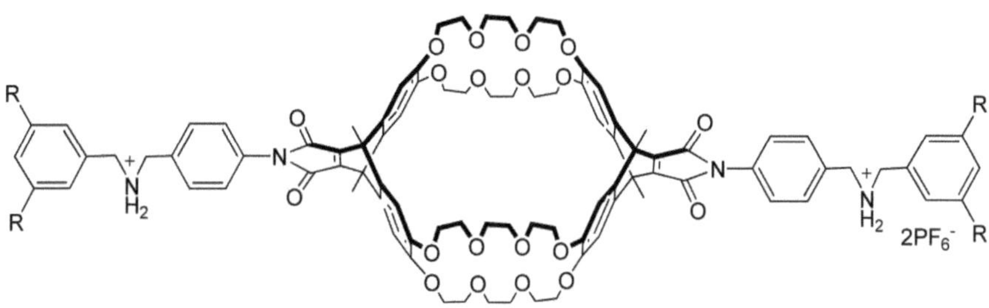

69 R = H
70 R = OCH_3

Fig. 10.36 Structures of self-complementary monomer **69** and model molecule **70**

In 2011, Hu and Chen [35] reported a couple of novel oxacalixarene-based isomeric [2]rotaxanes **68a** and **68b**, which were synthesized in 56 and 25 % yield, respectively, by the reaction of the mixture of macrocycles **66a**, **b** and the paraquat derivative **67** with 3,5-di-*t*-butylbenzoic anhydride in the presence of a catalytic amount of tri(*n*-butyl) phosphane. This was the first example of oxacalixarenes used as useful wheels for the synthesis of a new family of mechanically interlocked molecules (Scheme 10.9).

More recently, Su et al. [36] also designed and synthesized a novel kind of self-complementary monomer **69** (Fig. 10.36), which combined both a macrotricyclic polyether and two dibenzylammonium ions. It was found that this monomer **69** could self-assemble into a supramolecular polymer network by the host–guest interactions, which was revealed by the ^{1}H NMR spectrum, the MALDI-TOF MS spectrum, viscometry, and dynamic light-scattering (DLS) experiments. As a comparative study,

Fig. 10.37 Supramolecular gel formed from **69 a** in CH_3CN and its thermo-induced gel–sol transitions and **b** in $CDCl_3/CD_3CN$ (1:1, v/v) and its pH-induced gel–sol transitions. (Reproduced from [36], with the permission of John Wiley and Sons)

model molecule **70** was also prepared. As expectedly, molecule **70** could not form the self-assembly similar to that of **69,** because the sterically hindered end groups in the "guest moieties" of **70** would prevent the complexation. Moreover, it was also found that the decomposition and reformation of the supramolecular polymer of **69** could be chemically controlled by the use of triethylamine and trifluoroacetic acid. Interestingly, it was further found that the supramolecular polymer of monomer **69** could form a responsive organogel both in acetonitrile or acetonitrile/chloroform solution, which could achieve the reversible gel–sol transitions under the thermo- and pH-induced stimuli (Fig. 10.37).

References

1. Bashir-Hashemi A, Hart H, Ward DL (1986) Tritriptycene: a D_{3h} C_{62} hydrocarbon with three U-shaped cavities. J Am Chem Soc 108(21):6675–6679
2. Venugopalan P, Burgi HB, Frank NL, Baldridge KK, Siegel JS (1995) The crystal-structure of a heptiptycene-chlorobenzene clathrate. Tetrahedron Lett 36(14):2419–2422
3. Lu J, Zhang JJ, Shen XF, Ho DM, Pascal RA (2002) Octaphenylbiphenylene and dodecaphenyltriptycene. J Am Chem Soc 124(27):8035–8041
4. Hashimoto M, Takagi H, Yamamura K (1999) Three dimensional supramolecules of triptycenequinone and its 6,7-dimethyl derivative formed by weak intermolecular π–π interactions and C–H$\cdots$O hydrogen bonds. Tetrahedron Lett 40(33):6037–6040
5. Hashimoto M, Yamamura K, Yamane J (2001) Characteristic ribbon-like supramolecular structure and intermolecular D–A interactions in the crystals of triptycenequinones, triptycene-TCNQs and their clathrates as studied by X-ray investigations. Tetrahedron 57(52):10253–10258
6. Yamamura K, Kawashima T, Eda K, Tajima F, Hashimoto M (2005) Solid solution of triptycenequinone and triptycenehydroquinone as a non-stoichiometric quinhydrone. Bathochromic changes in color caused by local intermolecular interaction between *p*-benzoquinone and hydroquinone moieties. J Mol Struct 737(1):1–6

7. Veen EM, Postma PM, Jonkman HT, Spek AL, Feringa BL (1999) Solid state organisation of C60 by inclusion crystallisation with triptycenes. Chem Commun 17:1709–1710
8. Chong JH, MacLachlan MJ (2007) Synthesis and structural investigation of new triptycene-based ligands: en route to shape-persistent dendrimers and macrocycles with large free volume. J Org Chem 72(23):8683–8690
9. Yang JS, Liu CP, Lee GH (2000) Anomalous crystal packing of iptycene secondary diamides leading to novel chain and channel networks. Tetrahedron Lett 41(41):7911–7915
10. Yang JS, Liu CP, Lin BC, Tu CW, Lee GH (2002) Solid-state molecular folding and supramolecular structures of triptycene-derived secondary dicarboxamides. J Org Chem 67(21):7343–7354
11. Zhang C, Chen CF (2010) Synthesis and analysis of hydroxyl substituted triptycene adducts: the competitive recognition between the hydroxyl substituted triptycenes with 4,4′-bipyridine and solvent molecules. Crystengcomm 12(10):3255–3261
12. Zhang C, Chen CF (2007) Triptycene-based expanded oxacalixarenes: synthesis, structure, and tubular assemblies in the solid state. J Org Chem 72(10):3880–3888
13. Zong QS, Zhang C, Chen CF (2006) Self-assembly of triptycene-based cylindrical macrotricyclic host with dibenzylammonium ions: construction of dendritic [3] pseudorotaxanes. Org Lett 8(9):1859–1862
14. Guo JB, Han Y, Cao J, Chen CF (2011) Formation of 1:2 host–guest complexes based on triptycene-derived macrotricycle and paraquat derivatives: anion–π interactions between PF_6^- and bipyridinium rings in the solid state. Org Lett 13(20):5688–5691
15. Tian XH, Hao X, Liang TL, Chen CF (2009) Triptycene-derived calix [6] arenes: synthesis, structure and tubular assemblies in the solid state. Chem Commun 2009(44):6771–6773
16. Tian XH, Chen CF (2010) Triptycene-derived calix [6] arenes: synthesis, structures, and their complexation with fullerenes C_{60} and C_{70}. Chem Eur J 16(27):8072–8079
17. Xue M, Chen CF (2011) Aromatic single-walled organic nanotubes self-assembled from *NH*-bridged azacalix [2] triptycene[2]pyridine. Chem Commun 47(8):2318–2320
18. Li PF, Chen CF (2011) Triptycene-derived calix [6] resorcinarene-like hosts: synthesis, structure and self-assemblies in the solid state. Chem Commun 47(44):12170–12172
19. Xue M, Su YS, Chen CF (2010) Isomeric squaraine-based [2] pseudo-rotaxanes and [2]rotaxanes: synthesis, optical properties, and their tubular structures in the solid state. Chem Eur J 16(28):8537–8544
20. Zhang C, Chen CF (2007) Synthesis and structure of a triptycene-based nanosized molecular cage. J Org Chem 72(24):9339–9341
21. Yang JS, Lee CC, Yau SL, Chang CC, Leu JM (2000) Conformation and monolayer assembly structure of a pentiptycene-derived α,ω-alkanedithiol. J Org Chem 65(3):871–877
22. Wolpaw AJ, Aizer AA, Zimmt MB (2003) Synthesis of self-orienting triptycene adsorbates for STM investigations. Tetrahedron Lett 44(41):7613–7615
23. Fernandez-Torrente I, Franke KJ, Henningsen N, Schulze G, Alemani M, Roth C, Rurali R, Lorente N, Pascual JI (2006) Spontaneous formation of triptycene supramolecules on surfaces. J Phys Chem B 110(41):20089–20092
24. Satrijo A, Swager TM (2005) Facile control of chiral packing in poly(*p*-phenylenevinylene) spin-cast films. Macromolecules 38(10):4054–4057
25. Zu SZ, Sun XX, Liu Y, Han BH (2009) Supramolecular surface modification and solubilization of single-walled carbon nanotubes with cyclodextrin complexation. Chem Asian J 4(10):1562–1572
26. Han T, Chen CF (2008) Efficient potassium-ion-templated synthesis and controlled destruction of [2] rotaxanes based on cascade complexes. J Org Chem 73(19):7735–7742
27. Jiang Y, Guo JB, Chen CF (2010) A bifunctionalized [3] rotaxane and its incorporation into a mechanically interlocked polymer. Chem Commun 46(30):5536–5538
28. Jiang Y, Guo JB, Chen CF (2010) A new [3] rotaxane molecular machine based on a dibenzylammonium ion and a triazolium station. Org Lett 12(19):4248–4251
29. Guo JB, Xiang JF, Chen CF (2010) Synthesis of a bis-macrotricyclic host and its complexation with secondary ammonium salts: an acid-base switchable molecular handcuff. Eur J Org Chem 2010(26):5056–5062

30. Guo JB, Jiang Y, Chen CF (2010) Self-assembled interwoven cages from triptycene-derived bis-macrotricyclic polyether and multiple branched paraquat-derived subunits. Org Lett 12(24):5764–5767
31. Ashton PR, Chrystal EJT, Glink PT, Menzer S, Schiavo C, Stoddart JF, Tasker PA, Williams DJ (1995) Doubly encircled and double-stranded pseudorotaxanes. Angew Chem Int Ed 34(17):1869–1871
32. Zhu XZ, Chen CF (2005) A highly efficient approach to [4] pseudocatenanes by threefold metathesis reactions of a triptycene-based tris [2]pseudo-rotaxane. J Am Chem Soc 127(38):13158–13159
33. Jiang Y, Zhu XZ, Chen CF (2010) Multivalency-directed magic-ring [2] (3)catenane by olefin metathesis. Chem Eur J 16(48):14285–14289
34. Zhu XZ, Chen CF (2006) Efficient synthesis of a chiral [4] pseudocatenane and its derivatives: a novel ship's wheel-like interlocked structure. Chem Eur J 12(21):5603–5609
35. Hu SZ, Chen CF (2011) Triptycene-derived oxacalixarenes as new wheels for the synthesis of [2] rotaxanes: acid-base- and metal-ion-switchable complexation processes. Chem Eur J 17(19):5423–5430
36. Su YS, Liu JW, Jiang Y, Chen CF (2011) Assembly of a self-complementary monomer: formation of supramolecular polymer networks and responsive gels. Chem Eur J 17(8):2435–2441

Chapter 11
Iptycenes and Their Derivatives in Coordination Chemistry

11.1 Triptycene-Based Ligands

It is the rigid structures that make the triptycenes promising candidates as the building blocks for the construction of metal carbonyl complexes. Pohl and Willeford [1] prepared the first triptycene π complex, tricarbonyl(η^6-triptycene) chromium, via the reaction between triptycene and hexacarbonyl-chromium in the refluxing n-butyl ether (**1**, Fig. 11.1) in 1970. This complex could be soluble in most organic solvents, but decomposed quickly under sunlight.

In 1985, Gancarz et al. [2] firstly determined the crystal and molecular structures of tricarbonyl(η^6-triptycene)chromium **1** and nonacarbonyl(η^6-triptycene)tetracobalt **2**. As shown in Fig. 11.1, the deformation of the triptycene skeletons was observed in both molecular structures due to the effect of nonbonded repulsions between one of the triptycene rings and the $Cr(CO)_3$ or $Co_4(CO)_9$ tripods, and complex **2** displayed greater deformation since the much more bulky tripod. Moreover, they [3] further found that the relative rates of the bridging–equatorial, bridging–axial, and axial–equatorial transposition processes for complex **2** showed obvious temperature dependence, according to the variable-temperature ^{13}C nuclear magnetic resonance (NMR) studies.

Besides the formation of the complexes between the triptycene and carbonyl chromium, Moser and Rausch [4] obtained the triptycene-based complex with $(NH_3)_3Co(CO)_3$ in 1974. Several years later, Fung et al. [5] reported that the triptycene could react with mercury^{II} trifluoroacetate in trifluoroacetic acid to form the corresponding complex, while they also found that 9-methoxy and 9,10-dimethoxy substituted triptycene derivatives could not form the corresponding complexes under the same conditions. In 1989, Fagan et al. [6] reported the synthesis of the mono-substituted complex $[Cp^*Ru(\eta^6\text{-triptycene})]^+(OTf)^-$ (**3**) by the reaction of complex $Cp^*Ru(CH_3CN)_3{}^+(OTf)^-$ with triptycene in CH_2Cl_2. As shown in Scheme 11.1, when the triptycene was excess, the dications *syn*-**4** and *anti*-**4** could be obtained. Moreover, due to the steric effect, it was found that only the reaction between *anti*-**4** and another equivalent of $Cp^*Ru(CH_3CN)_3{}^+(OTf)^-$ could form the triangular trication $[Cp^*Ru(\eta^6, \eta^6, \eta^6\text{-triptycene})]^{3+}(OTf)^{3-}$, which was in an approximate triangular prism shape.

C.-F. Chen, Y.-X. Ma, *Iptycenes Chemistry*,
DOI 10.1007/978-3-642-32888-6_11, © Springer-Verlag Berlin Heidelberg 2013

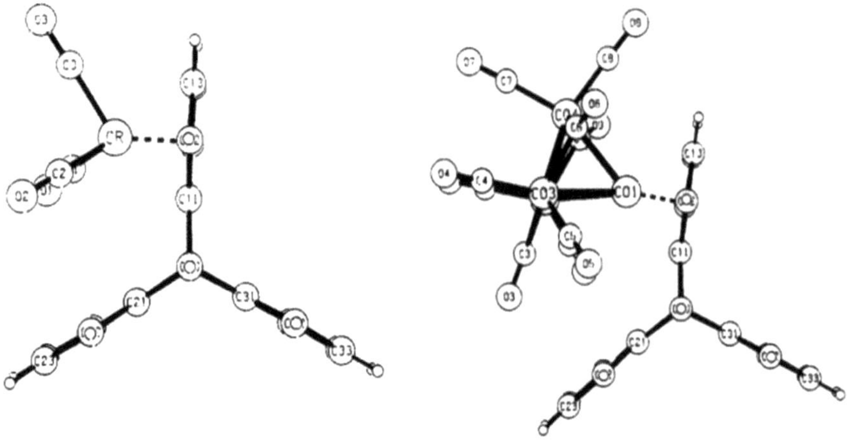

Fig. 11.1 The crystal structures of **1** (*left*) and **2** (*right*). (Reprinted with permission from [2]. Copyright 1985 American Chemical Society)

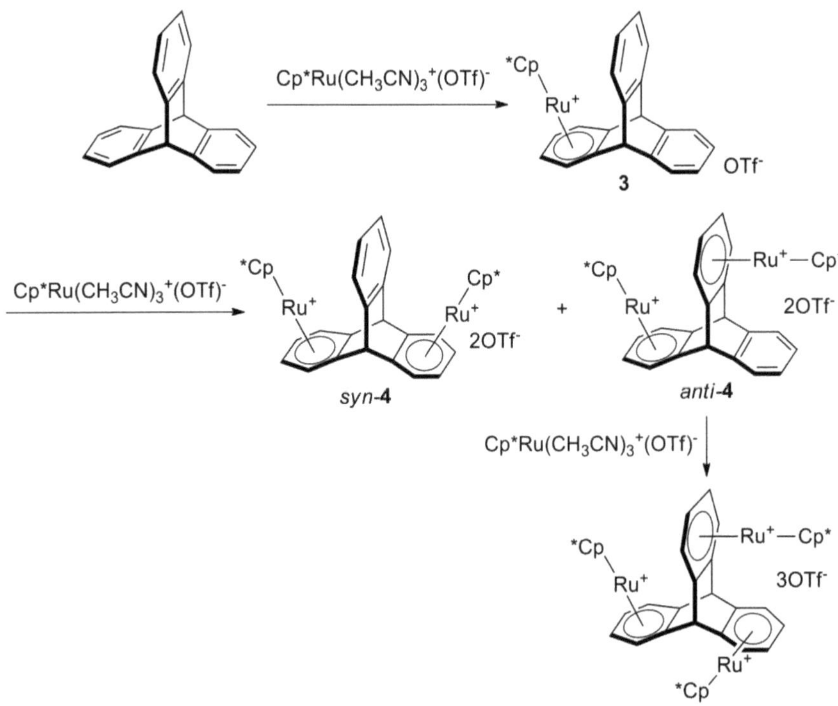

Scheme 11.1 Reaction of complex $Cp^{*}Ru(CH_3CN)_3{}^{+}(OTf)^{-}$ with triptycene

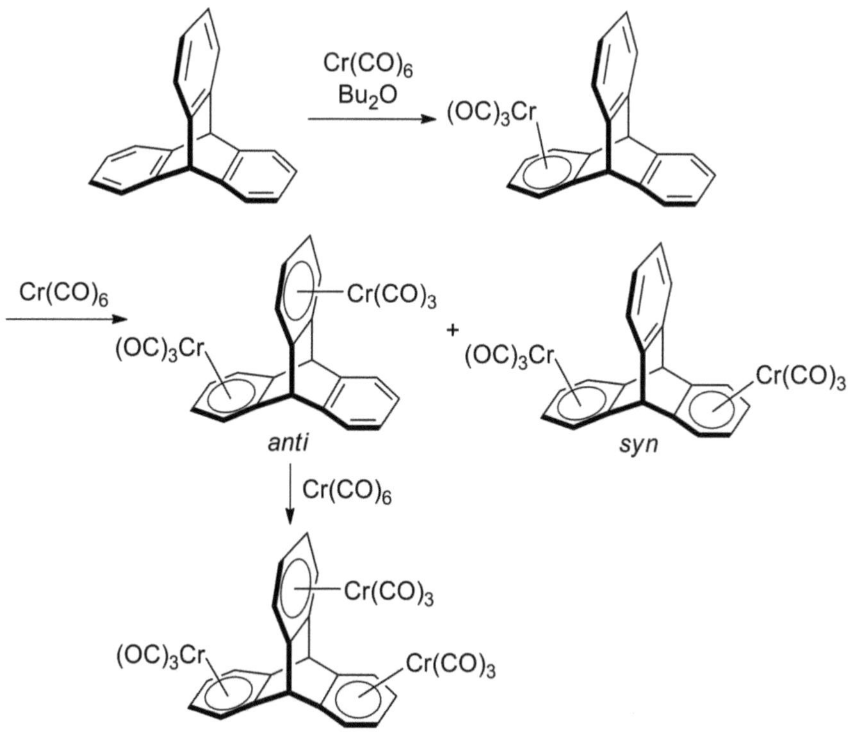

Scheme 11.2 Synthesis of triptycene-based mono-, bis-, and tris-(tricarbonylchromium) complexes

The unique structure of triptycene was also suitable for the bis(arene) coordination with the large dications such as Sn(II) and Pb(II). Thus, Schmidbaur et al. [7] obtained the crystalline complex **5**, $[(\text{triptycene})SnCl(AlCl_4)]_2$•2benzene by the reaction of triptycene with the equivalent anhydrous $SnCl_2$ and $AlCl_3$ in benzene. The resulting complex **5** showed the discrete centrosymmetrical dinuclear features, which was revealed by the X-ray crystal structure.

In 1997, Toyota et al. [8] reported the synthesis of a series of triptycene-based mono-, bis-, and tris-(tricarbonylchromium) complexes via the reaction of triptycene with $Cr(CO)_6$, which was outlined in Scheme 11.2. Similarly, there was a little deformation to compensate the narrow notch in the mono- and bis(tricarbonyl-chromium) complexes, when the $Cr(CO)_3$ groups lied in all notches of the triptycene.

In 1999, Munakata et al. [9] took advantage of the reaction between triptycene and silver(I) perchlorate, and synthesized the novel hydrocarbon-bridged polymeric complexes **6** ($[Ag_3(\text{triptycene})_3\ (ClO_4)_3](\text{toluene})_2$). Firstly, it was found that the triptycene preferred to interact with the smaller metal atom such as Co, by one arene ring (Fig. 11.2), meanwhile for the larger metal atom such as Sn, the mode of bis(arene) coordination (Fig. 11.2b) seemed to be taken. In the case of the complex with Ag, an unprecedented $\mu\text{–}\eta^2\text{–}\eta^2$ or $\mu\text{–}\eta^2\text{–}\eta^2\text{–}\eta^1$ coordination mode could

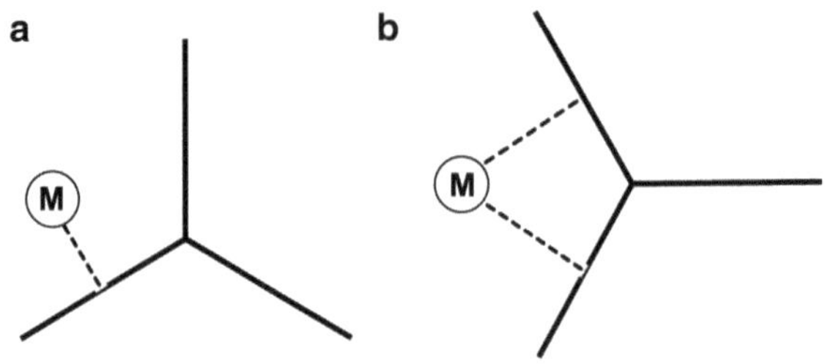

Fig. 11.2 The complexation modes of metal centers with triptycene

be found in the solid state of complex **6** containing a three-dimensional architecture with the less deformation (Fig. 11.3a). Subsequently, Wen et al. [10] further found this porous three-dimensional structure of complex **6** displayed the desorption and adsorption behavior of guest molecules, and the structure of **6** seemed to be unstable along with the structure changes. In spite of this deficiency, the features of desorption and absorption made it a potential inorganic functional material for wide applications. Afterward, they [11] further reported the synthesis of two novel silver(I) coordination complexes **7** and **8** based on triptycene-derived ligands with different anions and solvents. Unlike the complex **6** (Fig. 11.3a), the complex **7** [Ag(triptycene)$(THF)_2$](ClO_4) displayed a one-dimensional chain structure in a μ–η^1–η^1 mode. While the other complex **8** [Ag_6(triptycene)$_4$ $(CF_3SO_3)_2(H_2O)_6$] bearing the larger size anion showed a two-dimensional sheet structure with six metal ions bridged by ligands and oxygen atoms (Fig. 11.3b). Consequently, both the anion and solvent molecule could affect the structure of triptycene-based complex to a certain extent.

In 2008, Vagin et al. [12] prepared the first two-dimensional and three-dimensional metal-organic frameworks (MOFs) of 9,10-triptycenedicarboxylic acid with zinc nitrate under solvothermal reaction conditions. The reaction conditions, including the reactant ratios, reaction temperature, and time could affect the sizes and morphologies of MOFs. Only in a few cases could the smaller regular crystals with hexagonal column modes be prepared. Moreover, it was found that with the increased van der Waals interactions between triptycene units as well as coordinated solvent molecules, the two-dimensional framework further possessed the three-dimensional MOFs containing the hexagonal channels, which were filled with guest molecules (Fig. 11.4).

11.2 Substituted Triptycene-Based Ligands

11.2.1 Selenium Substitution

In 1999, Ishii et al. [13] designed and synthesized a series of the triptycene-derived selenenic acids (Fig. 11.5). As shown in Scheme 11.3, the reduction of the sulfide **9** by lithium 4,4′-di-tert-butylbiphenilide (LiDBB) gave the thiophenetriptycyllithium

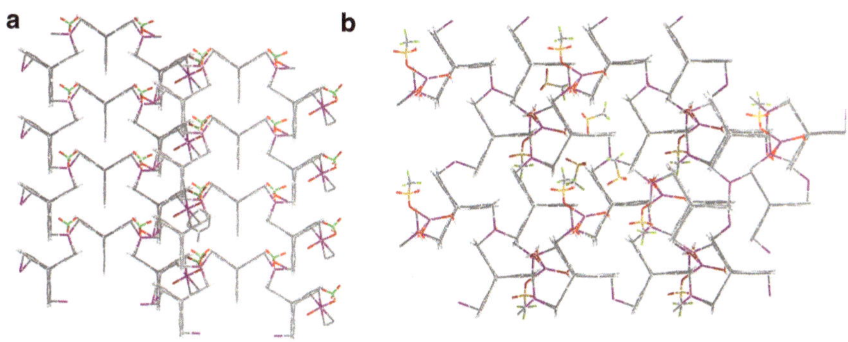

Fig. 11.3 Two-dimensional sheet structures of complexes **6** (**a**) and **8** (**b**)

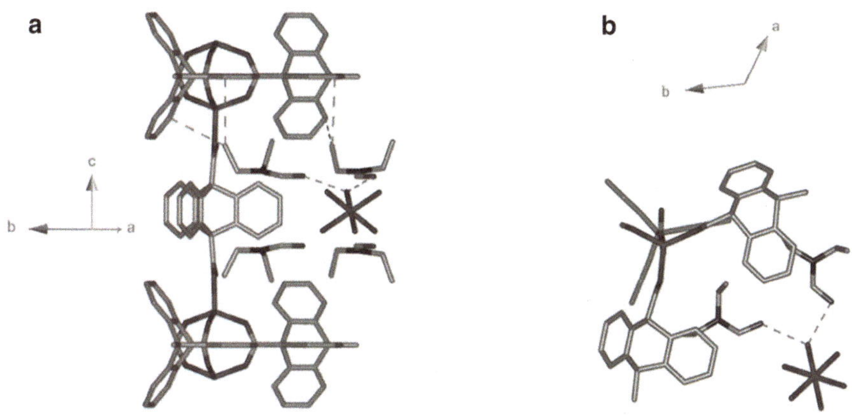

Fig. 11.4 Two views of the structural fragment of MOFs. (Reproduced from [12], with the permission of John Wiley and Sons)

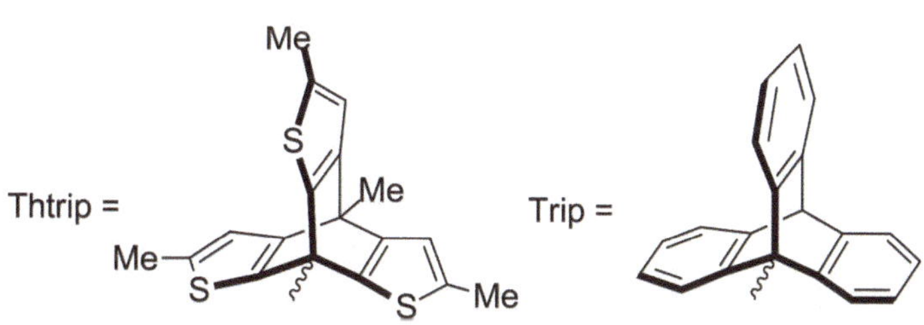

Fig. 11.5 Structures of triptycene ligands in selenenic acids

10, which then reacted with dibutyl diselenide to afford the target compound **11**, butyl thiophenetriptycyl selenide. With compound **11** as the precursor, the selenoseleninate **12** could be further synthesized via the intermediate **13**. Starting from the di-9-triptycyl triselenide **14**, the 9-triptyceneselenol **15** and butyl 9-triptycyl selenide

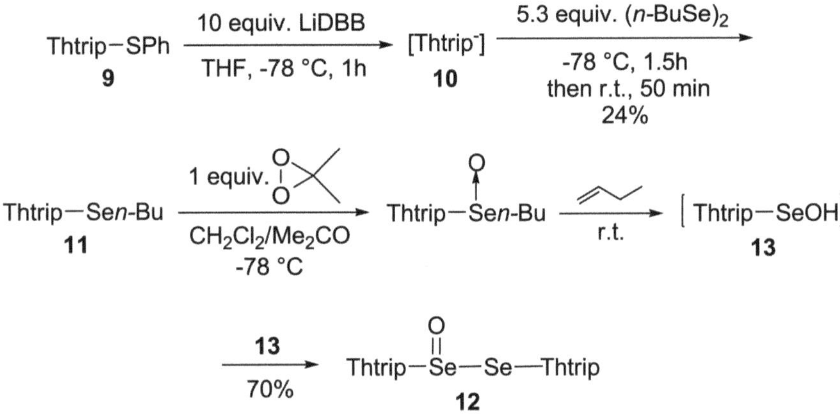

Scheme 11.3 Synthesis of triptycene-derived selenenic acids

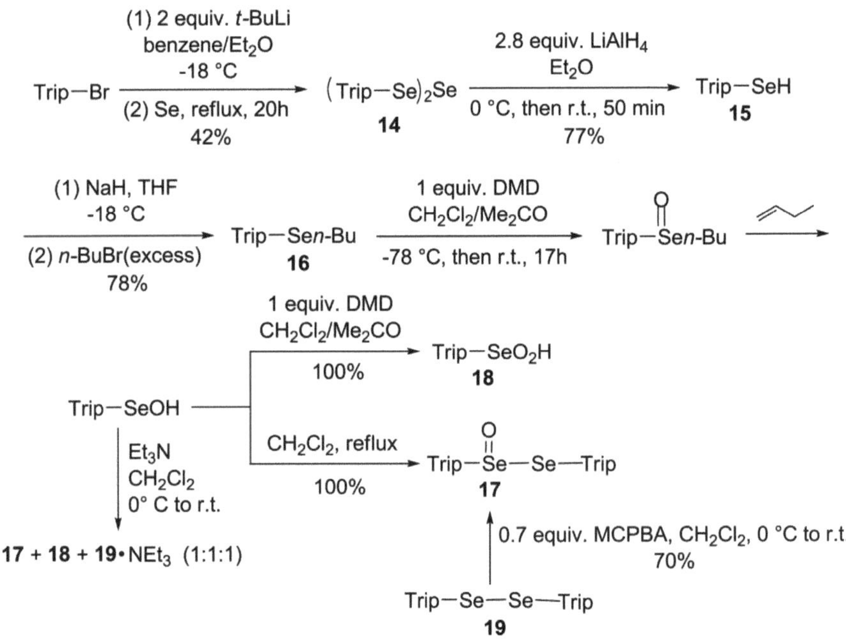

Scheme 11.4 Synthesis of selenoseleninate **17**

16 could be synthesized (Scheme 11.4). Moreover, according to the similar method as the synthesis of **12**, the selenoseleninate **17** was also obtained from the precursor of **16** (Scheme 11.4). It was noteworthy that both compounds **12** and **17** were the first examples of isolable selenoseleninates, and they displayed the *s-trans* conformation about the Se–Se single bond with the central symmetry revealed by the

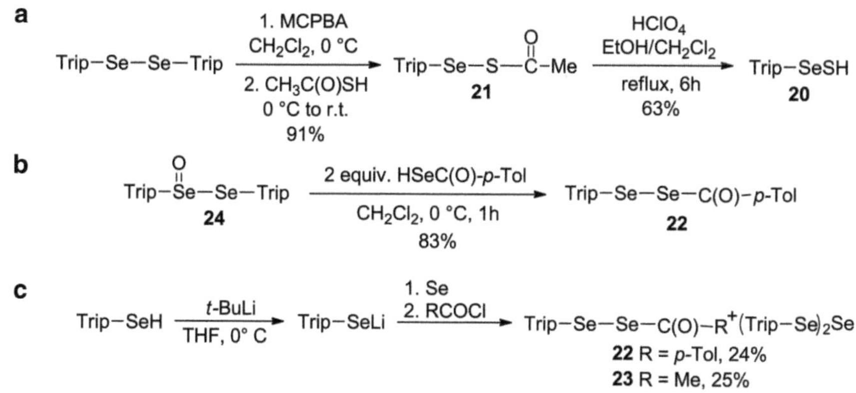

Scheme 11.5 Synthesis of triptycene-9-thioselenenic acid **20** (**a**), diselenides **22** (**b**), and **23** (**c**)

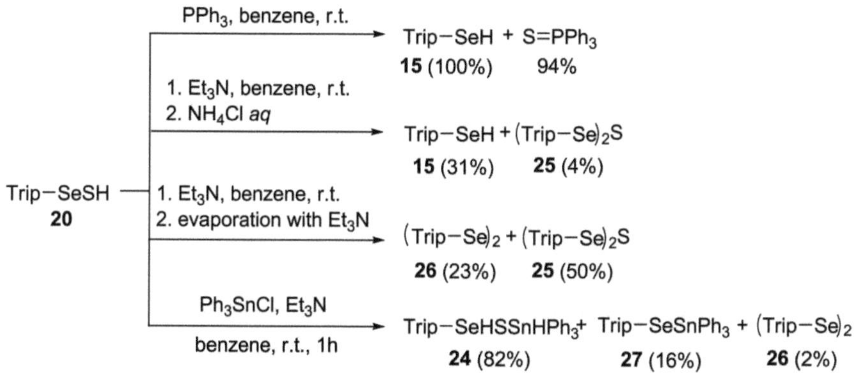

Scheme 11.6 Synthesis of triptycene-9-selenic derivatives

single-crystal structures. Moreover, it was found that treatment of the selenoseleninate **17** with 2 M $HClO_4$ in 1,4-dioxane gave the acid-catalyzed hydrolyzation product, selenenic acid (TripSeOH), while the alkaline hydrolysis products were the diselenide (TripSeSeTrip) and the seleninic acid ($TripSeO_2H$) [14].

Furthermore, Ishii et al. [15] reported the synthesis of the triptycene-9-thioselenenic acid **20** by the hydrolyzation of the acetyl thioselenenate **21** (Scheme 11.5a), which was stable under acidic conditions. However, when compound **20** was treated with the different reagents in different conditions, it would decompose or convert to the series of triptycene-9-selenic derivatives (Scheme 11.6). Moreover, the diselenides **22** and **23** [16] could be afforded from Se-9-triptycyl triptycene-9-selenoseleninate **24** and triptycene-9-selenenoselenolate salt, respectively (Scheme 11.5b and c).

In 2008, Ishii et al. [17] reported that these triptycene-derived selenium compounds could react with the platinum^0 complex (**28**, $[Pt(PPh_3)_2(\eta^2\text{-}C_2H_4)]$) via the intramolecular C–H bond activation to afford a five-membered selenaplatinacycle,

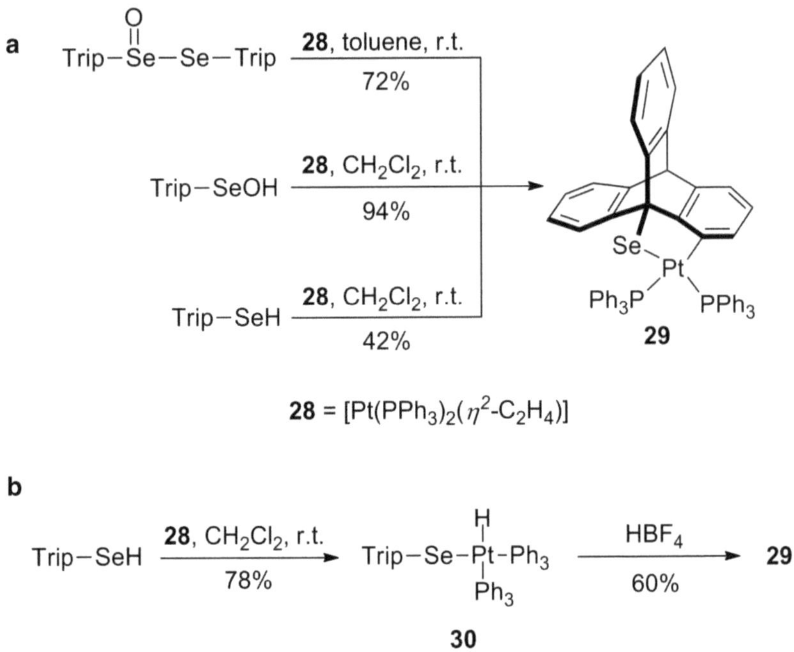

Scheme 11.7 Synthesis of selenaplatinacycle **29**

which is shown in Scheme 11.7a. It was suggested that the bulky substituent groups played an important role in the formation of selenaplatinacycle **29**. Moreover, the reaction between selenol and complex **28** could afford the hydrido Pt^{II} complex **30**, which was stable in the air at room temperature. The successful preparation of selenaplatinacycle and (hydrido) (selenolato) $platinum^{II}$ complex **30** obviously provided a novel strategy to the low-valent transition-metal complexes with selenium compounds. However, complex **30** treated with strong acid such as HBF_4 would form the complex **29** via the cationic intermediate (Scheme 11.7b).

Afterward, Ishii et al. [18] further investigated the addition reaction of the (hydrido)(selenolato)platinum(II) complexes with electron-deficient alkynes as well as the carboselenation reactions. The reaction between PtH(SeTrip)-$(PPh_3)_2$ (**30**) with dimethyl acetylenedicarboxylate (DMAD) or methyl propiolate (MP) provided the hydroselenation *syn*-adducts **31** and **32** (Scheme 11.8a); while for the [PtH(SeTrip)(dppe)] containing the stronger phosphane σ-donor ligand dppe, 1,2-bis(diphenyl phosphanyl)ethane, both *syn*- and *anti*-adducts (**33a** and **b**) could be produced (Scheme 11.8b). In addition, these addition reactions did not occur in the presence of PPh_3, which suggested that the dissociation of ligand PPh_3 played a key role in the hydroselenation. This result also offered an insight into the Pt-catalyzed hydroselenation reaction mechanism (Scheme 11.9).

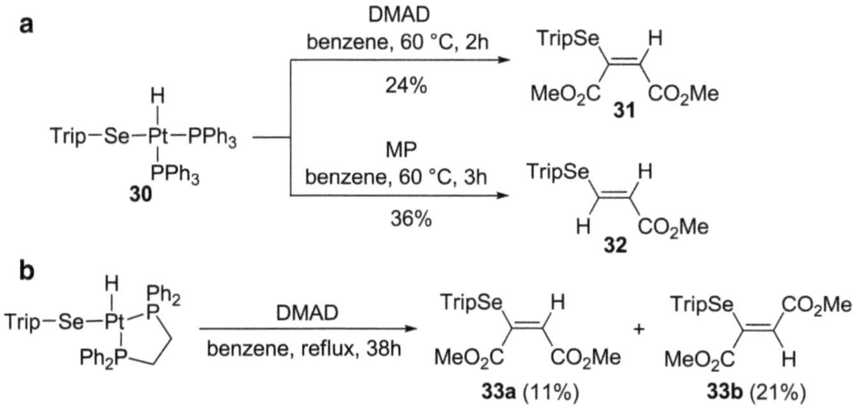

Scheme 11.8 Reaction between $PtH(SeTrip)$-$(PPh_3)_2$ (**30**) with DMAD or methyl propiolate

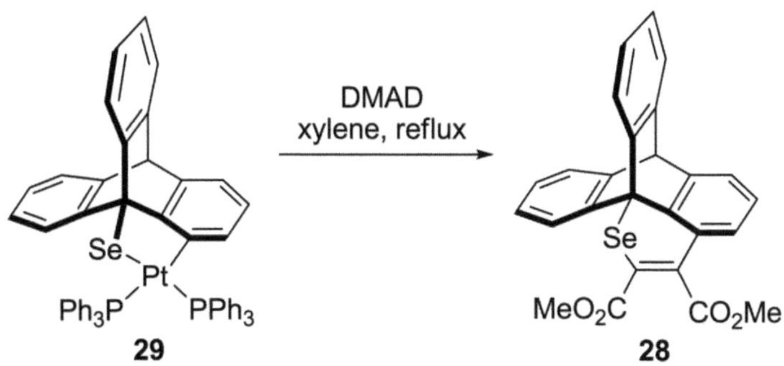

Scheme 11.9 9 Synthesis of compound **28**

Recently, Nakata et al. [19] further synthesized a series of novel Pd^0 complexes by the reactions of the selenol **15** with Pd^0 complexes $[Pd(PPh_3)_4]$. Unexpectedly, the novel selenolato-bridged dinuclear Pd^I complex $(Pd(PPh_3)_2(\mu\text{-SeTrip})_2$ **34**, Fig. 11.6a) rather than Pd^0 complex was obtained in 47 % yield by the reactions of the selenol **15** with $[Pd(PPh_3)_4]$ in toluene (Scheme 11.10a). Similarly, the five-membered selenapalladacycle, $Pd(\eta^2$-(C, Se)–Trip)(dppe) **35** (Fig. 11.6b) was obtained in 88 % yield in toluene by the reaction of **15** with $[Pd(dppe)_2]$ (Scheme 11.10b).

11.2.2 Germanium and Silicon Substitution

In general, the primary germanes are air-sensitive liquid (or gaseous) compounds at room temperature with relatively high activity. In 1999, Brynda et al. [20] took advantage of the surrounding phenylene hydrogen atoms of the triptycyl moiety to stabilize the high-reactivity bond, and obtained the air-stable hydride complex **36**

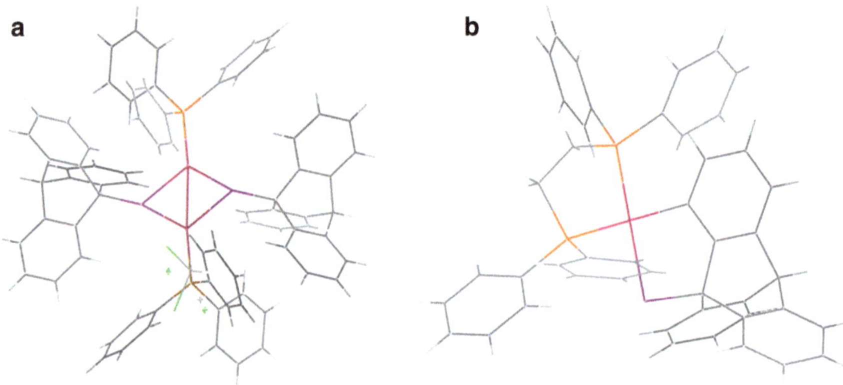

Fig. 11.6 ORTEP drawing of complexes **34** (**a**) and **35** (**b**).

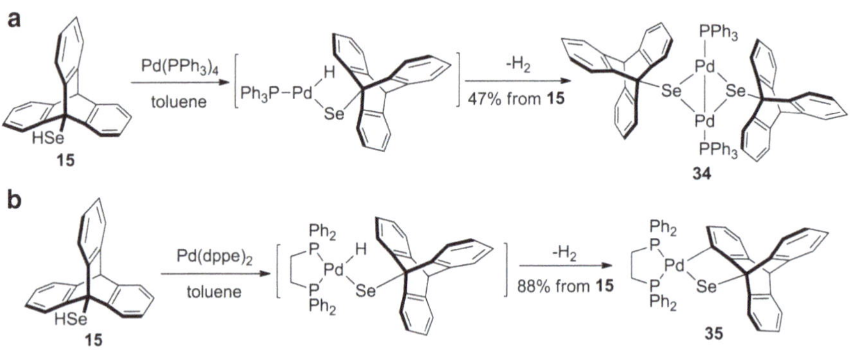

Scheme 11.10 Synthesis of Pd^0 complexes

[(Trip)GeH_3] by the reduction of the corresponding triptycenyl complexes of germanium halides, [(Trip)$GeCl_3$], which was prepared by the reaction of 9-lithiotriptycene with $GeCl_4$. Also, the "user-friendly" derivatives of triptycene-containing SiH_3 fragments, [(Trip)SiH_3] could be obtained in 82 % yield by the in situ reaction of the corresponding $SiCl_4$ with the lithium derivative of triptycene in tetrahydrofuran (THF) following the reduction with $LiAlH_4$ [21].

In addition, Nakata et al. [22] found that by the reaction of the triptycene-derived germanium compound **36** instead of selenium compounds with [Pt(PPh_3)$_2$(η^2-C_2H_4)] in toluene at room temperature, the stable dihydrogermyl (hydrido) platinum(II) complex **37** as colorless crystals could also be obtained in 99 % yield (Scheme 11.11). In the case of bidentate phosphine ligands dppe, the corresponding dihydrogermyl(hydrido) complex [(dppe)Pt(H)-(GeH_2Trip)] and bis(germyl) complex [(dppe)Pt(GeH_2Trip)$_2$] could be obtained by the reaction of **36** with varying amounts of [(dppe)Pt]. While the complex [(dcpe)Pt(H)-(GeH_2Trip)] with another bidentate phosphine ligands dcpe was obtained via the ligand-exchange reaction of complex **37** with free dcpe in toluene at room temperature. Moreover, the complex

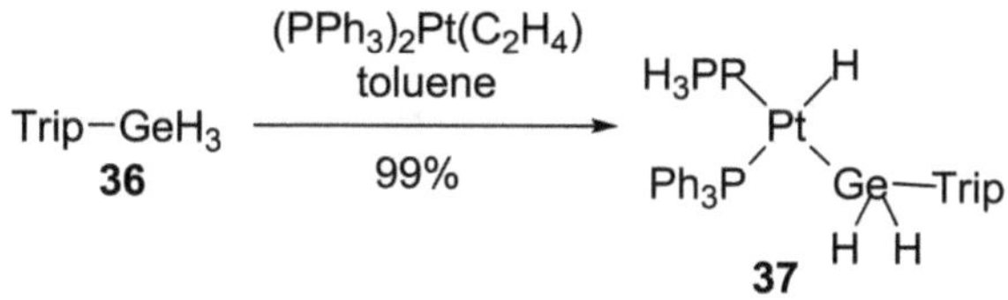

Scheme 11.11 Synthesis of dihydrogermyl(hydrido) platinum(II) complex **37**

38 would convert to the corresponding complex (dppe)Pt(H)[Ge(HTrip)GeH$_2$Trip] **39** in 88 % yield by heating to 60 °C for 1 day, and the possible formation mechanism is shown in Scheme 11.12. These results obviously provided an effective strategy to prepare the novel classes of germyl(hydrido) complexes.

Based on the above work, Ishii's group [23] further designed and prepared the palladium(II) hydrido complexes containing the dihydrosilyl or dihydrogermyl ligands, which had not been reported before. In consequence, the complex **40** [PdH(SiH$_2$Trip)(dcpe)] and [PdH(GeH$_2$Trip)(dcpe)] **41** with a distorted-square-planar geometry around the palladium centers could be efficiently produced in good yields by the oxidative-addition reactions of an overcrowded primary silane or germane with the complex [(μ-dcpe)Pd]$_2$ in toluene (Scheme 11.13). It was noteworthy that the Si–H σ-complex intermediate could be observed by the variable-temperature NMR experiments, which revealed the intramolecular interchange of coordination environments between the silyl and hydrido ligands for complex **40**, as shown in Scheme 11.14.

11.2.3 Phosphorus Substitution

Since the first stable phosphaalkyne, P ≡ C*t*-Bu, was prepared in 1981 [24], this type of compounds showed the potentialities for the versatile and widely applications in the field of coordination chemistry. Thus, Brym and Jones [25] designed and synthesized the first triptycene-based diphosphaalkyne **42** in a good yield by the reaction of the diphosphaalkene with a catalytic amount of KOH in 1,2-dimethoxyethane (DME). The diphosphaalkyne **42** showed the similar behavior to the monophosphaalkyne, and it could react with two equivalents of [Pt(C$_2$H$_4$)-(PPh$_3$)$_2$] or [RuHCl(CO)(PPh$_3$)$_3$] to afford the bis-platinum(0) coordination complex **43** and bis-ruthenium phosphaalkenyl complex **44**, respectively (Fig. 11.7).

Inspired by the synthesis of the diphosphaalkyne, Baker et al. [26] further attempted to prepare the more 9-triptycenyl ligands with the more group 13 and 15 elements. In fact, the primary phosphane [(Trip)PH$_2$] [27] and arsane [(Trip)AsH$_2$] [21] had been obtained by the reaction of 9-lithiotriptycene with either PCl$_3$ or AsCl$_3$, followed by the reduction with LiAlH$_4$ in situ before. Both of the compounds were air stable, because the triptycene framework could protect the highly reactive P–H or As–H bonds. However, the triptycenyl complexes of Sb could not be prepared in the same way. Based on this method, the reactions between 9-lithiotriptycene and

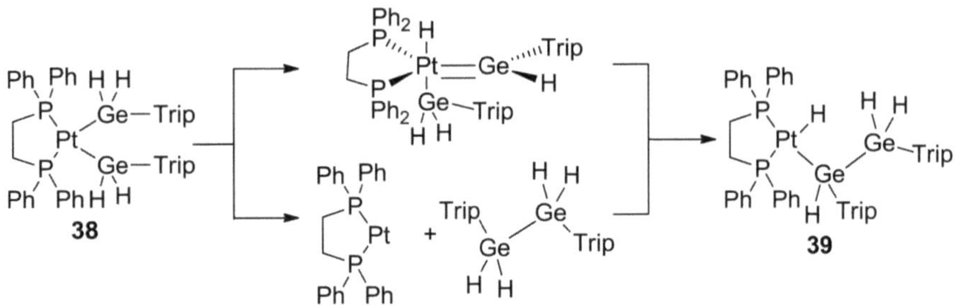

Scheme 11.12 Synthesis of complex **39**

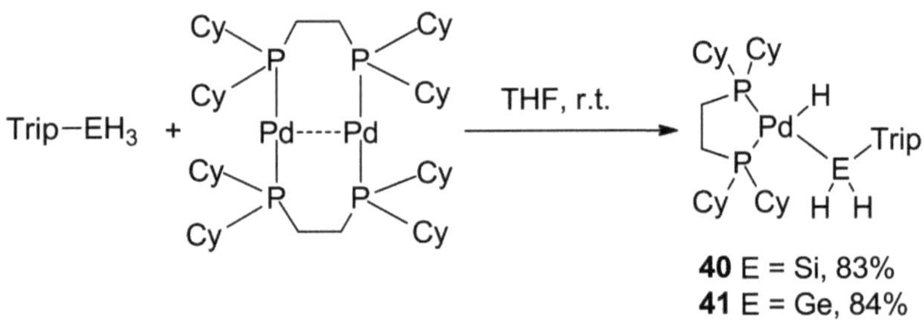

Scheme 11.13 Synthesis of complex **40** [PdH(SiH_2Trip)(dcpe)] and [PdH(GeH_2Trip)(dcpe)] **41**

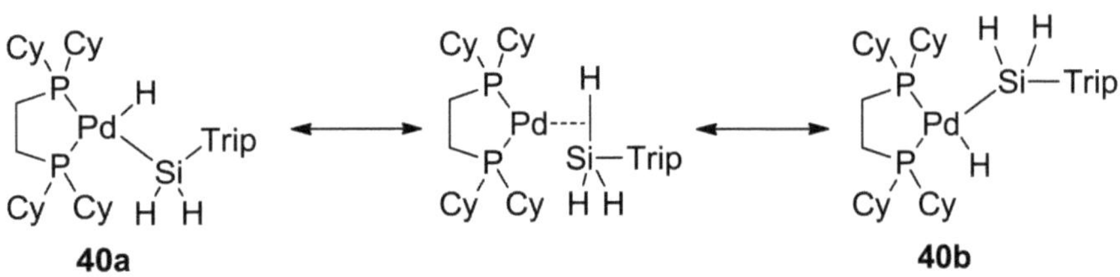

Scheme 11.14 Synthesis of complex **40b**

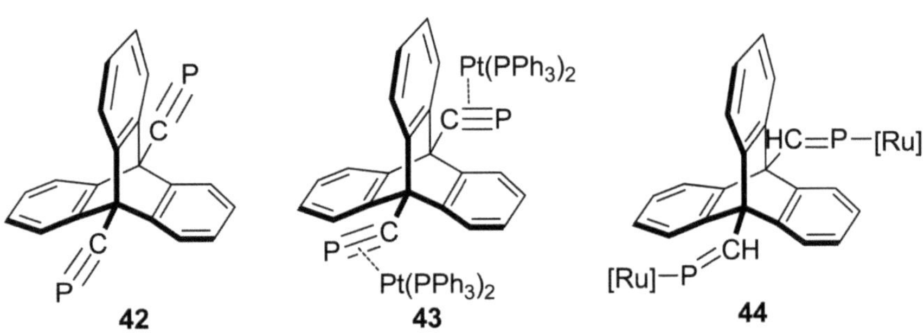

Fig. 11.7 Structures of triptycene-based diphosphaalkyne **42** and complexes **43** and **44**

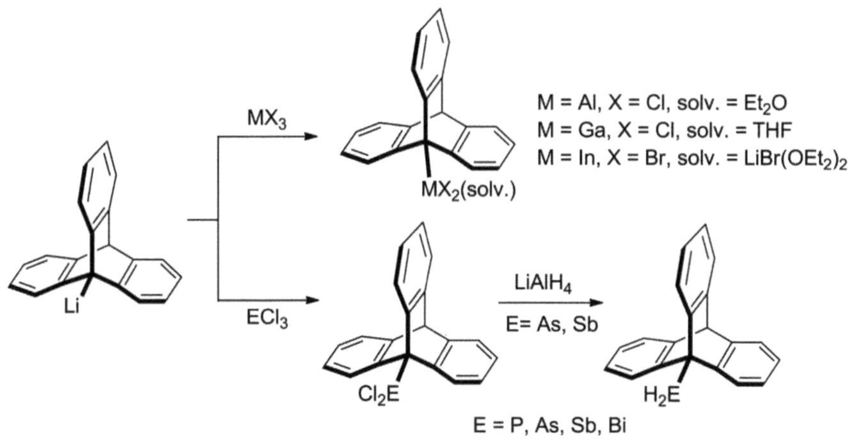

Scheme 11.15 Synthesis of triptycenyl complexes of group **13** halides

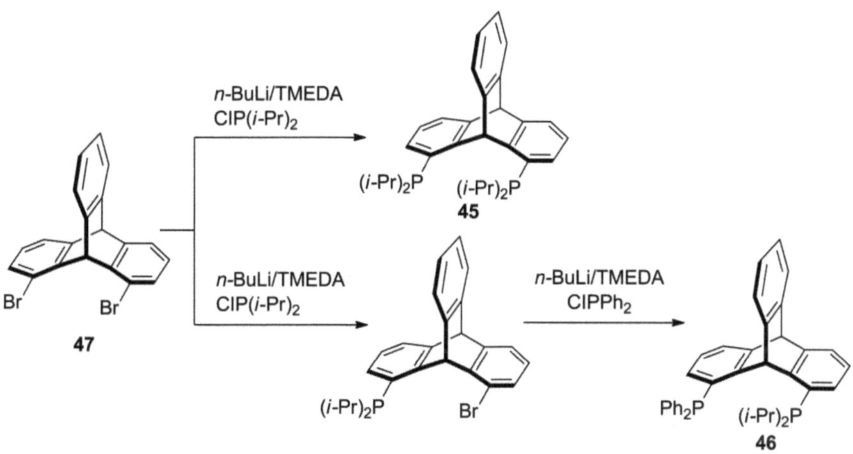

Scheme 11.16 Synthesis of chelating diphosphine ligands **45** and **46**

$AlCl_3$, $GaCl_3$, or $InBr_3$ could produce the air-stable triptycenyl complexes of group 13 halides in either diethyl ether or THF in good yields. Similarly, the 9-triptycenyl complexes of group 15 halides could also be afforded from the 9-lithiotriptycene. Moreover, the primary pnictanes [(Trip)AsH_2] and [(Trip)SbH_2] could be obtained by the reduction of the corresponding halide complexes with $LiAlH_4$ (Scheme 11.15).

In view of the unique rigid structural properties of triptycenes making them promising building blocks for the "wide bite angle" and "*trans*-chelating" diphosphines ligands, Grossman et al. [28] designed and synthesized a series of the chelating diphosphine ligands containing the triptycene subunit. The synthetic route to the ligands **45** and **46** with the readily accessible 1,8-dibromotriptycene **47** as starting material was outlined in Scheme 11.16. With these strongly bent triptycene-based diphosphines ligands, the *trans*-spanned complexes **48** and **49** could be obtained

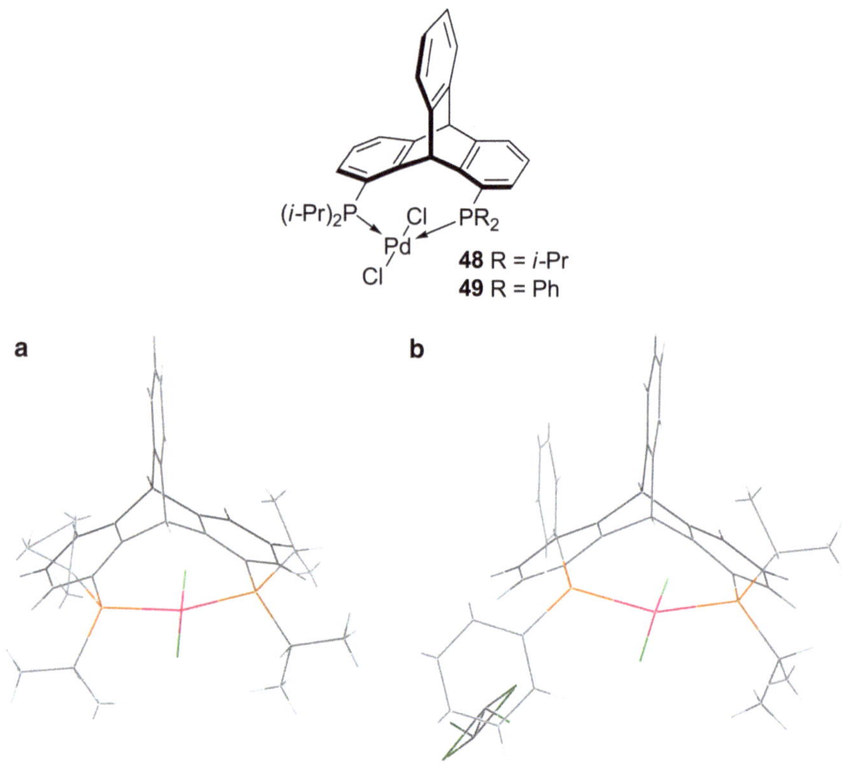

Fig. 11.8 Molecular structures, and ORTEP drawing of the crystal structures of complexes **48** (**a**) and **49** (**b**)

in good yields. The crystal structures showed a strong distortion of the palladium centers to a butterfly-like environment instead of the local square-planar geometry (Fig. 11.8). The novel ligands and the corresponding complexes displayed the potential in the catalysis of Suzuki–Miyaura cross-coupling reactions with excellent catalytic activity. Moreover, Grossman and Gelman [29] also synthesized the similar *trans*-spanned palladium complex **50**, which showed the efficient and selective catalytic activity in the mild palladium-catalyzed cyanation of aryl bromides in the mild conditions under the noninert atmosphere. Also, it was further found that the catalyst **50** could be applied for a large range of substrates without the assistance of the amine coligands (Table 11.1).

To reveal the coordination diversity of triptycene-based bidentate ligands, Azerraf et al. [30] synthesized and characterized a series of complexes between the triptycene-based ligands and varied d^8 transition metals (Ni, Rh, Pt; Fig. 11.9). According to the structural analysis of these complexes, the ligand **45** exhibited the high degree of flexibility with the varietal bite angles and geometries to achieve the most stable structure. For example, in the solid-state structure of complex **51**, it was revealed that the nickel center in **51** was in a heavy distorted square planar geometry with

Table 11.1 Suzuki–Miyaura cross-coupling reactions catalyzed by complex **50**

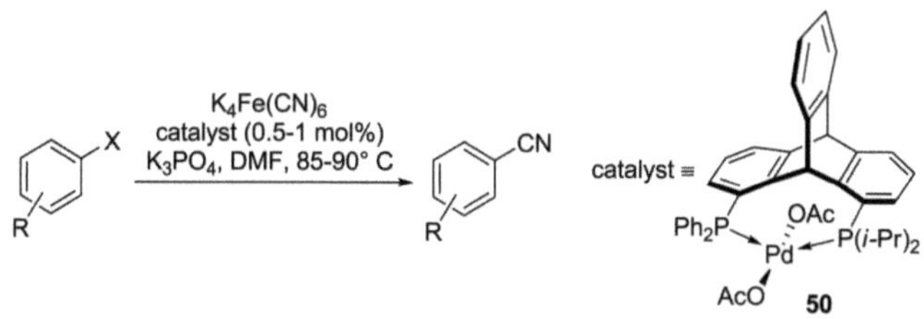

Entry	Ar-X	Ar-CN	Conversion (%)	Isolated yield (%)
1	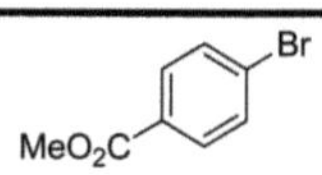	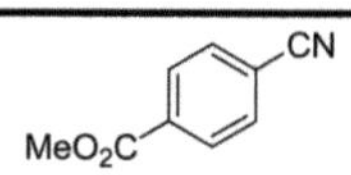	>99	96
2	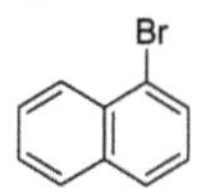	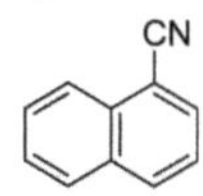	>99	98
3	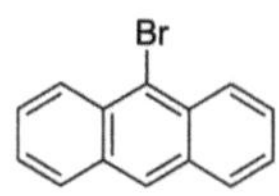	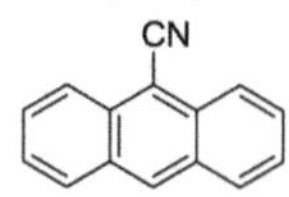	>99	84
4	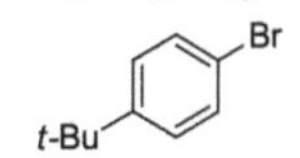	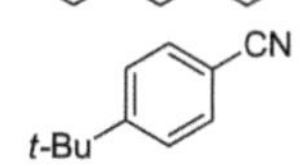	>99	94
5	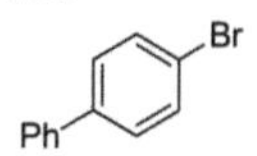	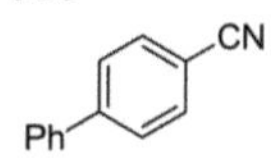	>99	93
6	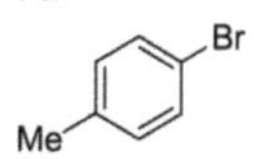	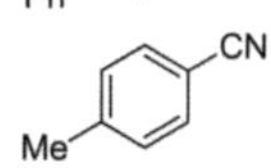	>99	96
7	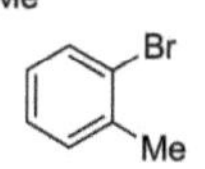	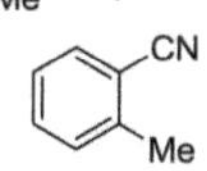	>99	88
8	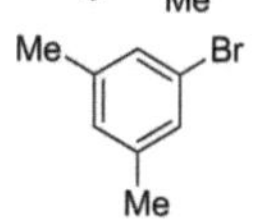	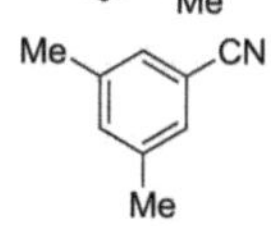	>99	91
9	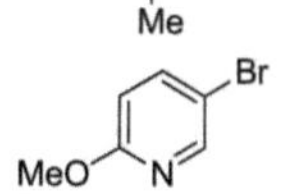	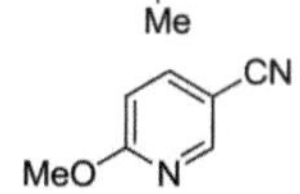	>99	81
10	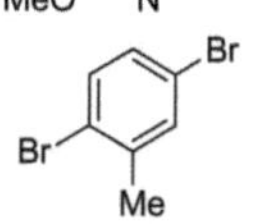	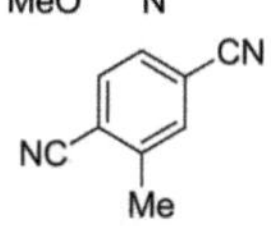	>99	95
11	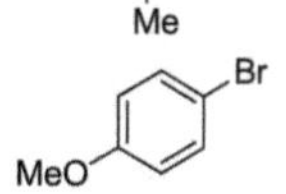	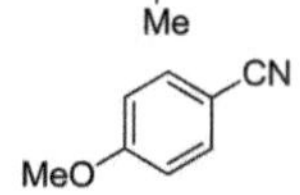	71	69
12	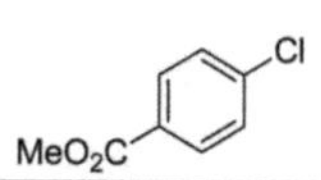	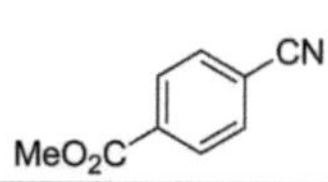	34	21

Reaction conditions: 1 equiv. of aryl halide (1 mmol), 0.3 equiv. of $K_4Fe(CN)_6$, 2 equiv. of K_3PO_4, 1 mol % of $Pd(OAc)_2$, 1.5 mol % of **50**, 3 mL of DMF, 85 or 90 °C, 24 h.

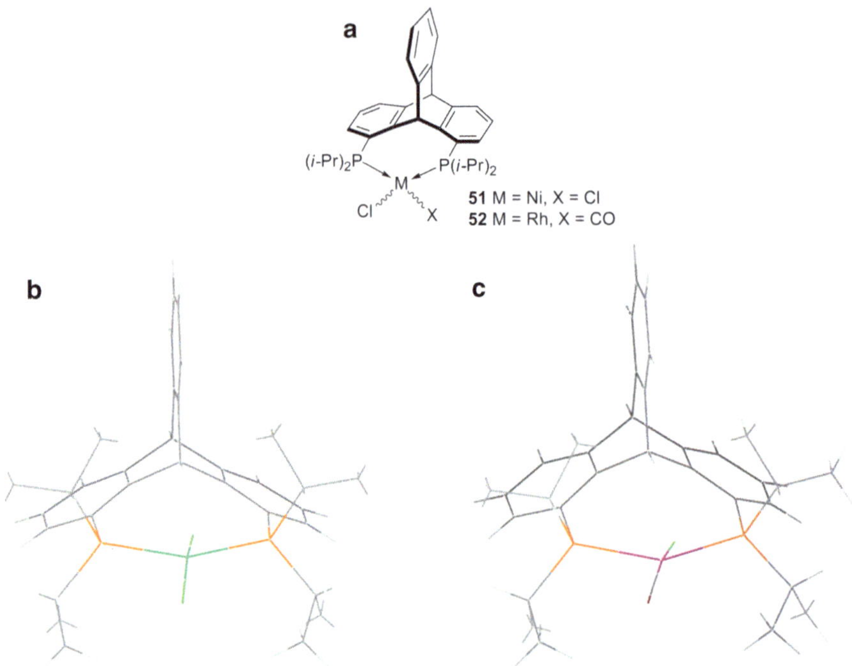

Fig. 11.9 Molecular structures of **51–52** (**a**), ORTEP drawing of the structures of complexes **51** (**b**), and **52** (**c**)

a 152.51(5)° P(1)–Ni–P(2) angle, and the angle of Cl(1)–Ni–Cl(2) was 162.60(2)° (Fig. 11.9b); while for complex **52**, the rhodium center in **52** was in butterfly-type geometry with P(1)–Rh–P(10) and C(25)–Rh–Cl(1) angles of 149.79° and 170.77°, respectively (Fig. 11.9c).

Moreover, the extraordinary *quasi*-closed halogen-bridged dipalladium complex **53** could be also synthesized by the ligand **45** as a bidentate ligand, which led to a nonplanar Pd_2Cl_4 site with a very short intermetallic distance (Fig. 11.10). With the further studies of the bi-nucleating behavior of rigid triptycene-based ligands, Azerraf et al. [31] demonstrated that the formation of the rare bimetallic *quasi*-closed chloride-bridged complexes was achieved from the *trans*-spanned transition-metal mononuclear complexes bearing the triptycene-based ligands via the ring-expansion progress. Based on this mechanism, the di-rhodium *quasi*-closed chloride-bridged complex **54** (Scheme 11.17) and the dissymmetric di-palladium complex **55** (Scheme 11.18) containing the $M_2(\mu\text{-}Cl_2)$ cores were successfully obtained from the corresponding monometallic complexes **56** and **57** by the ring-expansion reaction. In addition, these halogen-bridged compounds were stable in both solution and solid state, and they could not be dissociated even in the presence of the limited quantities of Lewis base such as pyridine.

Based on the robust triptycene scaffold, more and more of the PC(sp^3) P-based cyclo-metalated complexes have been subsequently synthesized. For example, Azerraf and Gelman [32] synthesized and characterized a novel robust

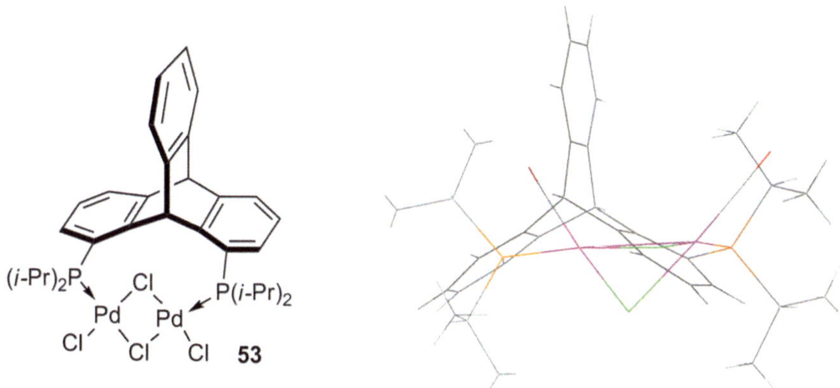

Fig. 11.10 Molecular structure and ORTEP drawing of crystal structure of **53**

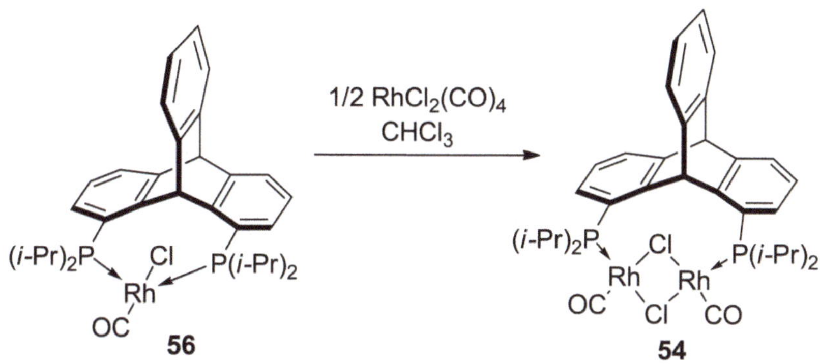

Scheme 11.17 Synthesis of di-rhodium *quasi*-closed chloride-bridged complex **54**

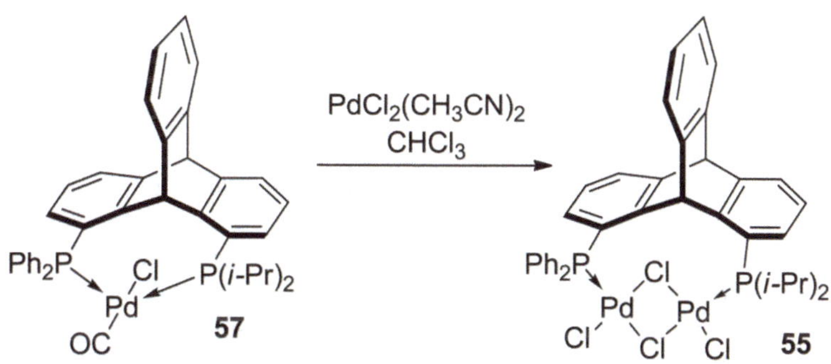

Scheme 11.18 Synthesis of dissymmetric di-palladium complex **55**

$Ir^{III}PC(sp^3)P$ complex **58** by the reaction of the triptycene-based ligand **46**, 1,8-bis(diisopropylphosphino)triptycene, with $IrCl_3(H_2O)_n$ in *N,N*-dimethylformamide (DMF). Complex **58** also displayed good thermal and conformational stability, and had no labile α- or β-hydrogen. Moreover, it was found that complex **58** showed the high catalytic activity in transfer dehydrogenation of ketones in acetone under air (Table 11.2) [33].

Additionally, Azerraf and Gelman [34] also reported the synthesis of the C(sp^3)-metalated pincer complexes containing platinum(II), ruthenium(II), and iridium(III) via the C–H activation, in which the *trans*-chelating bis(diisopropyl phosphino)triptycene **59** was used as the platform. However, the pathway to the complexes (**60–63**) varied slightly depending on the different transition metal or the precursor of the complexes, which are shown in Scheme 11.19 and Scheme 11.20, respectively. As mentioned before, both of these pincer complexes have not the labile α- or β-hydrogen, thus, there was little chance to occur elimination of H. As a result, they all displayed an excellent thermal and conformational stability even under very harsh condition. Moreover, all of them showed an unusual three-dimensional roof-shaped bent structures (Fig. 11.11), and these features made them promising scaffold for designing the PC(sp^3)P-based compounds.

In 2007, Bini et al. [35] synthesized a triptycene-based diphosphine ligand **64** in 20 % overall yield by an improved route from the difluorotriptycene moiety **65** (Scheme 11.21). Then, the new complexes could be obtained by the reactions between **64** and $PtCl_2$ and $Ni(cod)_2$. Also, the complex **64•**Ni(cod) was found to display the excellent conversion and high selectivity (98 %) toward 3-pentenenitrile in the hydrocyanation of butadiene.

Grossman et al. [36] also found that the bidentate ligand **45** (1,8-bisdi-*i*-propyl-phosphinotriptycene) could be used in a new protocol for the palladium-catalyzed Buchwald–Hartwig aminations of the aryl halides with benzophenone imines as ammonia surrogate in the mild conditions, which may be applied for the relatively sensitive such as base-sensitive substrates (Table 11.3).

11.2.4 Miscellaneous Substitutions

Besides the diphosphine ligands, Kodanko et al. [37] developed a new family of dinucleating ligands, the *syn* *N*-donor diethynyltriptycene ligands **66**. As shown in Table 11.4, the resulting triptycene-based ligands could be efficiently obtained by the dual Sonogashira cross-coupling of 2,3-diethynyl triptycene with a variety of functionally diverse 5-bromopyridines as well as an imidazole and a quinoline derivative. They [38] further prepared the heterodinuclear iron–sodium complex **67** starting from the ligand **66a** (Scheme 11.22). They also studied the metal-for-metal substitution of sodium by iron in complex **67**, and the kinetic experiments revealed the metal-for-metal exchange might be in a dissociative mechanism.

As mentioned before, Chong and MacLachlan [39] synthesized the shape-persistent ligands **68** and **69** (Fig. 11.12a) based on triptycene quinoxalines, and then constructed the noninterpenetrating coordination frameworks from the ligands and CuI (Fig. 11.12b). According to further analysis, these coordination frameworks containing the hydrophobic void spaces showed good thermal stability and water

Table 11.2 The catalyzed dehydrogenation of ketones

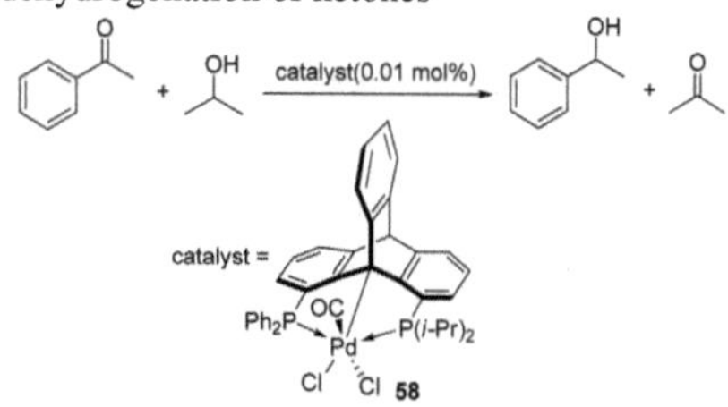

Entry	Ketone	Time	Conversion (%)	Isolated yield[a] (%)
1	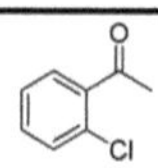	5 sec	99	98
	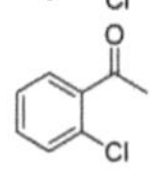	5 min[b]	99	99
2	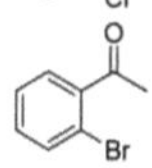	5 sec	99	98
	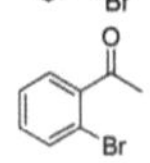	5 min[b]	99	99
3	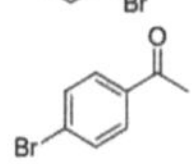	5 sec	96	95
	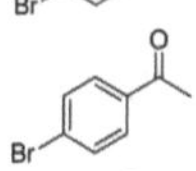	5 min[b]	96	95
4	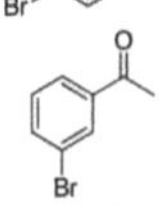	30 min	99	98
5	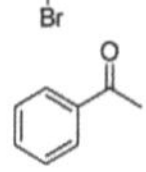	30 min	94	93
	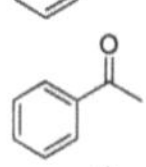	12 h[b]	96	95
	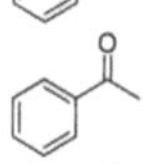	48 h[c]	96	93
	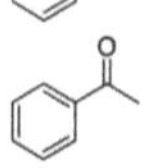	12 h[d]	45	-
6	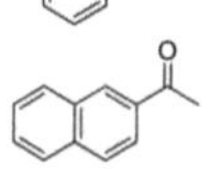	30 min	95	94
7	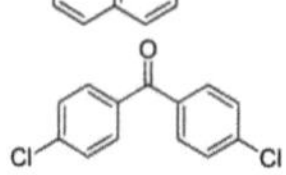	30 min	97	96
8	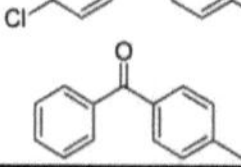	30 min	98	98

[a] Isolated yields of at least 95% pure compounds.
Reaction conditions: Ratio of catalyst/substrate/base = 1:2000:100, *i*-PrOH (1 M), 82 °C under air; [b] Ratio of catalyst/substrate/base = 1:10000:500, *i*-PrOH (4 M), 8 °C under air; [c] Ratio of catalyst/substrate/base = 1:100000:5,000, *i*-PrOH (4 M), 8 °C under air; [d] Ratio of catalyst/substrate/HCO_2Na = 1:100:500, CH_3CN/ H_2O (2:1), 82 °C.

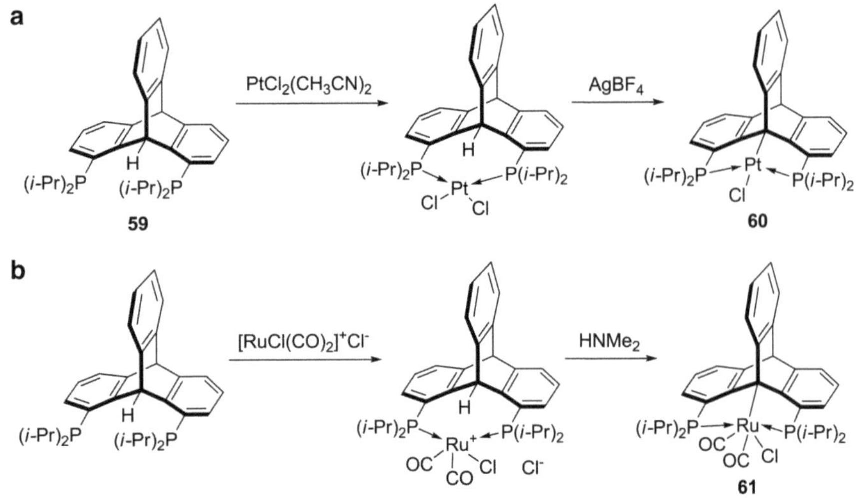

Scheme 11.19 Synthesis of complexes a **60** and b **61** from *trans*-chelating bis(diisopropylphosphino)triptycene **59**

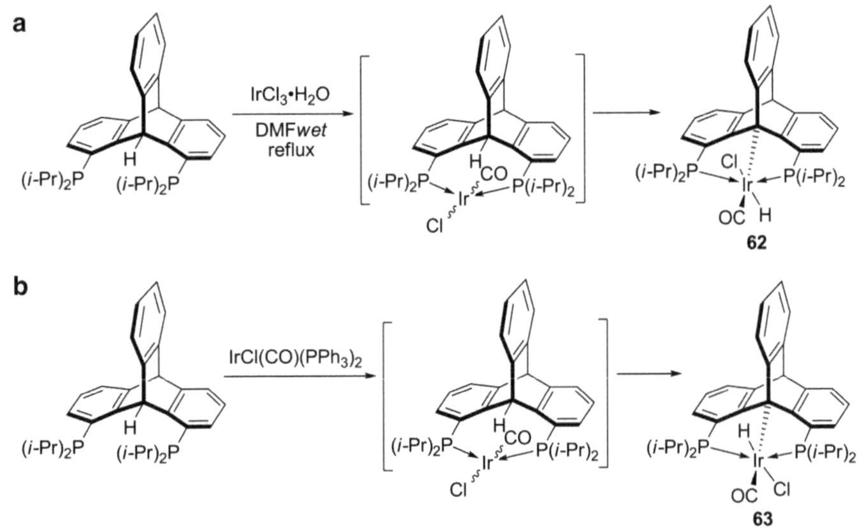

Scheme 11.20 Synthesis of complexes a **62** and b **63** from *trans*-chelating bis(diisopropylphosphino)triptycene **59**

solubility. Moreover, the frameworks displayed the ability of reversible solvent adsorption, and the Cu-**68** and its frameworks with the good sorption capacity could be used to remove organic pollutants such as benzene from contaminated water. Similarly, they [40] also found that there was a dimer with two molecules of **70**, which was bridged by Cu_2I_2. In addition, the extended framework in the dark red crystals

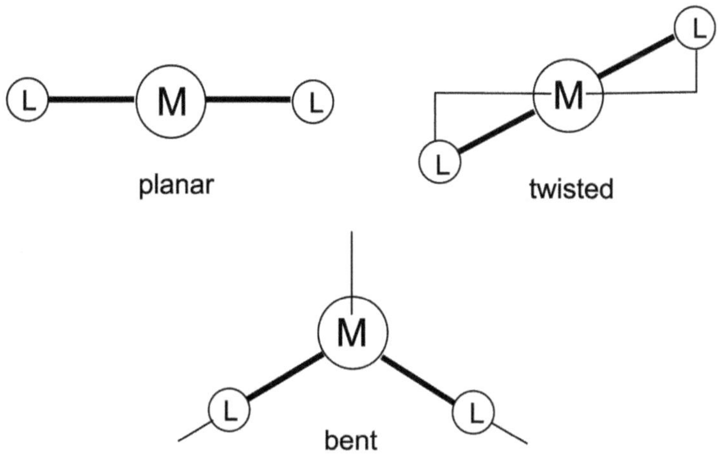

Fig. 11.11 Schematic diagrams of the three-dimensional structures of complexes

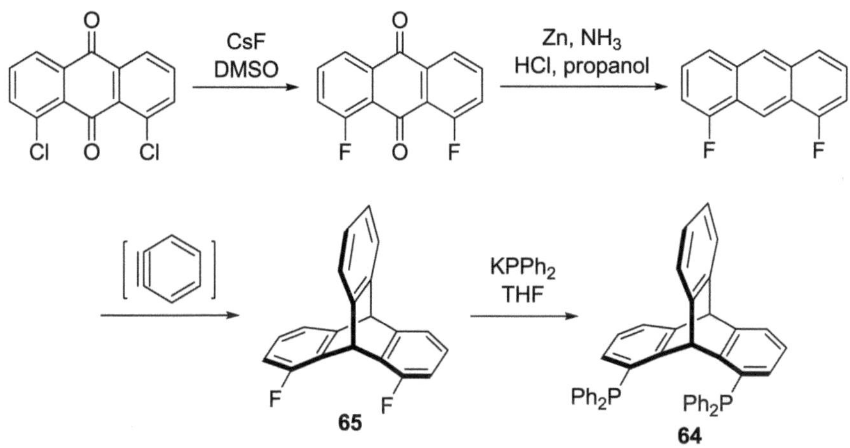

Scheme 11.21 Synthesis of triptycene-based diphosphine ligand **64**

of Cu-**70**, and the dimers further formed the herringbone-layered structure with π–π stacking, which contained the large solvent-filled voids as well (Fig. 11.12c).

In 2011, MacLachlan's group [41] further investigated the coordination complex of triptycenyl phenanthroline **71** with Cu^I ion, and found that the complex of **71** could be converted into a triptycene-scaffolded Cu^I Prussian blue analogues with $K_3[Fe(CN)_6]$ in methanol under solvothermal conditions, and the different coordination geometries with Cu^I centers in an unusual T-shaped complexation could be obtained (Fig. 11.13).

Additionally, the metal complexes based on triptycene bridging ligands could act as a rigid platform for the investigation of photo-physical properties. Thus, Beyeler

Table 11.3 The palladium-catalyzed Buchwald–Hartwig aminations of the aryl halides

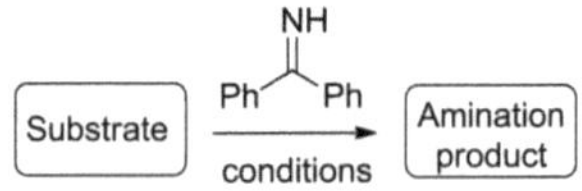

Entry	Substrate	Conditions	Product	Yield (%)
1	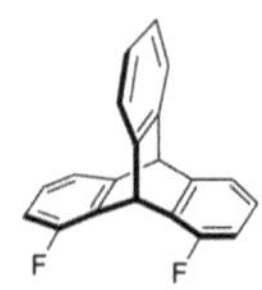	$Pd_2(dba)_3$/**45** (2 mol%), *t*-BuONa: 5 g scale	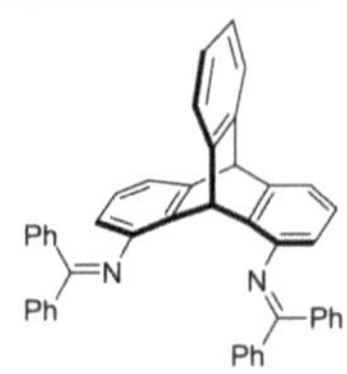	87
2	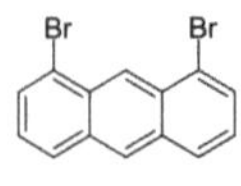	$Pd_2(dba)_3$/**45** (2 mol%), *t*-BuONa: 3 g scale	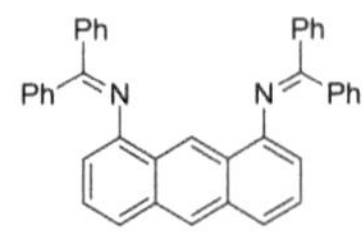	93
3	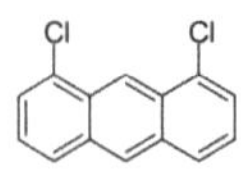	$Pd_2(dba)_3$/**45** (2 mol%), *t*-BuONa: 3 g scale	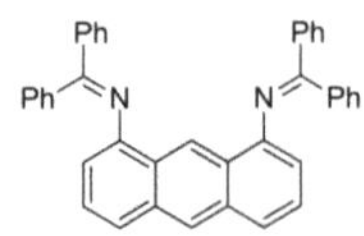	87
4	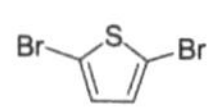	$Pd_2(dba)_3$/**45** (3 mol%), *t*-BuONa	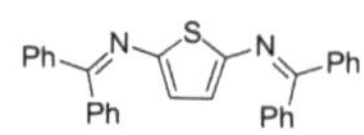	77
5	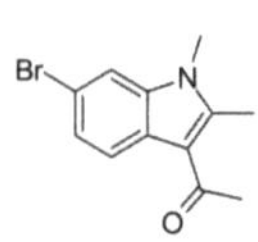	$Pd_2(dba)_3$/**45** (0.5 mol%), *t*-BuONa	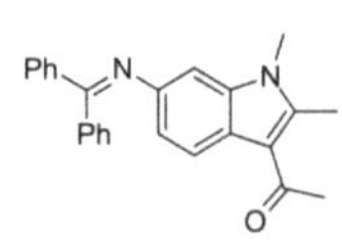	99
6	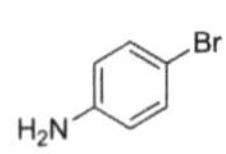	$Pd_2(dba)_3$/**45** (1.5 mol%), *t*-BuONa	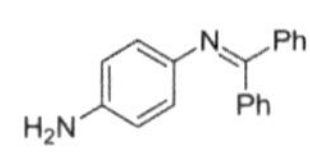	87
7	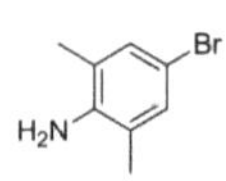	$Pd_2(dba)_3$/**45** (1.5 mol%), *t*-BuONa	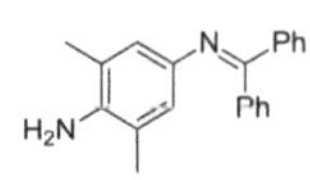	84
8	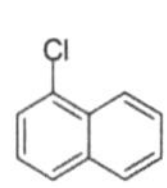	$Pd_2(dba)_3$/**45** (1.5 mol%), K_3PO_4	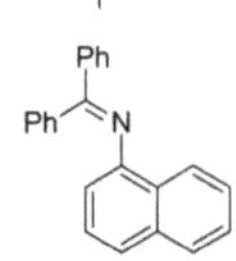	99
9	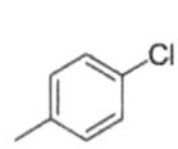	$Pd_2(dba)_3$/**45** (1.5 mol%), K_3PO_4	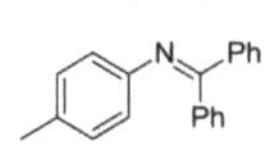	99
10	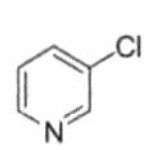	$Pd_2(dba)_3$/**45** (1.5 mol%), K_3PO_4	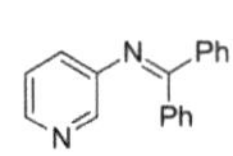	82
11	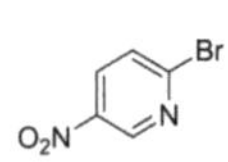	$Pd_2(dba)_3$/**45** (1.5 mol%), K_3PO_4	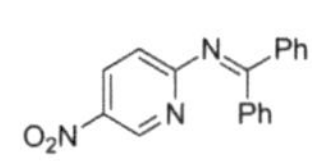	92
12	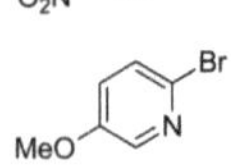	$Pd_2(dba)_3$/**45** (1.5 mol%), K_3PO_4	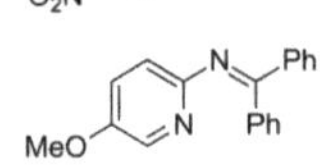	81

Table 11.4 The synthesis of triptycene-based dinucleating ligands

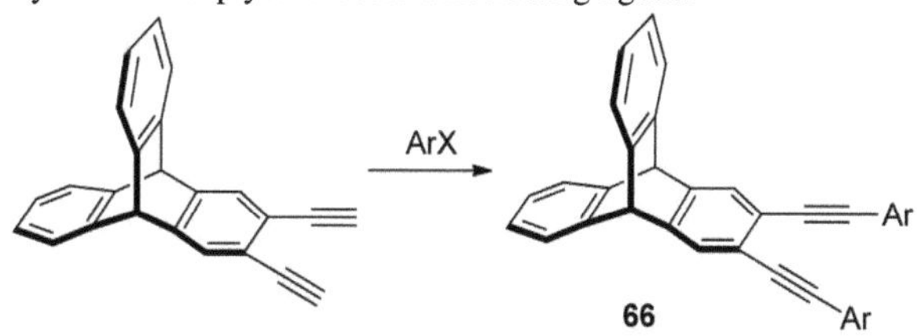

66

Entry	Ar-X	Product	Yield (%)
1	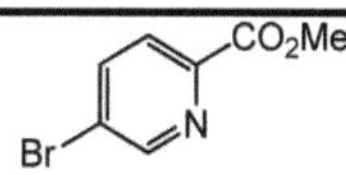	**60a**[a]	93
2	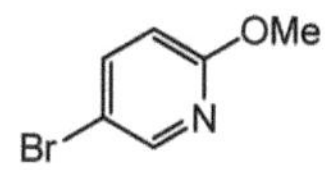	**60b**[a]	79
3	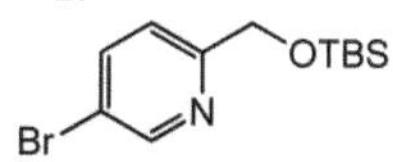	**60c**[a]	97
4	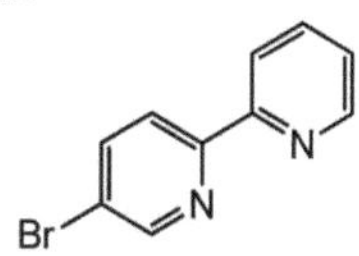	**60d**[a]	93
5	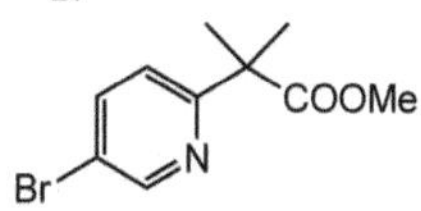	**60e**[a]	80
6	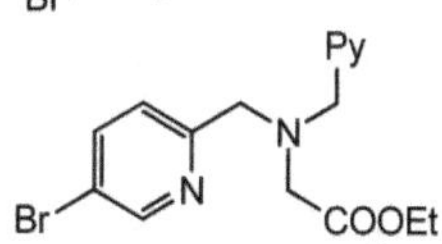	**60f**[a]	99
7	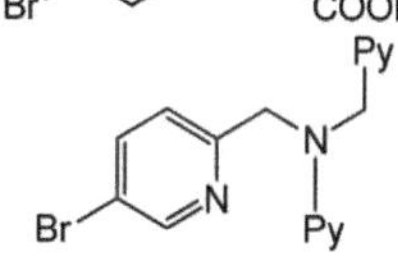	**60g**[a]	87
8	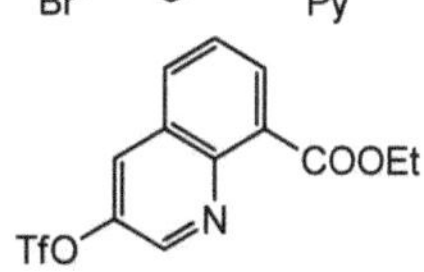	**60h**[b]	67
9	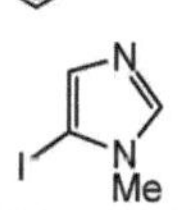	**60i**[c]	63

Reaction conditions: [a]10 mol % of $Pd(PPh_3)_4$, Et_3N, THF, 55 °C; [b] 10 mol % of $PdCl_2(PPh_3)_2$, 5 mol % of CuI, Et_3N, THF, r.t.; [c] 10 mol % of $PdCl_2(PPh_3)_2$, piperidine, 65 °C.

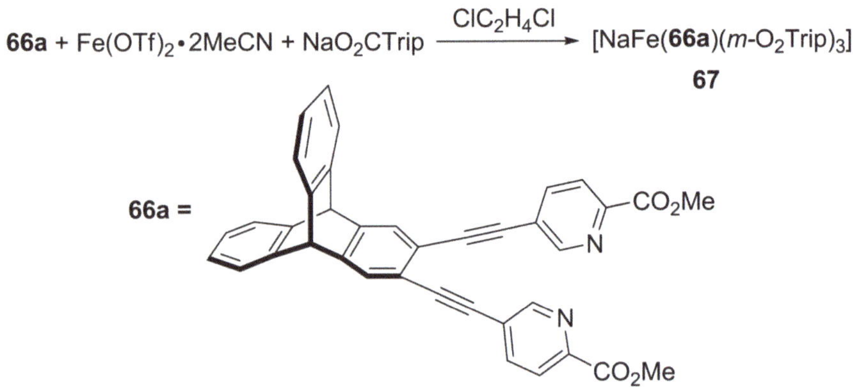

Scheme 11.22 Synthesis of hetero dinuclear iron-sodium complex **67**

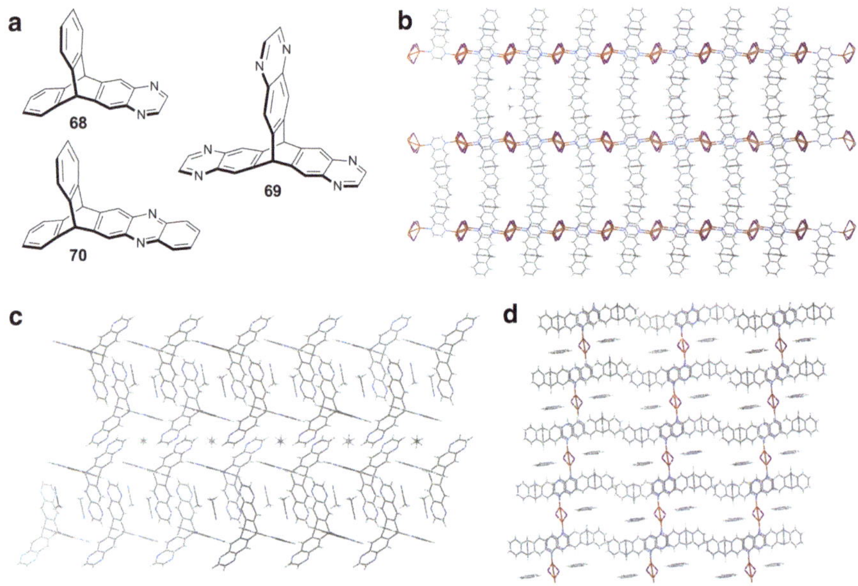

Fig. 11.12 **a** Structures of **68-70**. The viewed coordination frameworks of **b** Cu-**68** along the c-axis, **c** Cu-**69** along the b-axis, and **d** Cu-**70** along the b-axis.

and Belser [42] synthesized a series of rigid triptycene spacer connected to two bipyridine ligands by the Horner–Emmons-type reaction. With the new triptycene bridging ligands in hand, the corresponding dinuclear metal complexes **72–74** (Fig. 11.14) could be obtained. According to the metal-to-ligand charge-transfer bands in UV spectra of these complexes, they found that the donor effects of methoxy substituents on the triptycene led to the stronger emission than the unsubstituted triptycene ligand, whereas the ligand with quinone moiety strongly quenched the emission.

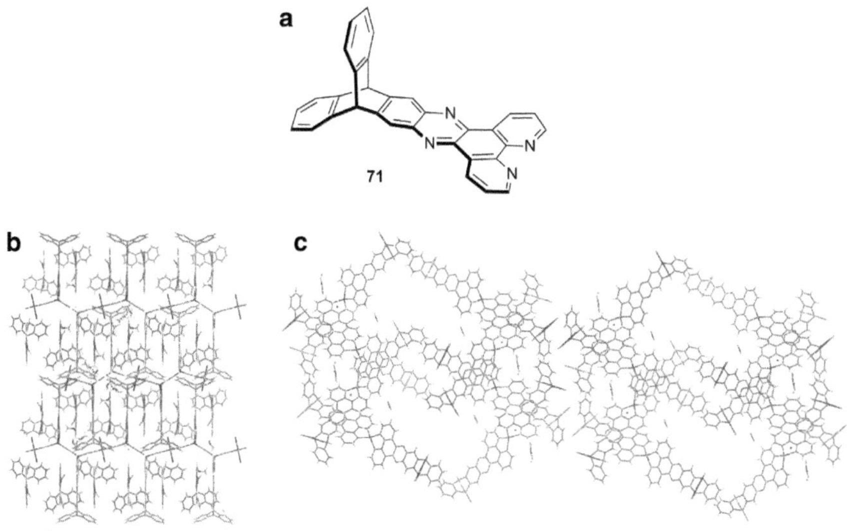

Fig. 11.13 **a** Structure of triptycenyl phenanthroline **71** and **b**, **c** the ORTEP structures of the two different complexes of **71** with $CuCl_2$, and $K_3[Fe(CN)_6]$

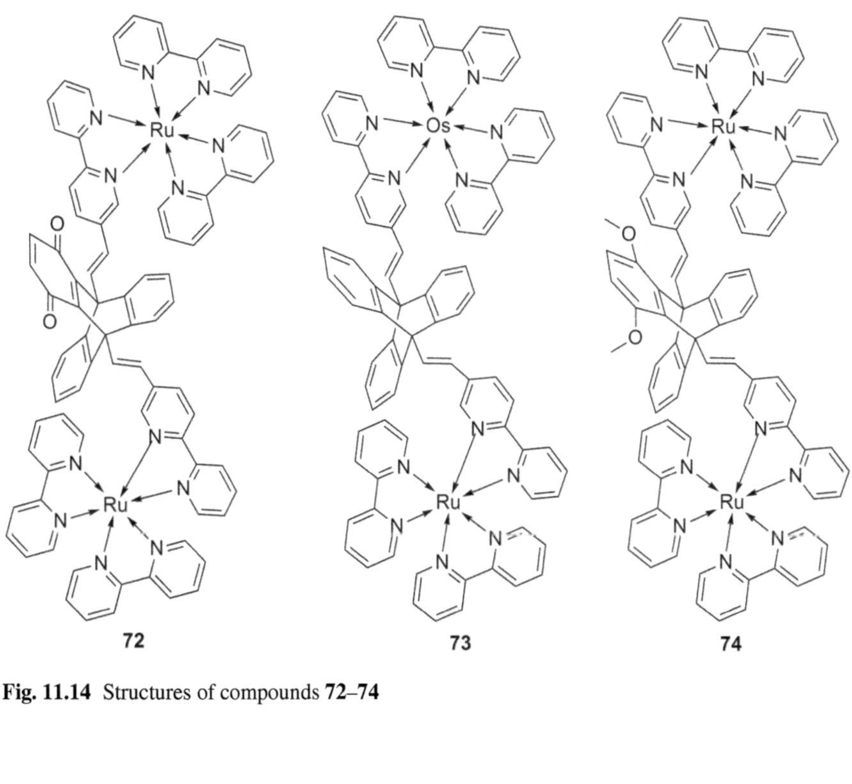

Fig. 11.14 Structures of compounds **72–74**

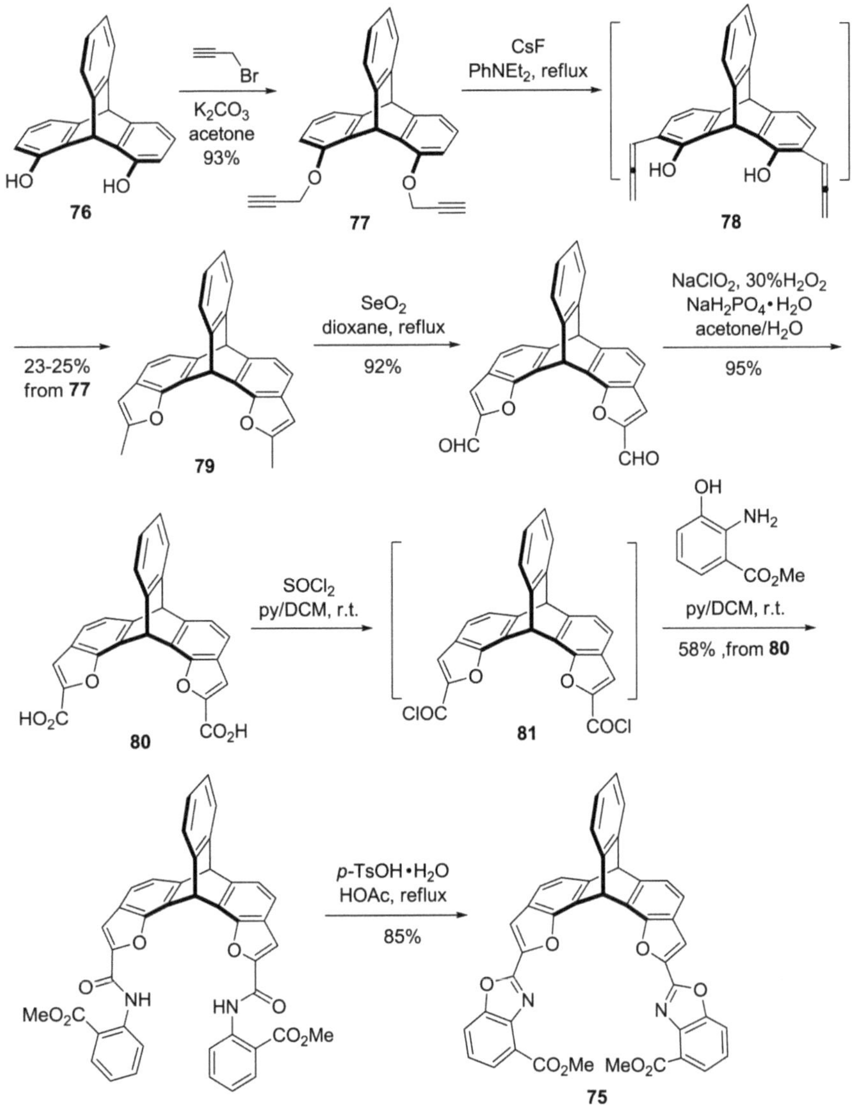

Scheme 11.23 Synthesis of bis-furan-fused triptycene backbone **75**

More recently, Li et al. [43] designed and synthesized a novel triptycene-based ligand with a preorganized framework. As shown in (Scheme 11.23), through a multistep synthesis, the target ligand with a bis-furan-fused triptycene backbone (**75**) along with the well-chosen benzoxazole *N*, *O*-donor arms could be efficiently synthesized. Firstly, 1,8-dihydroxytriptycene (**76**) was treated with propargyl bromide in an acetone solution in the pressure of K_2CO_3 to give the alkylation product **77** in 93 % yield. Then, the treatment of compound **77** with CsF in $PhNEt_2$ afforded the 2-methyl benzofuran **79** in 23–25 % yield via the highly reactive allene species (**78**). This step was one of the key steps in the process for the synthesis of target **75**. Another key step was the introduction of benzoxazole *N*, *O*-donor arms into the triptycene backbone. Thus, the diacid **80** was treated with excess amounts of pyridine (40 equivalents) and thionyl chloride (40 equivalents) in anhydrous methylene dichloride (DCM) to give the key benzoyl chloride intermediate **81**, which further reacted with methyl (3-hydroxy)anthranilate, and finally followed by the treatment with acetic acid and toluene sulfonic acid to afford the target ligand **75** in a reasonable overall yield.

References

1. Pohl RL, Willeford BR (1970) Triptycenetricarbonylchromium(0). J Organomet Chem 23(2):C45–C46
2. Gancarz RA, Blount JF, Mislow K (1985) Triptycene–π complexes: molecular-structures of $(C_{20}H_{14})Cr(CO)_3$ and $(C_{20}H_{14})Co_4(CO)_9$. Organometallics 4(11):2028–2032
3. Gancarz RA, Baum MW, Hunter G, Mislow K (1986) Dynamics of internal motion in nonacarbonyl(η^6-mesitylene)tetracobalt and nonacarbonyl(η^6-triptycene)tetracobalt. Carbonyl scrambling and rotation about the triptycene-cobalt axis. Organometallics 5(11):2327–2332
4. Moser GA, Rausch MD (1974) The utility of triammin-etricarbonylchromium, $(NH_3)_3Cr(CO)_3$, in the synthesis of arenetricarbonyl-chromium complexes. Syn React Inorg Met 4(1):37–48
5. Fung CW, Khorramdel-Vahed M, Ranson RJ, Roberts RMG (1980) Kinetics and mechanism of mercuriation of aromatic compounds by mercury trifluoroacetate in trifluoroacetic acid. J Chem Soc Perkin T 2(2):267–272
6. Fagan PJ, Ward MD, Calabrese JC (1989) Molecular engineering of solid-state materials: organometallic building-blocks. J Am Chem Soc 111(5):1698–1719
7. Schmidbaur H, Probst T, Steigelmann O (1991) A triptycene complex of Tin(II): $(C_{20}H_{14})SnCl(AlCl_4)_2$. Organometallics 10(9):3176–3179
8. Toyota S, Okuhara H, Oki M (1997) Synthesis and structures of all possible $Cr(CO)_3$ complexes of triptycene. Organometallics 16(18):4012–4015
9. Munakata M, Wu LP, Sugimoto K, Kuroda-Sowa T, Maekawa M, Suenaga Y, Maeno N, Fujita M (1999) Silver(I) complex assemblies with nonplanar aromatic compounds. Inorg Chem 38(25):5674–5680
10. Wen M, Munakata M, Suenaga Y, Kuroda-Sowa T, Maekawa M (2002) Porous silver(I) organometallic coordination polymer of triptycene, and the guest desorption and absorption. Inorg Chim Acta 340:8–14
11. Wen M, Munakata M, Li YZ, Suenaga Y, Kuroda-Sowa T, Maekawa M, Anahata M (2007) 1D chain and 3D framework of silver(I) organo-metallic polymers self-assembled with triptycene. Polyhedron 26(12):2455–2460
12. Vagin S, Ott A, Weiss HC, Karbach A, Volkmer D, Rieger B (2008) Metal-organic frameworks (MOFs) composed of (triptycenedicarboxylato)zinc. Eur J Inorg Chem 2008(16):2601–2609
13. Ishii A, Matsubayashi S, Takahashi T, Nakayama J (1999) Preparation of a selenenic acid and isolation of selenoseleninates. J Org Chem 64(4):1084–1085
14. Ishii A, Takahashi T, Nakayama J (2001) Hydrolysis of an isolable selenoseleninate under acidic and alkaline conditions. Heteroat Chem 12(4):198–203
15. Ishii A, Takahashi T, Tawata A, Furukawa A, Oshida H, Nakayama J (2002) First synthesis and characterization of isolable thioselenenic acid, triptycene-9-thioselenenic acid. Chem Commun 2002(23):2810–2811

16. Ishii A, Mori Y, Uchiumi R (2005) Synthesis and acetyl a study on and hydrolysis of *p*-toluoyl 9-triptycyl diselenides: generation of triptycene-9-selenenoselenoic acid. Heteroat Chem 16(6):525–528
17. Ishii A, Nakata N, Uchiumi R, Murakami K (2008) Reactions of a ditriptycyl-substituted selenoseleninate and related compounds with a platinum(0) complex: formation of selenaplatinacycle and hydrido selenolato platinum(II) complexes. Angew Chem Int Ed 47(14):2661–2664
18. Ishii A, Kamon H, Murakami K, Nakata N (2010) Hydroselenation and carboselenation of electron-deficient alkynes with isolable (hydrido) (selenolato)platinum(II) complexes and a selenaplatinacycle bearing a triptycene skeleton. Eur J Org Chem 2010(9):1653–1659
19. Nakata N, Uchiumi R, Yoshino T, Ikeda T, Kamon H, Ishii A (2009) Reactions of 9-Triptyceneselenol with palladium(0) complexes: unexpected formations of the dinuclear palladium(I) complex $\{Pd(PPh_3)\}_2(\mu\text{-SeTrip})_2$ and five-membered selenapalladacycle $Pd(\eta^2$(C, Se)-Trip)(dppe). Organo-metallics 28(7):1981–1984
20. Brynda M, Geoffroy M, Bernardinelli G (1999) Air-stable crystalline primary phosphines and germanes: synthesis and crystal structures of dibenzobarellenephosphine and tribenzobarellenegermane. Chem Commun 1999(11):961–962
21. Brynda M, Bernardinelli G, Dutan C, Geoffroy M (2003) Kinetic stabilization of primary hydrides of main group elements. The synthesis of an air-stable, crystalline arsine and silane. Inorg Chem 42(21):6586–6588
22. Nakata N, Fukazawa S, Ishii A (2009) Synthesis and crystal structures of the first stable mononuclear dihydrogermyl(hydrido) platinum(II) complexes. Organometallics 28(2):534–538
23. Nakata N, Fukazawa S, Kato N, Ishii A (2011) Palladium(II) hydrido complexes having a primary silyl or germyl ligand: synthesis, crystal structures, and dynamic behavior. Organometallics 30(17):4490–4493
24. Becker G, Gresser G, Uhl W (1981) Acylidenphosphines and alkylidenephosphines. 15. 2,2-dimethylpropylidynephosphine, a stable compound with a phosphorus atom of coordination number-1. Z Naturforsch B 36(1):16–19
25. Brym M, Jones C (2003) Synthesis, characterisation and reactivity of the first diphosphaalkyne. Dalton Trans 2003(19):3665–3667
26. Baker RJ, Brym M, Jones C, Waugh M (2004) 9-Triptycenyl complexes of group 13 and 15 halides and hydrides. J Organomet Chem 689(4):781–790
27. Ramakrishnan G, Jouaiti A, Geoffroy M, Bernardinelli G (1996) 9-substituted triptycene as a probe for the study of internal rotation around a C–PH bond in the solid state: a single crystal EPR study at variable temperature. J Phys Chem 100(26):10861–10868
28. Grossman O, Azerraf C, Gelman D (2006) Palladium complexes bearing novel strongly bent trans-spanning diphosphine ligands: synthesis, characterization, and catalytic activity. Organometallics 25(2):375–381
29. Grossman O, Gelman D (2006) Novel trans-spanned palladium complexes as efficient catalysts in mild and amine-free cyanation of aryl bromides under air. Org Lett 8(6):1189–1191
30. Azerraf C, Grossman O, Gelman D (2007) Rigid trans-spanning triptycene-based ligands: how flexible they can be? J Organomet Chem 692(4):761–767
31. Azerraf C, Cohen S, Gelman D (2006) Roof-shaped halide-bridged bimetallic complexes via ring expansion reaction. Inorg Chem 45(17):7010–7017
32. Azerraf C, Gelman D (2008) Exploring the reactivity of C(*sp3*)-cyclometalated Ir^{III} compounds in hydrogen transfer reactions. Chem Eur J 14(33):10364–10368
33. Levy R, Azerraf C, Gelman D, Rueck-Braun K, Kapoor PN (2009) Cyclometalated phosphine-based Ir(III) pincer complex as a catalyst for Oppenauer-type oxidation of alcohols. Catal Commun 11(4):298–301
34. Azerraf C, Gelman D (2009) New shapes of PC(*sp3*)P pincer complexes. Organometallics 28(22):6578–6584
35. Bini L, Mueller C, Wilting J, von Chrzanowski L, Spek AL, Vogt D (2007) Highly selective hydrocyanation of butadiene toward 3-pentenenitrile. J Am Chem Soc 129(42):12622–12623

36. Grossman O, Rueck-Braun K, Gelman D (2008) Trans-spanned palladium catalyst for mild and efficient amination of aryl halides with benzophenone imine. Synthesis 2008(4):537–542
37. Kodanko JJ, Morys AJ, Lippard SJ (2005) Synthesis of diethynyltriptycene-linked dipyridyl ligands. Org Lett 7(21):4585–4588
38. Kodanko JJ, Xu D, Song DT, Lippard SJ (2005) Iron substitution for sodium in a carboxylate-bridged, heterodinuclear sodium-iron complex. J Am Chem Soc 127(46):16004–16005
39. Chong JH, MacLachlan MJ (2006) Robust non-interpenetrating coordination frameworks from new shape-persistent building blocks. Inorg Chem 45(4):1442–1444
40. Chong JH, MacLachlan MJ (2007) Synthesis and structural investigation of new triptycene-based ligands: en route to shape-persistent dendrimers and macrocycles with large free volume. J Org Chem 72(23):8683–8690
41. Roy X, Chong JH, Patrick BO, MacLachlan MJ (2011) Molecular scaffolding of prussian blue analogues using a phenanthroline-extended triptycene ligand. Cryst Growth Des 11(10):4551–4558
42. Beyeler A, Belser P (2002) Synthesis of a novel rigid molecule family for the investigation of electron and energy transfer. Coord Chem Rev 230(1–2):29–39
43. Li Y, Cao R, Lippard SJ (2011) Design and synthesis of a novel triptycene-based ligand for modeling carboxylate-bridged diiron enzyme active sites. Org Lett 13(19):5052–5055

Chapter 12
Iptycenes and Their Derivatives in Sensors

12.1 Sensors Based on Iptycene-Containing Polymers

Generally speaking, the higher performance fluorescent chemosensors based on the conjugated polymers could be achieved through the amplification of the inherent sensitivity of the fluorescent material [1, 2]. In the iptycene-incorporated conjugated polymers, the rigid structure of iptycene can block the π–π stacking and the excimer formation in the solid state and these features make them potential candidates for fluorescent chemosensors. In 1998, Yang and Swager [3] took advantage of the films of pentiptycene-based conjugated polymers **1**–**3** (Fig. 12.1) with high fluorescence quantum yields and stabilities to act as the sensing materials for fluorescent chemosensors. These fluorescent chemosensors displayed wonderful abilities for the trace detection of the electron-deficient unsaturated species such as 2,4,6-trinitrotoluene (TNT), 2,4-dinitrotoluene (DNT), and benzoquinone (BQ). Especially, the polymer **1** exhibited high sensitivity to high explosive vapors (TNT, DNT), and repetitively great spectroscopic stability along with a fast response time. All of the excellent performances made it a commercially available device called "Fido", which had been applied for the detection technology of robotic and handheld explosive sniffers. The electron transfer from the excited polymer to the electron acceptors leading to the fluorescence quenching seemed to be regarded as the detection mechanism (Fig. 12.2). Further studies [4] suggested that the electron-rich environment, good balance of the electrostatic interactions, and the porous structures in the films played the key roles in the high sensitivity. Moreover, there were also varied factors that could determine the degree of fluorescence quenching, including the porosity and thickness of the film itself, the vapor pressure and the diffusion ability of analytes, and the exergonicity of electron transfer, the strength of binding between the polymer and the analytes.

In 2001, Yamaguchi and Swager [5] also synthesized a novel pentiptycene-based π-conjugated polymer **4** (Fig. 12.3) containing dibenzochrysene moieties. It was found that the polymer **4** exhibited a longer fluorescence lifetime (2.5 ns) than the corresponding phenylene-based π-conjugated polymers with a subnanosecond lifetime. Also, this polymer even displayed higher sensitivity than the polymer **1** in the sensing experiments of TNT.

C.-F. Chen, Y.-X. Ma, *Iptycenes Chemistry*,
DOI 10.1007/978-3-642-32888-6_12,

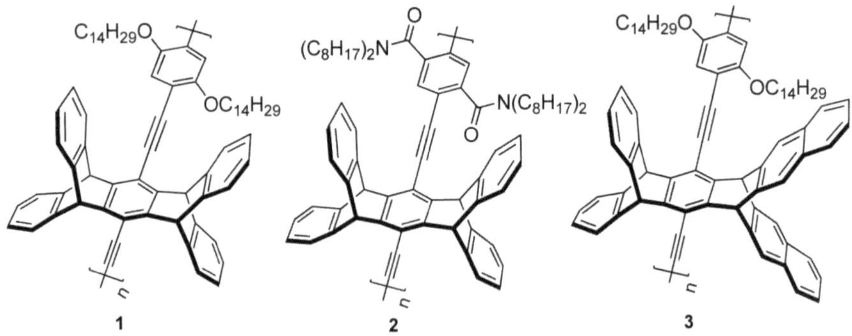

Fig. 12.1 Structures of pentiptycene-based conjugated polymers **1–3**

Fig. 12.2 Schematic representation of galleries defined between polymer chains that can host analytes. (Reprinted with permission from [3]. Copyright 1998 American Chemical Society)

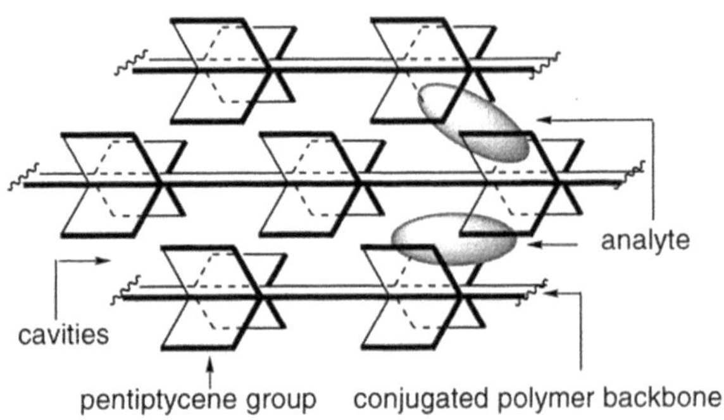

Fig. 12.3 Structure of pentiptycene-based π-conjugated polymer **4**

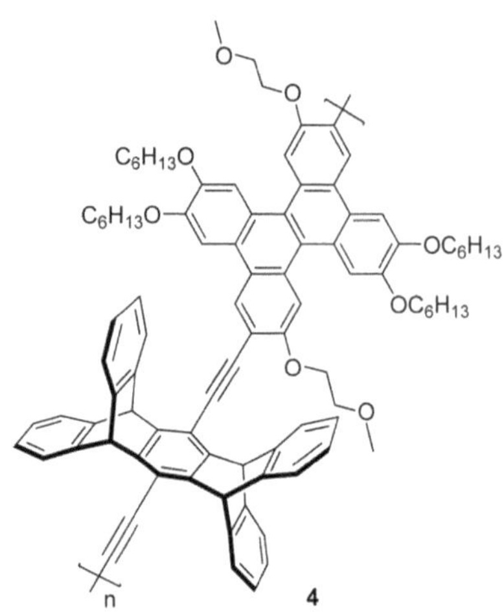

Soon afterwards, Zahn and Swager [6] further introduced the chiral alkyloxy side chains into the pentiptycene-based polymer **5** (Fig. 12.4). As a result, a helical

Fig. 12.4 Structure of pentiptycene-based conjugated polymer **5**

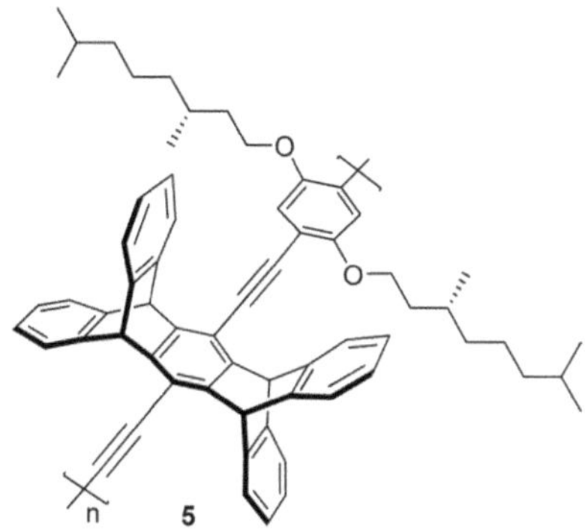

Fig. 12.5 Structure of triptycene-based PPVs **6**

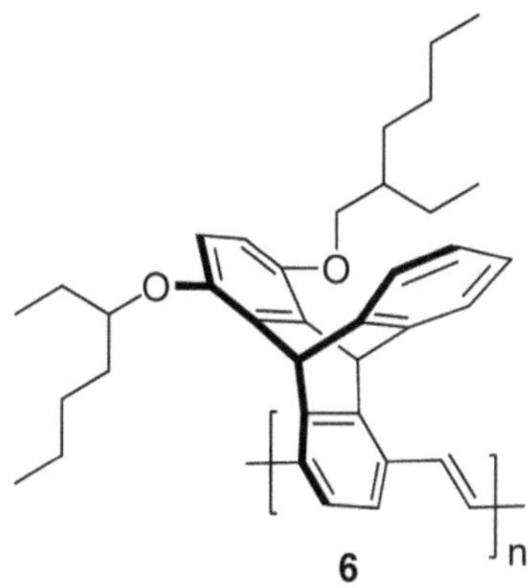

grid structure could be obtained by the aggregation of polymer **5** with the longer exciton diffusion length and conjugation length. The rigid pentiptycene moieties blocked the collinear aggregation of chains, and the chirality of the side chains could systematically create a strong electronic coupling without decreasing of fluorescence efficiency. This unique chiral structure and chiral, three-dimensional interactions resulted in the enhancements of sensitivity toward nitroaromatics with high quantum yields.

Recently, Malval and Leray [7] reported the photo-physical properties of the triptycene-based poly(phenylene vinylene)s (PPVs) **6** (Fig. 12.5) and its sensing ability for DNT. According to the studies, the strong fluorescent quenching seemed to be reasoned by a dynamic electron-transfer process from the polymer to the quencher. Moreover, it was found that there was a threshold value of film thickness for polymer **6**, and the fluorescence quenching rate of **6** displayed the thickness dependence when the films were thinner than 55 Å.

In 2005, Zhao and Swager [8] also reported a class of novel poly(iptycene-butadiynylene)s, which displayed excellent static quenching constants. Especially, for **7** (Fig. 12.6), the value could be up to 185 M^{-1} with TNT. It was suggested that there existed the ground-state associations, which was derived from the electrostatic

Fig. 12.6 Structure of poly(iptycene-butadiynylene)s **7**

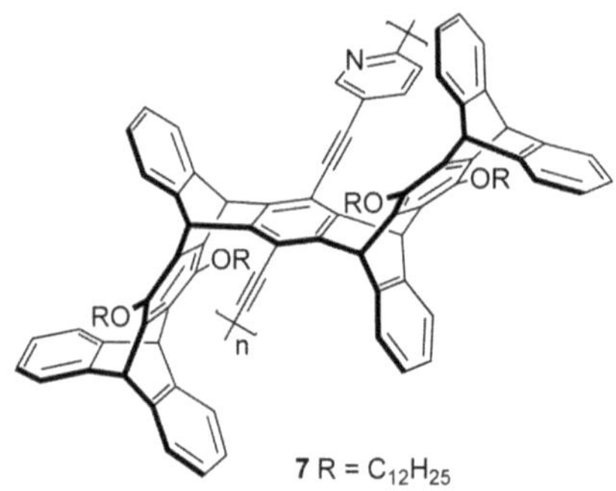

and π–π interactions between the electron-rich polymer and the electron-deficient analytes. However, the unique quencher-sequestering properties of these poly(*p*-phenylenebuta diynylene)s (PPDs) could lead to a relatively longer restoring time after the removal of TNT or DNT vapor.

In 2005, Amara and Swager [9] reported that the conjugated polymers **9a** and **9b** (Fig. 12.7) containing the strongly hydrogen bond-donating hexafluoro-2-propanol (HFIP) groups displayed higher fluorescence response than the control **8** to the hydrogen bond-accepting analytes, such as pyridine and 2,4-dichloropyrimidine. The incorporation of HFIP moieties not only increased the binding interactions between the analytes and the polymer, but also facilely activated the charge-transfer reactions with the analytes. A combination of these features resulted in the stronger fluorescence-quenching responses.

In 2006, Thomas and Swager [10] also reported a new kind of conjugated PPEs that could be used for the detection of trace hydrazine vapor in a turn-on typed method with a sensitivity below the permissible exposure limit of hydrazine (1 ppm). In addition, it was found that the highly electron-rich polymers such as the polymer **10** (Fig. 12.8) showed the better fluorescence response to the hydrazine vapor. In a certain sense, the greater magnificance fluorescence quenching of polymers in essence was, the higher selectivity levels of this type sensor had.

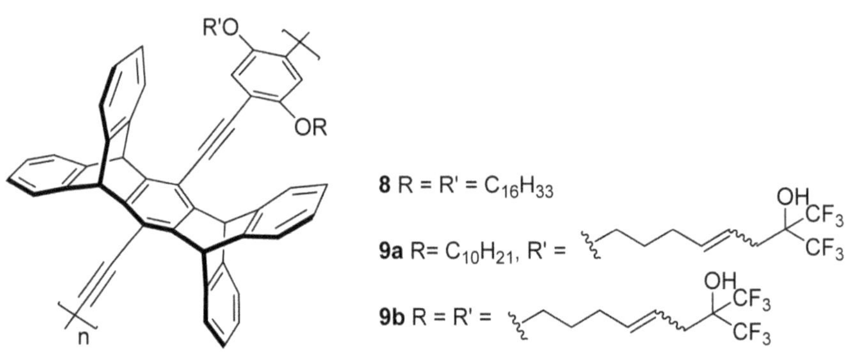

Fig. 12.7 Structures of polymers **8** and **9**

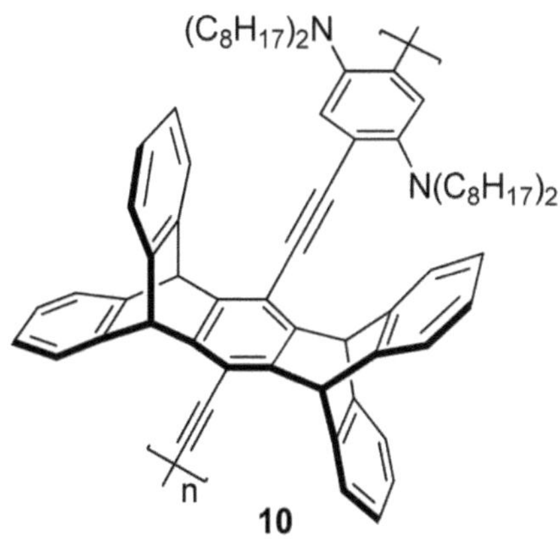

10

Fig. 12.8 Structure of highly electron-rich polymer **10**

Interestingly, in 2004, Kwan et al. [11] reported a kind of rotaxane-type conjugated polymers, and found that they had the sensing capability for alcohol vapor. As shown in Scheme 12.1, the reaction of rotaxane **11** with the aryl diacetylene gave the corresponding PPEs (**12a**, **b**) by the microwave-assisted Sonogashira–Hagihara-coupling method. In the presence of acidic alcohol vapor, the emission of thin films of **12** showed conspicuous quenching, which suggested to be the contribution of the hydrogen bonding between the rotaxane groups with acidic alcohols moieties, according to the nuclear magnetic resonance (NMR) experiments. In addition, the thin films of polymers **12a**, **b** doped with Zn^{2+} also displayed the sensitivity to alcohol vapor such as methanol, ethanol, 2-ethylhexanol, or cyclohexanol.

It was considered that the two-dimensional thin film with less chances of revisiting the same polymer segment probably resulted in the greater degree of migrational freedom of excitons. Therefore, Wosnick et al. [12] prepared the silica microspheres coating with the thin films of weakly anionic, nonaggregating PPE **13** (Fig. 12.9) as the polyanion, while the quaternary ammonium poly(diallyldimethylammonium chloride) (PDAC) as the polycation by the "layer-by-layer" (LbL) deposition method. Consequently, the resulting microspheres with fluorescence coat showed the similar fluorescence properties to the corresponding polymers in solution, and the higher sensitivity to nitro-aromatic quenchers with enhanced quenching behavior, due to both surface and electronic effects.

In addition, Joly et al. [13] also used the host–guest system to construct the sensing polymers. The hydrophobic interactions and the host–guest complexation of the crown ether-functionalized PPE **14** (Fig. 12.10), which introduced the pentiptycene-based monomer, made the system high sensitivity toward detecting viologens under the aqueous environments. Both the combination of hydrophobic interactions and the host–guest complexation between the crown ether moiety and cationic viologens made contributions to the high sensitivity of this pentiptycene-based PPE **14** with increased porosity. It was suggested that this hydrophobic, water-insoluble, conjugated polymer could form thin film with the rigid iptycene skeleton, which had potential

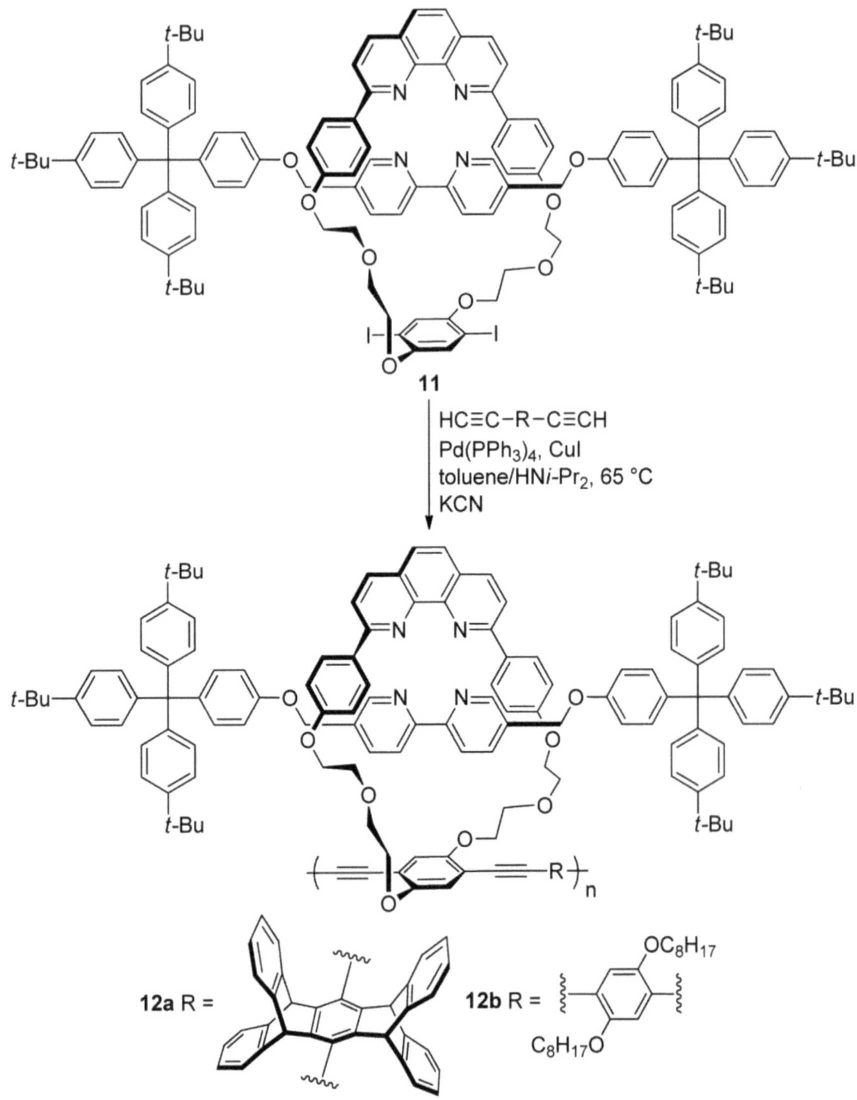

Scheme 12.1 Synthesis of rotaxane-type conjugated polymers

applications in the solid-state sensory devices in aqueous environments, rather than being confined to the solution-based sensors.

It was considered that when the iptycene-based amplifying fluorescent polymers (AFPs) were introduced with the biomolecules, the resulting polymers probably exhibited the biosensing. Thus, Zheng and Swager [14] designed and synthesized the organic solvent-soluble biotinylated PPE **15** (Fig. 12.11), which contained the pentiptycene moieties and showed greater thin-film quantum yield. According to the

Fig. 12.9 Structure of nonaggregating PPE **13**

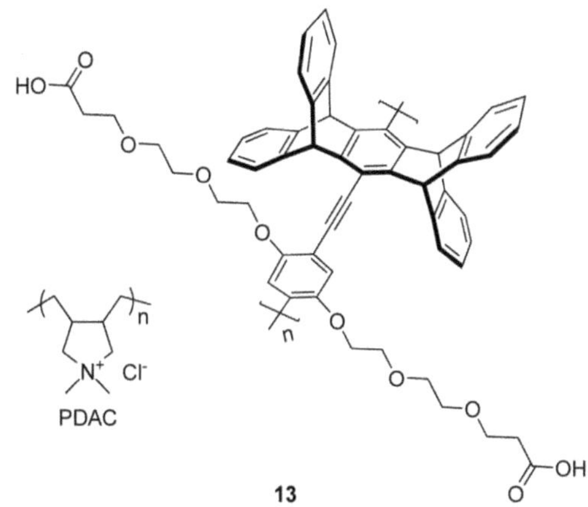

13

Fig. 12.10 Structure of crown ether-functionalized PPE **14**

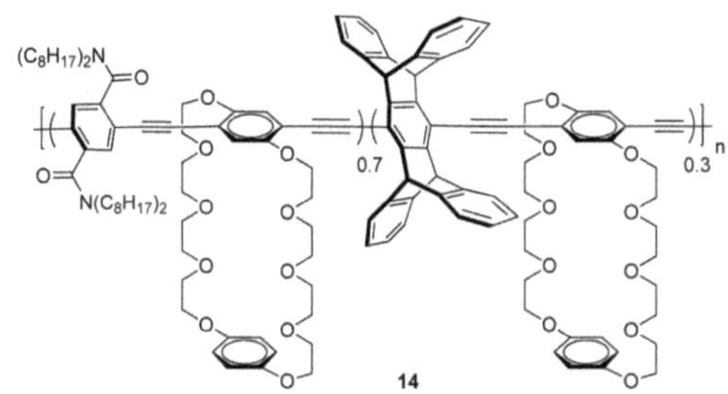

14

Fig. 12.11 Structures of organic solvent-soluble biotinylated PPEs **15**

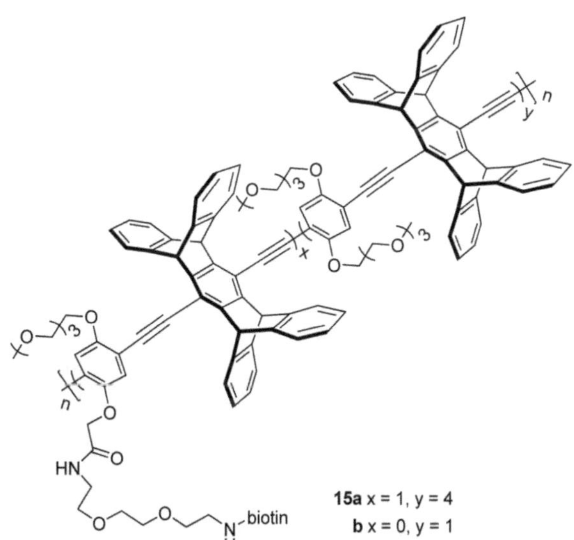

15a x = 1, y = 4
b x = 0, y = 1

Fig. 12.12 Structure of pentiptycene-bispyrenyl system **16**

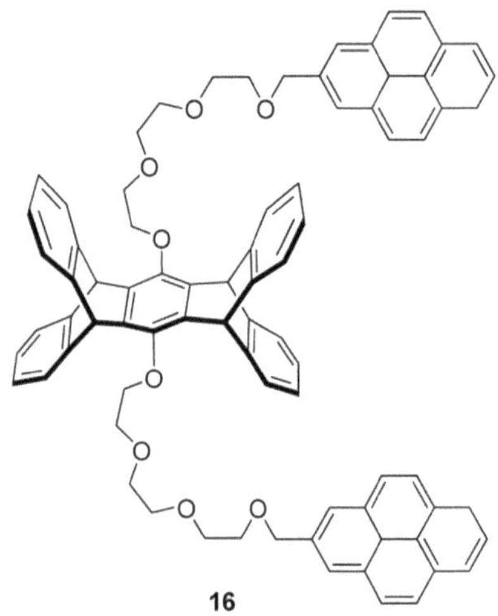

16

data of fluorescence spectra, it was suggested that the RhB dye had greater energy–transfer (ET) interactions than that of T-red-strept, which was probably caused by the more intimate interactions between the smaller molecule and the polymer chains. Further fluorescence spectral results revealed that the ET between the biotinylated polymer and the dye-labeled streptavidin played the important role in the biosensing application of the PPE **15**.

12.2 Other Iptycene-Based Sensors

In 2001, Yang et al. [15] designed and synthesized a pentiptycene-bispyrenyl system **16** (Fig. 12.12) starting from pentiptycene hydroquinone. They found that in the presence of Cu^{2+} ion, the compound **16** displayed a blue shift along with an intensity enhancement of the pyrene excimer emission. With further analysis, it was found that dynamic excimer of this system was in a sandwich-like shape, while the static one contained partially overlapping. The conversion between the different excimers led to this fluorescence responses.

In 2005, Gong et al. [16] reported a novel electrochemical method for the sensitive determination of biological thiols including homocysteine, cysteine, and glutathione. This method was based on the rational attachment of synthetic triptycene orthoquinone onto the single-walled carbon nanotubes (SWNTs; Fig. 12.13). Consequently, the system not only totally mimicked the redox properties inherent in the SWNTs themselves, but also increased the catalytic sites onto the SWNTs, which thus made this new electrochemical method promising as a highly sensitive and stable detector for biological thiols.

Recently, Zyryanov et al. [17] designed and synthesized a series of 1,4-diarylpentiptycenes and then tested their sensing abilities toward TNT and DNT.

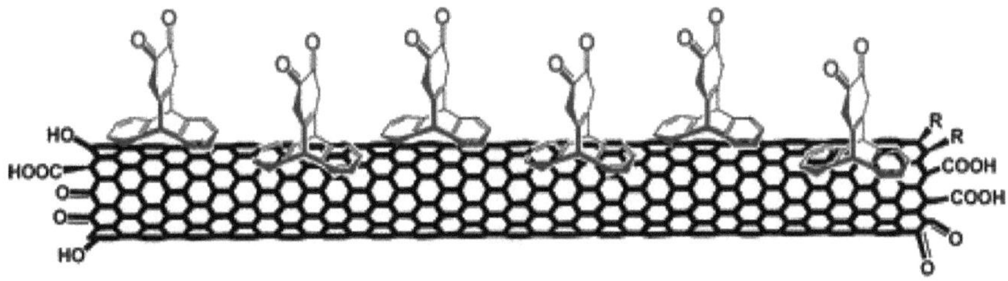

Fig. 12.13 Schematic illustration of attachment of triptycene orthoquinone onto SWNTs. (Reprinted with permission from [16]. Copyright 2005 American Chemical Society)

Fig. 12.14 Structure of 1,4-diarylpentiptycenes **17**

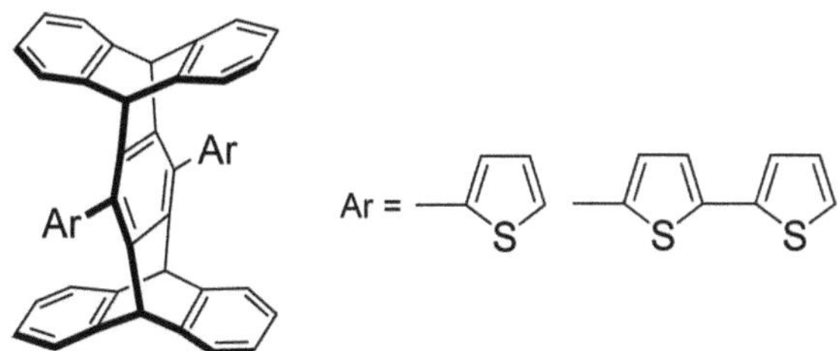

17

Consequently, it was found that the 1,4-diarylpentiptycenes (**17**, Fig. 12.14) displayed high sensing ability with the considerably fluorescence quenching for TNT in both solution-cast films and solid states.

In 2010, Zhao et al. [18] reported a new kind of selective supramolecular fluorescence Ba^{2+} probe based on the triptycene-derived macrotricyclic host (**18**). First, it was found that the host **18** could selectively form a 1:2 stable complex (**19**) with two (9-anthracylmethyl) benzyl-ammonium salts (**20**), and in the complex the 9-anthracyl groups of the two guests were all selectively positioned inside the lateral crown ether cavities of **18**. Then, the study revealed that the presence of Ba^{2+} would considerably induce the fluorescence enhancement of complex **19** in $CHCl_3/CH_3CN$ (1:1, v/v), but under the same tested conditions, no obvious spectral changes of **19** upon the addition of the other tested metal ions were observed (Fig. 12.15). These results suggested that Ba^{2+} ions might bind much stronger with host **18** than those of other tested metal ions. This feature made this complex potential in the supramolecular fluorescence probe for Ba^{2+}.

The further fluorescence and ^{1}H NMR spectroscopic titration experiments showed that the respond behavior of probe **19** toward Ba^{2+} was in a two-step process. As shown in Scheme 12.2, upon the addition of Ba^{2+} ion to complex **19** system, one of the guest molecules, (9-anthracylmethyl)benzylammonium ions (**20**), would be displaced to form the complex **21**, then as more and more amount of Ba^{2+} ions were added into the system, the remaining guest molecule was replaced by Ba^{2+} ion to give the complex **22**, which showed a higher fluorescence intensity than the local complex **19**.

More recently, Hu and Chen [19] also synthesized two pairs of triptycene-derived heterocalixarenes (**23–26**, Fig. 12.16) and investigated their binding properties

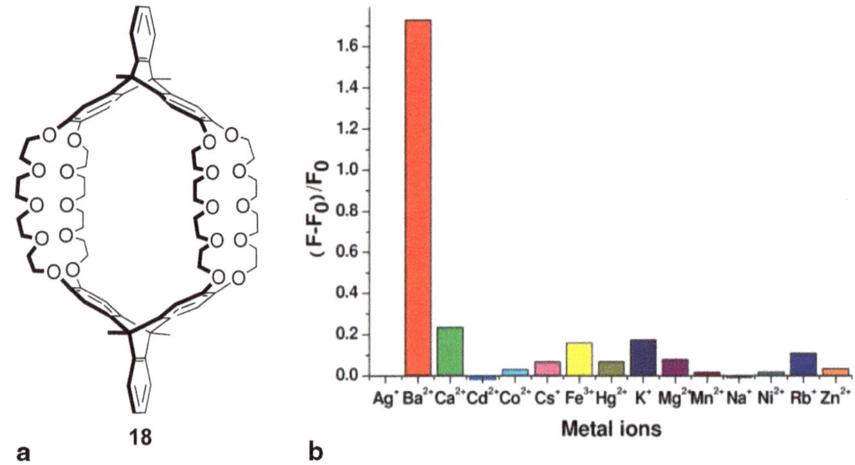

Fig. 12.15 **a** Structure of host **18**. **b** Fluorescence enhancement ratio of a solution of complex **19** in $CHCl_3/CH_3CN$ (1:1, v/v) in the presence of equimolar metal ions. (Reprinted with permission from [18]. Copyright 2010 American Chemical Society)

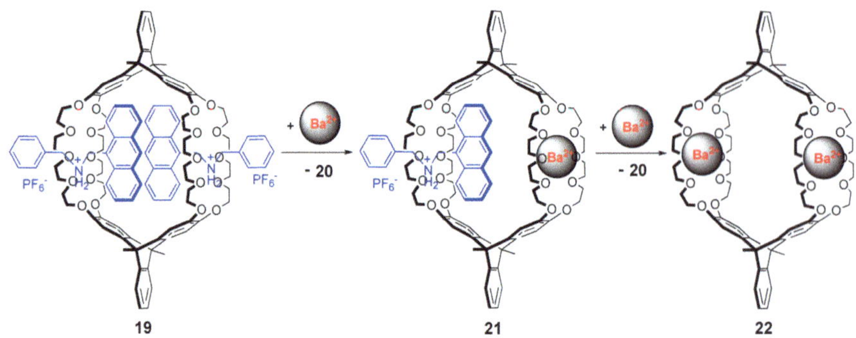

Scheme 12.2 Graphic representation of the displacement of guest in complex by Ba^{2+} ions

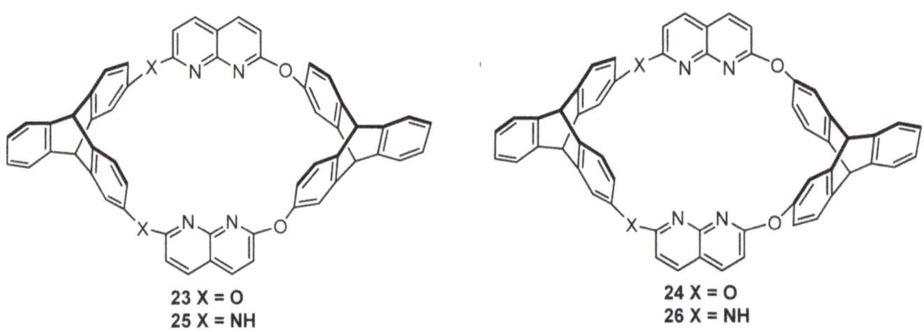

Fig. 12.16 Chemical structures of triptycene-derived heterocalixarenes **23–26**

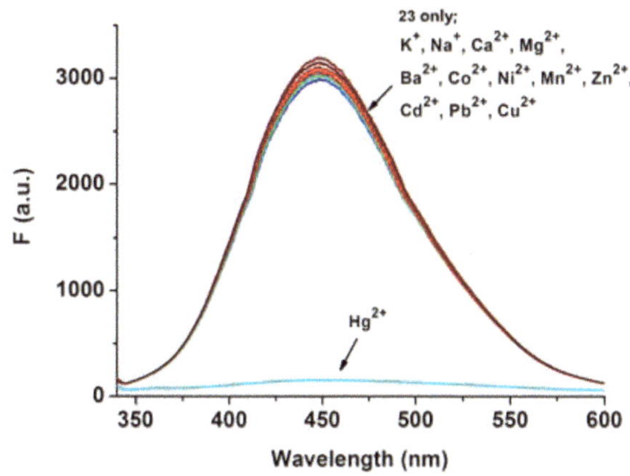

Fig. 12.17 Fluorescence spectra of **23** (10 μM in $CH_3CN–H_2O$, 98:2, v/v) in the absence and presence of 5 equivalent of metal ions

toward various metal ions in detail by spectroscopic methods. They found that among the four macrocycles only oxacalixarene **23**, a *cis* isomer with a boat-like *1,3-alternate* conformation and a symmetrical cavity structure, showed high selective response toward Hg^{2+} (Fig. 12.17). Moreover, it was further found that such selectivity was controlled by the bridging heteroatoms and cavity structures of macrocycles. This feature made the macrocycle **23** potential as a fluorescent probe for Hg^{2+} ion.

References

1. Zhou Q, Swager TM (1995) Fluorescent chemosensors based on energy migration in conjugated polymers: the molecular wire approach to increased sensitivity. J Am Chem Soc 117(50):12593–12602
2. Zhou Q, Swager TM (1995) Methodology for enhancing the sensitivity of fluorescent chemosensors: energy migration in conjugated polymers. J Am Chem Soc 117(26):7017–7018
3. Yang JS, Swager TM (1998) Porous shape persistent fluorescent polymer films: an approach to TNT sensory materials. J Am Chem Soc 120(21):5321–5322
4. Yang JS, Swager TM (1998) Fluorescent porous polymer films as TNT chemosensors: electronic and structural effects. J Am Chem Soc 120(46):11864–11873
5. Yamaguchi S, Swager TM (2001) Oxidative cyclization of bis(biaryl)acetylenes: synthesis and photophysics of dibenzo[*g*, *p*]chrysene-based fluorescent polymers. J Am Chem Soc 123(48):12087–12088
6. Zahn S, Swager TM (2002) Three-dimensional electronic delocalization in chiral conjugated polymers. Angew Chem Int Ed 41(22):4225–4230
7. Malval JP, Leray I (2010) Photophysical properties of poly(triptycene vinylene) derivatives and effect of nitrotoluene exposure. Chem Phys Lett 501(1–3):54–58
8. Zhao DH, Swager TM (2005) Conjugated polymers containing large soluble diethynyl iptycenes. Org Lett 7(20):4357–4360
9. Amara JP, Swager TM (2005) Synthesis and properties of poly(phenylene ethynylene)s with pendant hexafluoro-2-propanol groups. Macromolecules 38(22):9091–9094
10. Thomas SW, Swager TM (2006) Trace hydrazine detection with fluorescent conjugated polymers: a turn-on sensory mechanism. Adv Mater 18(8):1047–1050

11. Kwan PH, MacLachlan MJ, Swager TM (2004) Rotaxanated conjugated sensory polymers. J Am Chem Soc 126(28):8638–8639
12. Wosnick JH, Liao JH, Swager TM (2005) Layer-by-layer poly(phenylene ethynylene) films on silica microspheres for enhanced sensory amplification. Macromolecules 38(22):9287–9290
13. Joly GD, Geiger L, Kooi SE, Swager TM (2006) Highly effective water-soluble fluorescence quenchers of conjugated polymer thin films in aqueous environments. Macromolecules 39(21):7175–7177
14. Zheng J, Swager TM (2004) Biotinylated poly(*p*-phenylene ethynylene): unexpected energy transfer results in the detection of biological analytes. Chem Commun 2004(024):2798–2799
15. Yang JS, Lin CS, Hwang CY (2001) Cu^{2+}-induced blue shift of the pyrene excimer emission: a new signal transduction mode of pyrene probes. Org Lett 3(6):889–892
16. Gong KP, Zhu XZ, Zhao R, Xiong SX, Mao LQ, Chen CF (2005) Rational attachment of synthetic triptycene orthoquinone onto carbon nanotubes for electrocatalysis and sensitive detection of thiols. Anal Chem 77(24):8158–8165
17. Zyryanov GV, Palacios MA, Anzenbacher P (2008) Simple molecule-based fluorescent sensors for vapor detection of TNT. Org Lett 10(17):3681–3684
18. Zhao JM, Zong QS, Chen CF (2010) Complexation of triptycene-based macrotricyclic host toward (9-anthracylmethyl)benzylammonium salt: a Ba^{2+} selective fluorescence probe. J Org Chem 75(15):5092–5098
19. Hu SZ, Chen CF (2011) Hg^{2+} recognition by triptycene-derived heteracalixarenes: selectivity tuned by bridging heteroatoms and macrocyclic cavity. Org Biomol Chem 9(16):5838–5844

Chapter 13
Iptycenes and Their Derivatives in Molecular Balances

The molecular interactions play a great role in both chemistry and biology. To design better supramolecular systems and biologically active agents, it is indispensable to understand the mechanism involved in noncovalent interactions. In general, noncovalent interactions in natural complexes are overly complicated, and the precise geometry of an interaction in a conformationally dynamic biomolecule is hard to determine. Thus, the molecules, like the folding molecules, in which the small alterations in geometry lead to a considerable strength change of interaction, could be served as "molecular balances" to offer an attractive platform for the study of noncovalent interactions.

In the 1970s, Ōki [1] first investigated the restricted bond rotation in a series of triptycene derivatives. They found that the distance between the two substituents of 1,9-disubstituted triptycene (**1**) was unusually short, only about 2.7 Å between X and Y [2–4]. It was noteworthy that the characteristic of the substituent had an impact on the distance between X and Y. When the α-position of 9-substituent was a nonhydrogen atom, the distance between the β-atom of the 9-substituent was about 3.0 Å in the *syn*-configuration. In addition, the overlap of the substituents' orbitals, contributed effectively to the charge-transfer interactions between the electron donor and acceptor, which contributed effectively to the charge-transfer interactions between the electron donor and acceptor. These two structural features made it possible to detect a weak molecular interaction. In these systems, the small difference in the repulsive forces between the two rotameric positions could result in the change of configuration. Thus, 1,9-disubstituted triptycene **1** with typical structural characteristics could potentially act as the sensitive and convenient probes for the weak, yet attractive molecular interactions. Additionally, the substituents in the 9-(bridge-head) position brought an unusually high rotational barrier about the C–C bond [5]. Subsequently, Ōki [1] and Nakamura et al. [6] illustrated the conformational equilibrium in the 1,9-disubstituted triptycenes, as shown in Fig. 13.1. In the folded $\pm$ *syn*-conformation, the substituents at the 1- and 9-position could interact with each other; whereas, in the unfolded *anti*-conformation, they were diverged. Therefore, it could well determine the configuration simply via the evaluation of the intramolecular interaction between the two substituents. When the system has large enough substituents in 1- and 9-position, the distinct conformers with the increased rotation

C.-F. Chen, Y.-X. Ma, *Iptycenes Chemistry*,
DOI 10.1007/978-3-642-32888-6_13, © Springer-Verlag Berlin Heidelberg 2013

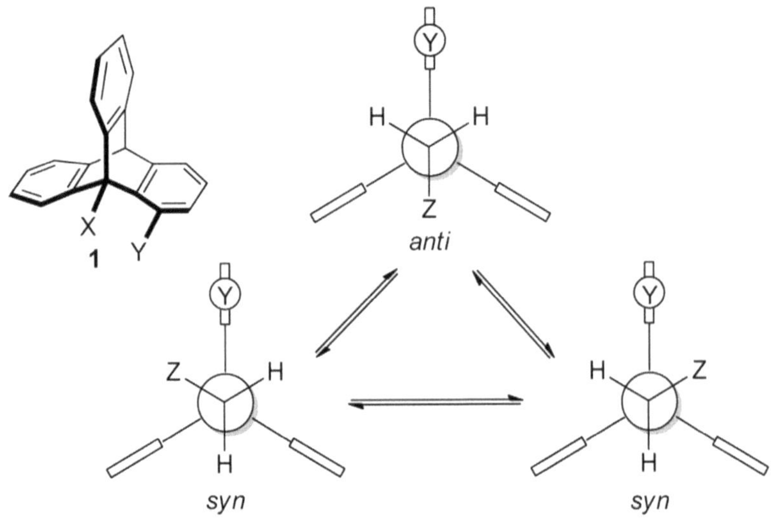

Fig. 13.1 The conformational equilibrium in the 1,9-disubstituted triptycene

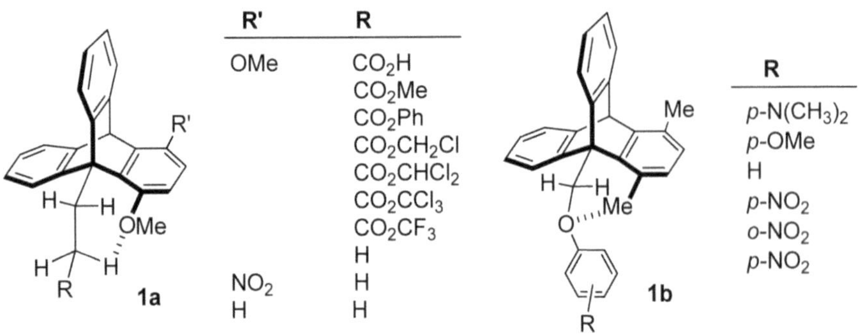

Fig. 13.2 Structures of 1,9-disubstituted triptycene derivatives

barrier could be simply distinguished through the ^{1}H NMR spectra at low temperatures [5]. On the basis of these pioneering studies, Ōki et al. developed a series of 1,9-disubstituted triptycene systems as molecule balances, as shown in Fig. 13.2. These systems with methoxyl or methyl group could detect many types of noncovalent interactions, including weak *n*–σ* [7–9], *n*–π* [10–12] charge-transfer interactions, and interactions involving a methyl group, like $CH_3 \cdots O$ hydrogen bond [7, 12–16] and $CH_3 \cdots$ arene interactions [17, 18]. It was noteworthy that the $CH_3 \cdots$ arene interactions had not been found in other systems before. For example, when the R-groups were electron-withdrawing, these systems preferred to adopt the folded *syn*-conformers. Thus, the large size and decreased electron density of chlorine instead of methoxyl group would result in the decreased *syn*/*anti*-ratios. Taking advantage of this behavior, the torsion balances for the $CH_3 \cdots O$ hydrogen bond could be prepared.

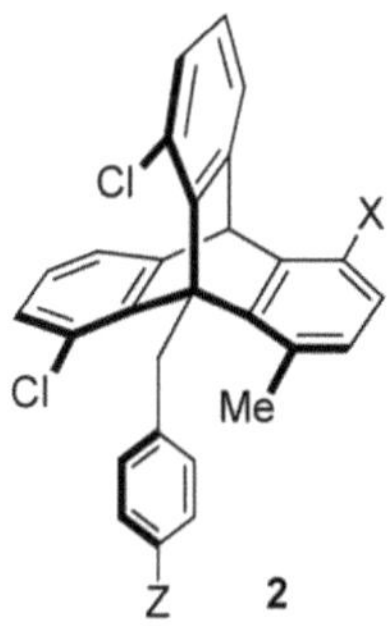

X	Z
Me	NO_2
	H
	NMe_2
H	H
CO_2Me	H
CN	H

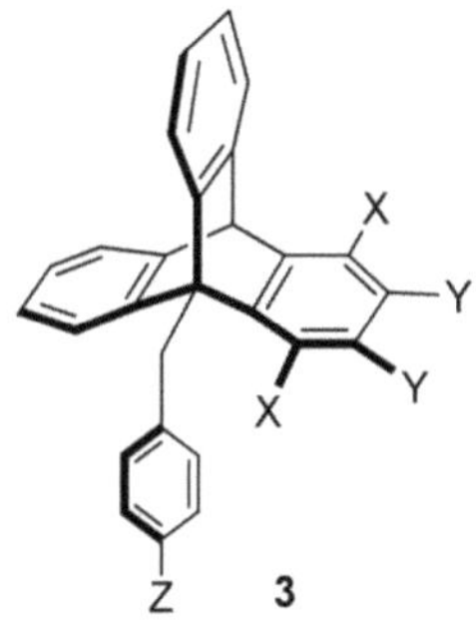

X	Y	Z
OMe	H	H
		Cl
		OMe
Cl	H	H
Br	Br	H

Fig. 13.3 Structures of 9-benzyl triptycene derivatives **2** and **3**

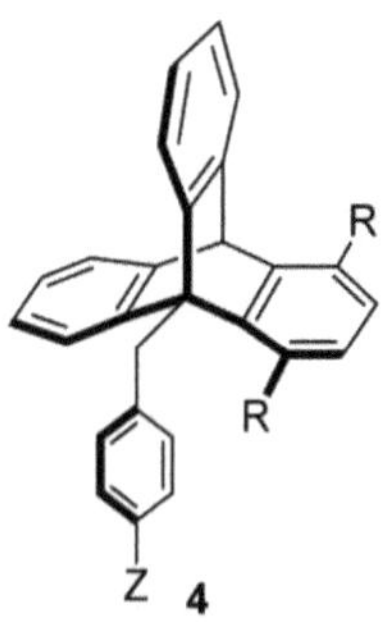

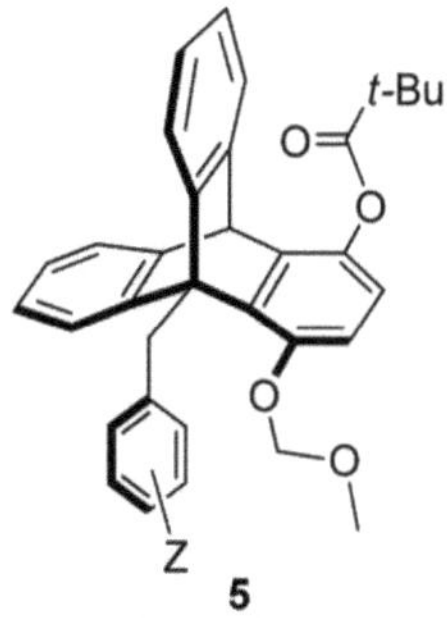

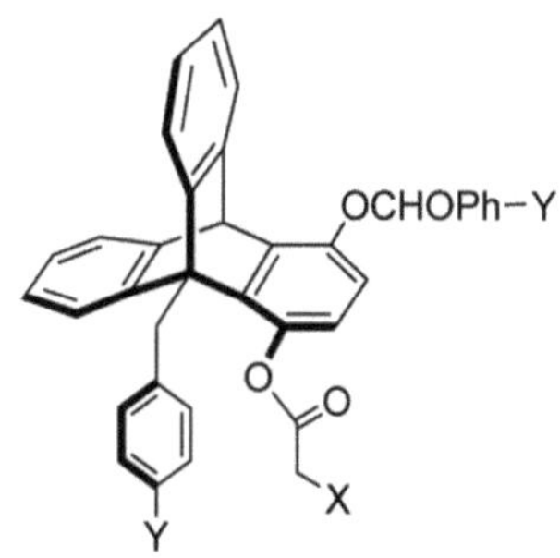

R = OMe, OCOMe, OCOEt, OCO*i*-Pr, OCOH, OCOCN, $OCOCF_3$
Z = MeO, Me, H, Cl, F, CN, CF_3

Z = *p*-NO_2, *p*-CN, *p*-CF_3, *p*-Br, *p*-Cl, *p*-F, H, *p*-Me, *p*-OMe, *p*-NMe_2, penta-fluoro

6a Y = H, X = CN, Cl, Br, I, H, Me, MeO
6b X = Me, Y = Me, H, F, Cl, Br, CN, NO_2

Fig. 13.4 Structures of triptycene–derived molecule balances **4**, **5**, and **6a, b**

As mentioned before, Ōki's molecule balances based on 9-benzyl triptycene derivatives could investigate the $CH_3 \cdots \pi$ interactions. Thus, Ōki [1] synthesized a series of 9-benzyl triptycene derivatives **2** and **3**, in which the substituents contained oxygen/halogen atoms (Fig. 13.3). These 9-benzyl triptycene derivatives could also be acted as molecule balances for studying the oxygen/halogen $\cdots$ arene interactions.

In 2005, Gung et al. [19, 20] designed and synthesized a series of 9-benzyl triptycene derivatives **4** and **5** containing the ester or methoxymethyl ether groups at 1-position (Fig. 13.4). For almost all cases, these unimolecule systems, especially for the methoxymethyl ether ones, preferred to be in folded *syn*-configuration. Moreover, it was noteworthy that the methoxymethyl ether triptycene system containing the electron-withdrawing Z-substituents of the arene showed the increased strength of the attractive interactions. These results revealed that both the dispersion forces and local dipole interaction in this system should be also considered, instead of being only treated the aromatic rings in the system as a simple quadrupolar functional

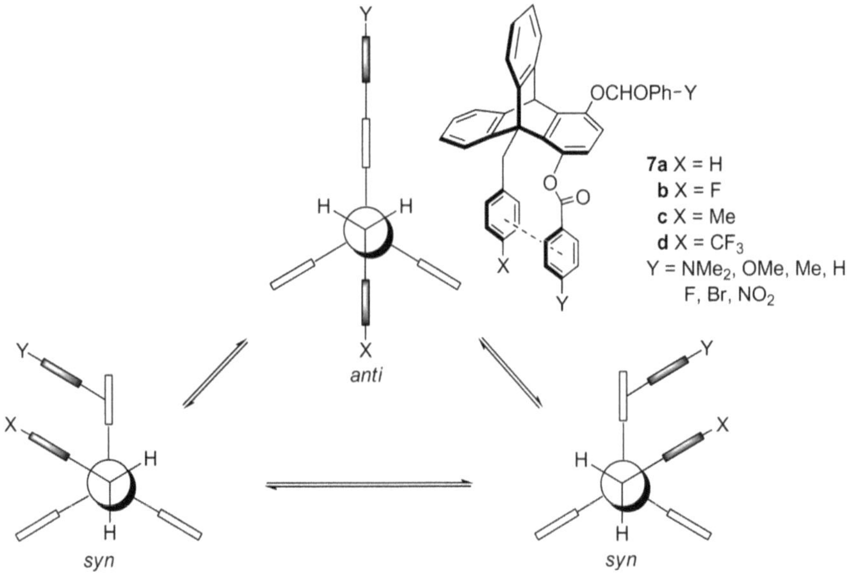

Fig. 13.5 The conformational equilibrium in the modified triptycene derivatives **7a–d**

group at Van der Waals' distance. Subsequently, Gung et al. [21] further introduced α-substituted acetates into the 1-position of triptycene, and obtained the new triptycene-derived molecule balances **6a, b** to quantify the CH $\cdots\pi$ interactions. Moreover, according to the results of the low-temperature ^{1}H NMR spectra, the free energies of the interactions and the equilibrium constants were also obtained from the ratios of the *syn*/*anti*-rotamers. In addition, it was found that the attractive interaction in these systems (**6**) were correlative with Hammett constants σ_m: the Hammett plot of these systems displayed the straight line for the substituent effect. However, these systems were not ideal probes to study X $\cdots\pi$ interactions, as the CH $\cdots\pi$ type interactions played the predominant role in these systems, and it was hard to exclude the CH $\cdots\pi$ interactions from X $\cdots\pi$ interactions.

Furthermore, Gung et al. [22] synthesized a series of modified triptycene derivatives **7a, d** as molecular balances for direct measurements of $\pi-\pi$ interactions in the offset-stacked orientation. These triptycene models allowed a stacking interaction between the two arene groups in the *syn*-conformation, whereas the interaction was absent in the *anti*-conformation (Fig. 13.5). Thus, the ratiosfor the *syn*- and *anti*-conformers could be determined by variable-temperature NMR spectroscopy, and the free energies could be calculated from the *syn*/*anti*-ratios, which ranged from slightly positive (0.2 kcal/mol) to considerably negative (0.98 kcal/mol) values in chloroform. Moreover, the interactions between the arenes bearing electron-donating groups or electron-withdrawing groups were fairly different: for the electron-donating groups, the interactions were either negligible or slightly repulsive; whereas, for electron-withdrawing groups, the interactions were attractive. For the majority of these systems, the temperature had a negligible influence on the *syn*/*anti*-ratios. In the

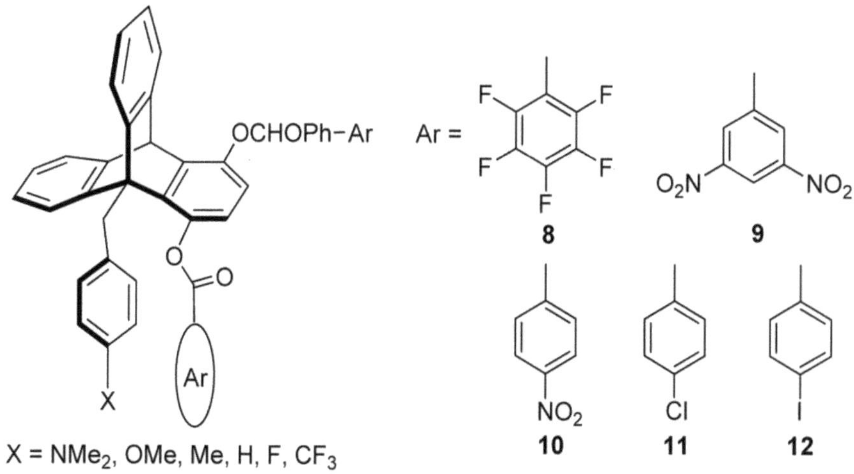

Fig. 13.6 Structures of triptycene model compounds **8–12**

same year, Gung et al. [23] also reported a series of triptycene model compounds **8–12** (Fig. 13.6) containing some strong donors and accepters. In consequence, the authors found that the two groups in the model compounds showed different aromatic interaction behaviors from the parallel stacked configuration. Among them, compounds **10–12** with monocyclic arenes showed a good correlation between the free energy attractions and Hammett parameters; however, in the models **8** and **9**, the multiple substituents strongly perturbed the aromatic rings which led to the exceptions to this correlation. These results revealed that the charge-transfer interactions might play a dominant role in the "abnormally behaved" compounds. Afterward, Gung et al. [24] found that the relative position of the arene substituents also influenced the $\pi-\pi$ stacking interactions in the 9-benzyl-substituted triptycene system. Moreover, the system with the methyl group in the *ortho* position showed more than 50 % increase in the strength of $\pi-\pi$ stacking interactions compared with the one in the *para* position.

In 2007, Gung et al. [25] further developed a series of triptycene-derived molecular models **13** (Fig. 13.7) and inserted an oxygen atom, which adopted a near-sandwich configuration. This system could be used to quantitatively study the near-perfect face-to-face π-stacking interactions between two aromatic rings. Compared with the models with arenes in the parallel-displaced configuration, the π-stacking interaction in **13** also preferred the *syn*-conformation in the sandwich configurations. However, the attractive interactions in the sandwich configurations seemed to be smaller than the parallel-displaced configuration. It was noteworthy that when one of the two arene rings was electron deficient enough, like pentafluorobenzoate, the attractive interactions could be observed; whereas, the character of the other arene ring almost had no influence on the interactions.

Recently, Gung et al. [26] also reported a series of triptycene-derived scaffolds **14** (Fig. 13.8) containing aromatic rings, pyridine or pyrimidine rings, which were used

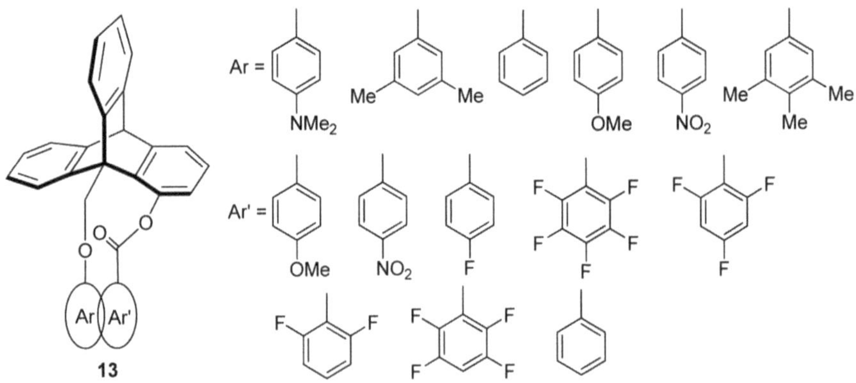

Fig. 13.7 Structures of triptycene-derived molecular models **13**

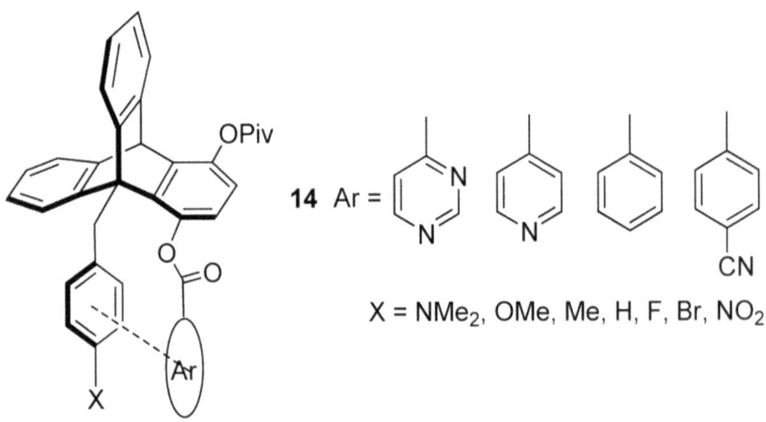

Fig. 13.8 Structures of triptycene-derived scaffolds **14**

as molecular balances to estimate the stacking interactions between aromatic rings and pyridine or pyrimidine rings. Consequently, the authors found that there were stronger attractive interactions between the pyridine or pyrimidine and benzene ring than the corresponding arene–arene interactions, as a result of the dominance of the *syn*-conformation. Additionally, the attractive interactions in the system of pyridine and pyrimidine were much less susceptible to the substituent effects compared with the ones in the corresponding carbocycles. Overall, when the two substituents were consistent with a predominant donor–acceptor interaction, like a pyrimidine ring and an *N*, *N*-dimethylaminobenzene, there were considerably attractive interactions between them.

References

1. Ōki M (1990) 1,9-Disubstituted triptycenes: an excellent probe for weak molecular interactions. Acc Chem Res 23(11):351–356
2. Mikami M, Toriumi K, Konno M, Saito Y (1975) Crystal-structure of 1,2,3,4-tetrachloro-9-*t*-butyltriptycene. Acta Crystallogr B 31(10):2474–2478
3. Nogami N, Ōki M, Sato S, Saito Y (1982) Implications of X-ray crystallographic results of 1,2,3,4-tetrachloro-9-(2-oxopropyl)triptycene rotamers. Bull Chem Soc Jpn 55(11):3580–3585
4. Ōki M, Takiguchi N, Toyota S, Yamamoto G, Murata S (1988) Rotamer populations and molecular-structure of 9-isobutyl-1,4-dimethoxytriptycene: further evidence for the presence of $CH_3 \cdots O$ hydrogen-bond. Bull Chem Soc Jpn 61(12):4295–4302
5. Ōki M (1976) Unusually high barriers to rotation involving tetrahedral carbon-atom. Angew Chem Int Edit 15(2):87–93
6. Nakamura M, Ōki M, Nakanish H, Yamamoto O (1974) Restricted rotation involving tetrahedral carbon. 10. Barriers to rotation of methyl-groups in 9-methyltriptycene derivatives. Bull Chem Soc Jpn 47(10):2415–2419
7. Nakanishi H, Yamamoto O (1978) Nuclear magnetic-resonance study of exchanging systems. 11. ^{13}C NMR-spectra of 9-ethyltriptycene derivatives and restricted rotation of ethyl group. Bull Chem Soc Jpn 51(6):1777–1783
8. Izumi G, Yamamoto G, Ōki M (1981) Intramolecular interactions between a group bearing normal-electrons and a $CH_2{-}X$ where X is an electronegative group. Bull Chem Soc Jpn 54(10):3064–3068
9. Tamura Y, Takizawa H, Yamamoto G, Ōki M, Murata S (1990) Molecular-structure of 9-chloromethyl-1,4-dimethoxytriptycene and its implications for the presence of $O \cdots CH_2Cl$ interactions. Bull Chem Soc Jpn 63(9):2555–2563
10. Suzuki F, Ōki M, Nakanishi H (1974) Restricted rotation involving tetrahedral carbon. 11. Barriers to rotation and conformational preferences of substituted 9-isopropyltriptycenes. Bull Chem Soc Jpn 47(12):3114–3120
11. Suzuki F, Ōki M (1975) Restricted rotation involving tetrahedral carbon. XIV. Conformational equilibria and attractive interactions in substituted 9-benzyltriptycenes. Bull Chem Soc Jpn 48(2):596–604
12. Ōki M, Izumi G, Yamamoto G, Nakamura N (1982) Attractive interactions between carbonyls and groups bearing lone-pair electrons in triptycene systems. Bull Chem Soc Jpn 55(1):159–166
13. Kikuchi H, Hatakeyama S, Yamamoto G, Ōki M (1981) Restricted rotation involving the tetrahedral carbon. 40. Barriers to rotation of 9-(1-methyl-2-propenyl)triptycenes. Bull Chem Soc Jpn 54(12):3832–3836
14. Tamura Y, Yamamoto G, Ōki M (1987) Effects of $CH_3 \cdots O$ hydrogen-bond on the rotamer populations in 9-(alkoxymethyl)-1,4-dimethyltriptycenes. Bull Chem Soc Jpn 60(10):3789–3790
15. Tamura Y, Yamamoto G, Ōki M (1987) $CH_3{=}O$ hydrogen-bond: implications of its presence from the substituent effects on the populations of rotamers in 4-substituted 9-ethyl-1-methoxytriptycenes and 9-(substituted phenoxymethyl)-1,4-dimethyltriptycenes. Bull Chem Soc Jpn 60(5):1781–1788
16. Tamura Y, Yamamoto G, Ōki M (1986) Substituent effects on the populations of rotational isomers in 9-(aryloxymethyl)-1,4-dimethyl triptycenes: implications for the presence of $CH_3 \cdots O$ hydrogen-bond. Chem Lett (9):1619–1622
17. Yamamoto G, Ōki M (1981) Restricted rotation involving the tetrahedral carbon. 36. Stereodynamics of 9-(2-methylbenzyl)triptycene derivatives. Bull Chem Soc Jpn 54(2):481–487
18. Nakai Y, Inoue K, Yamamoto G, Ōki M (1989) Implications of unusual population ratios in rotational isomers of 9-(4-substituted benzyl)-8,13-dichloro-1,4-dimethyltriptycenes and 4-substituted 9-benzyl-8,13-dichloro-1-methyltriptycenes $CH_3 \cdots \pi$ hydrogen-bond. Bull Chem Soc Jpn 62(9):2923–2931
19. Gung BW, Xue X, Reich HJ (2005) Off-center oxygen–arene interactions in solution: a quantitative study. J Org Chem 70(18):7232–7237

20. Gung BW, Zou Y, Xu Z, Amicangelo JC, Irwin DG, Ma S, Zhou HC (2008) Quantitative study of interactions between oxygen lone pair and aromatic rings: substituent effect and the importance of closeness of contact. J Org Chem 73(2):689–693
21. Gung BW, Emenike BU, Lewis M, Kirschbaum K (2010) Quantification of CH$\cdots\pi$ interactions: implications on how substituent effects influence aromatic interactions. Chem Eur J 16(41):12357–12362
22. Gung BW, Xue XW, Reich HJ (2005) The strength of parallel-displaced arene–arene interactions in chloroform. J Org Chem 70(9):3641–3644
23. Gung BW, Patel M, Xue XW (2005) A threshold for charge transfer in aromatic interactions? A quantitative study of π−stacking interactions. J Org Chem 70(25):10532–10537
24. Gung BW, Emenike BU, Alverez CN, Rakovan J, Kirschbaum K, Jain N (2010) Relative substituent position on the strength of $\pi-\pi$ stacking interactions. Tetrahedron Lett 51(13):1648–1650
25. Gung BW, Xue X, Zou Y (2007) Enthalpy (ΔH) and entropy (ΔS) for π−stacking interactions in near-sandwich configurations: relative importance of electrostatic, dispersive, and charge-transfer effects. J Org Chem 72(7):2469–2475
26. Gung BW, Wekesa F, Barnes CL (2008) Stacking interactions between nitrogen-containing six-membered heterocyclic aromatic rings and substituted benzene: studies in solution and in the solid state. J Org Chem 73(5):1803–1808

Chapter 14
Miscellaneous Applications of Iptycenes and Their Derivatives

14.1 Medicinal Chemistry

In 1999, Perchellet et al. [1] reported a class of triptycene bisquinones containing the *para*-quinone moieties, and found that they could be used as the bi-functional anticancer drugs. These drugs could not only block the synthesis of nucleic acid and protein but also inhibit the proliferation of cancer cells in vitro. Especially, the bisquinone **1** and its C-2 brominated derivative **2** (Fig. 14.1) exhibited the strong anti-proliferative activities, which were at the same level as the anti-proliferative ability of daunomycin (DAU). In 2002, Wang et al. [2] further designed and synthesized the amino-functionalized triptycene bisquinones **3a–b** (Fig. 14.1), which had the potent anticancer activities. It is noteworthy that these triptycene-based anticancer drugs could maintain their activity in multidrug-resistant tumour cells (daunorubicin-resistant HL-60 cell lines) with the ability to induce poly(ADP-ribose) polymerase-1 (PARP-1) cleavage. On the basis of further investigations, Wang et al. [3] found that these synthetic triptycene analogs could inhibit both DNA topoisomerase (topo) I and II activities, which could be used as dual inhibitors. In particular, the amino-functionalized triptycene showed the same level of power to the existing topo I inhibitor, such as camptothecin. Simultaneously, the triptycene derivative also exhibited the stronger inhibited effect on topo II than amsacrine. It was noteworthy that these synthetic triptycene analogs were cytostatic and cytotoxic, which could inhibit the L1210 leukemic cell growth and mitochondrial metabolism. Moreover, they could also block the cellular transport of purine and pyrimidine nucleosides, which was superior to that of DAU [1, 4, 5]. The potency and unusual mechanism of action made the triptycene bisquinones retaining their abilities to induce apoptosis in the multidrug-resistant (MDR) HL-60-RV and HL-60-R8 sublines [4]. Furthermore, Perchellet et al. [6] found that these bisquinones and their derivatives could trigger a caspase-independent release of cytochrome c and a caspase-2-mediated activation of caspase-9 and caspase-8, and keep their efficacy in DAU-resistant HL-60 cells without Fas signaling.

In 2006, Wang et al. [7] investigated whether these anti-tumour triptycene bisquinones might directly target mitochondria in cell and cell-free systems to cause the collapse of mitochondrial membrane potential, which was linked to

C.-F. Chen, Y.-X. Ma, *Iptycenes Chemistry*,
DOI 10.1007/978-3-642-32888-6_14, © Springer-Verlag Berlin Heidelberg 2013

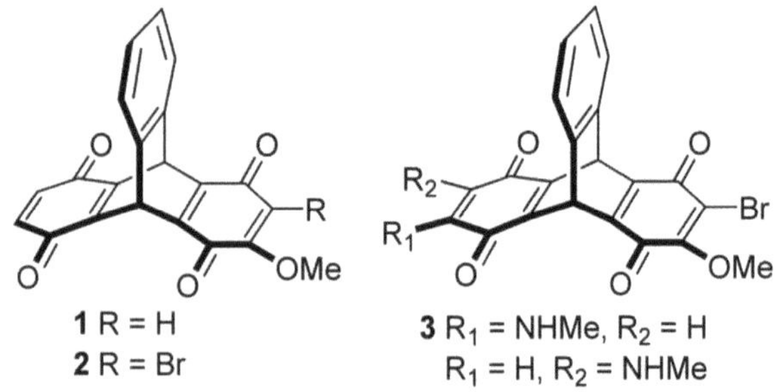

Fig. 14.1 Structures of triptycene bisquinones **1–3**

permeability transition pore (PTP) opening with the fluorescent probes of trans-membrane potential. It turned out that anti-tumour triptycene analogs could directly target mitochondria to trigger the Ca^{2+}-dependent and cyclosporin A-sensitive mitochondrial membrane potential, which probably induced the PTP opening and the apoptosis via mitochondrial pathway without nuclear signals.

Besides anti-cancer activities, Hua et al. [8] reported the unusual addition reactions among aliphatic primary, secondary amines and a substituted triptycene bisquinone **2**, and tested the biological activities of the triptycene bisquinone. According to the valuation of biological activities, these triptycene bisquinones showed not only the inhibition of L1210 leukemia cell viability, but also Plasmodium falciparum 3D7 with the IC50 values in the 0.11–0.27 μM and 4.7–8.0 μM range, respectively.

In 2003, Kourounakis [9] synthesized the triptycene quinones **4–6** (Fig. 14.2) by the improved synthetic pathways, and evaluated their anti-oxidant and anti-inflammatory activity with the standard and known protocols. In consequence, both of these triptycene quinones showed anti-oxidant activity, which could protect rat hepatic microsomal fraction against the lipid peroxidation; and they also could reduce the mouse paw edema induced by the Freund's complete adjuvant with anti-inflammatory activity.

14.2 Model for Jahn–Teller Systems

More than 50 years ago, the theory of the Jahn–Teller effect [10–13] had been worked out. In the early work, most manifestations of the Jahn–Teller effect were well known in metal trimer molecules, while the models in large molecules were comparatively

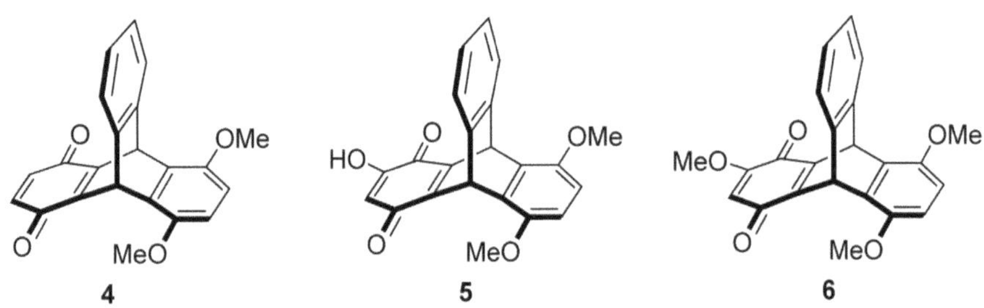

Fig. 14.2 Structures of triptycene quinones **4–6**

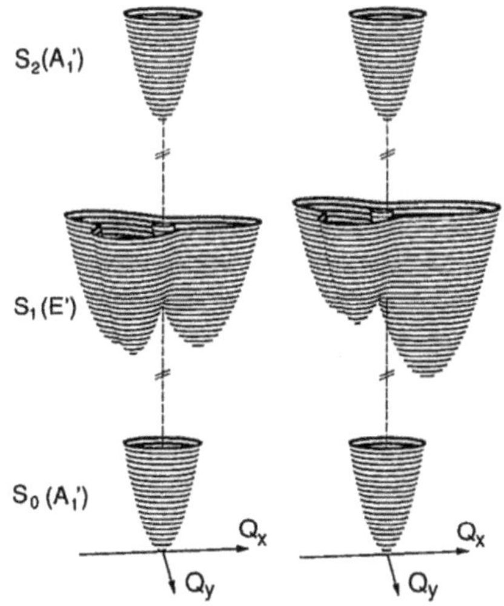

Fig. 14.3 A schematic view of (adiabatic) potential energy surfaces of uncomplexed triptycene (T, *left*) and T′ Ne (*right*) in the $S_0(A_1')$ ground state, $S_1(E')$ first excited state, and $S_2(A_1')$ second excited state as a function of displacement along the components Q_x and Q_y of the JT-active butterfly vibrational mode. (Reprinted with permission from ref. [17], Copyright 1994, American Institute of Physics)

rare. Until 1992, Furlan et al. [14] pointed out that the excited E' state of triptycene, which was measured by resonant two-photon ionization (R2PI), was a textbook example for the study of the Jahn–Teller effect. In fact, benzene wagging framework mode with low frequency vibronic levels and relatively large linear and quadratic coupling terms was the typical Jahn–Teller active vibrational mode. The first singlet excited state of triptycene showed the low vibronic level yields ($\nu_e' = 47.83\,\text{cm}^{-1}$), and the linear and quadratic terms were $\text{k} = 1.65$ and $\text{g} = 0.426$, which was precisely fitted to a strong $E' \otimes e'$ type Jahn–Teller effect.

Subsequently, Furlan et al. [14–17] further investigated a trimer vibronic coupling model for triptycene. The $E' \otimes e'$ Jahn–Teller effect, especially the higher order effects could be well applied to the interpretation of its highly resolved vibronic spectrum. The generalizations of the $E' \otimes e'$ model: $(A_1' \oplus E') \otimes e'$ and $(A_1' \oplus E') \otimes (a_2' \oplus e')$ trimer models could further be used for interpretation of the spectrum of triptycene. It was found that the $E' \otimes e'$ model was essentially correct, but the small details of this spectrum still gave an expression of momentum coupling with an a_2' vibration, which was confirmed by the data of the emission pattern of the laser induced fluorescence of the vibronic levels [15, 16]. This result was probably important to study and interpret the "molecular Barnett effect". According to the studies based on triptycene model, Furlan et al. [18] deemed that the wagging motions of the three benzene rings, which strongly coupled to the excited electronic state, led to the observed Jahn–Teller and Barnett effects in the e' and a_2' vibrations modes.

In addition, the triptycene•Ne_n ($n = 1$–3) van der Waals clusters with unique possibilities could be applied to investigate the intermolecular perturbation of a Jahn–Teller system through the supersonically cooled two-colour R2PI spectroscopy (Fig. 14.3) [17]. The complexes of triptycene with one to three Ne atoms exhibited a static distortion of the triptycene S_1 state Jahn–Teller potential energy surface and the weak intermolecular interactions on the intramolecular Jahn–Teller coupling. The $(A_1 \oplus E) \otimes e$ model with a uniaxial external strain component was suitable to

Fig. 14.4 Structures of donor–acceptor molecules with fixed distances **7a–b**

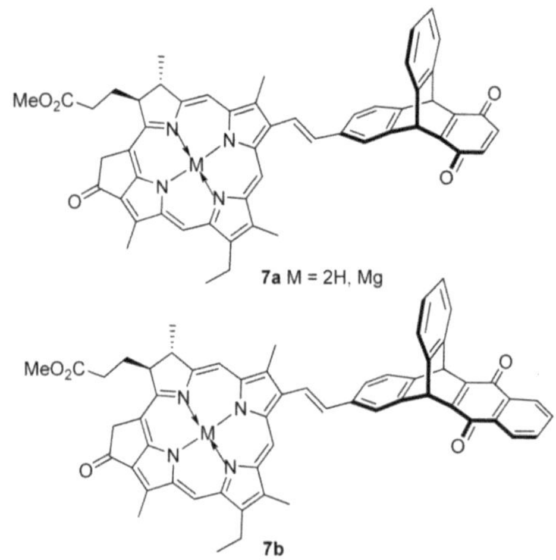

calculate the S_1 state levels and $S_1 \leftarrow S_0$ electronic spectra of all these complexes. It was found that the spectra of T•Ne and T•Ne_2 showed splitting of the E vibronic levels in C_{2v} symmetry and the additional transitions to levels, which was entirely different to the spectra of bare triptycene. However, there was a highly symmetric D_{3h} structure without inclusion of strain in the spectrum of T•Ne_3, since each of the three V-shaped compartments of triptycene were occupied by a single Ne atom. The 9-hydroxytriptycene, which was an isolated molecule, its $S_1 \leftarrow S_0$ R2PI spectrum and the fluorescence from various excited state vibronic levels showed the pseudorotation of the Jahn–Teller vibration with strong coupling to the torsional motion of the hydroxyl group [19]. The tunneling splitting both in the ground and excited states resulted in this special torsional motion. The spectra of 9-hydroxytriptycene exhibited a full C_{3v} vibronic point group, which was symmetry forbidden in the bare triptycene molecule.

14.3 Artificial Photosynthesis Models

In 1980s, Wasielewski [20] synthesized a series of donor–acceptor molecules with fixed distances (**7a–b**, Fig. 14.4), which were composed of methyl pyropheophorbide-a or methyl pyrochlorophyllide-a and triptycene quinone moiety. In these systems, electron transfer from the lowest excited singlet state of the donor to the acceptor was in a mechanism of fluorescence quenching, which was revealed by the picosecond transient absorption.

Soon after, Wasielewski et al. [21–23] further designed and synthesized a series of fixed-distance porphyrin–pentiptycene quinone molecules (**8** and **9**, Fig. 14.5). The unique structure of pentiptycene moiety gave a special polycyclic hydrocarbon spacer, which resulted in the porphyrin donor and the quinone acceptor to keep a fixed distance and orientation, and without the generation of electronically excited states of the donor or acceptor. In these systems, the presence of two positional isomers

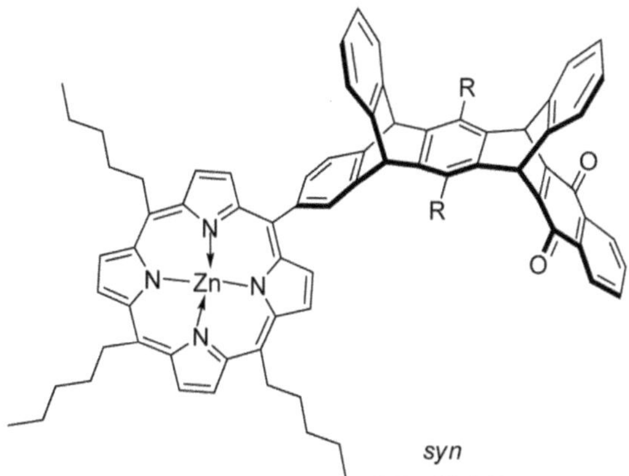

syn
8 R = H, OMe

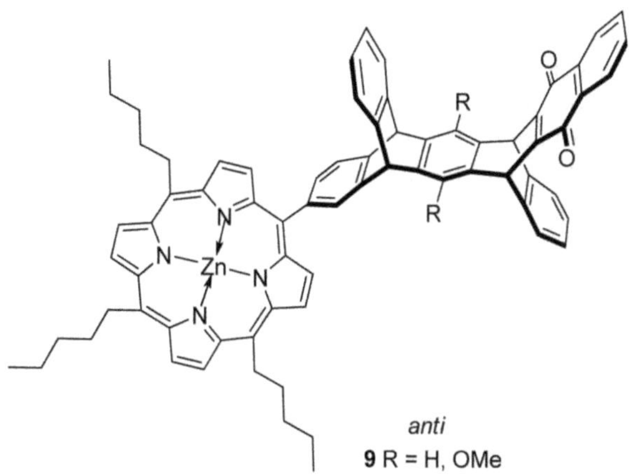

anti
9 R = H, OMe

Fig. 14.5 Structures of porphyrin–pentiptycene quinone molecules **8** and **9**

provided a platform for studying the photo-induced electron transfer reactions at a fixed distance with different orientation (*syn*/*anti*). Moreover, the presence of a central benzene ring of pentiptycene also allowed the modulation orbital energies of the spacer without conformational alteration, which finally resulted in a super-exchange mechanism of electron transfer.

Additionally, Wasielewski et al. [22] also designed a new kind of donor–acceptor molecules (**10a–b**, Fig. 14.6) with fixed distances, which involved Zn *meso*-triphenylporphine or Zn *meso*-tripentylporphine as donor, and a part of triptycene derivatives as high electron affinity acceptor. Likewise, the unique structure of triptycene afforded a rigid spacer. They attempted to use these systems to investigate the rates and energetics of these electron-transfer reactions at very low temperatures by the picosecond transient absorption spectra and EPR spectroscopy at 10 K. On the basis of the experimental results, they deemed that for the radical ion pairs of **10a–b** in the solid matrix at 10 K, the enthalpy was only about 0.9 eV, if its energy in fluid polar solvent at 294 K served as the standard value.

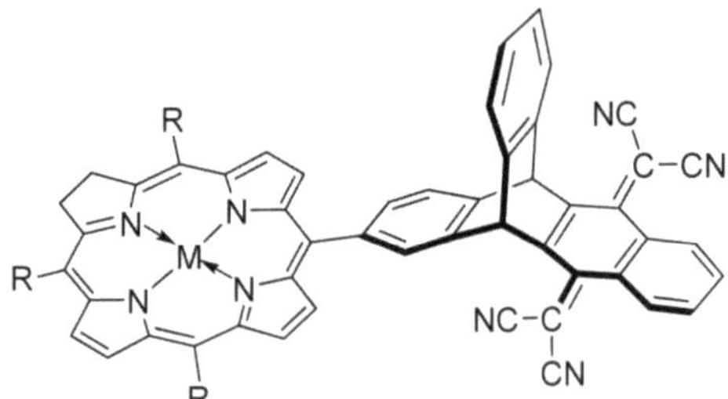

10a R = phenyl
10b R = *n*-pentyl

Fig. 14.6 Structures of donor–acceptor molecules **10a–b** with the fixed distance

Furthermore, Wasielewski et al. [24, 25] utilized the electron transfer rate data of 14 porphyrin–triptycene acceptor molecules to build a complete and quantitative picture of the dependence of the charge separation rate on free energy of reaction in a rigid glass [24]. These results showed that the porphyrin–triptycene molecules possessed ion-pair states that fluctuated with a range of 0.8 eV in rigid glasses, in contrast with their energies determined from electro-chemical measurements in polar liquids. This number displayed the dependence on the spacer structure of the porphyrin–triptycene acceptor molecules. On the basis of these results, they [25] subsequently designed the supramolecular systems as shown in Fig. 14.7, and aimed to separate charge efficiently in the solid state.

In 1999, Röder et al. [26] reported a class of porphyrin–triptycene–bisquinone systems. However, due to the electronic coupling, the porphyrin linked with triptycene and bis-quinone moieties could be conscribed as one acceptor unit as an integral whole in these systems. Thus, the bridge between the donor and the acceptor could be determined by the free enthalpy of the ET to a certain extent.

14.4 Preparation of Carbene

In 1995, Tomioka et al. [27] reported a kind of stable triplet carbene by utilizing the triptycene group as blocking group. Consequently, it was found that the presence of triptycyl group effectively enhanced the lifetime of some arylcarbenes. For the purpose of investigating the influence of electronic and steric effects on the structure and reactivity of triplet carbenes, they prepared a series of triplet 9-triptycyl(aryl)carbenes by photolysis of the corresponding diazomethanes, which was shown in Scheme 14.1. By the matrix isolation techniques at low temperature and time-resolved spectroscopies in solution at room temperature, they demonstrated that π-delocalization of the spin to aromatic rings led to the thermodynamic structure in the corresponding singlet state; however, extensive delocalization tended to a triplet arylcarbene. Moreover, it was also found that the steric factors affected the lifetime of triplet carbenes much more, rather than their structures.

In 2002, Iiba et al. [28] further designed and prepared a di(triptycyl)carbene by the similar route as above from the precursor di(triptycyl)diazomethane (Scheme 14.2).

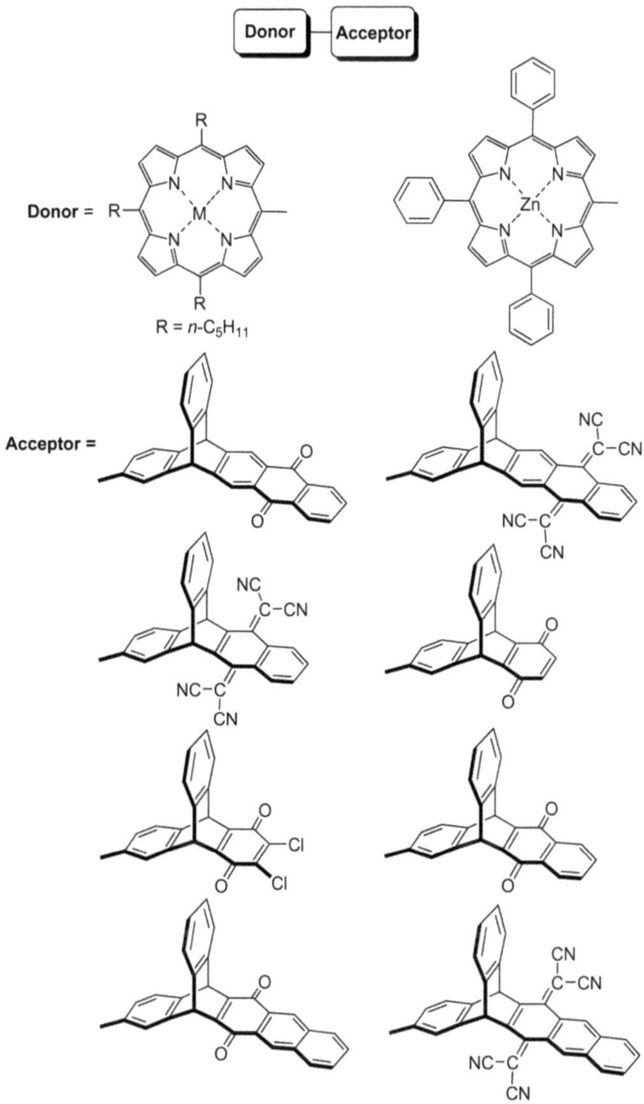

Fig. 14.7 Structures of donors and acceptors in supramolecular systems

The spectroscopical analysis and theoretical calculations indicated that this triplet bis(triptycyl)carbene was likely to be the most stable triplet dialkylcarbenes among the reported ones.

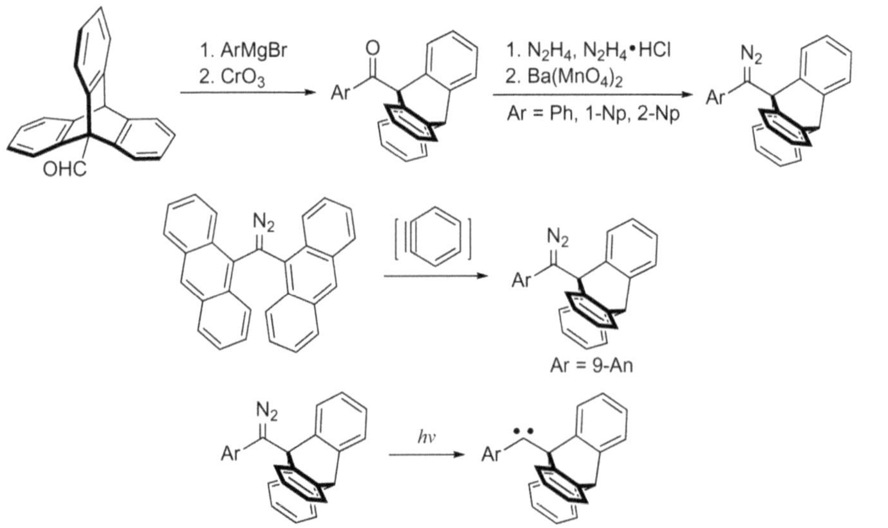

Scheme 14.1 Synthesis of triplet carbenes by utilizing the triptycene group as blocking group

Scheme 14.2 Synthesis of di(triptycyl)carbene from di(triptycyl)diazomethane

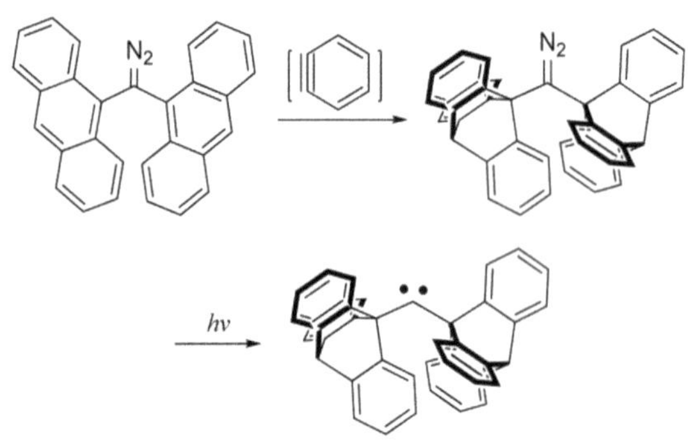

References

1. Perchellet EM, Magill MJ, Huang XD, Brantis CE, Hua DH, Perchellet JP (1999) Triptycenes: a novel synthetic class of bifunctional anticancer drugs that inhibit nucleoside transport, induce DNA cleavage and decrease the viability of leukemic cells in the nanomolar range in vitro. Anticancer Drugs 10(8):749–766
2. Wang Y, Perchellet EM, Tamura M, Hua DH, Perchellet JP (2002) Induction of poly(ADP-ribose) polymerase-1 cleavage by antitumor triptycene bisquinones in wild-type and daunorubicin-resistant HL-60 cell lines. Cancer Lett 188(1–2):73–83
3. Wang BN, Perchellet EM, Wang Y, Tamura M, Hua DH, Perchellet JPH (2003) Antitumor triptycene bisquinones: a novel synthetic class of dual inhibitors of DNA topoisomerase I and II activities. Anticancer Drugs 14(7):503–514
4. Wang BN, Wu MF, Perchellet EM, McIlvain CJ, Sperfslage BJ, Huang XD, Tamura M, Stephany HA, Hua DH, Perchellet JP (2001) A synthetic triptycene bisquinone, which blocks nucleoside transport and induces DNA fragmentation, retains its cytotoxic efficacy in daunorubicin-resistant HL-60 cell lines. Int J Oncol 19(6):1169–1178

5. Perchellet EM, Sperfslage BI, Wang Y, Huang XD, Tamura M, Hua DH, Perchellet JP (2002) Among substituted 9,10-dihydro-9,10-[1,2] benzenoanthracene-1,4,5,8-tetraones, the lead antitumor triptycene bisquinone TT24 blocks nucleoside transport, induces apoptotic DNA fragmentation and decreases the viability of L1210 leukemic cells in the nanomolar range of daunorubicin in vitro. Anticancer Drugs 13(6):567–581
6. Perchellet EM, Wang Y, Weber RL, Lou KY, Hua DH, Perchellet JPH (2004) Antitumor triptycene bisquinones induce a caspase-independent release of mitochondrial cytochrome c and a caspase-2-mediated activation of initiator caspase-8 and-9 in HL-60 cells by a mechanism which does not involve Fas signaling. Anticancer Drugs 15(10):929–946
7. Wang Y, Perchellet EM, Ward MM, Lou KY, Zha HP, Battina SK, Wiredu B, Hua DH, Perchellet JPH (2006) Antitumor triptycene analogs induce a rapid collapse of mitochondrial transmembrane potential in HL-60 cells and isolated mitochondria. Int J Oncol 28(1):161–172
8. Hua DH, Tamura M, Huang XD, Stephany HA, Helfrich BA, Perchellet EM, Sperfslage BJ, Perchellet JP, Jiang SP, Kyle DE, Chiang PK (2002) Syntheses and bioactivities of substituted 9,10-dihydro-9,10-[1,2] benzenoanthracene-1,4,5,8-tetrones. Unusual reactivities with amines. J Org Chem 67(9):2907–2912
9. Xanthopoulou NJ, Kouronakis AP, Spyroudis S, Kourounakis PN (2003) Synthesis and activity on free radical processes and inflammation of 9,10-dihydro-5,8-dimethoxy-triptycene-quinones. Eur J Med Chem 38(6):621–626
10. Longuet-Higgins HC, Opik U, Pryce MHL, Sack RA (1958) Studies of the Jahn−Teller effect. 2. The dynamical problem. Proc Roy Soc Lond 244(1236):1–16
11. Liehr AD (1960) Semiempirical theory of vibronic interactions in some simple conjugated hydrocarbons. Rev Mod Phys 32(2):436–439
12. Liehr AD (1962) Quantum theory: an essay on higher-order vibronic interactions. Annu Rev Phys Chem 13:41–76
13. Liehr AD (1963) Topological aspects of conformational stability problem. 1. Degenerate electronic states. J Phys Chem 67(2):389–471
14. Furlan A, Riley MJ, Leutwyler S (1992) The Jahn−Teller effect in triptycene. J Chem Phys 96(10):7306–7320
15. Furlan A, Leutwyler S, Riley MJ, Adcock W (1993) The Jahn−Teller effect in 9-fluorotriptycene. J Chem Phys 99(7):4932–4941
16. Riley MJ, Furlan A, Gudel HU, Leutwyler S (1993) A trimer vibronic coupling model for triptycene: the Jahn−Teller and Barnett effects. J Chem Phys 98(5):3803–3815
17. Furlan A, Leutwyler S, Riley MJ (1994) Intermolecular perturbation of a Jahn−Teller system: the triptycene•Ne_n ($n = 1$–3) van-Der-Waals clusters. J Chem Phys 100(26):840–855
18. Furlan A, Fischer T, Fluekiger P, Gudel HU, Leutwyler S, Luthi HP, Riley MJ, Weber J (1992) Low-frequency vibrations of triptycene. J Phys Chem 96(26):10713–10719
19. Furlan A, Leutwyler S, Riley MJ (1998) Coupling of a Jahn−Teller pseudorotation with a hindered internal rotation in an isolated molecule: 9-hydroxytriptycene. J Chem Phys 109(24):10767–10780
20. Wasielewski MR (1992) Photoinduced electron transfer in supramolecular systems for artificial photosynthesis. Chem Rev 92(3):435–461
21. Wasielewski MR, Niemczyk MP, Svec WA, Pewitt EB (1985) High-quantum-yield long-lived charge separation in a photosynthetic reaction center model. J Am Chem Soc 107(19):5562–5563
22. Wasielewski MR, Johnson DG, Svec WA, Kersey KM, Minsek DW (1988) Achieving high quantum yield charge separation in porphyrin-containing donor-acceptor molecules at 10 K. J Am Chem Soc 110(21):7219–7221
23. Wasielewski MR, Niemczyk MP, Johnson DG, Svec WA, Minsek DW (1989) Ultrafast photoinduced electron transfer in rigid donor-spacer-acceptor molecules: modification of spacer energetics as a probe for superexchange. Tetrahedron 45(15):4785–4806
24. Gaines GL, Oneil MP, Svec WA, Niemczyk MP, Wasielewski MR (1991) Photoinduced electron-transfer in the solid-state: rate vs free-energy dependence in fixed-distance porphyrin acceptor molecules. J Am Chem Soc 113(2):719–721

25. Wasielewski MR, Gaines GL, Oneil MP, Svec WA, Niemczyk MP (1990) Photoinduced spin-polarized radical ion-pair formation in a fixed-distance photosynthetic model system at 5 K. J Am Chem Soc 112(11):4559–4560
26. Korth O, Wiehe A, Kurreck H, Roder B (1999) Photoinduced intramolecular electron transfer in covalently linked porphyrin-triptycene-(bis)quinone diads and triads. Chem Phys 246(1–3):363–372
27. Tomioka H, Nakajima J, Mizuno H, Sone T, Hirai K (1995) Triptycyl(aryl)carbenes. a remarkably effective kinetic stabilizer of triplet carbenes. J Am Chem Soc 117(45):11355–11356
28. Iiba E, Hirai K, Tomioka H, Yoshioka Y (2002) Di(triptycyl)carbene: a fairly persistent triplet dialkylcarbene. J Am Chem Soc 124(48):14308–14309

MIX
Papier aus verantwortungsvollen Quellen
Paper from responsible sources
FSC® C105338

If you have any concerns about our products,
you can contact us on
ProductSafety@springernature.com

In case Publisher is established outside the EU,
the EU authorized representative is:
Springer Nature Customer Service Center GmbH
Europaplatz 3, 69115 Heidelberg, Germany

Printed by Libri Plureos GmbH
in Hamburg, Germany